DESIGN WITH MICROCONTROLLERS

McGraw-Hill Series in Electrical Engineering

Consulting Editor

Stephen W. Director, *Carnegie-Mellon University*

CIRCUITS AND SYSTEMS
COMMUNICATIONS AND SIGNAL PROCESSING
CONTROL THEORY
ELECTRONICS AND ELECTRONIC CIRCUITS
POWER AND ENERGY
ELECTROMAGNETICS
COMPUTER ENGINEERING
INTRODUCTORY
RADAR AND ANTENNAS
VLSI
Previous Consulting Editors

Ronald N. Bracewell, Colin Cherry, James F. Gibbons, Willis W. Harman, Hubert Heffner, Edward W. Herold, John G. Linvill, Simon Ramo, Ronald A. Rohrer, Anthony E. Siegman, Charles Susskind, Frederick E. Terman, John G. Truxal, Ernst Weber, and John R. Whinnery

Computer Engineering
Consulting Editor

Stephen W. Director, *Carnegie-Mellon University*

Bartee: *Digital Computer Fundamentals*
Bell and Newell: *Computer Structures: Readings and Examples*
Garland: *Introduction to Microprocessor System Design*
Gault and Pimmel: *Introduction to Microcomputer-Based Digital Systems*
Givone: *Introduction to Switching Circuit Theory*
Givone and Roesser: *Microprocessors/Microcomputers: Introduction*
Hamacher, Vranesic, and Zaky: *Computer Organization*
Hayes: *Computer Organization and Architecture*
Kohavi: *Switching and Finite Automata Theory*
Lawrence-Mauch: *Real-Time Microcomputer System Design: An Introduction*
Levine: *Vision in Man and Machine*
Peatman: *Design of Digital Systems*
Peatman: *Design with Microcontrollers*
Peatman: *Digital Hardware Design*
Peatman: *Microcomputer-Based Design*
Ritterman: *Computer Circuit Concepts*
Sze: *VLSI Technology*
Taub: *Digital Circuits and Microprocessors*
Wiatrowski and House: *Logic Circuits and Microcomputer Systems*

DESIGN WITH
MICROCONTROLLERS

John B. Peatman

Professor of Electrical Engineering
The Georgia Institute of Technology

McGraw-Hill Book Company

New York St. Louis San Francisco Auckland Bogotá Hamburg
London Madrid Mexico Milan Montreal New Delhi
Panama Paris São Paulo Singapore Sydney Tokyo Toronto

ABOUT THE AUTHOR

John B. Peatman is professor of electrical engineering at the Georgia Institute of Technology. He is the author of three previous McGraw-Hill texts, *The Design of Digital Systems* (1972), *Microcomputer-Based Design* (1977), and *Digital Hardware Design* (1980). In addition to microcontroller design experience gained with his students over the past eight years, Dr. Peatman has been involved in various design projects with Hewlett-Packard during five summers in Colorado Springs, Loveland, and Palo Alto, and a year in Scotland. He has been a director of Intelligent Systems Corporation for the past six years.

This book was set in Times Roman by Bi-Comp, Inc.
The editors were Alar E. Elken and David A. Damstra;
the copyeditor was Rita T. Margolies;
the designer was Kao & Kao Associates;
the production supervisor was Friederich Schulte.
Drawings were done by J & R Services, Inc.
R. R. Donnelley & Sons Company was printer and binder.

DESIGN WITH MICROCONTROLLERS

2 3 4 5 6 7 8 9 0   DOC DOC   8 9 2 1 0 9 8

ISBN 0-07-049238-7

Library of Congress Cataloging-in-Publication Data

Peatman, John B.
  Design with microcontrollers.

  (McGraw-Hill series in electrical engineering)
  Bibliography: p.
  Includes index.
  1. Electronic controllers—Design and Construction.
2. Electronic instruments—Design and construction.
3. Microprocessors. I. Title. II. Series.
TK7895.M5P43   1988   621.398'1   87-3135
ISBN 0-07-049238-7

To the parents in my life:
Lillie Burling Beisel
John Gray Peatman
Vivian Kendeigh Sutton
Edward Henry Sutton

CONTENTS

PREFACE

The evolutionary tree representing the growth of microprocessor technology is marked by a major fork. One branch is represented by microprocessors which have evolved in the direction of enlarged word widths (i.e., 16 or 32 bits) and increasingly powerful CPUs. Another branch is represented by microprocessors which have been combined with RAM, ROM, and various I/O facilities *on a single chip,* and which are commonly referred to as *microcontrollers*. It is this microcontroller branch of the microprocessor family tree which is addressed in this book.

This book is intended for the engineer who is interested in learning how to use a microcontroller in the design of an instrument or device. It opens the door to the many applications of microcontrollers. It differs from many books in that it develops design capability while keeping this breadth of application in mind.

To utilize microcontrollers effectively, a designer should develop at least three distinct capabilities. First, he or she must have a fundamental understanding of available components. This begins with the microcontroller itself, with its CPU register structure, its instruction set, its addressing modes, and its on-chip resources. It extends to keyswitches for setting up an instrument or device, displays for showing this setup information to the user as well as user output, transducers for sensing inputs, and actuators for control. Furthermore, for a designer to be effective this understanding must extend beyond the framework of only simplified and idealized devices. For example, a state-of-the-art microcontroller can juggle many real-time activities simultaneously using interrupt control. To achieve this without error, the designer must understand how the microcontroller handles interrupts and timing issues related to them.

Second, the designer must thoroughly understand the algorithmic processes required by each aspect of the design and be able to translate them into the language of the microcontroller. For example, the design of an antilock braking system for an automobile involves an understanding of both the brake system and the implementation of its control via a sequence of microcontroller instructions.

Third, the designer must understand how the extensive requirements of an instrument or device can be broken down into manageable parts. Almost any project can be likened to the process of jumping from boulder to boulder to cross a stream. Each boulder may represent the design ideas needed to understand and use a device like a liquid crystal display. But in addition to studying boulders, the designer must pay attention to how streams are crossed. For example, the use of hardware versus software to implement a function can be thought of as providing alternative routes across the stream.

This book attempts to organize and unify the development of these three capabilities: to understand and use components, to exploit powerful algorithmic processes, and to realize an effective organization of hardware and software so as to meet the specifications for an instrument or device.

From another point of view, this book is directed toward a specific goal of engineering studies—the development of creative design capability. With the availability of powerful, low-cost microcontroller chips, we are able to focus the design process on the microcontroller chip itself, the development tools needed to develop and debug the code which makes it run, and a variety of I/O devices which gives zest to the process of learning. Development tools have been greatly simplified by the availability of the ubiquitous personal computer. In fact, many designers have gained their initial experience with a specific microcontroller by using a low-cost emulator board (such as one of those discussed in Chap. 8) together with a personal computer. To encourage such activities, some microcontroller manufacturers are supporting their products with a dial-up facility which permits users to download an assembler and other useful software over the phone line. Never has an engineering professor been better able to put together a microcontroller design laboratory than now. And never has the basement entrepreneur been better supported with low-cost tools than now! To help readers take advantage of the opportunities for developing microcontroller design capability, most chapters close with an assortment of problems, many having a design flavor.

This book will typically be used in a one-semester or two-quarter course at the senior level. Alternatively, it might be used at the junior level if it is deemed worthwhile to trade off the increased engineering experience of seniors for the opportunity to follow this course with other design-oriented courses and individual project activities. Although the context of the book is electrical, each component is sufficiently explained to permit the book to be used in a variety of curricula as an introduction to design using a microcontroller. The incentive to so use the book lies in the diverse applications made possible by the availability of a "controller on a chip." Many of these applications are described in the first chapter.

This text uses two specific microcontroller chips, Intel's 8096 and Motorola's 68HC11, to illustrate the ideas which arise in the process of carrying out the design of an instrument or device. Each of these microcontrollers is treated somewhat generically throughout the text and then in concrete detail in an appendix. Nevertheless, the text has been written to support other microcontrollers. For example, the user's manual for Zilog's Super-8 microcontroller or National Semiconductor's HPC16040 microcontroller might be used as an "external appendix" to the book. Copy services like Kinkos, which are available on many college campuses, have made the use of such materials, which support the teaching of a course, painless and inexpensive.

There has been an attempt to make many parts of the book self-contained. Consequently, one way in which the book might be used is to study the bare bones of each chapter in order to accelerate the process of getting an overall view of how a microcontroller is designed. The remaining sections of each chapter can be studied at a later time as the need arises or in a subsequent course. Such an accelerated route through the features of a microcomputer might draw from the first half of Chap. 2 as well as the interpretation of Chap. 4 in terms of a specific microcontroller. An accelerated route through microcomputer software organization and ideas might be gained from the first four sections of Chap. 5 together with examples drawn from Sec. 7.5 on D/A conversion and Sec. 7.7 on position encoding. A microcontroller's interrupt structure, which plays a dominating role in providing fast real-time control, can be studied in Sec. 3.2. The programmable timer, which is probably the most distinguishing I/O feature of a microcontroller, can be studied in Secs. 3.3 and 3.4 as well as the appendixes. The I/O serial port, representing an important, unique feature of microcontrollers, can be studied in Sec. 6.6. This study can be augmented with examples drawn from Secs. 7.2, 7.3, 7.5, and 7.6, where various I/O chips which are driven from an I/O serial port are discussed.

One mark of good designers is their conservative handling of the timing considerations which arise in a design. Sections 6.3 and 6.4 deal with this topic for memory chips used to augment a microcontroller operated in a mode in which its internal busses have been made accessible externally. Sections 3.5 and 3.6 deal with the timing considerations which ensure reliable interrupt handling.

A study of microcontrollers would not be complete without a consideration of development tools. The variety of options available has never been greater. This includes some very low-cost development boards, and software which is supported by the ready availability of personal computers. The various alternatives are considered in Chap. 8.

It has been my good fortune to have the counsel and support of Dr. Demetrius T. Paris, director of the school of electrical engineering at the Georgia Institute of Technology. My students and I have been the grateful recipients of top-of-the-line development system equipment and support given to us by Hewlett-Packard Company. I am particularly grateful to Alan Herrmann and Randy Abler, who have served as system administrators in my laboratory and from whom I have gained a variety of learnings.

I am grateful to the following people for their support in the preparation of this book. Pat Heath, Motorola's technical marketing engineer for the

68HC11, reviewed Appendix A. Denis Regimbal, Intel's senior product marketing engineer for the 8096, and David Ryan, applications engineer for the 8096, reviewed Appendix B. In other capacities, I want to acknowledge the support of:

Skip Addison, Group Product Manager, Ungermann-Bass, Inc.
Tom Blakeslee, Vice President, R&D, Orion Instruments
Ed Bleichner, Field Applications Engineer, National Semiconductor
    Corp.
Michael Davidson, Field Applications Engineer, Motorola Inc.
John Hotchkiss, Electrical Engineer, Process Control Corp.
Jim King, Field Applications Engineer, Intel Corp.
Ed Klingman, President, Cybernetic Micro Systems, Inc.
Ashley Miller, Field Applications Engineer, Hewlett-Packard Co.
Charles Muench, President, Colorocs Corp.
Don Osburn, Chief Engineer, Electronic Systems International
Bill Pherigo, Digital Design Engineer, Hewlett-Packard Co.
Joe Putnal, Field Applications Engineer, Advanced Micro Devices, Inc.
Tom Saponas, R&D Manager, Hewlett-Packard Co.
Tim Settle, Sales Manager, Hewlett-Packard Co.
Mike Stipick, Field Applications Engineer, Hewlett-Packard Co.
Leland Strange, President, Intelligent Systems Corp.
Jeff Zurkow, President, Avocet Systems Inc.

I would also like to express my thanks for the many useful comments and suggestions provided by colleagues who reviewed this text during the course of its development, especially to Genevieve Cerf, Columbia University; Keith L. Doty, University of Florida; Mansur Kabuka, University of Miami; Kenneth W. Kolence, Kolence Associates; Michael A. Schuette, Carnegie-Mellon University and Ronald D. Williams, University of Virginia.

Finally, I am grateful to my wife, Marilyn, for her support, encouragement, editing, and proofreading of this book.

John B. Peatman

DESIGN WITH MICROCONTROLLERS

CHAPTER
1

THE ROLE
OF THE
MICROCONTROLLER

The last decade has seen an exciting evolution in the capabilities of micro-processors. Manufacturers have produced 16- and 32-bit processors (which can operate upon 16- and 32-bit operands in a single instruction and which can address megabytes of memory) to answer the need for ever more power-ful CPU (central processing unit) processing capability. These processors have found their niche in the development of ever more powerful personal computers. They have become the backbone of the workstations which are becoming the revolutionary tool of the engineer. And the architect. And the . . .

Because of their processing power and speed, these 16- and 32-bit processors have also found their way into the design of stand-alone products such as electronic instruments which require sophisticated control capability.

In this book we will explore another branch in the evolution of micro-processor capability. It is a branch which semiconductor manufacturers have been developing with equal vigor. Instead of focusing upon larger word widths and address spaces, the emphasis here has been upon exceedingly fast real-time control. It has focused upon the integration of the facilities needed to support fast control into a single chip.

In the past, the highest performance real-time control applications have employed 16- and 32-bit micros together with interrupt handler chips, pro-grammable timer chips, and ROM (read-only memory) and RAM (random access memory) chips to achieve what can now be achieved in a single state-

FIGURE 1.1
Microcontrollers are finding roles in even the most diverse jobs! (*Rand Renfroe*)

of-the-art microcontroller chip. Even for those real-time control applications for which the resources of a "single-chipper" are not adequate, such a chip still generally offers *the* optimal design approach. Its on-chip resources provide an integrated approach to a variety of real-time control tasks. By operating the chip in an *expanded* mode, not only do we gain the on-chip features, but also we can augment these features as we see fit. All the power available in *any* peripheral chip becomes available to our designs based upon such a microcontroller chip.

In this new world of the microcontroller, the applications which are appearing are limited only by our collective imaginations. As shown in Fig. 1.1, some designers' imaginations seemingly know no bounds!

In this chapter, we will look at some of this diversity of application. We will try to see what it is about the microcontroller which recommends its choice for such applications.

1.1   HAND-HELD INSTRUMENTS

As radio receivers have shrunk in size, it has become possible to make very small "pagers." Such a unit is an individually addressable receiver. The unit shown in Fig. 1.2 is $5\frac{1}{2}$ in long, permitting it to be carried in a pocket alongside a pen or pencil. It includes a CMOS (complementary metal-oxide semiconductor) microcontroller chip, which extends its battery operation as long as possible. With a nickel-cadmium (NiCd) battery, it will operate for an average of 32

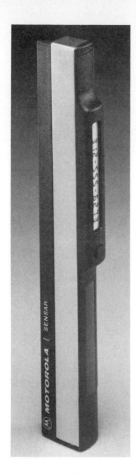

FIGURE 1.2
Numeric display radio pager. (*Motorola, Inc.*)

hours between charges. With a nonrechargeable zinc-air battery, it will operate for an average of 480 hours.

The microcontroller interprets the characters received, alerts a user with an audible tone, and presents any of the last five messages on its low-power liquid crystal display. Each message of up to 24 digits can indicate a phone number to call, a code indicating who the caller is, another code indicating the urgency of the request, etc.

Another battery-powered application is represented by the electronic planimeter shown in Fig. 1.3. This 14-in-long device is used to read areas and line lengths from maps and photographs. The microcontroller is able to employ sophisticated numerical algorithms and integrate the output of two small rotary encoders into either a line length or an area. The encoders permit the changing position of the optical tracer point to be resolved to better than 0.05 mm. The unit includes a "point mode" for computing the line length or area of a straight line figure simply by setting the tracer point at each intersection of the figure and recording a measurement. For more

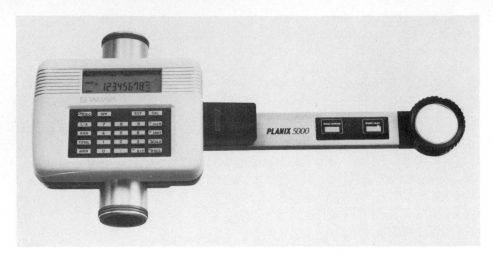

FIGURE 1.3
Electronic planimeter. (*Tamaya.*)

general figures, it includes a "stream mode" for which the curved outline of the figure is traced. The little keypad permits scale factors and any of several commands to be entered. It is even possible to pass the output to a computer over the built-in RS-232 interface. Its rechargeable NiCd batteries permit 15 hours of operation between charges.

The 6.3-in-long "levelmeter" shown in Fig. 1.4 measures the angle from the level position with an accuracy of 0.5° over a range of ±120°. Alternatively, it can measure grade (i.e., inches of rise divided by inches of run). An audible beeper may be turned on to notify a user when a horizontal

FIGURE 1.4
Digital levelmeter. (*Soar Corp.*)

position is attained. Two low-power liquid crystal displays, one on the top of the unit and one on the side, show the continuously updated angle or grade. Disposable batteries provide for approximately 50 hours of continuous operation.

Figure 1.5 shows an IC (integrated circuit) tester that is the size of a hand-held calculator. It includes a universal zero-insertion-force socket which will accept a DIP (dual in-line package) with up to 40 pins. The user need only align pin 1 of the chip in the upper-left-hand-corner of the socket. The up/down keys can be used to scroll through part numbers for a family of parts (e.g., the 74LSxxx TTL family). When the desired part number is reached, pressing the "test" key applies power to the part for approximately 0.1 second while an extensive diagnostic procedure is carried out. If the part fails the test, the offending pins are identified, along with the nature of the failure.

This tester includes an "auto" mode for testing an unknown part. Since it carries out a new test every millisecond, the unit can instantly identify a working part while also verifying its proper operation. This feature can even be used to bypass the need to scroll through the internal list of part numbers before testing. In this mode, the unit becomes the IC tester's equivalent of a "point and shoot" camera! The probability of classifying a bad IC as good is estimated to be less than 0.6 percent.

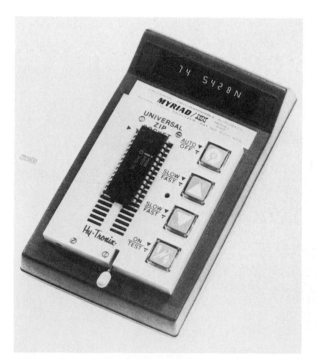

FIGURE 1.5
Universal integrated circuit identifier and tester. (*Hy-Tronix Instruments, Inc.*)

1.2 PERIPHERAL DEVICES

Another class of application for microcontrollers is represented by the variety of devices which do not stand alone but rather derive their usefulness by augmenting the capability of a personal computer. In fact, the keyboard for a personal computer, such as that in Fig. 1.6, employs a microcontroller to scan the keys and then serially transmit keycodes to the computer. This is an excellent job to unload from the computer's processor.

The keyboard's microcontroller continuously looks for the pressing of new keys. It "debounces" the keys in software, simplifying the design requirements for the mechanical keyswitches and simplifying their interface to the microcontroller. It also implements "N-key-rollover" capability. If several keys are depressed at the same time, their keycodes are *all* sent to the computer's processor in the order in which they were pressed. Furthermore, when a key is held down the microcontroller generates a series of "autorepeat" keycodes which make it appear as if the key is being repeatedly tapped.

FIGURE 1.6
Keyboard controller. (*IBM Corp.*)

At power-on time, the keyboard's microcontroller carries out a diagnostic procedure and informs the system unit if it detects a problem. While this procedure does not check for bad keyswitches, it goes a long way toward identifying other assorted problems, such as a loose keyboard cable.

Finally, by serializing the data, the keyboard's microcontroller simplifies the cabling to the computer. The resulting cable bears a closer resemblance to a sturdy telephone cord than to a ribbon cable (which would tend to fail after extensive flexing). It supplies power to the keyboard as well as providing two bidirectional signal lines.

The modem shown in Fig. 1.7 is used to connect a personal computer or terminal to another computer over the telephone lines. Its design employs two microcontrollers: one serves as a command interpreter while the other handles the actual data transmission.

The printer buffer shown in Fig. 1.8 can make a significant difference in the performance of a personal computer. It permits a file to be dumped at high speed, using a parallel port *or* a serial port from the computer, into the buffer. Then the buffer will send these characters to the printer at the printer's speed while the computer is free to do something else. This one box can drive a printer with either a serial port or a parallel port. It can be configured as a minimum 64K-byte buffer, or as large as a 2-Mbyte buffer.

FIGURE 1.7
Modem. (*Hayes Microcomputer Products, Inc.*)

FIGURE 1.8
Printer buffer. (*Quadram Corp.*)

As graphics, and in particular, color graphics, become more and more important to personal computer and workstation users, color plotters will become an increasingly pervasive peripheral. These put new requirements on a buffer box because of the extensive handshaking commands which go back and forth between computer and plotter. The buffer box shown in Fig. 1-8 supports these handshaking commands and thereby can serve both as a printer buffer and as a plotter buffer.

Because of the requirement for relatively low-cost RAM, a printer buffer typically uses dynamic RAM chips which must be refreshed by the microcontroller. Its ability to handle the refresh task without requiring extra parts makes it particularly well suited to the job. Also its ability to implement both serial and parallel ports with a minimum of extra parts helps to keep the design simple, low cost, and trouble-free.

The eight-pen plotter shown in Fig. 1.9 requires extensive real-time control capability of its microcontroller. The microcontroller must accept long strings of serial input data and handle the handshaking protocols which arise. It must also monitor the little 12-key array of keyswitches on the front panel. It must control the eight-pen carousel which is responsible for automatic pen changing and pen capping (to prevent the ink from drying up and clogging the pen). It includes incremental encoders for both paper position and pen position as well as dc motors on each of these drives, providing low-cost fast positioning with tight servo loops.

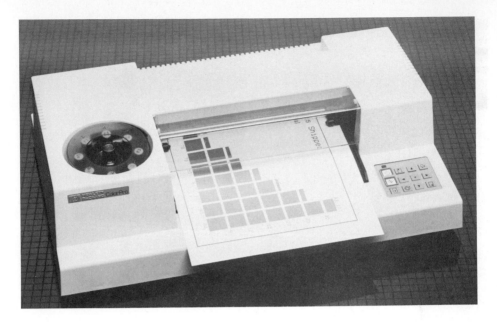

FIGURE 1.9
Color plotter. (*Hewlett-Packard Co.*)

1.3 STAND-ALONE DEVICES

The color copier shown in Fig. 1.10 uses the Intel 8096 microcontroller discussed in this book to carry out a wide array of control tasks. It uses the 8096's high-speed inputs in conjunction with four dc servo loops for paper positioning. It uses the 8096's high-speed outputs for phase control of 60-Hz ac power to control the color exposure. It uses the 8096's serial port to expand the number of output lines with shift registers, as will be discussed later in this book. The 8096 includes a multiplexed A/D converter which is used to measure temperature via thermistors. Also assorted optical sensors are monitored.

The 8096 chip does not include all of the resources needed for the overall job. Consequently, off-chip timers are used to generate pulse-width-modulated outputs. This is a simple way to generate an output with a dc component proportional to a digital number. Also an off-chip UART (universal asynchronous receiver transmitter) communicates with a separate processor which monitors the front panel keyswitches and generates control commands to the 8096. The 8096 is also supported with external RAM and ROM.

The designers of the color copier of Fig. 1-10 started out intending to use a 16-bit microprocessor designed for fast, general purpose computing. They knew that the micro would have a lot to do, and they knew that they

FIGURE 1.10
Color copier. (*Colorocs Corp.*)

would need fast multiplication. As they got into the design, they found that
the general purpose microprocessor just was not fast enough for the job. In
contrast, the 8096 attends to all of the jobs required with something like 35
percent of its CPU time left over.

The typewriter shown in Fig. 1.11 employs a cartridge printwheel to
print true letter-quality characters. It uses two microcontrollers. One of the
two microcontrollers controls printwheel selection, the printer and index
function, and the carrier escapement (for paper movement). Depending upon
which of 38 possible printwheels are selected, this microcontroller redefines
the spacing, or indexing, for each character. It handles any of four different
pitches, including proportional spacing (where the letter i takes less horizontal space than the letter m).

FIGURE 1.11
"Wheelwriter 5" typewriter. (*IBM Corp.*)

The other microcontroller in this typewriter carries out the scanning of the keyboard, controls the indicator lights, and implements all text functions. These functions include centering, underscoring, and sub- and superscripts. It also includes the storage, revision, and playback of about 7000 characters of text material allocated into a maximum of 99 storage areas. Text can include a "stop" code for the insertion of variable information and for changing the printwheel. Battery backup supports memory retention in the event of a power outage.

The cable TV terminal shown in Fig. 1.12 uses a microcontroller chip to support addressability functions. Not only does the unit permit a TV viewer to select cable channels, it also supports premium channel descrambling as well as pay-per-view descrambling.

The fish finder shown in Fig. 1.13 combines a sonar transducer with a sophisticated signal processor and a liquid crystal display. The finder displays the contour of the bottom and any intervening fish and can resolve fish as close as 6 inches apart. It uses a CMOS microcontroller so that virtually all of the power drawn from a 12-V automobile battery is used to operate the 200-kHz sonar transducer.

The lawn sprinkler control box shown in Fig. 1.14 uses a microcontroller chip to control lawn sprinkling automatically. It controls up to six solenoid valves buried in the lawn to provide optimal sprinkling, even when some parts of a lawn may require more watering than other parts. Each valve can be turned on for up to 99 min, up to 3 times per day, and on selected days during a 7-day cycle. The unit includes battery backup and a

FIGURE 1.12
Cable TV terminal. (*Scientific Atlanta*.)

FIGURE 1.13
Fish finder. (*Humminbird*.)

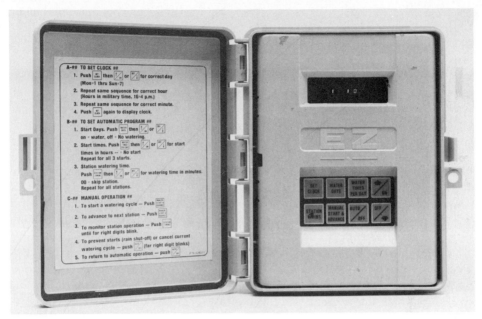

FIGURE 1.14
Lawn sprinkler. (*Rain Bird Sales, Inc.*)

FIGURE 1.15
Charge-card phones. (*R-Tec Systems Division of Reliance Electric Company.*)

24-hour clock that keeps time even during a power outage. The microcontroller chip allows this device to be within the financial reach of many home owners.

The telephones shown in Fig. 1.15 use a microcontroller to read a credit card and to handle phone number entry. They also permit access to a long-distance carrier: the user's billing code is entered followed by a phone number. In this way, they support long distance phone service without the hassle and vandalism associated with a coin-operated phone.

1.4 INSTRUMENT SUBFUNCTIONS

Microcontrollers are commonly used to implement subfunctions within an instrument design. For example, consider the precision digitizing oscilloscope shown in Fig. 1.16. This instrument can capture 16,000 samples of a waveform, with 10-bit resolution for each sample and with a maximum sample rate of 20 megasamples per second.

This instrument not only captures and displays the data, but can also analyze it. For pulse measurements, the oscilloscope can automatically compute and display rise and fall times, pulse width, duty cycle, amplitude, and overshoot and undershoot. For waveforms, it converts them to the frequency domain, carries out frequency domain analysis, and displays magnitude and phase spectrums on the same CRT display as the original time domain waveform.

This instrument can create a permanent copy of the original display and any derived displays on a floppy disk. It can also record the entire state of the instrument so that complex sequences of instrument setups can be recalled and used again. The data on the floppy disk is formatted for compatibility with Hewlett-Packard's 9000 series of computers. Consequently, the instrument can serve as a data gatherer and data analyzer; the captured records can be used to track waveform variations over an extensive period.

A microcontroller is employed in the design of this digitizing oscilloscope to handle front panel functions. It scans the keyboard to detect pressed keyswitches. It drives LED annunciators built into the keyswitches to indicate which ones have been pressed (or which ones have been turned on by replaying a previously recorded measurement). The CRT display includes touch-screen input capability which is operated by the microcontroller. Finally, parameter values can be entered via the numeric keypad or "tweeked" with an RPG.* The front panel microcontroller handles the RPG output.

Dedicating front panel functions to a separate microcontroller makes sense in the same way that dedicating keyboard scanning in Fig. 1.6 to a separate microcontroller makes sense. For their support, these functions will

---

* Rotary pulse generator, discussed in Chapter 7.

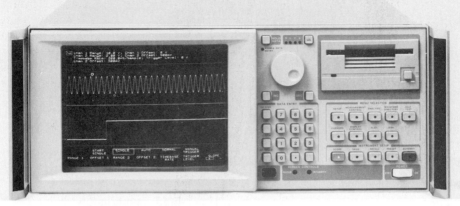

(a)

(b)

FIGURE 1.16
Front-panel handler for a digitizing oscilloscope. (*a*) Oscilloscope; (*b*) front-panel circuitry.
(*Hewlett-Packard Co.*)

FIGURE 1.17
Touch-sensitive liquid crystal display.
(*Kiel Corp.*)

require a separate printed circuit board mounted immediately behind the front panel. If the PC (printed circuit) board also includes a microcontroller, then the front panel control task can not only be broken out to this microcontroller, but the *communication* to the main processor in the instrument will require no more than a couple of wires.

Another instrument subfunction is illustrated in Fig. 1.17: a multiplexed liquid crystal display which also incorporates touch-screen input. The manufacturer of this unit is able to offer instrument designers an easy-to-use solution to the problems of user setup and the display of measured results. The display is composed from an array of 64 × 240 pixels. The touch screen has 4 × 10 touch areas. Instrument operation can be designed so that the instrument can be set up by the user, who presses labeled touch areas to respond to a sequence of menus.

To the instrument designer, the unit shown in Fig. 1.17 looks like a peripheral chip which is accessed over an 8-bit data bus. Control and display operations are executed with writes to either of two addresses. Status and touch-screen output information is gathered with reads from either of two addresses. An interrupt output signal can be used to gain the attention of the instrument's controller when a finger pressing is detected.

To achieve this performance, a microcontroller can carry out the continuous multiplexing of the LED display. It can monitor the touch-sensitive input mechanism. And it can handle the interactions which take place over the 8-bit data bus.

This touch-sensitive display is typical of a large class of applications. The user of the device never sees the interactions required to make a complex device work. The microcontroller can handle these complex interactions, but presents an easily managed interface to the user.

Some instruments are available with optional features. For example, the microwave counter shown in Fig. 1.18a can be operated as a stand-alone device. Alternatively, the "HP-IB interface" shown in Fig. 1-18b can be included as an option. This permits the instrument to be used in an automatic measurement system and controlled over the Hewlett-Packard Interface Bus, which is HP's implementation of a standard embraced by the

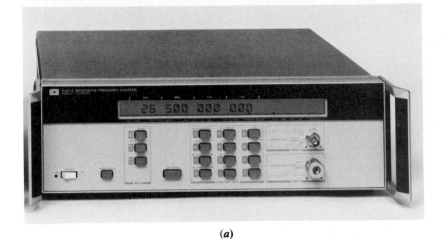

(*a*)

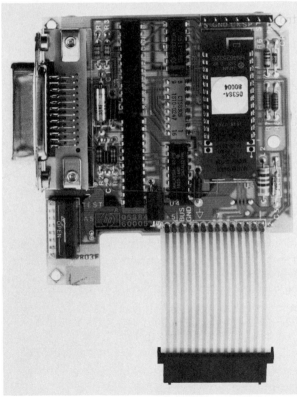

FIGURE 1.18
HP-IB interface option for a microwave counter. (*a*) Counter; (*b*) interface. (*Hewlett-Packard Co.*)

(*b*)

entire instrumentation industry. The 3 × 4 in board includes a microcontroller to support data transfers consisting of 21 bytes per reading and 130 readings per second. This simplifies the job of the main instrument controller, which can now handle a complete 21-byte reading at a time.

FIGURE 1.19
Multimeter which uses a microcontroller to help isolate and control its front end. (*Hewlett-Packard Co.*)

The high-performance multimeter shown in Fig. 1.19 uses serial communication between two microcontrollers to support an optically isolated front end. The multimeter includes a multiplexer so that any one of several inputs can be measured. Furthermore, the multiplexer and A/D converter combination *floats*, so that small differential inputs can be measured even when superimposed upon a *common-mode* voltage* that is large, perhaps hundreds of volts. The front end of the instrument needs to be controlled for channel selection, for type of measurement (i.e., voltage, current, or resistance), and for range. The measured output from the front end needs to be obtained by the instrument's main controller for display and possibly for transmission over the HP-IB interface to another instrument. By transferring data serially between these two controllers, only two lines are required to handle any amount of information transfer. Thus only two optoisolators are required to isolate these two parts of the multimeter from each other electrically.

The spectrum analyzer shown in Fig. 1.20 consists of a *family* of modules. They permit the unit to be configured in any of several ways, depending upon the needs of the user. For example, the spectrum analyzer shown employs the separate box on top for the display and control function. An alternative configuration would employ a smaller display and control module to take up three-eighths of the lower system mainframe. Either of these display and control modules employs a microcontroller to handle front-panel functions, just like the instrument in Fig. 1.16.

A spectrum analyzer configuration always includes a 3.0- to 6.6-GHz local oscillator module which also serves as the master controller for the system and includes a 16-bit microprocessor and extensive system software. The remaining RF (radio frequency) and IF (intermediate frequency) modules are selected according to the frequency range required of the system (e.g., 100 Hz to 2.9 GHz or 50 kHz to 26.5 GHz). Each of these remaining modules contains a microcontroller chip which communicates with the mas-

---

* That is, the average voltage relative to the multimeter's ground potential.

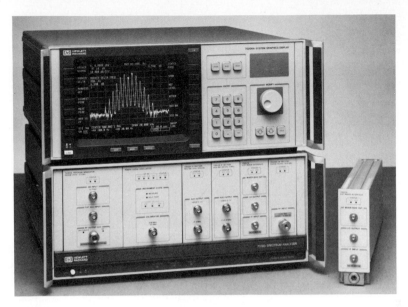

FIGURE 1.20
Modular spectrum analyzer. (*Hewlett-Packard Co.*)

ter controller over a serial bus. In this way, the interconnections are reduced to just a few wires while permitting extensive and fast interactions between modules.

1.5 AUTOMOTIVE APPLICATIONS

A major driving force behind the development of microcontroller capabilities has been the automobile industry. In fact, the Motorola 68HC11 microcontroller discussed in this book is an outgrowth of work between Motorola and the Delco Electronics Division of General Motors to specify and produce microcontroller chips for General Motors automobiles.

The needs of the automobile industry for microcontrollers are tremendous. A major application has been for engine control modules, such as that shown in Fig. 1.21. This unit maintains closed-loop engine control to boost fuel economy and to control exhaust emissions. It employs sensors for oxygen, coolant temperature, throttle position, manifold pressure, and barometric pressure. It controls the air-fuel mixture, spark timing, idle speed, and torque conversion.

Microcontrollers are beginning to serve other applications in automobiles. An antilock braking system monitors the traction on each tire. If the driver slams on the brakes, the system gains maximum traction from each tire short of skidding. In effect, the brake on each tire is pressed and released many times a second. The car comes to a halt with a speed and

FIGURE 1.21
Automobile engine control module. (*Delco Electronics Division of General Motors Corp.*)

with a degree of control which greatly exceeds what the driver could achieve without the system.

Another automotive application is dynamic ride control. It can adjust the automobile suspension system of a small car during cornering and braking so as to create the luxury ride normally associated with a larger automobile.

Electronic instrument panels, CRT displays, navigation systems, transmission controllers, multiplexed wiring systems, cellular telephones, electronic entry and security systems, electronic power steering, and collision avoidance systems represent further applications of microcontrollers to automobiles. The automobile of the future will indeed be a technological marvel!

CHAPTER
2

MICROCONTROLLER RESOURCES

2.1 OVERVIEW

The distinguishing characteristic of a microcontroller chip is the inclusion, on one chip, of *all* the resources which permit it to serve as the controller in a device or instrument. Furthermore, to support the expansion of resources beyond what is available on the chip itself, a microcontroller may include the "hooks" to make such expansion easy. In this chapter, we will look at the resources which permit a chip to carry out the controller function by itself. We will delay consideration of those resources which support expansion until Chap. 6. We will also delay until the next chapter a discussion of interrupts and programmable timers and the variety of issues which arise in the real-time control of events.

We will begin the chapter with a look at two popular state-of-the-art microcontroller families. Each is available in several packages. Each is also available with several alternative reduced feature sets, giving designers a lower-cost part if a certain feature* is not needed.

In the following section we will look at the role played by the widths of a microcontroller's data bus and address bus. We will find that the bus structures of the two microcontrollers studied extensively in this book differ markedly from each other. This will have implications relating to how data is stored, how quickly instructions are executed, and how the chips are expanded for adding more resources.

---

* For example, an analog-to-digital converter.

A microcontroller's program memory requirements and its data memory requirements can be met in any of several ways. The program memory alternative which is most useful during the development of a product is not what is least costly in production. The data memory alternative which is most convenient to use for normal processing is not the data memory which will retain the integrity of data when power is switched off or otherwise lost. In Secs. 2.4 and 2.5 we will examine the alternatives.

The parallel ports which consume most of the pins of a microcontroller chip possess features especially peculiar to the needs of a single-chip microcomputer. Each bit of a parallel port typically shares the pin with some other microcontroller resource. In addition, some bits can only be used as inputs, others only as outputs, and still others as *either* inputs or outputs. In Sec. 2.6 we consider these alternatives.

A microcontroller which can sense analog voltage inputs can interface directly to the many analog transducers available. A microcontroller which can generate analog voltage outputs can deal directly with a variety of actuators such as dc motors. Section 2.7 considers how these facilities are implemented, and how they can be used, in the two microcontrollers discussed in this book.

The manner in which a microcontroller is reset has received increasing attention as these chips have become critical to the operation of automobiles and other devices which affect life and limb. We certainly need to be able to force the chip into a known initial state when power is first applied. We also need to be able to kick it back to a known state with the help of a push button. Now we even have the ability to let the microcontroller monitor itself. With an on-chip timer that will reset the chip if it ever times out, the microcontroller need only reinitialize this timer periodically to maintain normal operation. A microcontroller which has lost control of itself will thus be restarted automatically. These considerations are developed in Sec. 2.8.

The final section of Chap. 2 deals with the events which take place when power is removed from a microcontroller, either by intention or accidentally. When power is lost, we can do better than have the microcontroller flail around on the I/O lines, causing momentarily strange and erratic behavior of a device. Thus a printer buffer should not write a few spurious characters to the printer every time a computer system is turned off. If some RAM is maintained through a power outage (either with battery backup or by using a nonvolatile technology), then again we do not want the microcontroller to lose control, flail around, and inadvertently write into this nonvolatile RAM. We will see the support which can be given to this issue, both by the microcontroller and by an external chip.

2.2 MICROCONTROLLER FAMILY MEMBERS

In order to describe microcontrollers within the constraints of present-day technology, we will consider two outstanding examples of microcontrollers in this book:

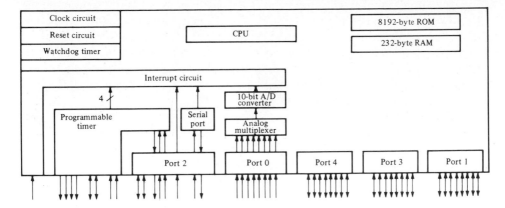

FIGURE 2.1
Intel 8096 block diagram (for single-chip operation).

Intel 8096

Motorola 68HC11

A block diagram of the Intel 8096 is shown in Fig. 2.1. That of the Motorola 68HC11 is shown in Fig. 2.2. Actually, both of these represent *families* of parts. Family members may start from the same internal chip but differ in the number of internal pads which are brought out to external pins. Thus the Intel 8096 is available in the 48-pin DIP (dual in-line package), the 68-lead PLCC (plastic-leaded chip carrier), the 68-pad LCC (leadless chip carrier) package, and 68-pin PGA (pin-grid array) package shown in Fig. 2.3. Furthermore, the production yield on the part is boosted by offering different versions of the chip, depending upon whether the internal ROM (read-only memory, for program storage) is functional or not. Obviously, if it is not, then the chip cannot

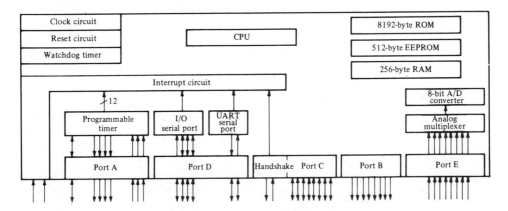

FIGURE 2.2
Motorola 68HC11 block diagram (for single-chip operation).

FIGURE 2.3
Intel 8096 packages. A 68-pad LCC package is shown in the lower left; two 68-pin PGA packages are shown in the upper left and upper center; two 48-pin DIP packages are shown on the right; a 68-lead PLCC package is shown in the lower center. (*Intel Corp.*)

be used as a single-chip microcontroller, but rather must access its program from off-chip memory. Also the part is offered in different versions depending upon whether the internal A/D (analog-to-digital) converter circuitry meets specifications. A user can obtain a lower-cost part if A/D conversion capability is not needed. All of these packaging and resource alternatives are discussed at the end of Appendix B.

The Motorola 68HC11 likewise comes in several versions. The 68HC11 can be obtained in the 48-pin DIP package or in the 52-lead quad pack (also called a PLCC) shown in Fig. 2.4. The 48-pin package requires a user to get along without four of eight possible inputs to an analog multiplexer, the output of which feeds into an on-chip A/D converter.

These two families of chips include ROM versions for program storage (necessarily programmed by the manufacturer at the time the chip is made). Each is also available as a user-programmable part. These options are discussed in Sec. 2.4.

2.3 BUS WIDTHS

Some early microcontrollers dealt exclusively with 4-bit operands. Thus, in adding two numbers together, the numbers had to be added 4 bits at a time. More recent microcontrollers have typically dealt with 8-bit operands (or possibly 16-bit operands for selected instructions).

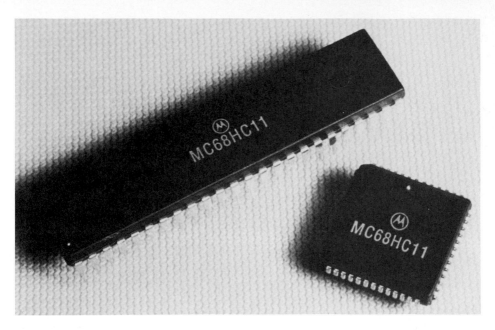

FIGURE 2.4
Motorola 68HC11 packages. (*Motorola Semiconductor Products Inc.*)

The Intel 8096 is inherently a 16-bit microcontroller in that the *data path* for operands is 16 bits wide. That is, when data is transferred between RAM or ROM and the CPU, it is transferred 16 bits per internal memory cycle. Because of this, we must *align* all 2-byte variables as we assign memory to them so that they begin at *even* addresses. When the CPU accesses a 2-byte variable in RAM, it accesses this even address and the immediately following address. The bus is organized to access these bytes simultaneously. (It is *not* organized to access the content of an odd address and the immediately following address simultaneously.)

The fetching of a 3-byte instruction (that is, 24 bits) will take two cycles. The addition of two 16-bit words takes the same time as the addition of two 8-bit bytes (assuming that the operands are accessed using the same addressing mode). As we will see in Chap. 4, the Intel 8096 also has some instructions for dealing with 32-bit operands (double words) as well as single-bit operands (for bit manipulations). These options are illustrated in Fig. 2.5.

The Motorola 68HC11 is an 8-bit microcontroller. That is, the data path for operands is 8 bits wide, and the vast majority of the instructions deal with 8-bit operands. While some instructions deal with 16-bit operands, each transfer of a 16-bit operand over the data bus takes two cycles. As in the case of the Intel 8096, the Motorola 68HC11 also includes instructions for dealing with single-bit operands. These options are illustrated in Fig. 2.6.

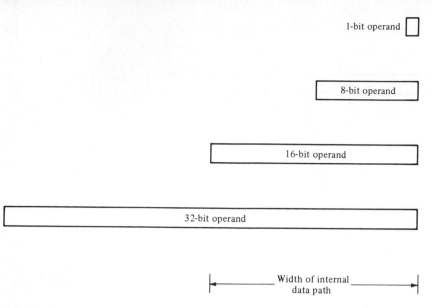

FIGURE 2.5
Length of operands handled by various Intel 8096 instructions.

The Motorola 68HC11 has an internal 16-bit address bus to access 2^{16} addresses, as shown in Fig. 2.7. In contrast, the Intel 8096 has the internal *eight*-bit address bus shown. The designers of this chip recognized that all on-chip RAM and register accesses can be identified by an 8-bit address, since there are fewer than 2^8 RAM and register locations. Jump instructions require an extra cycle (compared with what would be required if the address bus were 16 bits wide) to modify the latch contents which drive the upper byte of the ROM address.

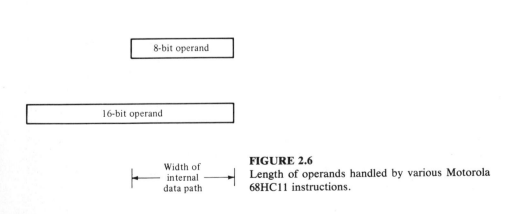

FIGURE 2.6
Length of operands handled by various Motorola 68HC11 instructions.

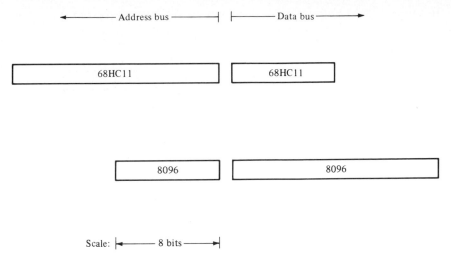

FIGURE 2.7
Motorola 68HC11 and Intel 8096 bus widths.

Both the Intel 8096 and the Motorola 68HC11 are designed to be able to access any of 65,536 addresses. While this represents a lot more on-chip resources than either chip really has, both chips can also be run in an *expanded* mode and can access off-chip resources (i.e., extra RAM or ROM or I/O ports) via these addresses.

The Intel 8096 includes an interesting option when operated in the expanded mode. While all internal accesses take advantage of the speed inherent in a 16-bit data bus, external accesses can be made to use *either* a 16-bit data bus or an 8-bit data bus. In the interest of minimizing parts count, an external 8-bit data bus can be chosen, at least tentatively. However, if the resulting speed of operation is unsatisfactory, then the circuitry can be changed to use an external 16-bit data bus. Refer to Sec. B.2 in Appendix B for more information on these two alternatives.

2.4 PROGRAM MEMORY

The memory used to hold the microcontroller's program is almost invariably *nonvolatile* memory. That is, it retains its content when power to the chip is turned off and then subsequently turned on again. One option available to designers is to have the program put into a *ROM*, or read-only memory. This option entails going to the manufacturer and paying a one-time mask charge of $3000 to $4000, and then placing a minimum order for maybe 1000 pieces. The completed order will follow in 12 to 15 weeks. Follow-on orders for parts do not require a repeat of the mask charge (assuming that the program can remain unchanged). The microcontroller parts cost might typically be $14 (1000 quantity) down to $7.50 (100,000 quantity).

Because of these constraints, designers want to be quite sure that they have a bug-free program, with all of the desired features working correctly, before committing themselves to the mask charge and the minimum quantities which a ROM implementation entails. One way to achieve this is to use a development system employing an *in-circuit emulator* to carry out debugging and development. We will discuss this in Chap. 8.

If a manufacturer of the microcontroller offers a version of the chip which the *designer* can program, then this presents a very attractive alternative to mask-programmed ROMs. This is especially true early in the production phase of a project when bugs may still be present in the program and features may be found wanting.

A microcontroller family of related parts will often include an EPROM (erasable, programmable read-only memory) version. Thus, the Intel 8096 family includes the several EPROM versions shown in Fig. 2.8. The quartz lid permits a user to erase the program stored in the EPROM with ultraviolet light; the light is supplied by an EPROM eraser, examples of which are shown in Fig. 2.9. Programming the chip requires a PROM programmer. The unit shown in Fig. 2.10 is a leading manufacturer's answer to having one top-of-the-line programmer for virtually all programmable chips.

The Motorola 68HC11 has at least 512 bytes of EEPROM (*electrically erasable, programmable read-only memory*) and is available in more expensive versions with either 2K (actually 2048 bytes) or 8K of EEPROM. EEPROM is

FIGURE 2.8
Intel 8096-family EPROM parts. (*Intel Corp.*)

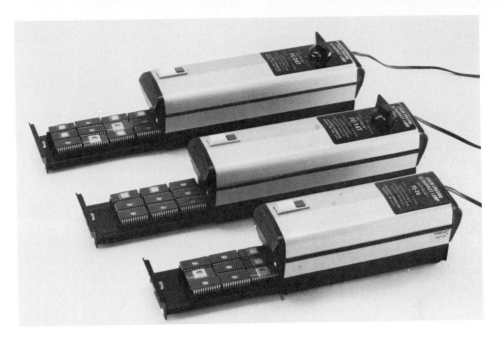

FIGURE 2.9
EPROM erasers. (*Spectronics Corp.*)

even easier to use than EPROM because it erases quickly and does not require the use of a separate device (an ultraviolet eraser) before the chip can be reprogrammed.

It represents too much of a digression at this point to discuss the programming and erasing of EPROM and EEPROM parts in this chapter. However, in Chap. 8 we will consider extremely simple programming alternatives which each of these chips makes possible. Then in the appendixes we will provide more thorough information on programming and erasing.

2.5 DATA MEMORY

A microcontroller is a device for which the designer feels the constraints of limited resources much faster than for a multiple-chip microcomputer. With the latter, the need for more memory can usually be handled by adding another memory chip. In the case of a (single-chip) microcontroller, "what you see is what you get." For example, the Motorola 68HC11 chip includes three kinds of memory:

> ROM (or possibly EEPROM) for program
>
> RAM (random access memory—really, read-write memory) for variables
>
> EEPROM (electrically erasable, programmable read-only memory) for

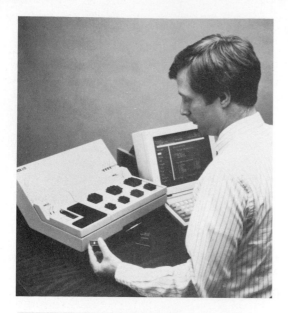

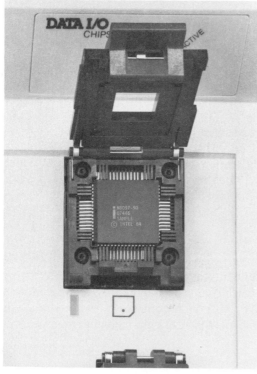

FIGURE 2.10
EPROM programmer. (*a*) Unit; (*b*) socket for Intel 8096-family part, PLCC package. (*Data I/O Corp.*)

nonvolatile storage of variables (i.e., storage which can survive power being turned off and on again)

The "normal" data memory (256 bytes of RAM), takes up an amount of space on the chip which is of the same order of magnitude as for the 8192 bytes of ROM taken by the program. This is the reason for the disparity in the amount of RAM and ROM on the chip; each byte of RAM "costs" much more in terms of chip area than each byte of ROM.

The Motorola 68HC11's RAM can be written into or read from during a single internal clock cycle of 0.50 μs. That is, the STAA instruction which stores the content of an accumulator to RAM requires several cycles for its execution, but only one of those cycles is used for the actual storing of data into RAM.

The Motorola 68HC11's EEPROM looks just like the chip's RAM or ROM as far as the time required for *reading* is concerned. Actually, this is true *only* if the EEPROM is undergoing neither an erase operation nor a program operation at the time when we want to read from the EEPROM. It is this potential conflict between reading EEPROM and modifying EEPROM which deters a designer from casually using the EEPROM as more read-write memory, in addition to the on-chip RAM.

The designers of the Motorola 68HC11 have given us a versatile repertoire of ways to modify the EEPROM, given that these modifications are going to be slow, requiring 10 ms for an erase operation or for a program operation. First, it is necessary to *erase* a location before writing to it. There are three erase modes, each taking 10 ms:

A single byte of EEPROM can be erased

A "row" of 16 bytes can be erased

The entire EEPROM can be erased

The programming of the EEPROM also requires 10 ms/byte. Thus, the programming of 512 bytes of EEPROM takes about 5 s.

Providing for nonvolatility is a tricky subject. It is not enough simply to have EEPROM or battery-backed-up RAM available. It is also necessary to make sure that at the moment when power is lost, the CPU is not permitted to "go berserk," writing crazily to random locations in memory. The 68HC11 provides the "hooks" to protect its EEPROM at such a time. We will discuss these hooks at the end of this chapter.

The Intel 8096 has 256 addresses which it divides between 232 bytes of RAM for variables and 24 registers for monitoring and controlling the assorted auxiliary functions on the chip (e.g., the programmable timer function). To provide for nonvolatility, 16 bytes of RAM can be "kept alive" with a battery supplying about one milliampere of current to a separate V_{PD} "power-down" pin on the chip when +5 V power is removed from the normal V_{CC} pin. During normal operation, or if this feature is not needed, these 16 bytes of RAM derive their power through the normal V_{CC} pin. If more nonvolatile memory is needed, it can be obtained by using an off-chip EEPROM or a battery-backed-up CMOS RAM. Augmenting RAM by adding extra RAM chips on the microcontroller's expansion bus will be discussed in Chap. 6.

2.6 PARALLEL PORTS

In many, if not most, applications of a microcontroller, its most limited resource is the small number of input and output ports. Accordingly, the pins for these ports tend to have a versatility that lets each one serve more than one purpose, depending upon the needs of the designer. Earlier generations of microcontrollers tended to employ 40-pin packages and were thus even more tightly constrained than either the Motorola 68HC11 (with 48 or 52 pins, depending upon the packaging) or the Intel 8096 (with 48 or 68 pins). In any case, a commonly used technique to make the I/O ports go as far as possible is to *share* their pins with other optionally used hardware facilities on the chip. For example, in the next chapter we will discuss how a programmable timer supports real-time control. Then in Chap. 6, we will see how the resources of a microcontroller can be expanded with the help of either a UART serial port or an I/O serial port. If we choose not to use one or more of these resources, we can gain, instead, extra lines on general-purpose I/O ports.

Figure 2.1 shows this sharing of the pins making up ports 0 and 2 of the Intel 8096. Any lines which are not needed as analog inputs to the A/D converter can be used as general-purpose digital inputs to port 0. Likewise, if the chip's serial port capability is not needed, then this frees up an input line and an output line on port 2. As can be seen in Fig. 2.1, several other lines of port 2 are also shared. While that figure does not show it, the entirety of ports 3 and 4 are shared with drivers and receivers which let them become a multiplexed address and data bus if we wish to operate the chip in its expanded mode rather than in its single-chip mode. We will discuss this in Chap. 6.

The *quasi-bidirectional* lines of the Intel 8096's port 1 can serve as outputs simply by writing to them. The pullup capability of the output is very weak except for the one clock cycle following a "write" to the port which changes the output from low to high. This is done to speed up the output transition when significant capacitive loading is present on the output. To use one of these lines as an input, it is only necessary to write a 1 to the line first. This can be overridden by an external device driving the line either high or low.

The Intel 8096's ports 3 and 4, which have open-drain outputs, can be used in much the same way. Since the open-drain output can only pull the output low, any line to be used as an output needs the addition of a pullup resistor, as shown in Fig. 2.11, to pull the line high when a 1 is written to that line of the port. Any line to be used as an input need only have a 1 written to it first (to disable the pulldown driver).

The Motorola 68HC11 has the five parallel ports shown in Fig. 2.2. Any

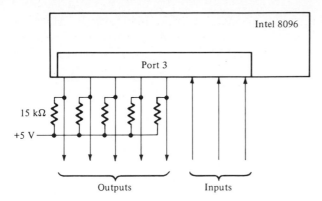

FIGURE 2.11
Use of pullup resistors to create output lines from open-drain outputs.

lines which are not serving the specialized alternate use shown in that figure can serve a more general function. Two of these ports provide lines which can be individually configured as either inputs or outputs, while the other three ports are fixed in their configuration (one as an input port, one as an output port, and one as a mixture). Associated with the two bidirectional ports are two "data direction registers." The 1s and 0s written to a data direction register configure the corresponding lines of the associated port as outputs and inputs, respectively, as illustrated in Fig. 2.12.

The Motorola 68HC11 has one further feature associated with one of its general-purpose parallel ports. Two handshake lines are available so that port C can be set up as either a handshaking input port or as a handshaking output port. For example, as shown in Fig. 2.13, if port C is set up as a handshaking input port, then the external device presents a byte of data to port C and strobes the Strobe line (which can be made to respond to either edge of Strobe). A latch in port C captures the 8 bits of data (so that the external device's output need only be valid at the precise moment that the edge of Strobe occurs). At the same time, a flag is set (and optionally, an interrupt can be made to occur) in the 68HC11 signifying that data has been captured at the port. Also in response to the Strobe input edge, the Ready output automatically goes low. When the

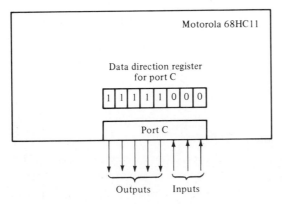

FIGURE 2.12
Role of Motorola 68HC11's data direction registers.

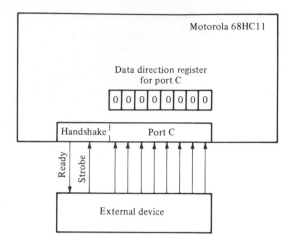

FIGURE 2.13
Handshaking input port.

CPU reads the data from the latch, the Ready line is automatically raised. This signals the external device that its last byte of data has been accepted by the 68HC11 and that the external device can now present the next byte for input.

2.7 A/D AND D/A CONVERTERS

In the interest of making microcontrollers which can be "all things to all people," chip designers have found the way to include analog input and output capability on-chip. The addition of *A/D* (analog-to-digital) conversion capability, in particular, means that a large variety of transducers for temperature, pressure, strain, position, etc., can be used *directly* with a microcontroller. Any transducer which generates an output voltage proportional to a desired physical parameter can be brought into the realm of the microcontroller, without the need for any additional parts.

The conversion of an analog voltage to a digital number can be implemented with several very different approaches:

Dual-slope A/D converters. These converters use a counter, analog integrator, voltage comparator, reference voltage, and control circuitry to build relatively slow but accurate A/D converters. By integrating an input voltage over an interval equal to one cycle of the power line frequency, they can be made to have excellent *normal-mode rejection.* That is, they can be made to measure a small analog voltage (e.g., a thermocouple's output in the millivolt range) and yet ignore a much larger 60-Hz noise signal impressed upon this small dc voltage.

Successive-approximation A/D converters. These converters use a D/A converter, voltage comparator, reference voltage, and control circuitry to carry out fast conversions.

Flash A/D converters. These converters use a tapped resistor to divide a reference voltage into 2^n equal parts, 2^n voltage comparators to compare the input voltage against each of these, and combinational circuitry (i.e., a 2^n-line-to-n-line priority encoder) to generate the digital output. This is the fastest of the three approaches, depending solely upon the propagation delays through the circuitry. However, it requires a vast amount of circuitry relative to either of the other two approaches.

The Intel 8096 and the Motorola 68HC11 each includes a successive-approximation A/D converter. Each is configured with an analog multiplexer to permit the A/D converter to convert the voltage on more than just one input pin.

The successive-approximation conversion process is analogous to making successive comparisons of an unknown weight using known weights of 2^n, 2^{n-1}, $\cdots$, 4, 2, and 1 oz. Such a process is illustrated in Fig. 2.14. Note that one comparison is required for each bit of the binary result, which is 10110001 (binary) or $128+32+16+1 = 177$ (decimal).

If all we want to do is to make isolated measurements on analog inputs, then it really does not matter whether the input changes during the conversion process. The converter output will be equal to what the input was *at some time* between the first comparison and the last comparison. That is, we will never get a wildly incorrect result like 00000000 because the input was equivalent to 01111111 when the first comparison was made and equivalent to 10000000 for the subsequent comparisons. In fact, in this case the result will be 01111111.

If, on the other hand, we want to carry out *fast sampling* of an analog input voltage, then we need to be more concerned about an input that changes during the conversion process. Since the output will be equal to a conversion of what the input was at some time during the conversion process, the effect of a changing input is to produce *jitter* into the sampling process; that is, the samples do not appear to be equally spaced in time. If the conversion time is a significant portion of the sample time, then the effect will be more profound than if the conversion time is essentially instantaneous relative to the sample time. For example, the A/D converter in the first generation of Intel 8096 parts carried out a conversion in 42 μs. If it were used to sample a waveform every millisecond, then the jitter of 42 parts in 1000, or 1 part in 24, was not very significant. On the other hand, if a waveform were sampled at the maximum rate of once every 42 μs, then the jitter became significant. This is illustrated in Fig. 2.15.

To understand why the successive-approximation conversion process produces the effect of jitter into the sampling process, consider the 4-bit converter shown in Fig. 2.16*a*. In Fig. 2.16*b* we see three analog inputs plotted versus time during the conversion process. Also on the vertical scale, each digital output value is listed at the voltage level which the D/A converter of Fig. 2.16*a* would produce, given this digital value as input. For example, note that the two consecutive values 1100 and 1101 are shown bounding the range of dc inputs which will actually produce an output coding of 1100.

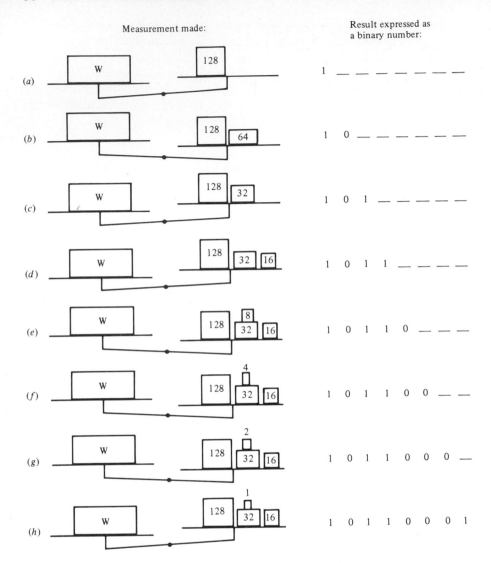

FIGURE 2.14
Successive-approximation A/D conversion analogy. (*a*) $W > 128$; (*b*) $W < 128+64$; (*c*) $W > 128+32$; (*d*) $W > 128+32+16$; (*e*) $W < 128+32+16+8$; (*f*) $W < 128+32+16+4$; (*g*) $W < 128+32+16+2$; (*h*) $W > 128+32+16+1$.

The conversion requires four consecutive comparisons, each taking place at one of the times shown (e.g., second comparison). The "compare" voltage of Fig. 2.16a at each comparison depends upon the previous measurement. The possible values are marked by small rectangles. For example, the first comparison of the analog input will be made against the voltage produced when the D/A converter has 1000 as its input. The second comparison will be made against

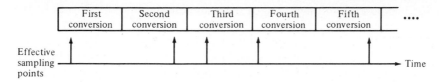

FIGURE 2.15
Effective sampling points when *not* using a sample and hold circuit.

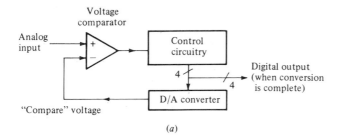

(a)

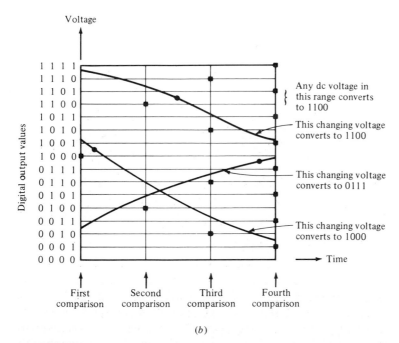

(b)

FIGURE 2.16
Successive-approximation A/D conversion. (a) Circuit; (b) conversion of changing voltages.

the voltage produced by either 1100 or 0100, depending upon the result of the first comparison.

Now consider the top example of a changing analog input. Its conversion results in comparisons against 1000, then 1100, then 1110, and finally 1101. The resulting output is 1100. Note the encircled point on the curve, placed halfway between the 1100 and the 1101 threshold levels, corresponding to the average voltage level which will produce an output of 1100. Since this point occurs about halfway between the second and third comparison, it is as if we had sampled the changing input waveform at this point. The other two examples of changing analog inputs produce digital outputs which, if aligned with *when* the analog input equaled this value, appear to have been sampled at different times in the conversion process. That is, the input which produces the output 0111 appears to have been sampled just before the fourth comparison whereas the one which produces 1000 appears to have been sampled just after the first comparison.

A *sample-and-hold* circuit in effect takes a snapshot of the analog input voltage and then holds this value during the conversion process. The result is shown in Fig. 2.17 in the case where a new conversion is carried out immediately after the previous one has been completed. A block diagram of the complete multiplexer—sample-and-hold—A/D converter combination is shown in Fig. 2.18a. This figure also illustrates that the A/D converter really converts the analog input voltage *relative to* the reference voltage range defined by the voltage inputs labeled

$$V_{ref} \text{ and } V_{gnd}$$

In fact, both the Intel 8096 and the Motorola 68HC11 bring these two voltage inputs to their A/D converters out to pins so that their voltages can be closely controlled to +5 V, independent of any voltage drop occurring in the lines supplying the normal +5 V power to the chip.

The Intel 8096 and the Motorola 68HC11 designers have produced A/D converters which possess somewhat different performance characteristics. The second-generation 8096 converter is a 10-bit converter, with sample-and-hold, which carries out individual conversions in 22 μs. The 68HC11 converter is an 8-bit converter, with sample-and-hold, which carries out *four* conversions in a total of 64 μs. The four conversions can be four samples of a single input

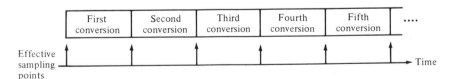

FIGURE 2.17
Effective sampling points when using a sample-and-hold circuit.

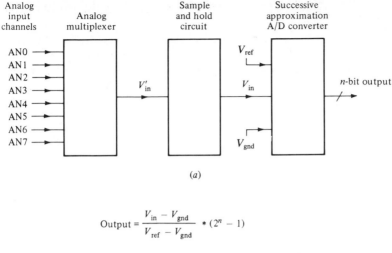

(a)

$$\text{Output} = \frac{V_{in} - V_{gnd}}{V_{ref} - V_{gnd}} * (2^n - 1)$$

If $V_{in} = V_{gnd}$, then output $= 00 \cdots 000$

If $V_{in} = V_{ref}$, then output $= 11 \cdots 111$

(b)

FIGURE 2.18
Multiplexer—sample-and-hold—A/D converter combination. (a) Circuit; (b) input-output relationship.

spaced 16 μs apart. Alternatively, they can be four samples taken one right after the next, each from a different input channel and sampled 16 μs apart. These two possibilities are illustrated in Fig. 2.19.

The Motorola 68HC11's A/D converter also offers the possibility of *continuous scanning*, as shown in Fig. 2.20. After each batch of four samples is taken, a flag is set, and the CPU must grab the samples quickly before they are overwritten with new samples. Unfortunately, the sample rate of this continuous scanning is not made available under program control. Either we must use the maximum rate or we must collect four samples, wait, collect four more samples, wait, etc.

The Intel 8096's A/D converter supports the continuous collection of samples by letting the A/D converter be triggered from one output of its programmable timer (a facility to be discussed in the next chapter on real-time control). Consequently, we can easily set an arbitrary sample rate (e.g., 100 samples per second) and have the timer initiate each conversion. Furthermore, since the completion of each conversion can trigger an interrupt, the CPU need only be involved each time a new sample is ready.

Features of these two A/D converter capabilities are listed in Fig. 2.21. It might be noted that the specification on absolute accuracy for Intel's 10-bit

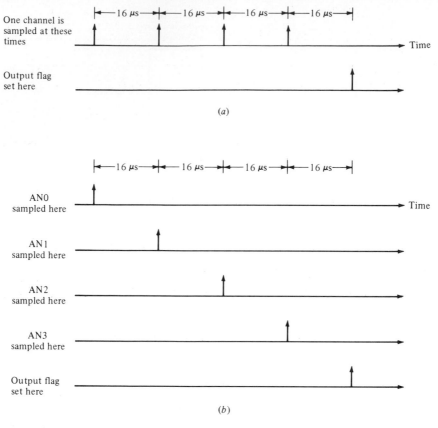

FIGURE 2.19
Motorola 68HC11 single scan. (*a*) One channel; (*b*) four channels.

converter is the same as that for Motorola's 8-bit converter. In many applications, particularly feedback control applications and applications involving small differences between two measurements, *resolution* is even more important than absolute accuracy. In such cases, the extra 2 bits of the Intel converter are valuable for the fourfold increase in resolution which they provide.

In comparing the contribution which an A/D converter can make to the capability of a microcontroller with that which a D/A converter can make, it is probably fair to say that the former is more valuable in more applications. An A/D converter can make large inroads into what might otherwise be rather intractable transduction problems. A D/A converter helps with *actuation*; that is, with making something happen (e.g., a movement or a temperature change). Actuation is a job which is often amenable to a digital solution. For example, position control can be handled with stepper motors which can be driven with the digital output from a microcontroller. A heating coil for controlling temper-

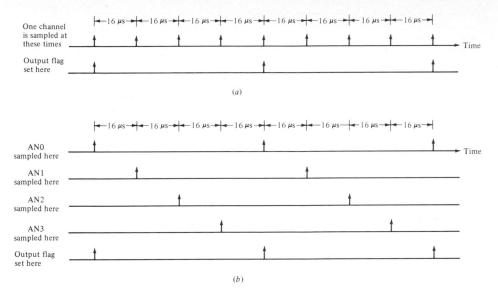

FIGURE 2.20
Motorola 68HC11 continuous scan. (*a*) One channel; (*b*) four channels.

ature can be driven with a pulse-width modulated output from a microcontroller, giving an average dc value which is proportional to the duty cycle of the output, as shown in Fig. 2.22.

Any microcontroller can generate a pulse-width modulated output. If this output is used to drive a heating coil, then the period of the output waveform can be very long (e.g., 0.1 s) relative to the clock rate of the microcontroller. In this case, almost any way we implement the function in software will give great accuracy and take only a tiny percentage of the microcontroller's execution time.

The Intel 8096 includes a pulse-width modulator as one of its on-chip resources. Once it has been set up, it requires no further CPU intervention to continue generating a waveform with a period of 64 μs and a duty cycle of any value between 0 and 255/256, depending upon the 8-bit number written to a PWM register. Because of this, it provides a reasonable way to carry out a D/A conversion if all we need is a waveform with a *dc component* which is proportional to a digital quantity. This might be used directly (e.g., in a heating coil application). Alternatively, as shown in Fig. 2.23, a low-pass filter can be used to eliminate the ac component of the output and thereby obtain a dc output.

The Motorola 68HC11 includes a feature of its programmable timer which supports up to *four* pulse-width modulated outputs. Once set up, each requires no further CPU intervention to continue generating a waveform whose dc component takes on any one of 65,536 values. We will discuss this capability in further detail in Chap. 7 and also in the appendixes.

	Intel 8096	Motorola 68HC11
Number of A/D channels on large package	8	8
Number of A/D channels on 48-pin DIP package	4	4
Sample and hold	Yes	Yes
Resolution	10 bits $0.001V_{ref}$	8 bits $0.004V_{ref}$
Accuracy	$\pm0.004V_{ref}$	$\pm0.004V_{ref}$
Minimum number of conversions made before a "Done" flag is set	1	4
Conversion time per conversion	22 μs	16 μs
Conversion time before "Done" flag is set	22 μs	64 μs
Alternatives for starting conversion:		
By setting a control bit	Yes	Yes
From programmable timer output	Yes	No
Continuous	No	Yes
Measurement mode alternatives:		
One measurement on one channel	Yes	No
Four measurements on one channel	No	Yes
One measurement on each of four channels	No	Yes
Continuous measurements on one channel	No	Yes
Continuous measurements on four channels	No	Yes
Number of buffer registers for storing successive measurements without program intervention	1	4
Alternatives for notification when done:		
By setting a flag	Yes	Yes
By generating an interrupt	Yes	No

FIGURE 2.21
Features of the Intel 8096 and Motorola 68HC11 A/D converters.

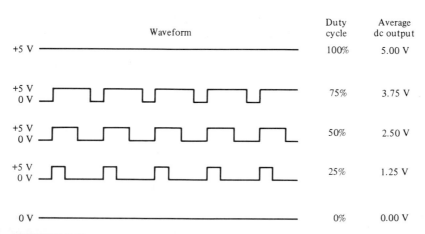

FIGURE 2.22
Pulse-width modulation to achieve variable average dc output.

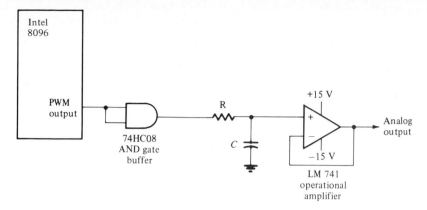

FIGURE 2.23
D/A conversion via the dc component of a pulse-width modulated waveform.

2.8 RESET CIRCUITRY AND WATCHDOG TIMERS

The reset input to a microcontroller can serve the reset function under as many as four circumstances:

Push-button reset

Power-on reset

Power-glitch reset

Watchdog-timer reset

The job of a reset circuit is to handle each possibility reliably. Push-button reset, if that were all the resetting capability we needed, is simple to implement, as shown in Fig. 2.24. This figure assumes that the reset input is active-low, executing the reset function when the pin is pulled low, as is usual for micro-controller chips.

When power is first turned on, we want to ensure that the microcontroller sees its active-low reset input held low until after the power to the chip reaches its specified minimum value for proper operation of the chip. For both the Intel

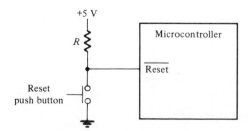

FIGURE 2.24
Push-button reset circuit.

Voltage

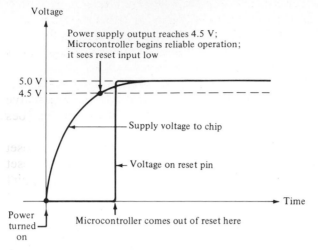

Power supply output reaches 4.5 V;
Microcontroller begins reliable operation;
it sees reset input low

5.0 V

4.5 V

Supply voltage to chip

Voltage on reset pin

Time

Power turned on

Microcontroller comes out of reset here

FIGURE 2.25
Power-on reset timing.

8096 and the Motorola 68HC11, this minimum operating voltage is 4.5 V (i.e., 5.0 V − 10 percent). If the reset input is *not* held low until after this time, then the CPU will not see it and will not act in a manner which we expect. The desired reset input signal is illustrated graphically in Fig. 2.25.

The simplest power-on reset circuit used with microcontrollers is the *RC* circuit shown in Fig. 2.26. For this to work correctly, the *RC* time constant must be long enough to hold the reset input below its threshold value until after the time specified in Fig. 2.25. This calls for a large *RC* value. On the other hand, the *R* value must be low enough to pull the reset input high in spite of the leakage current specification for the reset input. These requirements can lead to a large value for *C* (e.g., 1 to 100 μF).

If power is lost momentarily, the microcontroller (and other chips in the device being designed) can go awry when the power drops below 4.5 V. This is bad enough in and of itself. However, if the power loss is only milliseconds long, then we want to ensure that when power is restored above 4.5 V, the reset input will be pulled low. Otherwise the microcontroller's program counter might have had garbage loaded into it and it will not get restarted again correctly, if at all.

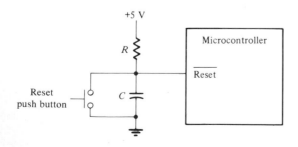

+5 V

Microcontroller

R

Reset

Reset push button

C

FIGURE 2.26
RC—power-on reset circuit.

Even if this *RC* reset circuit is designed properly to handle power-on resetting, it is probably unsatisfactory for handling power-glitch resetting. That is, when power is momentarily lost to the chip, the large capacitor in the reset circuit can hold the reset line high for the duration of the power glitch. Thus when power is restored to the chip, it will *not* be reset.

As we shall see in the next section, an *RC* reset circuit is also unsatisfactory for *halting* the microcontroller when power is turned off. The alternative of having the microcontroller wildly changing its output lines as power goes down is unsatisfactory for many applications.

The solution to all of these problems is to use a *voltage-threshold* reset circuit instead of a time-constant reset circuit. Such a circuit will hold the Reset input low until the power supply output rises above 4.70 V or so. At that point the Reset pin is snapped high, starting CPU operation. When power is lost, the voltage-threshold reset circuit waits until the power supply output drops below 4.65 V or so. At that point the Reset pin is snapped low, halting CPU operation. Such a circuit is shown in Fig. 2.27 using a low-cost part which is internally trimmed to give a turn-off threshold of 4.65 V.

The final circumstance which we might want to handle is resetting the microcontroller by a watchdog timer. A watchdog timer is a feature of some microcontrollers, including both the Intel 8096 and the Motorola 68HC11. Its function is to detect that, somehow or other, the microcontroller is "off in the weeds." That is, it has stopped executing its program correctly. As we will see in Chap. 4, the stack is a critical feature for maintaining order in the sequencing of a program. If the stack ever is inadvertently corrupted, the microcontroller might *never* get back on track, executing its program correctly. A watchdog timer can restore proper operation by doing the most drastic of things; it can pull the reset line low *from within the microcontroller*! It does this not only to

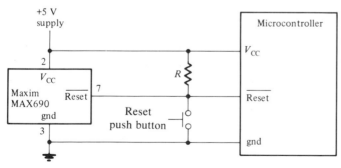

Maximum Integrated Products
MAX690
Reset generator
Lower threshold = 4.65 V
8-pin DIP

FIGURE 2.27
Voltage-threshold reset circuit.

reset the microcontroller, but also to reset any other chips tied to this same reset line.

Before looking at the specific features of the watchdog timers in the Intel 8096 and the Motorola 68HC11, let us consider the implication of having a watchdog timer upon the reset circuitry we have been discussing. Of the three variations we have been discussing, the only ones which will work satisfactorily when an internal driver tries to drive the reset line low is the push-button reset circuit of Fig. 2.24 and the voltage-threshold reset circuit of Fig. 2.27. The resistor in Fig. 2.24 has a high value chosen solely to ensure that the reset input is pulled high and out of the reset state, in spite of the effect of any leakage currents within the chip. The watchdog timer's driver necessarily has enough drive capability to pull the line low in spite of this resistor.

The capacitor in the circuit of Fig. 2.26 may have either of two bad

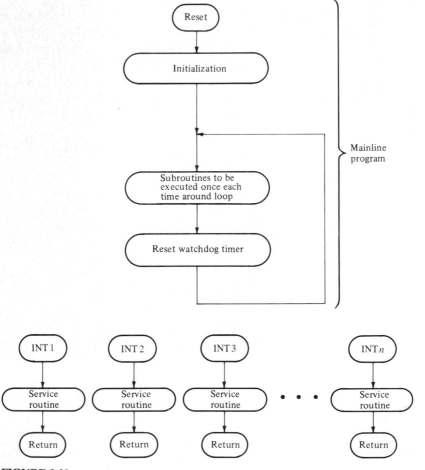

FIGURE 2.28
A common microcontroller program organization.

effects. If the watchdog timer's reset driver does not limit its current to a safe value, then it may burn itself out (or degrade its life expectancy) as it tries to discharge the capacitor. Even if the driver's current is safely limited, the timer may not be activated for long enough to discharge the capacitor. For example, the Intel 8096's watchdog timer drives the reset line for only 0.5 μs.

The voltage-threshold reset chip in Fig. 2.27 includes a *weak* pullup which can be safely overridden by a watchdog timer's driver. Thus, in a single chip we get all of the performance which is needed to reset the microcontroller under all circumstances which we have discussed.

Use of a watchdog timer presupposes that the microcontroller's program is built around a loop of routines which are repeatedly executed. For example, consider the organization shown in Fig. 2.28. This illustrates a program which, after reset, carries out whatever initialization is required. Then it proceeds to a loop of routines which are executed again and again. A variety of on-chip facilities (e.g., A/D converter and parallel port data transfers) may be set up to be handled under interrupt control, using one of the interrupt service routines shown at the bottom of the figure. Each time one of these facilities causes an interrupt, the execution of the mainline program is set aside and then the corresponding interrupt service routine is executed. When it has been completed, execution of the mainline program continues.

Within the loop of the mainline program, we imbed the instructions which will reset the watchdog timer. We depend on our knowledge of the routines we have written, and on our knowledge of the frequency and duration of the interrupt service routines, to select a "safe" timeout interval for the watchdog timer. Nobody gets points for being close. So, our tendency is to pick a timeout interval which is *considerably* longer than the longest time we estimate it will ever take to go around the loop.

The characteristics of the Intel 8096 watchdog timer are presented in Fig. 2.29. Those for the Motorola 68HC11 are shown in Fig. 2.30. Note that both can be set aside and never used. Note also that once either one is enabled, it is very difficult (and, in fact, impossible for the 8096) to disable it. This is done so that if the microcontroller goes awry, it will not inadvertently turn off the very mechanism which could get it started all over again.

Finally, note that resetting either watchdog timer is purposely done as a

Disabled: By resetting the microcontroller chip.

Enabled: By carrying out the first "reset watchdog timer" sequence described below. Once enabled, it cannot be disabled again (without resetting the chip).

Timeout interval (with 12-MHz crystal): 16.384 ms

To reset watchdog timer: First write 00011110 (binary) to address 000A (hex); then write 11100001 (binary) to address 000A (hex).

FIGURE 2.29
Intel 8096 watchdog timer characteristics.

Disabled: By a 1 stored in an EEPROM bit.

Enabled: By a 0 stored in an EEPROM bit.

Timeout interval (with 8-MHz crystal): Programmed by two bits in a register which can only be changed during the first 64 clock cycles after reset. The options are: 16.384 ms, 65.536 ms, 262.14 ms, and 1.048 s

To reset watchdog timer: First write 01010101 (binary) to address 003A (hex); then write 10101010 (binary) to address 003A (hex).

FIGURE 2.30
Motorola 68HC11 watchdog timer characteristics.

rather intricate, two-step process. Again, this is done so that a malfunctioning microcontroller will not inadvertently, and indefinitely, postpone the timing out of the watchdog timer by *accidentally* executing its reset sequence again and again.

2.9   POWER-DOWN CONSIDERATIONS

A microcontroller can take advantage of the low power dissipation of CMOS RAM to provide nonvolatile data storage. This is valuable for any setup data which defines the mode of an instrument or device. For example, a home automatic lawn sprinkler system can have a controller like that of Fig. 1.14 which keeps track of what days to sprinkle, what time of the day, and how long. A valuable feature for such a unit is its ability to retain this setup data through short power outages. Similarly, a complex test instrument may require an extensive setup procedure. If the most recent setup information is retained in nonvolatile memory, then the user is not burdened with the tedious task of reentering this setup information each morning or after a power outage.

Not only is it necessary to maintain power to the CMOS RAM chip retaining data, it is also necessary to ensure that the microcontroller which uses it does not do any wild flailing as it loses power. When power drops below the specified range for proper performance (e.g., 5.0 V − 10 percent = 4.5 V), if the microcontroller's clock is still running and it is trying to execute instructions, then it has the potential to destroy data in the CMOS RAM. In fact, it has so much potential to produce erratic performance that many instrument and device designs monitor the power supply voltage and shut down the microcontroller as power begins to fade, *whether or not* data retention is an issue. To support battery-backup capability, a prime building block is a battery-backup circuit which will smoothly switch from ac power over to battery power when the ac power dies. It must smoothly switch back again when power is restored. Finally, it should "trickle charge" the battery when ac power is present so that the battery will be fully charged when it is needed.

The same voltage-threshold-resetting chip which was used in Fig. 2.27 also supports battery backup of the 68HC11 CMOS chip, using the circuit of

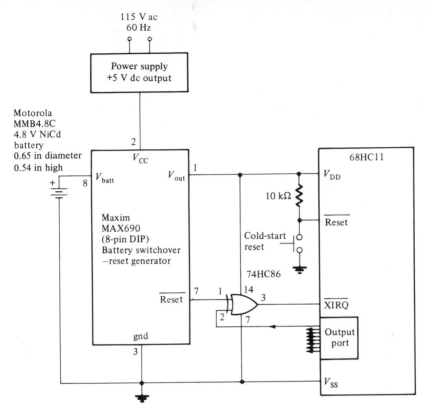

FIGURE 2.31
Battery backup of 68HC11.

Fig. 2.31. When power is lost, the MAX690's V_{out} pin is internally reconnected from V_{CC} to V_{batt}. When power is restored, the process is reversed. The tiny NiCd battery shown in Fig. 2.32 is one of a family, available in 1.2, 2.4, 3.6, and 4.8 V versions. The MAX690 will provide a small trickle charge to the 4.8 V version shown in Fig. 2.31 while the supply provides power.

Instead of using the MAX690's Reset output to control the Reset input to the 68HC11, its falling edge is used to interrupt the 68HC11, using a nonmaskable* interrupt input. The interrupt service routine uses one line of an output port going to an exclusive-OR gate to raise its XIRQ input. Then it executes a STOP instruction which halts all activity in the 68HC11, even stopping the clock. This puts the CMOS 68HC11 in a state where it dissipates hardly any power at all.

When power is restored, the MAX690's Reset output will go high, driving the 68HC11's XIRQ input low again. This will trigger the 68HC11 to exit from

---

* That is, an interrupt input which cannot be turned off and to which the CPU will respond immediately.

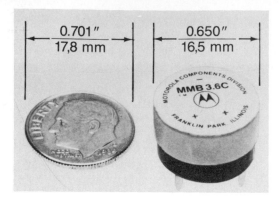

FIGURE 2.32
Rechargeable NiCd PC-board-mounted memory-backup battery. (*Motorola Inc., Components Div.*)

the STOP state and go on to the next instruction after the STOP instruction. In this way the 68HC11 can get going again in whatever manner is deemed appropriate, perhaps picking up where it left off when it was interrupted.

To see how the STOP instruction can have a disastrous effect upon the functioning of the watchdog timer, consider the following scenario. Completely independent of a power outage, the CPU might go awry, executing data as if it were program code. If the CPU were to fetch data which has the same value as the op code for the STOP instruction, its execution of the (enabled) STOP instruction would stop the clock. For this reason, the designers of the chip had the foresight to add an enable bit for the STOP instruction which is somewhat difficult to change accidentally. This is important because stopping the clock also stops the watchdog function, which would otherwise have gotten the CPU back on track by resetting the microcontroller.

After a power outage we may actually have no desire to continue as if nothing had happened. After all, the microcontroller is most likely serving as a controller in an instrument or device which has not had power for the rest of its circuitry supported by battery backup. The state of this other circuitry after a power outage is likely to be different from what it was beforehand, and continuing as if nothing had happened will lead to strange behavior. In this case, the CPU can recover by jumping to a Restart point in the program which initializes the state of the instrument to some reasonable condition. Then the instrument can begin again, based upon any setup information which has been retained.

A useful modification of this approach employs a state variable in RAM which can be updated as an ongoing activity during normal operation. Then when power is lost and subsequently restored, the microcontroller can use the state variable to decide how to begin again.

While the Motorola 68HC11 gives outstanding capabilities for handling a power outage, other microcontrollers have their corresponding capabilities. For example, the Intel 8096 has a pin which can be used to supply battery-backed-up power to 16 bytes of its internal RAM. With this approach, the heavier current load of the rest of the chip is permitted to drop to zero, letting the battery supply only the one milliampere or so needed by this section of RAM. The Reset pin can again be controlled by the MAX690 so that the chip will be reset as power drops away. This will keep the CPU from flailing around when power is lost even though the internal clock is not stopped.

2.10 SUMMARY

In this chapter we have looked at the resources available in a typical microcontroller chip. We have used the Motorola 68HC11 and the Intel 8096 as examples to give insight into two specific microcontroller implementations. More complete information on the features of these two microcontrollers is presented in the two appendixes at the end of the book.

The discussion of some microcontroller resources has been postponed to later chapters. In the next chapter we will consider factors which bear upon the use of a microcontroller for real-time control. Real-time control is usually the name of the game for a microcontroller. The microcontroller's *interrupt-handling mechanism* and its *programmable timer* provide for real-time interactions with external devices.

Another hardware feature of many microcontroller chips is the inclusion of some kind of a serial port. This facility permits the microcontroller to transfer data between itself and other devices efficiently, without using many pins on the chip. In Chap. 6, we will consider two kinds of serial ports. The I/O serial port will be used to expand I/O lines and to drive I/O devices which have a shift register input or output mechanism. The UART (Universal Asynchronous Receiver Transmitter) serial port provides an alternative standard interface to those devices which also have a UART.

Another major resource of a microcontroller is its instruction set and the manner in which it can access operands for its instructions. In Chap. 4 we will explore these for the Motorola 68HC11 and the Intel 8096.

PROBLEMS

Many of these problems are intended to be worked in terms of a specific microcontroller. Refer to the information on the Intel 8096 or the Motorola 68HC11 in the appendixes. Alternatively for another microcontroller, refer to a manufacturer's user's manual or equivalent information.

2.1. *Microcontroller family.* For a specific microcontroller, what are the different packages available? If some packages have fewer pins than others, then what functions are not brought out on the packages with fewer pins? Is there an EPROM version available for prototyping purposes?

2.2. *Microcontroller features.* Consider an earlier generation microcontroller (e.g., the Intel 8048) and depict its features in the same way as in Figs. 2.1 and 2.2. Use this to obtain a general overview of its on-chip facilities.

2.3. *Bus widths.* For a specific microcontroller:

(*a*) What is the width of the data bus for operands? If the chip supports any operations for both 2-byte operands and 1-byte operands, then do these take the same length of time?

(*b*) What is the width of the address bus? Does the address of an operand affect the access time for the operand?

2.4. *Program memory.* For a specific microcontroller, how many bytes of program memory does it have? Is it possible to get an EPROM part? Or an EEPROM part?

2.5. *Data memory.* For a specific microcontroller, how many bytes of data memory does it have?

2.6. *Memory.* For several microcontrollers, determine (and compare) the ratio of ROM memory bytes to RAM memory bytes. Is this ratio roughly equal among the microcontrollers you have looked at?

2.7. *Data memory.* For a specific microcontroller, is there any facility for nonvolatile storage of at least some variables (i.e., EEPROM, or a separate pin for battery backup of at least some RAM, or a low-power mode for battery backup of the entire chip)? If so, then what is the battery requirement (i.e., minimum voltage and maximum current drain)?

2.8. *EEPROM.* Consider a microcontroller which includes some EEPROM memory.

(*a*) Does programming or erasing require any unusual voltages (i.e., voltages other than +5 V)?

(*b*) What does an application program have to do to program the EEPROM?

(*c*) Can it be selectively erased, or can it only be erased in its entirety? How long does erasure take?

(*d*) How long does it take to program a byte? How does the CPU know when programming is done so that another byte can be programmed?

(*e*) How many erasure-reprogram cycles does the manufacturer guarantee?

2.9. *Data memory.* In some applications, all of the resources of a specific microcontroller may meet the requirements of the application except for the amount of RAM needed. Does the manufacturer support the addition of extra RAM in any way? For example, can the microcontroller be run in an expanded mode, where the internal busses are brought out so as to access additional external resources? Or are there RAM peripheral chips designed to work with this microcontroller which can be accessed through just a few pins?

2.10. *Parallel ports.* For a specific microcontroller:

(*a*) How are the parallel ports configured to be inputs? To be outputs?

(*b*) Are general-purpose parallel port lines shared with other I/O capabilities? How is the function of each pin determined in such a case?

2.11. *Parallel port control lines.* For a specific microcontroller:

(*a*) Is there a handshaking parallel port in which two control lines participate in handshaking automatically? If so, then can it be used as a handshaking input port? As a handshaking output port?

(*b*) Is there an "output strobe" which is automatically generated by writes to an output port? By reads from an input port? By either one, depending upon how it is set up?

(*c*) Is there an "input strobe" which either sets a flag or generates an interrupt (or both) whenever it sees an edge? Is there an input port which can latch its input data when this edge occurs? If so, can the latch be disabled (i.e., made transparent) if we do not want to synchronize input data on the port with an input edge?

2.12. *Parallel ports.* For a specific microcontroller, what is the drive capability of an output port for driving TTL parts?

(*a*) When an output is driven low by the microcontroller, how many milliamperes can it drive and still not be pulled up above 0.4 V? One 74LSxx input requires that this be at least 0.4 mA.

(*b*) When an output is driven high by the microcontroller, how many microamperes can it drive and still not be pulled down below 2.7 V? One 74LSxx input requires that this be at least 20 μA. .

(*c*) If an output port has an open-drain output, then what value of pullup resistor (to +5 V) will just barely supply the 20 μA at 2.7 V required by an 74LSxx input? What current load does this place on the output port when the microcontroller drives the line low, down to 0.4 V (or so)? Can the port handle this load?

2.13. *Parallel ports.* Microcontroller output ports have no trouble driving CMOS devices *low* since their input impedance is so high. On the other hand, CMOS parts like to have their inputs driven much higher than a TTL input needs. A typical "high" voltage input requirement is 4.0 V or higher (with a load current of zero). For a specific microcontroller, can its output ports satisfy this requirement? (If not, then pullup resistors will have to be attached to the outputs which are going to drive CMOS inputs.)

2.14. *A/D converter.* Consider the 8-bit successive approximation algorithm depicted in Fig. 2.14. Describe the operation of the nth weighing in terms of weight 2^{8-n}. That is, after this weight is put on the scale, what is done based upon the result of the weighing? What is the effect of this nth weighing upon the binary result?

2.15. *A/D converter.* In the discussion accompanying Fig. 2.16, we considered the effective "jitter" which is introduced into the sampling process when an A/D converter is not preceded by a sample-and-hold circuit. In that figure, three cases were illustrated. One case produced effective sampling early in the conversion. Another produced effective sampling late in the conversion.

(*a*) If the input is monotonically increasing during the conversion process, then does this affect where the effective sampling occurs? Explain.

(*b*) If the input is monotonically decreasing during the conversion process, then does this affect where the effective sampling occurs? Explain.

(*c*) If the input is constant during the conversion process, then does this affect where the effective sampling occurs? Explain.

2.16. *A/D conversion.* For a specific microcontroller (other than the two discussed here) which includes an A/D converter:

(*a*) Does it multiplex its inputs? If so, then how many channels are available?

(*b*) Does it include a sample-and-hold circuit?

(*c*) How long does a conversion take?

(*d*) What is the resolution of the converter (e.g., 8 bits)?

(*e*) If the converter includes a sample-and-hold circuit, then it has the potential for continuously sampling an input, using a highly accurate sampling period. If such is the case, then does it include its own timing circuitry to collect samples using an arbitrarily set sampling interval? Or can it be controllable *directly* by a programmable timer output?

(*f*) Does it include a sample-and-hold on *each* separate channel so that inputs can be sampled simultaneously? If not, then how much skew must necessarily occur between samples taken on different channels?

2.17. *D/A converter.* For a microcontroller which employs a successive approximation A/D converter, it would seem reasonable for the designers of the chip to permit user access to the control register which drives the internal D/A converter. It would also seem reasonable to bring out the (unbuffered) D/A converter output in such a way that it could drive an external operational amplifier (having the normal ±15 V supply needed to reproduce the output with both linearity and drive capability).

(*a*) If this were done, what advantage would it give over the pulse-width modulation scheme offered by the Intel 8096?

(*b*) A strobe pulse could also be made available to drive an external sample-and-hold circuit every time that the D/A converter were written to. If an input waveform were sampled and processed, and the resulting waveform made available on the output of a sample-and-hold circuit, then what would be an application for such a capability (taking into account the 8-bit or 10-bit resolution, or whatever)?

2.18. *Watchdog timer.* With its two-operand instructions, the Intel 8096 can execute each of the writes required by the watchdog timer (to hexadecimal address 000A) as a single, four-cycle instruction. If the loop time of Fig. 2.28 varies between 500 μs and 10 ms but averages out to about 2 ms, then what percentage of the CPU's time is spent servicing the watchdog timer? (Assume a 12-MHz crystal frequency.) What percentage of its time is available for doing its useful work?

2.19. *Malfunction avoidance.* A watchdog timer can fix gross malfunctions in the running of the CPU. However, all it does is make sure that execution of the loop of the mainline program continues. Many other potential malfunctions can be avoided by *reinitializing state information*, wherever possible. State information is used here in the most general sense. Any flip-flops in a system, whether in the microcontroller or elsewhere, hold state information. If any of this inadvertently changes, then until this state information is corrected the system is malfunctioning. In the software organization of Fig. 2.28, the implication is that state information is initialized once in front of the main loop. In actual fact, a lot of state information can be repeatedly initialized, especially if we are aware of this as a potential opportunity to avoid trouble.

(*a*) We can generate a pulse on an output line by toggling (i.e., changing a 0 to a 1, or a 1 to a 0) the line twice. What is the potential problem here and what is the solution?

(*b*) If we have a 1-byte variable that we are using as a counter so as to execute the instructions in a loop 15 times, then we might choose to count up from 0 toward 15 or down from 15 toward 0. Our choices for testing when we are done include comparing the variable for $=0$, <0, $=<0$, $=15$, >15, $>=15$. (Assume that we are aware enough to start or end at 14 or 16 if that is what it takes to loop 15 times.) What is our best choice so that we will terminate looping if the RAM bits holding the variable ever inadvertently "drop a bit," resulting in a grossly incorrect value?

(*c*) Some microcontrollers, like the Motorola 68HC11, have a data direction register to set individual bits of a port as either inputs or outputs. It is common to initialize these bits once and for all, in the initialization which precedes the main loop of Fig. 2.28. What is the risk of doing this and what can be done about it?

(*d*) Consider some other facility of the microcomputer that has been discussed in this chapter, and which has flip-flops holding state information. Describe a potential mode of use which is subject to malfunction, if state information is corrupted, which can be avoided by using it differently.

2.20. *Battery backup*. For a specific microcontroller that has a pin to support some degree of nonvolatility:

(*a*) What is supported (e.g., the 16 bytes of RAM in the Intel 8096 whose power can be maintained independently of the power for the rest of the chip)?

(*b*) What is the minimum backup voltage requirement?

(*c*) What is the worst-case current load on the backup battery?

2.21. *Powerdown and watchdog timers*. Consider a specific microcontroller which supports both battery backup of some of its internal resources and also watchdog timing.

(*a*) What provision is made (if any) to keep the CPU from flailing out and destroying battery-backed-up data as its power drops out of the specified range? You are to assume that the CPU has been notified of an imminent power loss.

(*b*) Does this same provision disable the functioning of the watchdog timer?

(*c*) If so, then what provision is made so that the same mechanism designed to be a *help* for the situation of part (*a*) does not actually become a *liability*, if the CPU ever goes awry? We want to be assured, if possible, that the runaway CPU is not likely to invoke the provision of part (*a*), thereby also disabling the watchdog timer.

2.22. *Cold-start reset*. The circuit of Fig. 2.31 shows a "cold-start reset" switch.

(*a*) What purpose does this serve? That is, what more could we ever want than being able to have the microcontroller begin again, after a power outage, in exactly the same state that it was in just before the power outage?

(*b*) Consider the home automatic lawn sprinkler system controller of Fig. 1.14, and discuss a time when the cold-start reset switch would be helpful.

REFERENCES

For a more detailed presentation of the hardware characteristics of the Intel 8096 and the Motorola 68HC11, refer to the appendixes on each one. The ultimate, definitive source of information on *any* microcontroller is the manufacturer's user's manual and application manual.

CHAPTER

3

REAL-TIME
CONTROL

3.1 OVERVIEW

This chapter will deal with the major role of a microcontroller chip. That is, it will deal with the real-time control of the resources which make up an instrument or device. This will sometimes require interactions to take place at times specified by the hardware outside of the microcontroller. The mechanism for having this happen, the microcontroller's *interrupt structure*, will be our first consideration in this chapter. Other interactions with external hardware require the generation of carefully timed output signals or the precise measurement of timed events (e.g., the width of a pulse). In support of these, a microcontroller typically includes a *programmable timer* facility. We will examine this next.

Some of the interactions between a microcontroller and external hardware need not take place very quickly, but rather with slow regularity. For example, an instrument which includes optics which need to be positioned or focused might use a relatively low-performance stepper motor to achieve low-cost positioning. Perhaps the stepper motor can start and stop "on a dime" with a stepping rate of 100 steps per second. We might organize its stepping with a *real-time clock* which produces interrupts at a 100-Hz rate. Interactions between a real-time clock and such devices will be discussed next.

The remainder of the chapter will deal with a variety of issues which arise in the application of real-time control. We need to know that the design which we carry out can actually meet the timing requirements of the application. We also need to explore several pitfalls which can arise if we are not both aware and careful.

3.2 INTERRUPT STRUCTURES

The goal of a microcontroller's interrupt structure is to permit real-time events to control the flow of program execution. To do this, two capabilities of this structure are paramount:

> The CPU should be able, at any time, to disable interrupts, execute a few instructions, and then reenable interrupts. This is a critical capability which is needed when a few instructions *must* be executed one right after another.
>
> Of all the interrupt capabilities designed into the microcontroller, a designer should be able to enable only those of value for a specific application, disabling the rest.

After these capabilities have been taken care of, the designer of a microcontroller has as paramount considerations:

> Fast servicing for interrupting devices
>
> Minimization of any *delay* which occurs before servicing any of several devices which request service at the same time

The fast servicing goal is really sought for its aid in resolving the minimization of delay goal. Typically a large part of the microcontroller's job gets done by having each device signal the CPU when it needs servicing. When the device gets the attention of the CPU, the service routine is quickly executed and then the CPU can go back to doing other tasks. The execution of some tasks will lead to the initiation of other tasks. Furthermore, some of these tasks cannot be carried out completely at the moment that they are initiated. For example, initiating the task of reading a 20-byte message from another device using the on-chip UART at 9600 baud will require 20 slices of CPU time, extending over about 20 ms. This is illustrated in Fig. 3.1. Each millisecond, the UART will interrupt the CPU's execution of the *mainline* program (i.e., the ongoing program which is being executed when no interrupt servicing is taking place) to

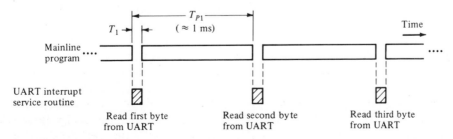

FIGURE 3.1
CPU time taken by an interrupt service routine.

have it take the byte of data and store it in memory. The role of the microcontroller designer in supporting such interrupt-driven applications is twofold:

Make the time T_1 short so that *other interrupts need not be postponed very long* if they should request service during T_1

Make *the percentage of CPU time taken by T_1 small*; that is, make the ratio of T_1/T_{P1} small

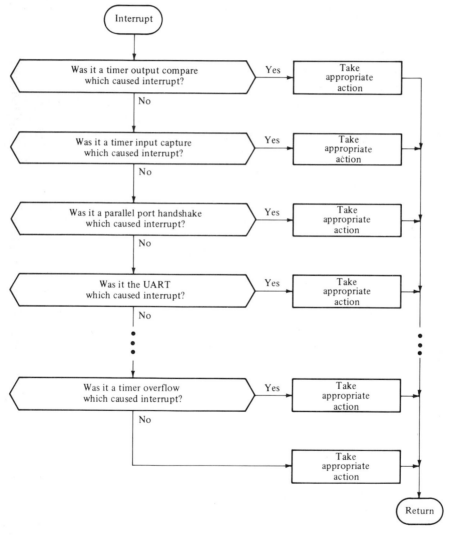

FIGURE 3.2
A polling routine to determine the source of an interrupt.

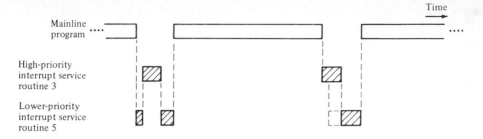

FIGURE 3.3
Use of CPU time by vectored priority interrupts.

Some microcontrollers must execute a *polling* routine whenever an interrupt occurs in order to determine the source of the interrupt. That is, the first part of the interrupt service routine is used to poll each possible source of the interrupt to determine which one caused this interrupt. This is illustrated in Fig. 3.2. Other microcontrollers support *vectored interrupts*. In this case, each source of an interrupt leads directly to the code which needs to be executed to service that specific source. Vectored interrupts reduce the time T_1 of Fig. 3.1. In so doing, they also reduce the ratio T_1/T_{P1}.

Still other microcontrollers support *vectored priority interrupts*. In this case, one source of an interrupt may be receiving service when a higher-priority source suddenly becomes ready for service. Rather than making the higher-priority source wait, some microcontrollers will let the higher-priority source preempt the lower one. The lower-priority service routine is put on hold until the higher-priority service routine is finished, at which point the lower one picks up again. This is illustrated in Fig. 3.3. On the left, interrupt service routine 5 is shown interrupting the mainline program. In the middle of its execution, the higher-priority interrupt service routine 3 butts in and gets serviced, whereupon 5 finishes up. The right part of the figure is intended to illustrate the higher-priority 3 source getting service and causing the servicing of the 5 source to be delayed until it is done.

Another possibility, which is almost as good as long as no interrupt service routine takes very long, is illustrated in Fig. 3.4. This shows one interrupt service routine getting serviced initially. It keeps interrupts disabled throughout its service routine. While it is being serviced, four other sources become ready for service. At the completion of the first interrupt service routine, the four interrupt sources which are *pending* (i.e., waiting) are automatically sorted out by the CPU, which immediately goes into the service routine for the highest priority source. The only difference between this case and the previous case is the duration of the longest interrupt service routine, since this is the maximum amount of *increased latency* (i.e., increased delay) which any source will see.

With all the many sources of interrupts in a microcontroller, a vectored

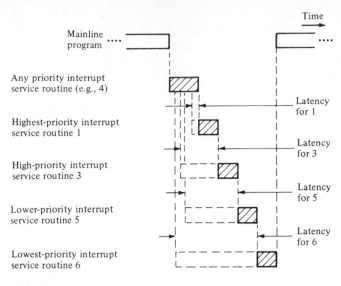

FIGURE 3.4
Priority of *pending* interrupts handled automatically by CPU.

priority interrupt structure raises another issue: Who sets the priority? If the priority is set by the manufacturer, then this will be no better than a vectored interrupt structure if the desired priority is just opposite to that desired in a specific application. That is, when an interrupt occurs, the interrupt service routine will simply not reenable interrupts for *any* other interrupt source until it is done.

   If a microcomputer's vectored priority interrupt structure permits the priority to be set by the application software, then a further question arises: Is the priority set during the microcomputer's initialization routine or is it done as an ongoing part of each interrupt service routine? If it can be done once and for all during initialization, then every device will be serviced as fast as possible. On the other hand, if each interrupt service routine is lengthened significantly as it reenables interrupts from some sources but not others, then the T_1/T_{P1} ratio of Fig. 3.1 will be increased significantly. In this case we may do better to forget about reassigning priority and live with short interrupt service routines which leave interrupts disabled throughout their execution.

   The designers of the two microcontrollers we have been considering have gone to great lengths to provide us with excellent interrupt structures. The Motorola 68HC11 has *21* vectors, with a separate one assigned to each source of an interrupt (almost). Except for the UART (which shares one vector between the receive function and the transmit function), any interrupt will lead right to the required code to be executed. The 68HC11 does not include a very convenient way to reenable interrupts on a priority basis within an interrupt service routine. However, it does sort out the priority of pending interrupts, so it is an excellent example of the scheme illustrated in Fig. 3.4. Furthermore, any *one* of the normal, maskable sources of interrupts can be elevated to the highest priority at initialization time, thereby providing extremely fast run-time performance for one interrupt source, for users who are happy with the remaining priorities. The 68HC11's interrupt sources are listed in Fig. 3.5 to illustrate that the problem of a fixed priority is not necessarily too bad. For example, the UART is near the bottom of the priority list since its interrupt can be held pending for as long as a millisecond, when it is receiving data at 9600 baud, and still carry out its function without error.

Shown at the top of Fig. 3.5 are the highest priority interrupts, associated with either resetting or with fault detection. The XIRQ pin interrupt is really a nonmaskable interrupt for power-fail detection (or some other external event requiring fast, drastic, certain action). Interrupts from this pin are initially

Interrupt source (highest priority first)	Vector location	CPU mask bit	Source* enable bit	Source* flag
External reset pin	FFFE-F	Nonmaskable	—	—
Clock failure	FFFC-D	Nonmaskable	0039/3	—
Watchdog timer	FFFA-B	Nonmaskable	003F/2	—
Illegal op code trap	FFF8-9	Nonmaskable	—	—
XIRQ pin	FFF4-5	X bit	—	—
Any *one* of the following I-bit maskable interrupt sources can be assigned the next priority	????-?	I	????/?	????/?
IRQ pin/port C control pin	FFF2-3	I	0002/6	0002/7
Real-time clock	FFF0-1	I	0024/6	0025/6
Timer input capture 1	FFEE-F	I	0022/2	0023/2
Timer input capture 2	FFEC-D	I	0022/1	0023/1
Timer input capture 3	FFEA-B	I	0022/0	0023/0
Timer output compare 1	FFE8-9	I	0022/7	0023/7
Timer output compare 2	FFE6-7	I	0022/6	0023/6
Timer output compare 3	FFE4-5	I	0022/5	0023/5
Timer output compare 4	FFE2-3	I	0022/4	0023/4
Timer output compare 5	FFE0-1	I	0022/3	0023/3
Timer overflow	FFDE-F	I	0024/7	0025/7
Pulse accumulator overflow	FFDC-D	I	0024/5	0025/5
Pulse accumulator input edge	FFDA-B	I	0024/4	0025/4
I/O serial port	FFD8-9	I	0028/7	0029/7
UART serial port—transmitter	FFD6-7	I	002D/7	002E/7
UART serial port—receiver	FFD6-7	I	002D/5	002E/5
Software interrupt instruction	FFF6-7	Nonmaskable	—	—

* Register addresses have been mapped to page zero, as suggested in Appendix A.

FIGURE 3.5
Motorola 68HC11 interrupt sources, listed in order of priority (highest at top, lowest at bottom).

disabled to ensure that the CPU will get to execute a few critical, initialization instructions (e.g., initializing the stack pointer) before the interrupt can occur. Once the X bit is changed to enable an "XIRQ" interrupt, such an interrupt cannot ever be disabled again without resetting the chip.

The interrupts which are intended to occur during proper, ongoing operation of the 68HC11 can all be quickly enabled or disabled by clearing or setting the I bit (with a CLI or SEI instruction). In addition, a *source enable bit* is associated with each of these interrupts, and its location is identified in Fig. 3.5. For example, the real-time clock's interrupt can be enabled (or disabled) by setting (or clearing) bit 6 of address 0024 (hex). Note that the source enable bits give us the hooks needed to reassign the priorities of most of the interrupts within each interrupt service routine by fiddling with the contents of addresses 0022 and 0024. At the beginning of an interrupt service routine, we could push the contents of these two addresses onto the stack. Next we could AND their contents with zeros in selected bit positions to disable lower-priority interrupts. Then we could reenable interrupts. After carrying out the job of the interrupt service routine we still have to restore the old contents of addresses 0022 and 0024 from the stack before returning. The trouble with this procedure is that we have probably increased the duration of the interrupt service routine significantly. We are probably better off having *short* interrupt service routines and somewhat less than optimum priority.

The final item to note in Fig. 3.5 is the column labeled *source flag*. When any of these interrupts occur, its source flag will be set. This is really a bit whose role is to keep a pending interrupt from getting lost. The interrupt service routine must clear it before returning (or, really, before reenabling interrupts, which is probably going to be delayed until returning). The clearing process for each bit is *not* carried out by writing a 0 to the bit in the register which contains it. Rather, it uses a mechanism which ensures that only that source flag will be cleared, and no other.

The designers of the Intel 8096 have taken an interesting approach to handling interrupts. All possible interrupt sources are divided into the nine *groups* shown in Fig. 3.6. Each group has its own vector. All but the software interrupt instruction can be enabled (or disabled) by setting (or clearing) the I bit with an EI (or DI) instruction. This might be thought of as the first, or highest, level of enabling or disabling. The interrupt mask register bits, shown in Fig. 3.6 as the "2d level enable" bits, form a second level of enabling or disabling for each group. Then within some of the groups, a third level of enabling or disabling can take place. For example, each of the programmable timer inputs, which can capture the time when an event occurs (as will be discussed in the next section) can be individually enabled or disabled.

When an interrupt occurs (with the I bit enabled), the CPU automatically vectors to the interrupt service routine, pushing the program counter onto the stack in the process. The flag which caused the interrupt is automatically cleared. All of these flags are in the interrupt pending register, shown in Fig. 3.6 as the "2d level flags." Interrupts are disabled only long enough for the inter-

Interrupt group (highest priority first)	Vector location	1st level enable bit	2d level enable bit (interrupt mask register)	2d level flag (interrupt pending register)
External interrupt	200F-E	I	0008/7	0009/7
Serial port	200D-C	I	0008/6	0009/6
Software timers	200B-A	I	0008/5	0009/5
High-speed input 0	2009-8	I	0008/4	0009/4
High-speed outputs	2007-6	I	0008/3	0009/3
High-speed input data ready	2005-4	I	0008/2	0009/2
A/D conversion complete	2003-2	I	0008/1	0009/1
Timer overflow	2001-0	I	0008/0	0009/0
Software interrupt	2011-0	Nonmaskable	—	—

FIGURE 3.6
Intel 8096 interrupt groups, listed in order of priority (highest at top, lowest at bottom).

rupt service routine to execute one instruction (which may, in turn, disable interrupts indefinitely).

At this point, the whole reason for the second level of enabling comes into play. A typical interrupt service routine is shown in Fig. 3.7. The first instruction, PUSHF, stacks the contents of the CPU flag register (i.e., the carry bit, the I bit, etc.) and the interrupt mask register. Also it then clears *all* of the bits in these two registers, which disables further interrupts indefinitely. This instruction, the POPF instruction at the end (which just undoes what PUSHF did), and the RET instruction (which executes the return from the interrupt) are almost necessarily a part of any interrupt service routine. The other two instructions (2 μs worth!) are all that are needed to reenable interrupts immediately *with a user-defined priority*!

A further level of astuteness was shown by the 8096 designers in their handling of interrupts within each group. Consider first the "high-speed input

```
ISRi:   PUSHF                           2 μs
        EI                              1 μs
        LDB    INT_MASK,#xxxxxxxxB      1 μs
           .
                      (service routine)
           .
        POPF                            2.25 μs
        RET                             3 μs

(Times shown assume a 12-MHz crystal)
```

FIGURE 3.7
Intel 8096 interrupt service routine.

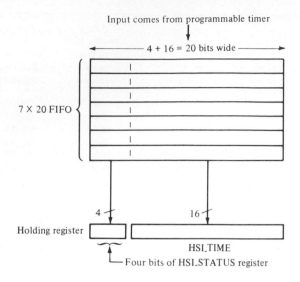

Input comes from programmable timer

◄——— 4 + 16 = 20 bits wide ———►

7 × 20 FIFO

4

16

Holding register

HSI_TIME

Four bits of HSI_STATUS register

Bit 7 of IOS1 register ☐   If set, FIFO has received at least one entry, which it has passed along to the holding register

FIGURE 3.8
Intel 8096 FIFO for timed input events.

data ready" interrupt of Fig. 3.6. This interrupt occurred because an input event, which occurred on any one of four inputs, was recorded by the programmable timer. The time of the event and the input that caused it are stored by the timer in a 7 × 20 *FIFO*, as shown in Fig. 3.8. This FIFO, or first-in–first-out memory, gives the CPU some breathing room as input events and other interrupts occur. Up to eight input events can occur before the CPU has handled the oldest one and the CPU still will not have lost an event. When the CPU finally services the interrupt, it reads the HSI_STATUS (high-speed input status) register to determine which input caused the oldest event. Then it reads the HSI_TIME register to determine when the event took place. This latter reading also dumps the entire holding register. If the FIFO contains the result of another input event, it will ripple down through the FIFO and load the holding register again. Each time that this happens, bit 7 of the IOS1 (input/output status register 1) is set. Consequently this bit can be tested within the interrupt service routine to see if there is anything else to do before returning.

The Intel 8096's programmable timer outputs benefit from an equivalent kind of buffering. As many as eight output events can be pending at a time. To generate a timed event, we write the nature of the event and its time into a holding register made up of a command register and a time register. This information is automatically transferred to the buffer; it then gets checked every 2 μs. When an "output compare" occurs for any of the up to eight events, the event is initiated and the buffer cleared of this entry. The kinds of timed events which can be initiated in this way are:

Set or clear any of six output lines at a specified time. We can also make a "high-speed outputs" interrupt (of Fig. 3.6) occur at that time if we choose.

Notify us with the setting of any of four flags when a specified time is reached. We can also make a "software timers" interrupt occur at that time if we choose.

Start an A/D conversion at a specified time.

From this it is apparent that if we reach an interrupt service routine by vectoring through address 2007-6, we need only sort out, by reading the appropriate status register, which output event(s) occurred in order to decide what to do next.

In this section we have omitted many details while attempting to give a good working idea of the interrupt mechanisms for the Motorola 68HC11 and the Intel 8096. More details of these two mechanisms are given in the two appendixes at the end of the book.

3.3   PROGRAMMABLE TIMERS

Input/output interactions of a microcontroller with an external device sometimes require the kind of *handshaking* described in Sec. 2.6. The handshaking may be automatic, as described there, or it might be implemented using two general purpose I/O lines, monitored and controlled by software.

Interactions with some other external devices may require the microcontroller to determine *when* the interactions should occur. For example, a thermal printer built into an instrument controlled by a microcontroller may require the microcontroller to output the data for each successive character, as depicted in Fig. 3.9. Then the microcontroller must turn on a print character pulse for 10 ms, to heat up the character-generating dot matrix in the print head to print the character on the heat-sensitive paper. Ten milliseconds is typical of the time required by many electromechanical devices in that it is a lot longer than the time it takes the microcomputer to execute instructions. In such circumstances, there are several approaches which a designer might take.

Delay-routine approach. We can write a little software routine which will waste time for 10 ms after turning the pulse on and before turning it off again. Such a routine for the Motorola 68HC11 is shown in Fig. 3.10. It is not our purpose here to be concerned about the 68HC11 instructions used and what they mean. Note only that this sequence of instructions does not introduce much code to achieve its purpose. However, it does tie up the microcomputer, which does nothing else during this time.

Interrupt service routine approach. We can turn the pulse on and ask an on-chip programmable timer to generate an interrupt when the 10 ms is up. Then while the timer is counting out 10 ms, the microcontroller is free

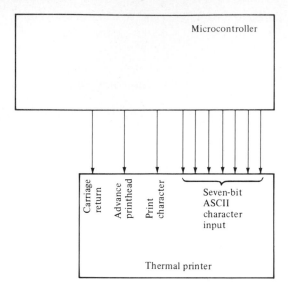

FIGURE 3.9
Thermal printer control.

to do other useful work. When the interrupt occurs, its service routine can contain the instruction(s) to turn the pulse off.

Virtually any microcontroller's programmable timer can provide the kind of control required by approach 2 above. A different kind of problem arises when the timed output event requires *precise* timing. Each of the approaches described above presents drawbacks.

The *delay-routine approach* requires that interrupts from other unrelated events be turned off during the timed operation. Otherwise an interrupt occurring while the delay routine is being executed will increase the duration of the timed event by the duration of the interrupt service routine. This is illustrated in Fig. 3.11.

The *programmable timer approach* can suffer from the same problem. Consider that we have two unrelated sources of interrupts, as in Fig. 3.12.

```
TIME    EQU  20000            ;Number of clock cycles to delay for 10 ms


        PSHX                  ;Set content of X register aside    (4 cycles)
        LDX #(TIME−12)/6      ;Load X with needed number of loops (3 cycles)
LOOP:   DEX                   ;Decrement loop count               (3 cycles)
        BNE  LOOP             ;Loop again if not done             (3 cycles)
        PULX                  ;Restore original content of X      (5 cycles)
```

FIGURE 3.10
Motorola 68HC11 ten-millisecond delay.

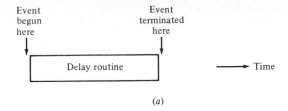

(a)

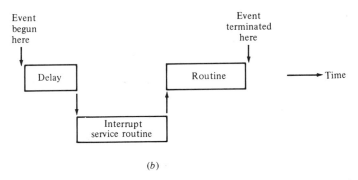

(b)

FIGURE 3.11
Effect of an unrelated interrupt upon timing achieved via a delay routine. (a) Desired timing; (b) actual timing, when an interrupting event occurs in the middle of the delay routine.

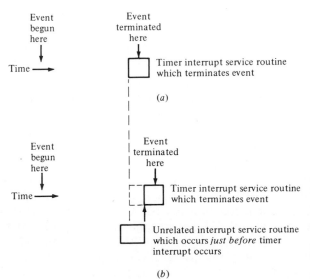

(a)

(b)

FIGURE 3.12
Effect of an unrelated interrupt upon timing achieved via a programmable timer. (a) Desired timing; (b) actual timing, when an unrelated interrupting event occurs *just before* the timer's interrupt would have occurred; the timer's interrupt service routine is placed on hold until the other routine is done.

	Intel 8096	Motorola 68HC11
Always available	4	4
Shared with another timer capability	2	1
Total possible outputs	6	5

FIGURE 3.13
Microcontroller outputs which are controlled directly by a programmable timer.

Before the interrupt that we are concerned with occurs, the other source causes an interrupt and postpones the servicing of our interrupt.

Again, this timing uncertainty is not a problem for many kinds of tasks which a microcontroller must control. An unrelated interrupt service routine which butts into the middle of the timing of a 10-ms event and throws the timing off by 50 μs is not going to upset too many tasks.

In the case where precise timing is a necessity, the programmable timer needs the ability to cause an *external* event to occur at some later time, *independent* of what the CPU is doing at that later time. Both the Intel 8096 and the Motorola 68HC11 have this capability, albeit implemented in somewhat different ways and summarized in Fig. 3.13.

The components needed to effect this kind of precise timing are shown in Fig. 3.14. The *free-running counter* keeps track of actual time. This is commonly a 16-bit counter, leading to a tradeoff between the resolution of time measured (i.e., the rate at which this counter is clocked) and the maximum amount of time which can be measured before the counter runs through a complete cycle of counts. To optimize this tradeoff, the designers of the microcontroller can add a *prescaler* between the internal clock and the clock input to this free-running counter. For example, the Intel 8096 has a divide-by-8 prescaler, making the time increment of the counter 2 μs (given a 12-MHz crystal). The Motorola 68HC11 permits the designer to make the tradeoff using a programmable divide-by-1, 4, 8, or 16 prescaler, making the time increment of the counter 0.5, 2, 4, or 8 μs (given an 8-MHz crystal).

While the free-running counter keeps track of actual time, the *output-compare register* and the *comparator* are the components of the timer in Fig. 3.14 which permit it to set the exact time when an event is to occur. Typically, we read the present time from the free-running counter, add an offset to make the desired event occur an exact amount of time later than this, and store the result in the output-compare register. The *output-level bit* is loaded with the desired output level (1 or 0). Then when the comparison occurs, the flip-flop will be clocked, transferring the output-level bit to an output pin of the micro-

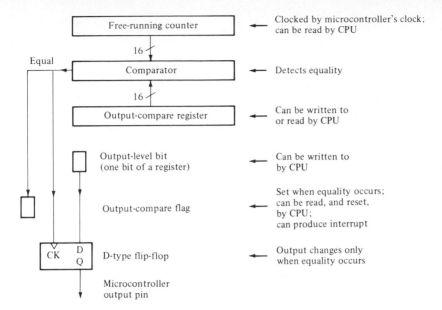

FIGURE 3.14
Programmable-timer output mechanism.

controller. In addition, the *output-compare flag* is set, with the option of send-ing an interrupt signal to the CPU, to indicate that the event has taken place.

An example is shown in Fig. 3.15*a*, where the timed event might be the generation of a 2.000-ms positive-going pulse. Initially the output-level bit is 0, corresponding to the output being low until the pulse is made to occur. The sequence of events is:

Disable interrupts

Read the free-running counter

Add a small "magic number" to this value to make the first comparison occur at the exact moment when we are ready for it to occur (we aren't ready yet!)

Store this number in the output-compare register

Set the output-level bit to 1

Reenable interrupts

The magic number mentioned above should be chosen to make the comparison occur shortly after the output-level bit has been set to 1. Interrupts were dis-abled at the beginning to ensure that between the reading of the free-running counter and the writing to the output-compare register and to the output-level

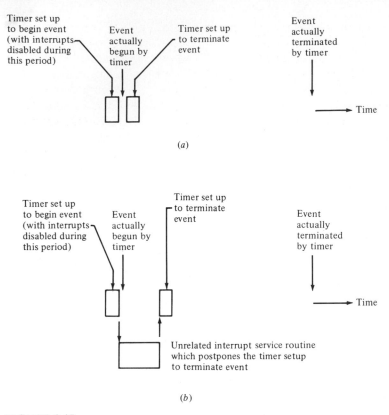

(a)

(b)

FIGURE 3.15
Use of a programmable timer to generate event timing which is unaffected by unrelated interrupts.
(a) Desired operation; (b) unrelated interrupt affects when timer is set up, *not* when output event
occurs.

bit, no extra time is inadvertently inserted by an interrupting device. (If this
were to happen, how would the time of occurrence of this event be affected?)

To set up the progammable timer to terminate the pulse, we must carry
out the following steps:

Read the *output-compare register* to regain the exact time when the output pulse was to begin.

Add 2000/INCREMENT to this value, where 2000 μs is the desired pulse
width and INCREMENT represents how often the free-running counter is
clocked (in microseconds).

Store the result back in the output-compare register.

Clear the output-level bit.

Note that we did not bother to disable interrupts here. Even if interrupts occur,
the timer will still perform its function reliably *as long as* this last sequence of

activities is carried out before the 2 ms transpires, as shown in Fig. 3.15*b*. For a much shorter pulse width, we would leave interrupts disabled during the entire setup process to ensure that the pulse gets turned off at the desired time, not a complete cycle of the free-running counter later!

A further role of a programmable timer is to permit the precise measurement of a time interval. For example, we might want to measure an input pulse width precisely. A programmable timer can make this measurement, again in a way which is independent of interrupts occurring during the measurement process. We first set up the programmable timer to capture the time of the starting event. Then we set it up to capture the time of the terminating event. The difference between these two values is the time interval of interest. Again, it is not necessary for the CPU to be involved at the time that either of these two measurements is made. It need only *set up* the measurements. The programmable timer input mechanism is shown in Fig. 3.16. The edge to be detected (i.e., a positive-going edge or a negative-going edge) is first selected and the *input-capture flag* cleared. Then when the input edge occurs, it will automatically transfer the content of the free-running counter to the input-capture register and set the input-capture flag (and generate an interrupt to the CPU, if that is desired).

To measure an input pulse, we simply capture the leading edge, save the time when it occurred (by storing the content of the input-capture register in RAM), reinitialize the programmable timer to capture the trailing edge, capture it, and subtract the first time from the second. This works fine as long as the input pulse is shorter in duration than the cycle time of the free-running counter (e.g., 2 μs $* 2^{16}$, or about 130 ms if the counter is being clocked every 2 μs). For longer times, we can use the *timer-overflow flag* to sense when the counter rolls

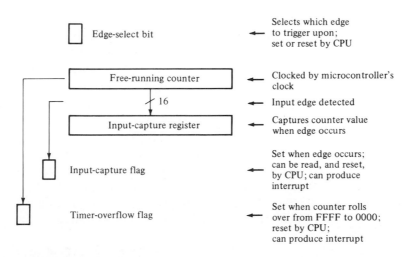

FIGURE 3.16
Programmable-timer input mechanism.

	Intel 8096	Motorola 68HC11
Always available	2	3
Shared with another timer capability	2	0
Total possible inputs	4	3

FIGURE 3.17
Microcontroller edge-sensitive inputs which can capture the time of occurrence of events.

over from all 1s to all 0s. We can increment a variable in RAM once per setting of the timer-overflow flag to keep track of how many counter cycles occur between the two edges of the input signal. This gives the information we need in order to reconstruct the time which elapsed between the two edges. An example is given in Appendix A.

The programmable timer in some microcontrollers permits triggering not only on

Positive-going edges
Negative-going edges

but also on

Both positive-going *and* negative-going edges

That is, the occurrence of *any* edge can be made to trigger a capture of the time. This capability is available in both the Intel 8096 and the Motorola 68HC11. A summary of the number of independent timer inputs available in these two microcontrollers is given in Fig. 3.17.

Another programmable timer capability found in some microcontrollers is the ability to make pulse train timing measurements. That is, instead of making timing measurements only from edge to edge of a waveform, some microcontrollers permit an additional measurement mode which measures the time duration for a specified number of pulses. For example, consider Fig. 3.18 which shows a variable-reluctance sensor generating a pulse each time a tooth of a gear passes beneath it. If we want to make a measurement of revolutions per minute, we could measure the time between two successive pulses and, taking into account the number of teeth on the gear, compute revolutions per minute. However, the measurement would be subject to tooth-to-tooth variations in the response of the sensor.

As another alternative, we could take a time-interval measurement over many pulses by having the CPU respond each time an edge occurred, incre-

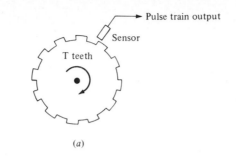

(a)

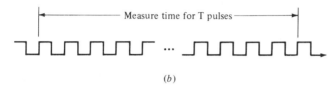

(b)

FIGURE 3.18
A pulse-accumulation measurement. (a) Mechanical gear and sensor; (b) measurement to be made.

ment a memory location, and stop only when T teeth have been counted, where T is the number of teeth on the gear. By keeping track of time to that point, we get a more accurate time measurement.

It is this latter measurement which some microcontrollers automate, including both the Intel 8096 and the Motorola 68HC11. Instead of requiring CPU interaction after each input edge, such a microcontroller has a *separate* counter, independent of the free-running counter discussed previously, configured more or less as shown in Fig. 3.19. This counter can be clocked directly from an input to the chip. The counting can be turned on and off by a *count-enable bit*. The counter can be preset to an arbitrary value, and when it overflows, a flag is set and an interrupt can be enabled to occur.

As an example, consider that we want to count the time duration for 144 pulses. We can preset the counter to its maximum count minus 144; that is, to $255 - 144 = 112$ for an 8-bit counter. For an exact measurement, the pulse train

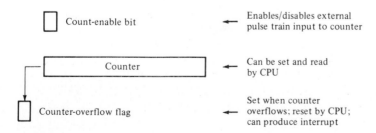

FIGURE 3.19
Programmable timer's pulse-accumulation mechanism.

input can be presented not only to the pulse-accumulator input, but also to an input-capture input. Now counting is enabled. After the first edge occurs, we need to note the "time" captured by the input-capture input. Further input-capture inputs are ignored until the counter-overflow flag is set (after a total of $144 + 1 = 145$ input edges). At that time the input-capture register is again read and the time interval since the first edge determined. This time interval between 145 edges is the time for 144 pulses, or one revolution of the gear. A detailed example is presented in Appendix A.

3.4 REAL-TIME CLOCK

One final programmable-timer capability found in some microcontrollers (e.g., both the Motorola 68HC11 and the Intel 8096) is that of a *real-time clock.** This is really just a simplification of the output-compare mechanism discussed in conjunction with Fig. 3.14. However, instead of driving an output pin, the function of a real-time clock is simply to generate periodic interrupts. Thus, the Motorola 68HC11 can be set up to divide the internal clock by 2^{13}, 2^{14}, 2^{15}, or 2^{16}. With an 8-MHz crystal (i.e., with an internal 2-MHz clock), these values correspond to about 4, 8, 16, or 33 ms between interrupts.

If we have several different kinds of events to time, each of which requires a time interval which is an integral multiple of 4 ms (for example), then we can ask the on-chip programmable timer to generate interrupts repeatedly—*every* 4 ms. The job of the timer's interrupt service routine is shown in Fig. 3.20, using pseudocode to represent the things to be done. CALL TIMER represents the calling of a TIMER subroutine. The subroutine is supposed to reinitialize the programmable timer to interrupt again in 4 ms. The FOR . . . NEXT I loop checks each task to be looked at every 4 ms. Each task has a counter COUNTER(I) whose state determines whether any action is to be taken during this 4-ms tick. If COUNTER(I) does not equal 0, then it is decre-

```
REAL_TIME_CLOCK_INTERRUPT_SERVICE_ROUTINE
     CALL TIMER
     FOR I=1 TO NUMBER_OF_TASKS
          IF COUNTER(I) .NE. 0 THEN
               DECREMENT COUNTER(I)
               IF COUNTER(I) .EQ. 0 THEN
                    CALL ACTION(I)
               END_IF
          END_IF
     NEXT I
     RETURN_FROM_INTERRUPT
```

FIGURE 3.20
Real-time clock interrupt service routine to serve many tasks.

* The Intel 8096 supports this with one of its "software timers."

mented. The action associated with this task is taken only if the decremented result equals 0.

As an example, we might use tasks 1 and 2 to generate a 40-ms pulse on bit 1 of port B. The pulse is started by storing the number 1 into COUNTER(1). Then when the next tick occurs, COUNTER(1) will be decremented to 0 and ACTION(1) must set bit 1 of port B as well as storing the number 10 (since 40 = 10 ∗ 4) in COUNTER(2). After 10 more ticks, COUNTER(2) causes the initiation of ACTION(2), which clears bit 1 of port B.

A real-time clock permits us to control the timing of several events to the nearest tick time, where one tick is much slower than the execution speed of the microcontroller. It does this while tying up no more than a single output compare facility of a programmable timer (or not even that for the Motorola 68HC11, with its separate real-time clock facility).

The timing of events which are controlled by a real-time clock are accurate only to the nearest tick. An unrelated interrupt (for example, from a UART which has accepted a byte of data serially from some other device) will extend the time of an output event being generated from within the real-time clock interrupt service routine. Furthermore, the real-time clock approach has an inherent difficulty in generating precise timing since it includes a *sequence* of tasks to be monitored, only some of which are actually going to lead to action during any given tick. This makes the time when any single task gets serviced somewhat inexact. That is, the time from the beginning of the interrupt service routine until action is taken for a certain task depends upon what actions are first taken for other tasks. On the other hand, relative to a 4 ms (or so) tick time, this variation is likely to be irrelevant.

In the next section we are going to be concerned with the latency experienced by the source of an interrupt. A real-time clock interrupt service routine of the form shown in Fig. 3.20 would appear to cause major damage to our attempts to minimize the latency experienced by *other* sources of interrupts (see also Sec. 5.4). We will find that we can easily reenable *all* interrupt sources early in the real-time clock interrupt service routine. Doing this generally makes the worst-case latency for any other interrupt source *independent* of the real-time clock interrupt service routine. It is also easy to do, even for a microcontroller like the Motorola 68HC11, which normally does not reenable interrupts while an interrupt service routine is being executed.

3.5 LATENCY

We are finally in a position to look at the effect of using a microcontroller's interrupt mechanism upon *latency*. Latency represents how long it will take from the time that the source of an interrupt says it is ready for service until the interrupt service routine is servicing that interrupt source. The latency issue has already been raised in conjunction with Fig. 3.4. In that figure we saw that a low-priority interrupt could be delayed by the time taken to service many other

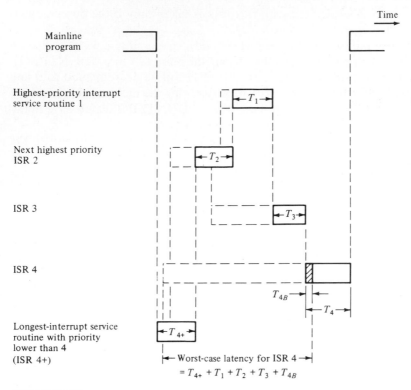

FIGURE 3.21
Worst-case latency for a low-priority interrupt.

interrupt service routines. This is an eventuality which must be expected and included in the design.

To examine latency more closely, consider Fig. 3.21. Here we assume that we have five or more *unrelated* sources of interrupts, all active at the same time. This figure shows the worst-case latency for interrupt service routine 4. Interrupt service routines 1, 2, and 3 have higher priority than 4 and have execution times of T_1, T_2, and T_3, respectively.

In this worst case, an even lower-priority interrupt service routine than 4 would have just begun execution when 4 requests service. If we assume that interrupts are turned off for the duration of the entire interrupt service routine (as they are with the Motorola 68HC11), then the worst case requires that we pick the lower-priority interrupt service routine which takes the longest to execute. We might identify this as ISR 4+ and its execution time as T_{4+}. On the other hand, if we assume that interrupts are turned on again in ISR 4+, then T_{4+} represents the time taken by any lower-priority interrupt service routine until interrupts are reenabled (when the CPU will switch to a higher-priority interrupt service routine).

The time marked T_{4B} represents that part of T_4 until ISR 4 begins execut-

	Intel 8096	Motorola 68HC11
Crystal clock frequency	12 MHz	8 MHz
T_{iB} (disabling interrupts for the duration of the interrupt service routine, and pushing CPU state onto stack)	8.25	7
T_{iB} (doing the above plus reassigning priorities and reenabling interrupts)	10.25	—

FIGURE 3.22
Time, T_{iB}, taken by Intel 8096 and Motorola 68HC11 interrupt service routines until they are ready to execute the instructions which actually service the interrupt source, expressed in microseconds.

ing the instructions which actually service the interrupt source. This includes the time which the CPU takes in its automatic response. It also includes the time which the CPU takes, either automatically or under program control, to push a minimum of CPU state information (e.g., at the very least, the program counter and the CPU flags) onto the stack. It includes any time taken to readjust interrupt priorities and reenable interrupts, if that is done. This time, called T_{iB} in general, is listed in Fig. 3.22 for both the Intel 8096 and the Motorola 68HC11.

With all of these factors taken into account, we see in Fig. 3.21 that the worst-case latency for interrupt service routine 4 equals the sum of the execution times for all higher-priority interrupt service routines plus the execution time for the longest lower-priority interrupt service routine* plus the time taken by ISR 4 to get to its servicing instructions.

Note that as far as interrupt service routine 4 is concerned, it does not matter whether the higher-priority interrupt service routines each reenable interrupts *within* their routines (as in Fig. 3.3) or not (as in this figure). The low priority sources just ask one thing of each of the higher-priority interrupt service routines: *keep it short!*

What determines the latency for the very highest priority interrupt source? First of all, while this source is active, we want to make sure that we do not ever disable interrupts any longer than necessary. Why, when we know we want to give good response time to active sources of interrupts, would we *ever* disable interrupts? Consider the code in Fig. 3.23. This is the Motorola 68HC11 code which uses the programmable timer to initiate a precisely timed output pulse, as discussed earlier in conjunction with Fig. 3.15. During the execution of this *critical region* of code, interrupts *must* be disabled. Otherwise, an interrupt occurring between *any* of the instructions would result in the pulse not being generated *at all*. This critical region takes 14 µs, during which even the most ardent of (maskable) interrupt sources is going to be put off.

---

* Refer to the discussion associated with Fig. 3-24 for a refinement of this factor.

```
SCALE    EQU 1               ;Timer's free-running counter is clocked every cycle.
MAGIC    EQU   23/SCALE      ;Makes compare occur after CLI instruction.


         SEI                 ;Disable interrupts                    (2 cycles)
         LDD   TCNT          ;Read timer                            (5 cylces)
         ADDD  #MAGIC        ;Add small number                      (4 cycles)
         STD   TOC2          ;Store result in output comp. reg. 2   (5 cycles)
         LDAA  TCTL1         ;Set                                   (4 cycles)
         ORAA  #11000000B    ;     output level                     (2 cycles)
         STAA  TCTL1         ;                        bit 2          (4 cycles)
         CLI                 ;Reenable interrupts                   (2 cycles)

                             ;Total time = 0.5 µs/cycle *           28 cycles
                             ;           = 14 µs
```

FIGURE 3.23
Protecting a critical region of Motorola 68HC11 code by temporarily disabling interrupts.

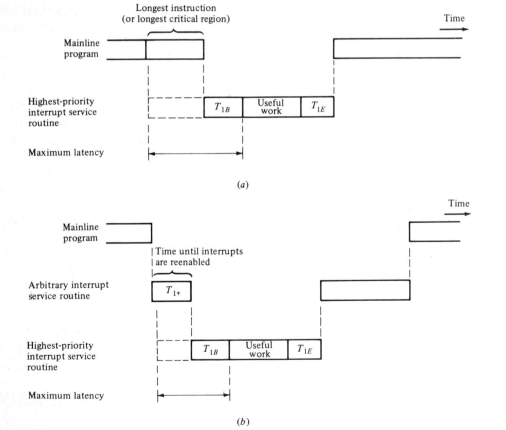

(a)

(b)

FIGURE 3.24
Latency time considerations. (a) Interrupt held pending by a long instruction; (b) interrupt held pending by temporarily disabled interrupts.

Another factor which affects the latency of the highest priority interrupt source is the execution of an instruction which takes a long time. The interrupt will be held pending until the completion of the instruction. For example, while most Motorola 68HC11 instructions take no longer than 2 or 3 μs to execute, a lurking "divide" instruction in our code can postpone an interrupt for 20.5 μs!

The effect of either of the above conditions upon latency is illustrated in Fig. 3.24a. Upon completion of a long instruction, or a critical region, the CPU sets aside whatever it was doing in the mainline program and vectors to the interrupt service routine. The time it takes to make this transition and to execute any preliminary "housekeeping" instructions in the interrupt service routine (e.g., the instructions to assign new interrupt priorities before reenabling interrupts) is labeled T_{1B} in Fig. 3.24a. We see its contribution to latency in that figure.

In Fig. 3.24b we again see the effect of any other interrupt service routine upon the latency of the highest priority interrupt source. Once any interrupt occurs, interrupts will be disabled for at least some minimal time. In Fig. 3.24b, this time is labeled T_{1+} and has exactly the same definition as was given for Figs. 3.21 and 3.22. In the worst case, the highest priority interrupt service routine will be put off by the longer of the times shown in Fig. 3.24a and b. That is, it will be delayed by whichever of the following three factors takes the most time:

The longest instruction in the code which will be executed while this highest priority interrupt source is active

The longest critical region which will be executed while this highest priority interrupt source is active

	Intel 8096	Motorola 68HC11
Crystal clock frequency	12 MHz	8 MHz
Worst-case latency of highest-priority interrupt service routine held pending by longest instruction	20.75	27.50
Worst-case latency of highest-priority interrupt service routine held pending by temporarily disabled interrupts within another interrupt service routine.	19.25	13.50 + X*

* X is the duration of the useful work instructions in the *longest* interrupt service routine. This assumes that 68HC11 interrupt service routines will leave interrupts disabled during their *entire* duration.

FIGURE 3.25
Worst-case latency time for Intel 8096 and Motorola 68HC11 highest-priority interrupt service routine, expressed in microseconds.

	Intel 8096	Motorola 68HC11
Crystal clock frequency	12 MHz	8 MHz
Minimum interrupt service routine execution time. This is given as the CPU time taken by an interrupt service routine *just* to enter and exit (i.e., not counting instructions to do useful work)	14.50*	13

* This includes the instructions shown in Fig. 3.7 (which reassign priority and reenable interrupts).

FIGURE 3.26
Minimum execution time for Intel 8096 and Motorola 68HC11 interrupt service routines, expressed in microseconds.

> The longest interval during which a lower-priority interrupt service routine will keep interrupts disabled

In fact, the T_{i+} factor used in conjunction with Fig. 3.21 should actually represent this same choice. If the lower-priority interrupt service routines are shorter than a long instruction or a critical region, then its time should be replaced by this value. Refer to the appendixes on the Motorola 68HC11 and the Intel 8096 for more information on instruction execution times. Meanwhile, Fig. 3.25 shows the worst-case latency for the highest-priority interrupt service routines for the Intel 8096 and the Motorola 68HC11, subject to two of the three factors listed above.

Since the latency for an interrupt service routine is a function of the duration of *other* interrupt service routines, it is useful to have some indication of how long these may take. Figure 3.26 lists these times.

3.6 INTERRUPT DENSITY AND INTERVAL CONSTRAINTS

In the last section, we were concerned with how long the interrupt service routine for a device might be put off. In this section, we will look at the relationship between the *frequency* of interrupts from a given source and the time required to service each one. Referring back to Fig. 3.1, we see the example provided by the reception of serial data by a UART. As we will see in Chap. 6, a steady stream of data being received at a baud rate of 9600 baud implies the reception of a new character roughly every millisecond. If no characters are to be lost, then each character must be accepted by the CPU from the UART before a new character is received to overwrite it. In this case, the

UART can be successfully used as long as the maximum latency for its interrupt service routine is no longer than 1 ms.

The ratio T_1/T_{P1} (the time taken by the interrupt service routine divided by the interval between interrupts) is used to determine whether a microcontroller is fast enough to handle all of its interrupting devices successfully. For each interrupting device, we first need to form this ratio. If the interval between interrupts varies, then the worst-case value (i.e., the shortest interval which will ever occur) will give us a properly conservative result. Likewise, if the duration of the interrupt service routine varies, then the worst-case value will be its longest duration.

Next, we need to sum these ratios for all devices which can be actively generating interrupts at the same time. This will give an *interrupt density* inequality which we will want to see satisfied in order to help ensure successful operation:

$$T_1/T_{P1} + T_2/T_{P2} + \cdots + T_n/T_{Pn} < 1.0$$

If this inequality is satisfied, then we can infer that even in the worst case the interrupt service routines do not take up all of the CPU time. Since we do get back to the mainline program regularly, even the lowest-priority interrupt service routine will be able to interrupt the mainline program and get serviced.

Successful interrupt handling, in the face of many independent interrupt sources, actually requires *more* than the above inequality. To see this, consider the example of Fig. 3.27. If T_{Pi} represents the worst-case period between inter-

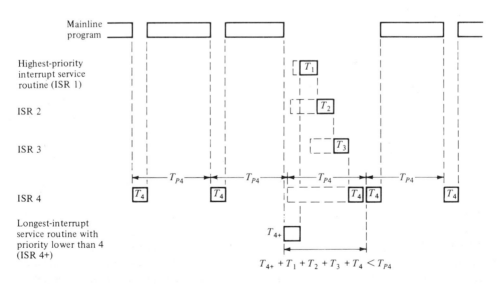

FIGURE 3.27
Interrupt interval constraint.

rupts from the device with the *i*th priority, then we must also satisfy an *interrupt interval* inequality for all values of *i*:

$$T_{i+} + N_1 T_1 + N_2 T_2 + N_3 T_3 + \cdots + T_i < T_{Pi}$$

where T_{i+} was defined at the end of the last section and where

$$N_1 = \text{INT}((T_{Pi} - T_i)/T_{P1}) + 1$$
$$N_2 = \text{INT}((T_{Pi} - T_i)/T_{P2}) + 1$$

etc.

In these equations, $T_{Pi} - T_i$ represents the maximum amount of time which the *i*th interrupt source can be put off, after it requests service, and still complete its interrupt service routine before interrupting again. The integer value N_1 represents the maximum number of interrupts which source 1 can invoke during this time. For simplicity, Fig. 3-27 shows the case in which this number for each of the higher priority interrupt sources is one.

The above inequality handles the case where the interrupt density may not be a problem but where one of the interrupting devices needs frequent service. This inequality ensures that it will be able to execute its interrupt service routine before its maximum latency time runs out.

Example 3.1. Assume that a microcontroller is used in a certain application with three interrupt sources, which we shall call sources 1, 2, and 3. Source 1 has the highest priority while source 3 has the lowest. Assume that the minimum time between interrupts for any of these sources is 100 μs. That is,

$$T_{P1} = T_{P2} = T_{P3} = 100\ \mu s$$

Furthermore, the CPU is diverted from what it was doing for 30 μs, as it services any one of these three sources. That is,

$$T_1 = T_2 = T_3 = 30\ \mu s$$

The longest critical region is 15 μs. The longest instruction used in code which can be interrupted takes 5 μs. Represent these last two conditions with

$$CR = 15\ \mu s \qquad INST = 5\ \mu s$$

Finally, assume that the servicing of each interrupt source must be completed before that source interrupts again and that interrupts are left disabled during the servicing of any interrupt.

(a) Is the interrupt density inequality satisfied? As we see below, it is.

$$T_1/T_{P1} + T_2/T_{P2} + T_3/T_{P3} < 1.00\ ?$$
$$30/100 + 30/100 + 30/100 < 1.00\ ?$$
$$0.30\quad + 0.30\quad + 0.30\quad < 1.00\ ?$$
$$0.90\quad < 1.00$$

(b) Will source 1 be certain to receive satisfactory service? The inequality, which says that it will be OK, is shown below.

$$T_{1+} + T_1 < T_{P1} ?$$
$$\text{(largest of } T_2, T_3, \text{ CR, INST)} + T_1 < T_{P1} ?$$
$$T_2 + T_1 < T_{P1} ?$$
$$30 + 30 < 100 ?$$
$$60 < 100$$

(c) Will source 2 be certain to receive satisfactory service? The inequality, which says that it will also be OK, is shown below.

$$T_{2+} + N_1T_1 + T_2 < T_{P2} ?$$
$$\text{(largest of } T_3, \text{ CR, INST)} + N_1T_1 + T_2 < T_{P2} ?$$
$$T_3 + N_1T_1 + T_2 < T_{P2} ?$$
$$30 + 30 + 30 < 100 ?$$
$$90 < 100$$

(d) Will source 3 be certain to receive satisfactory service? The inequality is shown below. It is *not* satisfied. Consequently, on occasion, source 3 will fail to be serviced in time.

$$T_{3+} + N_1T_1 + N_2T_2 + T_3 < T_{P3} ?$$
$$\text{(larger of CR, INST)} + N_1T_1 + N_2T_2 + T_3 < T_{P3} ?$$
$$CR + T_1 + T_2 + T_3 < T_{P3} ?$$
$$15 + 30 + 30 + 30 < 100 ?$$
$$105 \text{ is } not < 100$$

Example 3.2. Assume that a microcontroller is used in a certain application with just two interrupt sources, which we will call sources 1 and 2. Assume that the minimum time between interrupts for each of these sources is

$$T_{P1} = 60 \ \mu s \qquad T_{P2} = 100 \ \mu s$$

and that the CPU is diverted from what is was doing to handle each interrupt service routine for

$$T_1 = 20 \ \mu s \qquad T_2 = 35 \ \mu s$$

Also, assume that the longest critical region is 8 μs and that this is longer than the execution time of the longest instruction in our program. Finally, assume that the servicing of each interrupt source must be completed before that source interrupts again and that interrupts are left disabled during each interrupt service routine.

(a) Determine whether the interrupt density inequality is satisfied. As shown below, this inequality is indeed satisfied.

$$T_1/T_{P1} + T_2/T_{P2} < 1.00 ?$$
$$20/60 + 35/100 < 1.00 ?$$
$$0.33 + 0.35 < 1.00 ?$$
$$0.68 < 1.00$$

(b) If interrupt source 1 is given the higher priority, then determine whether it will invariably be serviced OK. For this case, the pertinent inequality is shown below. CR represents the longest critical region and INST represents the longest instruction which holds the possibility for putting off the CPU from servicing an interrupt. We see that the inequality is satisifed.

$$T_{1+} + T_1 < T_{P1} ?$$
$$\text{(largest of } T_2, \text{ CR, INST)} + T_1 < T_{P1} ?$$
$$T_2 + T_1 < T_{P1} ?$$
$$35 + 20 < 60 \ ?$$
$$55 < 60$$

(c) If interrupt source 1 is given the higher priority, then determine whether interrupt source 2 will invariably be serviced OK. Again, the pertinent inequality is shown below. We see that the inequality is satisfied.

$$T_{2+} + N_1 T_1 + T_2 < T_{P2} ?$$
$$\text{(larger of CR, INST)} + \quad 2T_1 + T_2 < T_{P2} ?$$
$$\text{CR} + \quad 2T_1 + T_2 < T_{P2} ?$$
$$8 + \quad 40 + 35 < 100 ?$$
$$83 < 100$$

(d) If interrupt source 2 is given the higher priority, then determine whether it will invariably be serviced OK. The pertinent inequality is shown below. We see that the inequality is satisfied.

$$T_{2+} + T_2 < T_{P2} ?$$
$$\text{(larger of } T_1, \text{ CR, INST)} + T_2 < T_{P2} ?$$
$$35 + 20 < 100 ?$$
$$55 < 100$$

(e) If interrupt source 2 is given the higher priority, then determine whether interrupt source 1 will invariably be serviced OK. We see from the inequality below that on occasion, interrupt source 1 will *not* receive service in time and that an error will occur.

$$T_{1+} + N_2 T_2 + T_1 \quad < T_{P1} ?$$
$$\text{(larger of CR, INST)} + \quad T_2 + T_1 \quad < T_{P1} ?$$
$$8 + \quad 35 + 20 \quad < 60 \ ?$$
$$63 \text{ is } not < 60$$

Example 3.3. Assume that a microcontroller is used in a certain application with two interrupt sources, which we shall call sources 1 and 2. Source 1 has the higher priority. The minimum time between interrupts for either of these sources is 100 μs. Let T_1 represent the time that the CPU is diverted from what it was doing to handle the source 1 interrupt. Likewise, let T_2 represent the time that the CPU is diverted from what it was doing to handle the source 2 interrupt. The longest critical region is 16 μs, and the longest instruction execution time which may be encountered is 20 μs. Finally, assume that the servicing of each interrupt source must be completed before that source interrupts again and that interrupts are left disabled within the interrupt service routines.

(a) Is there any combination of T_1 and T_2 for which source 2 is assured of receiving satisfactory service while source 1 is not? To answer this, consider the interrupt interval inequality which must be satisfied for each interrupt source:

For source 1:

$$T_{1+} + T_1 < 100$$
$$\text{(largest of } T_2, \text{ CR, INST)} + T_1 < 100$$

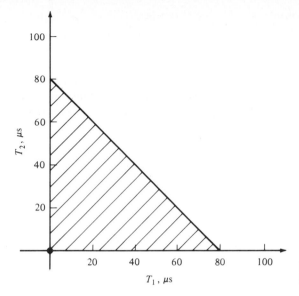

FIGURE 3.28
Solution to Example 3-3b.

For source 2:

$$T_{2+} + T_1 + T_2 < 100$$
$$(\text{larger of CR, INST}) + T_1 + T_2 < 100$$

Note that if the inequality for source 2 is satisfied, then the inequality for source 1 will *necessarily* also be satisfied.

(b) On a graph of T_2 versus T_1, shade in the area for which T_1 are T_2 are *both* assured of receiving satisfactory service. For the purposes of this plot, assume that T_1 and T_2 can be short as $0 \ \mu s$. T_1 and T_2 must satisfy both inequalities above. But satisfying the inequality for source 2 will also satisfy the inequality for source 1. Consequently the boundary between satisfactory and potentially unsatisfactory service is given by the equation

$$20 + T_1 + T_2 = 100$$

or

$$T_1 + T_2 = 80$$

The values of T_1 and T_2 which provide for satisfactory service of both interrupt sources are thus shown as the shaded area in Fig. 3-28.

3.7 SHARED RESOURCES AND CRITICAL REGIONS

When an interrupt occurs, the interrupted program is set aside. This interrupted program is returned to when the interrupt service routine has been executed. It is crucially important that once the interrupt service routine has been executed, the state of the CPU be exactly as it was at the time that the interrupt occurred; only then should execution of the interrupted program resume. If an accumulator or the carry flag were changed, for example, then this would destroy the reliable operation of an interrupted program that was right in the middle of an arithmetic computation.

All microcontrollers protect the program counter content, generally by automatically putting it onto a stack at the beginning of the interrupt service routine and restoring it at the end. All microcontrollers also provide instructions to set aside, and later restore, the CPU registers. The Motorola 68HC11 automatically puts the *entire state* of the CPU onto the stack at the beginning of an interrupt service routine. The RTI (return from interrupt) instruction at the end restores this CPU state from the stack. In this way, the interrupted program is *automatically* protected from changes in CPU state.

The Intel 8096 handles this problem in a very different way. The program counter is automatically stacked. Then by executing

PUSHF   and   POPF

instructions at the beginning and end of the interrupt service routine, respectively, the CPU flags (i.e., the carry bit, the Z flag, etc.) are set aside and subsequently restored. The CPU then, in effect, can carry out a *context switch*. That is, the registers which it uses in the interrupt service routine need not be the same ones which are used by the interrupted program. As we shall see in the next chapter, the Intel 8096 has addressing modes which permit *any* of its internal RAM locations to behave like an accumulator. It does not even have a dedicated CPU accumulator as such.

What does this mean for an interrupt service routine? Consider the job undertaken by the interrupt service routine for receiving UART data, represented in Fig. 3.1. Within the interrupt service routine, the following will happen:

The received byte of data is read out of the UART.

It is stored in RAM, perhaps at a location pointed to by RECEIVE_PTR, a 16-bit pointer.

RECEIVE_PTR is incremented, to be ready for the next interrupt.

Another variable, BYTE_COUNT, is incremented and tested to see if we have received all of the bytes of data required.

The UART is dealt with, either to turn off further interrupts or to make it ready for the next interrupt.

The return-from-interrupt instruction, or instruction sequence, is executed to get back to the interrupted program.

The Intel 8096 handles the two variables, RECEIVE_PTR and BYTE_COUNT, as variables which are dedicated to use by this interrupt service routine. The 8096 instructions deal with these variables directly, without having to move them into CPU registers first. In so doing, the 8096 drastically shortens the execution of interrupt service routines.

In contrast, the Motorola 68HC11 must move the content of a RAM variable called RECEIVE_PTR into a CPU register before using it since only the CPU register can serve as a pointer to memory. At the end of the interrupt service routine, it must save the content of the incremented CPU register back out to the RAM variable called RECEIVE_PTR so that it will be ready for the next interrupt. Note that the automatic stacking and unstacking of CPU registers by the 68HC11 does only *half* the job needed by an interrupt service routine. The other half, consisting of moving RAM variables needed in CPU registers by the interrupt service routine, must be done by extra instructions in the interrupt service routine.

Besides CPU registers, what other resources of the microcontroller are shared by an interrupt service routine and the mainline program? What resources are shared between interrupt service routines? What precautions must be taken to ensure successful operation in spite of these sharings?

Any variables which are used by both an interrupt service routine and the mainline program must be dealt with carefully. If the mainline program requires *more than one instruction* to update such a variable (or variables), then the instructions involved constitute a *critical region*. For reliable operation, interrupts must be disabled before the critical region and reenabled afterward.

Example 3.4. Suppose that a Motorola 68HC11 mainline program and a real-time clock interrupt service routine both deal with a 2-byte variable called NUM_STEPS. This variable represents a signed number, coded using twos-complement code, and keeps track of the number of steps remaining to be taken by a stepper motor. Each time the real-time clock causes an interrupt, it examines NUM_STEPS. If NUM_STEPS is 0, it does nothing. If NUM_STEPS is positive, it decrements NUM_STEPS and takes one step clockwise. If NUM_STEPS is negative, it increments NUM_STEPS and takes one step counterclockwise.

Whenever the mainline program wants to move the stepper motor, it simply adds the required number of steps to NUM_STEPS. It does not even matter whether any previously commanded stepping has been completed since NUM_STEPS acts as an accumulator for all stepping commands, past and present.

For the Motorola 68HC11, the code to do this follows:

```
LDD     #<number of steps required>
SEI                 ;Disable interrupts
ADDD    NUM_STEPS
STD     NUM_STEPS
CLI                 ;Reenable interrupts
```

In this code, a 2-byte accumulator (called D) is first loaded with a signed number representing the number of steps and direction of stepping to be taken. Then NUM_STEPS is read and added to this value. Finally, the result is stored back in NUM_STEPS. If a real-time interrupt were permitted to occur between the ADDD and STD instructions, and if NUM_STEPS contained a nonzero value, then the interrupt service routine would take a step and increment or decrement

NUM_STEPS. Upon returning, the mainline program would be oblivious to this change in NUM_STEPS (since it has *already* read out the old value and added it to D) and would update NUM_STEPS based on the old value rather than the new value. In this way the step taken by the interrupt service routine would go unaccounted for and an error in step position would occur. If the stepper motor were driving the pen along one axis of an *X-Y* plotter, then the remaining plot would be shifted one increment from where it should be.

Example 3.5. Reconsider the last example as it might be implemented by the Intel 8096. Now the code becomes

```
ADD NUM_STEPS,#<number of steps required>
```

The Intel 8096 instruction set permits the RAM variable, NUM_STEPS, to be updated with a single instruction. Consequently, a critical region does not arise in this case.

Next, consider the case of a resource which is shared by two interrupt service routines.

Example 3.6. For the Motorola 68HC11, suppose that the job of INT 2 involves updating bits 7 and 6 of an output port, generically denoted as PORT. For example, INT 2's job might be to drive the two lines for a stepper motor. The job of INT 1 involves updating bits 5 and 4 of PORT (perhaps to drive another stepper motor). For any Motorola 68HC11 output port, we can read the port to find out what is already there.

If bits 7 and 6 of accumulator B contain the desired bits to be updated by INT 2 (while the remaining bits of B are 0), then the following code might be used:

```
LDAA PORT          ;Read PORT into accumulator A
ANDA #00111111B    ;Force bits 7 and 6 of A to zero
ABA                ;Add B to A
STAA PORT          ;Restore updated value
```

In this case, *because interrupts are turned off throughout the entire interrupt service routine*, a critical region does not arise. The code in INT 1 is virtually the same, but deals with bits 5 and 4.

Example 3.7. Repeat the above PORT manipulation using the Intel 8096. Its output ports can also be read to determine their state. For this example, assume that bits 7 and 6 of a temporary variable called AL contain the information for updating PORT bits 7 and 6 and that the remaining bits of AL are 0. The code for INT 2 follows:

```
DI                    ;Disable interrupts
LDB   AH,PORT         ;Read PORT into another temp. var.
ANDB  AH,#00111111B   ;Force upper two bits to zero
ADDB  AL,AH           ;Combine into AL
STB   AL,PORT         ; and restore to PORT
EI                    ;Reenable interrupts
```

Here, we get a critical region assuming that INT 1 is a higher-priority interrupt service routine and assuming that INT 2 reenables interrupts. If INT 2 *did not* treat this as a critical region and disable interrupts during its execution, then the following could happen. If INT 1 were to request service after INT 2 had read PORT, then INT 1 would change PORT. However, upon the completion of INT 1, the remaining instructions in INT 2 would put bits 5 and 4 of PORT back the way it found them, completely undoing the action of INT 1.

In these examples, problems arose when the execution of several instructions dealing with a variable or a register was broken up by an interrupt service routine that dealt with the same variable or register. In every case, the solution lay in recognizing when a sequence of instructions constituted a critical region and then disabling interrupts while these instructions were executed.

Earlier, in conjunction with Fig. 3.23, we saw a different circumstance which can give rise to a critical region. In that case, we read a programmable timer's free-running counter. The code which followed depended for proper operation upon not being put off by an interrupt. If an intervening interrupt had been permitted to occur, then the interrupted code would have set up an event to take place at a time which had already passed.

3.8 SUMMARY

This chapter has explored the microcontroller resources which deal specifically with the real-time control of events in a device or instrument. Interrupts permit us to share the CPU among a variety of tasks. Each task gets CPU time only in response to its request for it. We have seen the sophisticated manner in which many sources of interrupts are handled by the Motorola 68HC11 and the Intel 8096. More detailed information is available in the two appendixes at the end of the book.

Both of these microcontrollers include state-of-the-art programmable timer facilities. They can be used to cause output events to occur at precise times, *even while the CPU is doing something else*. They can also be used to measure the time of occurrence of input events (e.g., when an input edge occurs), again even while the CPU is doing something else.

A real-time clock can be thought of as a special purpose programmable timer. Instead of causing an action directly, it causes all of the actions in its interrupt service routine to be executed periodically, perhaps every 10 ms. In this way, it permits relatively slow, periodic events to be handled automatically. Since these actions require CPU time for their execution, they can be held up somewhat by higher-priority interrupt service routines. For an event which occurs no more often than every 10 ms, a delay of even some hundreds of microseconds is not likely to be of concern. Under such circumstances, the event can be handled in this way.

The issue of latency is concerned with how long a requesting source of an interrupt will be put off before being serviced. If this time is longer than the source can manage, then faulty operation will arise. We have seen how the maximum permissible latency could be found for a typical input device, namely a UART which accepts new bytes of data no more often than every millisecond. As long as we handle each byte of data within the millisecond before *another* byte is received, then operation is satisfactory. Another source of interrupts would lead to similar considerations for its maximum permissible latency.

We have explored how a microcontroller's interrupt-handling mechanism affected the latency of each of its interrupt sources. We have seen that higher-priority interrupt sources are subject to shorter worst-case latency times. We have also seen how the Motorola 68HC11 permits any one of its interrupt sources to be elevated to the highest priority, in order to reduce its latency to a minimum. In addition, we have seen how the Intel 8096 permits the priority of its eight groups of interrupt sources to be rearranged for the same purpose. The appendixes explore these possibilities in detail.

Two constraints have been examined which, when satisfied, ensure reliable operation with interrupts. The first, an *interrupt density* constraint, makes sure that the CPU can keep up with the amount of processing required by the interrupt service routines. The second, an *interrupt interval* constraint, really addresses the issue of latency from a global point of view.

The chapter closes with an examination of shared resources. We have found a variety of situations in which interrupt service routines could cause faulty operation if a resource (i.e., a CPU register, a RAM variable, or a hardware facility like an output port) were acted upon both by the interrupt service routine and by the program code whose execution it interrupted. We have also seen how an interrupt service routine could mess up the desired use of a programmable timer. In all cases, the solution began with the recognition of a potential problem. It ended with the disabling of interrupts during the execution of a *critical region* of code.

PROBLEMS

3.1. *Interrupts*. If a microcontroller requires a polling routine, like that of Fig. 3.2, in order to sort out which source caused the interrupt, then what is the effect of this upon the times to service each of the sources of interrupts $T_1, \ldots, T_4$? Answer this by describing the difference between the length of servicing each source and the time required if each source had its own vector. Assume that each test in the polling routine takes 5 µs.

3.2. *Interrupts*. When the Intel 8096 vectors to its interrupt service routine for timed input events, described in conjunction with Fig. 3.8, then the interrupt service routine must include a polling routine to sort out which of four bits in a status register is set, in order to service the right device. When it is done servicing one source, it needs to check bit 7 of the IOS1 register to see if another one needs servicing. It continues this to service all sources of interrupts in this group before returning. The Intel 8096 includes a JBS three-operand instruction which tests an arbitrary bit of an arbitrary register and jumps to a specified address if the bit is set. The instruction takes five cycles to execute when the jump is not taken and nine cycles to execute when it is. Each cycle lasts 0.25 μs.

(a) Once the priority of the four sources has been set by the writing of the polling routine, how much longer does it take to begin servicing the second priority source than it takes to begin the first? The third than the first? The fourth than the first? In answering this question, assume that only one of these sources has interrupted.

(b) Assume that before beginning the polling routine, interrupts are reenabled only for selected *other* sources listed in Fig. 3.6, but not for high-speed input data ready interrupts, which got us into this interrupt service routine in the first place. What is the effect of selectively reenabling interrupts upon the servicing of the first, second, and third priority source when the fourth has caused an interrupt, has been polled, and its servicing has begun when all the other three put entries into the FIFO of Fig. 3.8 (signifying that they are each ready for service)?

3.3. *Interrupts.* The Intel 8096's FIFO, used to buffer captured input events, permits input pulse widths to be measured with a great deal of freedom. What determines the shortest input pulse width which can be measured when using a single input line?

3.4. *Programmable timer.* For a specific microcontroller, does it include a programmable timer? If so, then:

(a) What is the basic increment of time with which it deals? That is, how often is it clocked?

(b) Is this increment programmable? If so, then what are the choices?

(c) What is the maximum time interval which can be dealt with before the timer's counter cycles back to where it started?

(d) Does the programmable timer include provision for timing (or controlling) longer events, such as a flag which gets set and an interrupt which occurs each time the timer's counter rolls over?

(e) Can the programmable timer interrupt the CPU after a specified amount of time? Can it do this for two or more concurrent but independent time intervals?

(f) Can it change an output line when a specified time is reached? Can it do this for two or more concurrent but independent output events?

(g) Can it capture the time when an input line sees an edge? Can it do this for two or more concurrent but independent input events?

(h) Can it accumulate events (e.g., rising edges on an input line) over a specified time interval without program intervention at each input edge?

(i) Does it have any other unique measurement or control capabilities?

3.5. *Delay routine.* Consider the "waste time" approach to timing an event as exemplified by Fig. 3.10. The trouble with this approach is that it throws away CPU time. This is fine if the microcontroller has nothing better to do. On the other hand, if useful work could be going on, then it is of interest to know how much computing capability we throw away by using this approach. For a specific microcontroller, how many instructions, more or less, could the CPU be executing during 10 ms?

3.6. *Output timing.* For a specific microcontroller that includes set bit and clear bit instructions in its instruction set, what is the narrowest pulse (i.e., with no delay) that can be made to occur on one line of an output port (without the help of a programmable timer, if it could help)? That is, we want to set a bit and then clear a bit, leaving the other bits of the port unchanged.

3.7. *Programmable timer output.* Consider the magic number algorithm described in conjunction with Fig. 3.15. If interrupts were *not* disabled at the beginning, then:
   (a) Would it be possible that the output pulse might not even occur at all? When would an interrupt have to occur and how long would it have to last for this to happen, if it could happen?
   (b) Could an interrupt at the wrong time lengthen the pulse more than the intended time? Describe.
   (c) Could an interrupt at the wrong time shorten the pulse to less than the intended time? Describe.

3.8. *Programmable timer output.* For a specific microcontroller with a timer which can control an output line, what is the *narrowest* output pulse possible? Are very narrow pulses just as easy to generate as longer pulses?

3.9. *Programmable timer input.* For a specific microcontroller with a timer which can measure input events, what is the *narrowest* input that can be measured? Are very narrow pulses just as easy to measure as longer pulses?

3.10. *Measuring time intervals.* If we capture the time T_l when the leading edge of a pulse occurs as well as the time T_t when the trailing edge occurs, then we should be able to determine the pulse width. Assume that the free-running counter of Fig. 3.16 is a 16-bit counter, clocked every microsecond.
   (a) Assume that the free-running counter does not go through a complete count cycle. If the 16-bit number T_t is larger than the 16-bit number T_l, then will the 16-bit binary difference equal the pulse width in microseconds? Show an example.
   (b) Assume that the free-running counter still does not go through a complete count cycle. If the 16-bit number T_t is *smaller* than the 16-bit number T_l, then will the lower 16 bits of the binary difference equal the pulse width in microseconds? Show an example.
   (c) Assume that the free-running counter rolls over many times between the leading edge and the trailing edge of the pulse. Each time it overflows, it generates an interrupt and increments a variable in RAM called CYCLE. Figure out an algorithm for determining the pulse width. (*Hint:* "If $T_t > T_l$, then ...")

3.11. *High-resolution timing measurements.* We want to measure a time interval with *more* resolution than that of the free-running counter. We can do this for a time interval which repeats again and again.
   (a) For example, suppose that our free-running counter increments every microsecond (with crystal oscillator accuracy). Also suppose that we want to measure a pulse width of 26.xxx μs which repeats every 100 μs or so. As long as the pulse source is not *synchronized* to the microcontroller's clock, then we can *average* successive readings to get an accurate result having more resolution then 1 μs. Thus if we average 100 readings of 26 μs with 200 readings of 27 s, then what is our best estimate of the pulse width?
   (b) Time interval averaging like this is, of course, a statistical process. If we want the least-significant digit of the decimal result to be calculated for 1 standard deviation confidence level, then we need to make at least 25 measurements to justify adding the first digit and at least 2500 measurements to justify adding

two digits. Thus, if we want to measure the pulse in part (*a*) with a 10-ns resolution (e.g., 26.57 μs), then we might take something over 2500 measurements. To simplify the division by the number of measurements, we might take $2^{12} = 4096$ measurements. If we can collect and sum a new measurement every 100 μs, how long will this process take? If we want to measure intervals up to 1 ms long and if we want to sum them all together before dividing, then can we hold the sum as a 2-byte (16-bit) binary number? Or as a 3-byte (24-bit) binary number? Or as a 4-byte (32-bit) binary number? What do we have to do to divide this binary sum by 4096? We will leave the binary-to-decimal conversion as an exercise for another day.

3.12. *Pulse-accumulation measurement.* Consider a specific microcontroller which can carry out pulse-accumulation measurements (without CPU intervention after each pulse). In conjunction with Fig. 3.18, determine which registers and flags will be involved in the measurement of the time for a gear to make one revolution. Without considering the actual instructions to carry out the measurement, describe exactly what must be done.

3.13. *Real-time clock.* As was mentioned in Sec. 3.4, the Motorola 68HC11 includes a real-time clock which can be set up to generate an interrupt every 4 ms (for example). Once set up, it requires no ongoing activity on the part of the interrupt service routine to generate the next interrupt. Furthermore, it leaves all the rest of the programmable timer capability free for doing other things. For a specific microcontroller (other than the 68HC11) which has a programmable timer, how much of its timer capability will still be available if we decide to use the programmable timer *continuously* as a real-time, or tick, clock?

3.14. *Tick-clock timing.* Consider the approach to controlling external events described by the pseudocode of Fig. 3.20. Describe the functioning of the one or more tasks that could:

(*a*) Generate a 1-Hz square wave on bit 0 of port 1 of a microcontroller (to make an LED annunciator blink continuously on the microcontroller's PC board, giving a visual indication that it is alive and well).

(*b*) Generate a burst of five pulses on bit 1 of port 1, each pulse lasting 0.1 s, with 0.3 s between them. Also describe what a program would do to initiate this burst. This might be used for driving an LED on a user-held probe, to call the attention of the user to the occurrence of some event.

3.15. *Latency.* For a specific microcontroller other than the Intel 8096 or the Motorola 68HC11, determine:

(*a*) The worst-case latency time for its highest-priority interrupt service routine, when held pending by the longest instruction.

(*b*) The worst-case latency time for its highest-priority interrupt service routine, when held pending by temporarily disabled interrupts within another interrupt service routine. Describe any contingencies, just as was done in Fig. 3.25.

3.16. *Interrupt density and interval constraints.* As shown in Fig. 3.5, the Motorola 68HC11 has 15 different vectors to support 16 different sources of interrupts which are maskable by the CPU's I bit. If each interrupt service routine takes $T_i = 30$ μs and if the minimum time between interrupts T_{Pi} is the same for all sources and if the worst-case critical region lasts $T_{CR} = 13$ μs, then how short can T_{Pi} be and still allow reliable operation? Assume that each source of an interrupt must be serviced before it is ready to interrupt again.

3.17. *Interrupt density and interval constraints.* Consider a design which has four sources of interrupts, all of which can be active at the same time. The sources are identified by numbers 1, 2, 3, and 4, with 1 representing the highest priority source and 4 the lowest. The corresponding interrupt service routines take $T_1=31$, $T_2=29$, $T_3=37$, and $T_4=43$ μs, respectively. The minimum time between interrupts from the same source is $T_{P1}=120$, $T_{P2}=200$, $T_{P3}=150$, and $T_{P4}=1000$ μs, respectively. For reliable operation, each source must be serviced before its next interrupt occurs.

(a) Will operation be reliable, even if interrupts are turned off for the duration of each interrupt service routine? Explain.

(b) If the mainline routine is to be permitted to include critical regions, during which interrupts are temporarily turned off, then how long can these last and still maintain reliable operation?

(c) If each interrupt service routine were to reenable interrupts after 10 μs for higher-priority sources, then would operation be reliable? Explain.

(d) Answer the question in part (b) in this case.

(e) For case (a), given critical regions of no longer than 13 μs, then how much could T_1 be lengthened and still ensure reliable operation?

(f) Repeat for case (c).

(g) For case (a), could the priorities of the sources be rearranged so as to improve the maximum length of critical regions? If so, then how should they be rearranged, and what is the resulting maximum critical region?

(h) Repeat for case (c).

(i) For case (c), does the answer change if the critical region is in one of the interrupt service routines instead of the mainline program? If so, then does the priority of the interrupt service routine having the critical region matter?

(j) For case (a), if we added 8 μs to the duration of the interrupt service routines for sources 1, 2, and 3 so that interrupts could be reenabled within these service routines with only higher-priority interrupts enabled, then what would be the effect upon the worst-case critical region?

(k) If the T_{Pi} times were actually *average* times instead of minimum times between interrupts, then for case (a), what percentage of CPU time is available to the mainline routine, on the average?

(l) Is the answer to part (k) changed if we consider case (c)?

3.18. *Interrupt density and interval constraints.* In light of your answer to the previous problem, how would you set the priorities of interrupts from four sources, given the four interrupt service routine times, T_i, and the four minimum times between requests for service from each source, T_{Pi}, so as to maximize the acceptable duration of critical regions?

3.19. *Critical regions.* It is of paramount importance to be able to disable interrupts to protect critical regions in our code. Consider, for example, the common technique for changing a bit on an output port while leaving the other bits unchanged. We can read the port (even though it is an output port) into an accumulator. Then we can force the selected bit (e.g., bit 5) to 1 with an OR instruction (leaving the other bits unchanged). Finally, we can write the result back out to the port.

(a) Assuming that there is an interrupt service routine which changes bit 3 on this same port when it is called, why should interrupts be disabled for the above operation which is intended to change bit 5? To answer this, describe a scenario which leads to a malfunction.

(b) How often is such an error likely to recur if bit 3 is changed once per millisecond, with the three instructions taking 1 μs each, and if bit 5 is changed once every 10 ms or so? Assume that the two events are not synchronized to each other, but that at least one of them occurs in response to an unsynchronized, external event. (This part of the problem is for probability fanatics only.)

(c) Many malfunctions which occur rarely, are in fact the result of just this kind of unintended coupling between two unrelated activities. If you were going to characterize the *general* problem (i.e., more general than this one of two unrelated tasks dealing with one output port) so that you could avoid it, how would you describe it?

(d) Is there any benefit to be had in an instruction which sets (or clears or toggles) selected bits of a port *directly*? Explain. (Not all microcontrollers have such an instruction, but some do, including both the Motorola 68HC11 and the Intel 8096.)

CHAPTER
4

PROGRAMMING FRAMEWORK

4.1 OVERVIEW

In this chapter, we will consider the CPU register structure, the addressing modes, and the instruction set which are typical of a microcontroller. We will look at the features of an assembly language which support the writing of microcontroller programs. Finally, we will consider some ways for breaking the complexity of a big program development effort down into manageable tasks.

We will again look at the examples offered by the Intel 8096 and the Motorola 68HC11. We will see that the designers of these two chips approached the problem of selecting a microcontroller architecture from two very different points of view:

> The architecture of the Intel 8096 was chosen specifically to make it a *fast* controller. Whatever that takes, the designers have tried to do. The designers have subverted other laudable goals, such as compatibility with earlier products, in order to achieve this overriding goal. They have *taken advantage* of the limited resources available by giving access to those resources as efficiently as possible.

> The Motorola 68HC11 has an architecture which represents an evolution from Motorola's earlier microcontroller, the 6801. It has a somewhat upgraded CPU register structure and instruction set. However, its startling new features are located in the I/O arena. In addition, it is a CMOS part. Engineers familiar with the 6801 will love the enhancements incorporated into the 68HC11! All of their microcontroller programming skills will translate intact from the 6801 to the 68HC11.

4.2 CPU REGISTER STRUCTURE

The CPU of a microcontroller contains registers whose role is to support the execution of its instruction set. A typical register structure is shown in Fig. 4.1.

The *status register* contains assorted, unrelated bits. Some bits (e.g., the *carry* flag and the *zero* flag) are set or cleared in response to some instructions. They are tested by other instructions, thus providing a versatile way to *link* pairs of instructions together into super instructions. For example, consider the combination of the instructions to load the content of a memory location into the accumulator, set a flag if the accumulator contains a 0 in a certain bit position, and then jump to a specified address if that flag is set. In effect, this lets us carry out a conditional jump instruction based upon the state of one bit stored in an arbitrary memory location.

Other status register bits are used to enable or disable interrupts from selected sources. They may be set and cleared under program control. They may also be changed automatically by the CPU, as it responds to an interrupt.

Still other status register bits permit selected features of the microcontroller to be enabled or disabled. For example, the Motorola 68HC11's S bit is used to enable or disable the functioning of the STOP instruction, discussed in Chap. 2.

The *accumulator* is a register which serves a function similar to that of the carry flag or the zero flag. That is, it permits the construction of super instructions out of combinations of instructions. For example, we often want to read an input port and save what we read for later manipulation. In this case, what we need is an instruction to move data from the input port to a RAM location. If

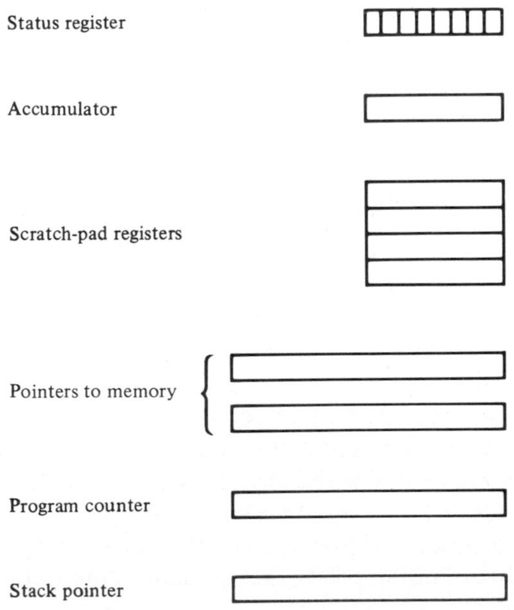

Status register

Accumulator

Scratch-pad registers

Pointers to memory

Program counter

Stack pointer

FIGURE 4.1
The register structure of a typical microcontroller's CPU.

we do not have such an instruction, then we can build one, using the accumulator as an intermediary. Data is moved from the input port to the accumulator during the execution of one instruction. It is then moved from the accumulator to a RAM location during the execution of the next instruction.

Scratch-pad registers are included in some microcontrollers. Once a scratch-pad register has been loaded with data, instructions can access that data using very few bits of the instruction for addressing. Thus, such instructions are fast. Unfortunately, the time taken, and the extra instructions needed, to move data between memory and scratch-pad registers seems to outweigh the advantage, once it is there. Neither the Intel 8096 nor the Motorola 68HC11 has scratch-pad registers.

Pointers to memory permit a program to vary the address of an operand. Consider the following example for adding together two 32-bit binary numbers using an 8-bit adder.

Example 4.1. Each of two 32-bit binary numbers is stored in RAM as a 4-byte-long array, as shown in Fig. 4.2a. The algorithm to add these two numbers together begins with the initialization instructions shown in Fig. 4.2b. If the CPU has two pointers to memory, called POINTER1 and POINTER2, then POINTER1 is initialized with the address of the least-significant byte (LSB) of the first number, while POINTER2 is initialized to point to the least-significant byte of the second number. A variable, SCRATCH,* is initialized to 0, to serve as a loop counter. The carry bit C is cleared.

The actual addition of bytes is shown in Fig. 4.2c. The first byte of the first number is loaded into accumulator A using POINTER1. Then the first byte of the second number is added to the content of A plus the 1-bit number represented by the carry bit. Eight of the nine bits of the result are put back into A. The ninth bit is automatically stored by the add instruction in the carry bit. This 9-bit result is represented by the expression

$$C,A$$

The lower 8 bits (i.e., the content of A) are stored back to memory using POINTER1.

Now we get to the rationale for this example. We want to use *the same instructions* over again on the second bytes of the two numbers. Since the bytes are identified by the content of CPU registers, we only need to increment these registers before looping around again. And again. And again. The loop counter, SCRATCH, terminates the algorithm after the 4-byte numbers have been added.

The *program counter* keeps track of the address of the next instruction to be executed. The CPU begins each instruction cycle by fetching the first byte of the next instruction and automatically incrementing the program counter. On the basis of the *opcode*† byte, the CPU knows whether it needs to fetch any

---

* Located either in a scratch-pad register or in RAM.

† Operation code: the code which tells the CPU which instruction is to be executed and whether it needs to fetch more bytes, for operands.

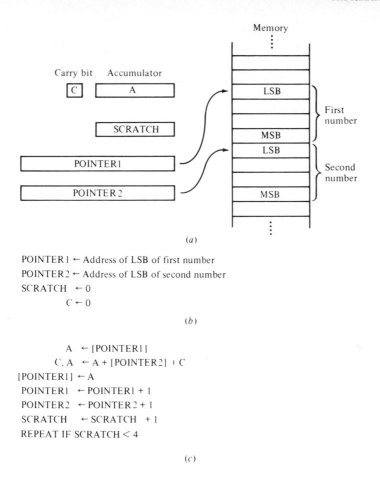

POINTER1 ← Address of LSB of first number
POINTER2 ← Address of LSB of second number
SCRATCH   ← 0
        C ← 0

(b)

        A   ← [POINTER1]
      C, A ← A + [POINTER2] + C
[POINTER1] ← A
 POINTER1   ← POINTER1 + 1
 POINTER2   ← POINTER2 + 1
 SCRATCH   ← SCRATCH + 1
REPEAT IF SCRATCH < 4

(c)

FIGURE 4.2
Four-byte binary addition using two pointers. (a) CPU registers and RAM; (b) initialization; (c) addition algorithm.

more bytes to complete the fetching of a multiple-byte instruction. Note that when it has completed fetching as many bytes as are called for, the program counter will be pointing at the first byte of the next instruction. Thus, when the execution of the present instruction has been completed, the program counter will be poised to get the next one.

The *stack pointer* is a pointer to RAM and is used automatically by the CPU to handle the temporary storage of return addresses during subroutine calls. That is, when a subroutine is called, the CPU will "push" the address of the next instruction (the address for which is in the program counter) onto the stack. Then it will load the program counter with the address of the subroutine. At the completion of the subroutine, the CPU executes a return from subroutine instruction which "pops" the address from the top of the stack and restores it to the program counter. This automatic mechanism is exactly what is needed to handle *nested* subroutine calls, where one subroutine is called from within another one. The return addresses are pushed onto the stack in the order in which they are called. As the most deeply nested subroutine is completed, its return address is popped off the stack, bringing the next return address to the top of the stack.

> **Example 4.2.** An illustration of the use of the stack is given in the example of Fig. 4.3. This is a skeletal view of the Motorola 68HC11 source code file for the algorithm of Fig. 2.28 (without showing the interrupt service routines). When the CPU executes the first subroutine call, to the INITIAL subroutine, it first pushes the return address onto the stack. This is the address of the first byte of the next instruction in line, JSR SUB1. The address is represented by the label LOOP. With the return address on the stack, the CPU now loads the address represented by the label INITIAL into the program counter. Consequently, the next instruction to be executed will be the first instruction of the INITIAL subroutine.
>
> The CPU completes the INITIAL subroutine by executing the RTS (return from subroutine) instruction and popping the 2-byte address off of the stack (one byte after the other) and putting this into the program counter. Since this is the address represented by the label LOOP, execution proceeds with this next instruction in the mainline routine. At this point, the stack has no entries in it.
>
> Now we come to the call of the SUB1 subroutine, which in turn will call the SUB2 subroutine. Calling SUB1 will cause its return address, labeled MAIN_1, to be pushed onto the stack. When the instruction in SUB1 calls SUB2, *its* return address, labeled SUB1_1, will be pushed onto the stack also. At that point, the stack holds two return addresses with the one on top being SUB1_1 and the other being MAIN_1. The RTS instruction at the end of the SUB2 subroutine pops SUB1_1 back into the program counter. Then the RTS instruction at the end of the SUB1 subroutine pops MAIN_1 back into the program counter. Once again the stack has been emptied out.

To use the stack pointer, it must first be initialized to point to the beginning of an area of RAM set aside for this purpose. Both the Intel 8096 and the Motorola 68HC11 have stack mechanisms which "grow down" in memory. Consequently, if 20 bytes of RAM are set aside for the stack, then the stack pointer must be initialized so that the first call of a subroutine will store the 2-byte return address into the highest two addresses set aside.

> **Example 4.3.** To initialize the Motorola 68HC11's stack pointer, we need to know where its RAM is located. As pointed out in Appendix A, this internal RAM can be *mapped* to any 256 addresses of the form
>
> I000H   to   I0FFH
>
> where I represents any hexadecimal digit (that is, 0, 1, 2, . . . , E, or F). The default mapping (which exists if we do nothing to change the mapping) is
>
> 0000H   to   00FFH

Since the stack grows toward lower addresses, and since the stack pointer is automatically decremented *after* each byte is pushed onto the stack, it makes sense to initialize the stack pointer by loading it with the address 00FFH, as shown at the very beginning of the listing of Fig. 4.3. This sets aside the top part of the available RAM for the stack.

Example 4.4. To initialize the Intel 8096's stack pointer, we need to know that RAM extends over the 230 hexadecimal addresses

$$001AH \quad to \quad 00FFH$$

However, the addresses from

$$00F0H \quad to \quad 00FFH$$

are the ones which are amenable to battery backup, as discussed in Chap. 2. Therefore it makes sense to have the stack grow down from address 00EFH. Since the stack pointer is automatically decremented *before* the first byte of a return address is pushed onto the stack, the stack should be initialized to 00F0H.

The stack pointer is also used by the CPU during interrupts, automatically pushing the program counter (at the very least) onto the stack before vectoring off to the interrupt service routine. Most microcontrollers also include instructions to push and pop CPU registers onto and off the stack, using it for temporary storage. Needless to say, such temporary storage *must* be used with care.

```
HIGHSTK EQU   00FFH              ;Upper boundary of stack


MAIN:    LDS   #HIGHSTK          ;Initialize stack pointer
         JSR   INITIAL           ;Initialization needed before entering loop
LOOP:    JSR   SUB1              ;Call of a subroutine, SUB1
MAIN_1:        ·                 ;Return address for SUB1 subroutine
               .
         JMP   LOOP              ;End of loop
INITIAL: ·                       ;Initialization subroutine
               .
         RTS                     ;Return from INITIAL subroutine
SUB1:          ·                 ;Code for SUB1 subroutine
               .
         JSR   SUB2              ;Nested call, of SUB2 subroutine
SUB1_1:        ·                 ;Return address for SUB2 subroutine
               .
         RTS                     ;Return from SUB1 subroutine
SUB2:          ·                 ;Code for SUB2 subroutine
               .
         RTS                     ;Return from SUB2 subroutine
```

FIGURE 4.3
Illustration of stack use to manage subroutine return addresses.

The inadvertent misuse of the stack is an easy way for a microcontroller to go awry. Fortunately, the malfunction is so gross that the designer will surely detect that a problem exists before customers ever see production versions of the instrument or device.

The CPU register structure of the Motorola 68HC11, shown in Fig. 4.4, is very close to the "typical" structure of Fig. 4.1. The status register contains

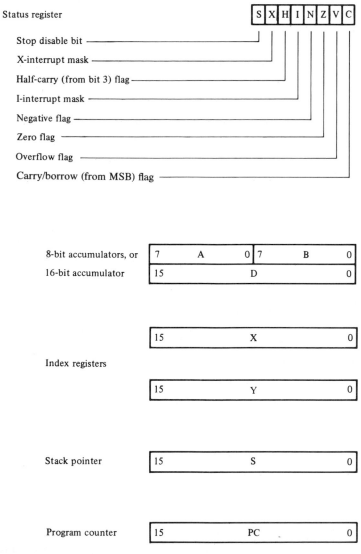

FIGURE 4.4
Motorola 68HC11 CPU register structure.

five flag bits (C, V, Z, N, H) which are set or cleared in conjunction with the execution of certain instructions. The function of each of these follows:

C, the *carry/borrow* flag, is the ninth bit of the result of an arithmetic operation on two 8-bit numbers. It is the seventeenth bit of the result of an arithmetic operation on two 16-bit numbers.

V, the *overflow* flag, is set if the result of an arithmetic operation on two signed numbers is too large to fit into the same size signed number. For example, the addition of $(+125) + (+125) = (+250)$. Each of the numbers to be summed can be expressed as the 8-bit 2s-complement signed number:

$$01111101$$

The *largest* 8-bit 2s-complement signed number is

$$01111111$$

or $+127$. If these two numbers are added in the 68HC11, the result will be

$$11111010$$

This is actually the 2s-complement code for the *negative* number, -6! The overflow flag V serves to flag this condition so that we know when to take appropriate action.

Z, the *zero* flag, is set when the result of an 8-bit operation is equal to

$$00000000$$

and when the result of a 16-bit operation is

$$0000000000000000$$

It is cleared when the result is anything else.

N, the *negative* flag, is set when the most-significant bit (MSB) of an operation is a 1, and cleared otherwise. For example, if the result of the sum of two 8-bit numbers is

$$10001111$$

then the N flag will be set, signifying that the result is negative, *if* we were really trying to add together 2s-complement numbers. If we were not, then we simply ignore the N flag.

H, the *half-carry* flag, is used to make a super instruction by helping to couple together two instructions. The usual way of dealing with decimal numbers is to use a binary-coded-decimal (BCD) representation of the numbers whereby each digit is coded by its 4-bit binary equivalent. Thus, the BCD representation of the decimal number 52 is

01010010

The 68HC11 uses an ADD instruction followed immediately by a DAA (decimal adjust accumulator) instruction to create a BCD instruction. The H flag is set if the ADD instruction produces a carry between bits 3 and 4 (i.e., between the lower-four bits and the upper-four bits). The DAA instruction looks at the content of A, C, and H to figure out what the correct BCD result should be and then stores the result back into A.

The Motorola 68HC11's status register also contains two interrupt mask bits, defined as follows:

X is the mask bit for the XIRQ interrupt pin on the chip. It disables (i.e., masks) interrupts from this pin when it is set and enables them when it is clear. At reset time, this X bit is set, disabling XIRQ interrupts. It can be cleared under program control to enable XIRQ interrupts. Once enabled, XIRQ interrupts cannot be disabled under program control.*

I is the mask bit for the IRQ interrupt pin on the chip, as well as all the other internal, maskable sources of interrupts (refer to Fig. 3.5).

Finally, the status register contains one bit that controls whether the STOP instruction is to be executed or treated as a NOP (i.e., a "no operation," or do nothing, instruction):

S is the STOP disable bit. If it is set, the STOP instruction is treated like a NOP instruction.

The effect of each Motorola 68HC11 instruction upon these status bits is shown in Fig. A.39 in Appendix A. Thus, the instruction

```
LDAA   TEMP
```

will affect the N and Z bits, clear the V bit, and leave the remaining bits unchanged.

The Motorola 68HC11 register structure of Fig. 4.4 includes two 8-bit accumulators A and B. Either one of these can be identified by two-operand instructions as the source of one (8-bit) operand and the destination of the result. For example, the instruction

```
ADDA TEMP
```

---

* However, the *occurrence* of an XIRQ interrupt sets the X bit, disabling further XIRQ interrupts while the XIRQ interrupt service routine is executed. Exiting from the XIRQ interrupt service routine reenables XIRQ interrupts by restoring the state of X from the stack.

says to add the content of the 1-byte variable called TEMP to the content of accumulator A and to put the result back into accumulator A. For instructions dealing with two 16-bit operands, A and B are concatenated together to form a 16-bit accumulator called D which holds one of the operands. It also serves as the destination for the 16-bit result.

The Motorola 68HC11's two 16-bit index registers, X and Y, serve as pointers to memory. As we will see in the next section, they each permit an operand in memory to be identified by the sum of the content of the register and an 8-bit offset which is included as part of the instruction. For example, the instruction

```
LDAB 4,X
```

says to add (temporarily) the number 4 to the content of the X register and use the result as the operand address. The operand is read from this address and then loaded into accumulator B.

The final CPU registers shown in Fig. 4.4 are the Motorola 68HC11's program counter and stack pointer. These are each 16 bits wide, permitting access to $2^{16} = 65,536$ different addresses. The 68HC11 uses nowhere near this number of addresses to identify each of its internal resources. On the other hand, when it is being operated in its *expanded* mode, with its internal address bus and data bus brought out to external pins, then these addresses can be used to access external devices (e.g., more RAM).

The Intel 8096 CPU includes a register structure that is simpler to describe and is shown in Fig. 4.5. The outstanding feature of the 8096's register structure is the absence of an accumulator, scratch-pad registers, and pointers to memory. As we will see, the *multiple-operand* instructions of the 8096 largely eliminate the need for accumulators and scratch-pad registers. However, whenever we need an accumulator (or scratch-pad register), be it for a 1-byte, 2-byte, or 4-byte operand, RAM can be used. In this sense, the 8096 has 230 1-byte accumulators, 115 2-byte accumulators, and 57 4-byte accumulators.

In like manner the 8096's *indirect* and *indexed* addressing modes, which can employ any of 115 2-byte pointers located in RAM, negate the need for pointers in the CPU. We will discuss the possibilities in the next section.

The Intel 8096's *program status word*, shown in Fig. 4.5, is 2 bytes long and consists of two parts. The lower byte is the interrupt mask register discussed in the last chapter in conjunction with Fig. 3.6. Each of its bits is referred to in that figure as a "2d level enable bit." The upper byte includes an interrupt enable bit I. This "1st level enable bit" must be set for *any* interrupts to be acknowledged by the CPU. The other six bits in this byte which have a defined function are:

Z, N, and V which are defined to serve the same function as the corresponding bits in the Motorola 68HC11.

C is a *carry/not-borrow* flag. Its carry definition is the same as for the 68HC11. However, its borrow definition for multiple-byte subtraction op-

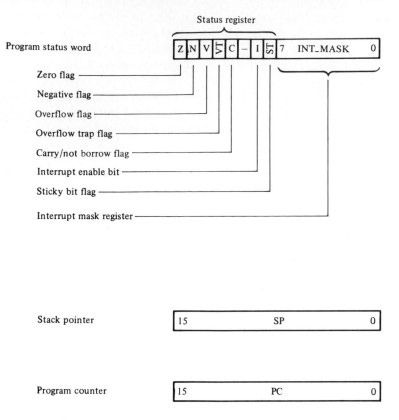

FIGURE 4.5
Intel 8096 CPU register structure.

erations is just the opposite. That is, C = 0 when a borrow occurs as a result of a subtraction. However, the subtraction circuitry in the Intel 8096 sorts this out so that a multiple-byte subtraction algorithm for the 8096 looks virtually identical to that for the Motorola 68HC11.

VT is an *overflow-trap* flag. This bit is set any time that the V flag is set. However, it is only cleared by an explicit instruction such as CLRVT (clear VT). Consequently, it is a way to replace repeated tests on the V flag (to see if an operation has set it) with a single test on the VT flag (to see if *any one* of the repeated operations has set it). Instead of testing the V flag *each time* it could be set, simply clear the VT flag first and test it when the entire sequence of operations is done.

ST is a "*sticky*" bit. The 8096 includes a right shift instruction which can be used to divide a binary integer (or a 2s-complement-coded integer) by 2^n, where n is specified by the instruction. After such an instruction, the carry flag represents the number 0.5. As such, if it is set, then a rounding operation on the quotient should increment the quotient. If the carry flag is clear, then rounding would leave the quotient alone. The sticky bit takes this one step further. After a right shift operation in which the operand was shifted more than one place, if the sticky bit equals 0, then *no* 1s were shifted into *and* out of the carry bit. Consequently, the dividend is *exactly* equal to the number remaining in the operand address and the carry bit. For example, a right shift of four places on the binary number

0010 0000 0001 1000

produces the *exact* result

0000 0010 0000 0001 . 1

That is, $(8192 + 16 + 8) / 16 = 513.5$ *exactly*. This can be used to give more information during a rounding operation.

4.3 ADDRESSING MODES

In this section we will look at a variety of addressing modes. We will use the Intel 8096 and the Motorola 68HC11 as examples, illustrating the *representation* of cach addressing mode within an assembly language program.

Immediate addressing identifies the operand as part of the instruction itself. An immediate operand is represented by a "#" prefix in both the Intel 8096 and the Motorola 68HC11 assemblers.

Example 4.5. The Intel 8096's three-operand instruction

```
ADD   HSO_TIME,TIMER1,#30
```

says to read the 16-bit content of the programmable timer's free-running counter (called TIMER1), add 30 to it, and store the result back out to one of the programmable timer's output-compare registers.*

Extended addressing permits the address of an operand to extend over the entire address space of the microcontroller. For both the Intel 8096 and the Motorola 68HC11, this address space goes from 0000H (hexadecimal) to FFFFH, giving 65,536 addresses (any one of which can be identified by a 2-byte number). The good feature about extended addressing is that it can reach *any* address. However, it does this at the expense of requiring 2 bytes of an instruction to hold this operand address.

---

* Which is actually a file of up to eight output-compare registers, each of which is loaded by a write to the address HSO_TIME. Refer to Appendix B.

Example 4.6. Motorola's default memory mapping allows the 64 registers associated with the 68HC11's internal facilities (e.g., UART and programmable timer) to be addressed in the range 1000H to 103FH. Any of them can be accessed with extended addressing. For example, the 3-byte instruction

LDD   TCNT

will read the content of the programmable timer's 16-bit free-running counter, located at address 100EH, into the 16-bit accumulator D.

Register addressing refers to the addressing of any of the CPU's scratch-pad registers. The purpose of a CPU's scratch-pad registers is to minimize the number of instruction bits required to identify an operand. For example, the popular microprocessor of the past, the Intel 8080, had six scratch-pad registers which could be represented by three bits of the opcode. Thus, a 1-byte 8080 instruction could add the content of any of these scratch-pad registers to the content of the accumulator and put the resulting sum back into the accumulator.

Direct addressing refers to the addressing of a specific *page* of the entire address space of the microcontroller, where a page refers to all addresses having the same value in the upper byte. Both the Motorola 68HC11 and the Intel 8096 treat *page 0* in this special way. That is, the page with addresses in the range

0000H   to   00FFH

can be accessed with instructions which require only 1 byte of the instruction for identifying any of 256 operands.

Example 4.7. Users of the 68HC11 normally employ this capability to access the 256 bytes of on-chip RAM. However, the designers of the chip made it possible to remap the RAM and the internal registers (associated with the internal facilities) so that each of these begins at an arbitrary address of the form

X000H

Consequently, we *could* set up the 68HC11 so that its registers are accessed with the following page 0 addresses

0000H to 003FH

while the RAM is accessed from some other page. Because the 68HC11's bit-manipulating instructions are so valuable for dealing with the registers, and because these instructions are supported by direct addressing but not extended addressing, we recommend this remapping of the registers to page 0 in Sec. A.2 of Appendix A.

The Motorola 68HC11 even permits us to map *both* RAM and registers to page 0. Accesses to any address between 0000H and 003FH are designed to access the registers in this case. The 64 bytes of RAM assigned to these same addresses are effectively disabled and therefore lost to our use. We are still left with 192 bytes of RAM, and we obtain faster, more efficient access to the registers, as well as RAM, by using direct addressing for everything.

This strategy of mapping both RAM and registers to page 0 and then supporting access to them with direct addressing is also used by the Intel 8096. However, it is even more powerful in this case since *any* operand field of *any* 8096 instruction can identify a page 0 operand.

Example 4.8. Consider the Intel 8096 instructions to determine a short pulse width

```
SUB     TEMP,HSI_TIME,TEMP
SHL     TEMP,#1
```

The first instruction assumes that a 2-byte variable in RAM, called TEMP, has previously been loaded with the start time for the pulse. The 2-byte register called HSI_TIME holds the programmable timer's input capture register content, which represents the stop time for the pulse. This three-operand instruction consists of 4 bytes:

The opcode

The destination operand's address on page 0

One source operand's address on page 0

The other source operand's address on page 0

The second instruction shifts this difference one place to the left to multiply it by two. This adjusts for the programmable timer being incremented every 2 μs. The result is therefore the pulse width, measured in microseconds.

Indirect addressing permits the address identified within an instruction to hold a *pointer* to an operand.

Example 4.9. The Intel 8096 can use *any* of its 2-byte RAM locations (starting at an *even* address) to hold a pointer to an operand. This means that an interrupt service routine can use a pointer without first loading up a CPU register, speeding up interrupt servicing significantly. For example, the instruction

```
STB     SBUF,[QUE_IN]
```

reads a byte from the UART's serial input register, SBUF, and stores this number into a queue at a RAM address pointed to by a 2-byte RAM variable called QUE_IN.

Indirect with autoincrement addressing permits the operation employing indirect addressing to follow this with the automatic incrementing of the pointer.

Example 4.10. The Intel 8096 supports this addressing mode. In fact, it supports *two versions* of it. If the instruction is dealing with 1-byte operands, then the pointer is incremented once. If the instruction is dealing with 2-byte operands, the pointer is incremented twice.

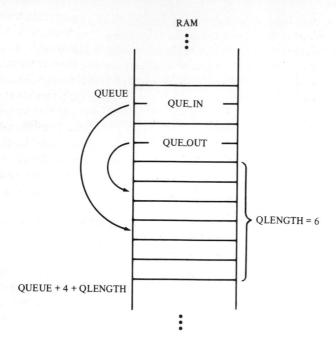

FIGURE 4.6
A queue structure, with two pointers and a data buffer.

Reconsider Example 4.9 in light of the actual operation desired. As we will discuss in the next chapter, a queue is a software version of a FIFO (first-in–first-out) memory. It uses two pointers, QUE_IN and QUE_OUT, to access some RAM set aside for buffering data. If the queue structure begins at an address labeled QUEUE and includes two pointers labeled QUE_IN and QUE_OUT plus a number of bytes of storage labeled QLENGTH, then we have the structure depicted in Fig. 4.6. The Intel 8096 instruction sequence for adding a byte into the queue from the UART is:

```
        STB     SBUF,[QUE_IN]+
        CMP     QUE_IN,#(QUEUE+4+QLENGTH)
        JNE     NEXT
        LD      QUE_IN,#QUEUE+4
NEXT:      .
```

The first instruction follows the storage of the single byte taken from SBUF into the queue (at the address pointed to by QUE_IN) with an automatic increment of the pointer in QUE_IN. This means that the next time the UART causes an interrupt, the pointer will be pointing to the right place in the queue for storing this new byte. The one exception occurs when we increment past the end of the RAM set aside for the queue.

The remaining instructions in the example above reset the pointer to the beginning of the RAM area set aside for the queue in this case. The CMP (compare) instruction will set the Z flag at that point. The JNE (jump if not equal) instruction normally causes execution to skip over the instruction which resets the pointer. However, when the 16-bit number stored in QUE_IN equals the 16-bit number represented by QUEUE+4+QLENGTH, then the Z flag will be set and the jump will not be taken. In this case, the content of QUE_IN will be reloaded with the address of the beginning of the queue's buffer, QUEUE+4. (At the beginning of our program listing we need to define labels like QUEUE and QLENGTH using the assembler directives to be discussed in Sec. 4.5.)

Example 4.11. In the last example, the queue dealt with single bytes. In this example, we want the queue to handle 2-byte words. This queue is to accept successive readings of the programmable timer's input-capture register. Each reading is a 2-byte number. The code is shown below.

```
              .
              .
       ST     HSI_TIME,[QUE_IN]+
       CMP    QUE_IN,#(QUEUE+4+(2*QLENGTH))
       JNE    NEXT
       LD     QUE_IN,#QUEUE+4
NEXT:         .
              .
```

We see that the only change is in the first two instructions. The STB instruction has been changed to an ST instruction. STB stores 1-byte variables whereas ST stores 2-byte variables. The CPU increments QUE_IN *twice* to complete the first instruction.

Indexed addressing temporarily adds a number in the instruction to a pointer to form the effective address of the operand.

Example 4.12. The Motorola 68HC11 uses indexed addressing as its way of accessing a variable using a pointer. For example, consider Fig. 4.7. The 2-byte 68HC11 assembly language instruction

```
ADDA    4,X
```

instructs the CPU to add 4 to the content of the X index register temporarily, to fetch a 1-byte operand from that address, to add it to the content of accumulator A, and to put the result back into accumulator A. The number in the instruction (4 in this case) can be any value between 0 and 255 (i.e., 00H to FFH). Hence, this number contributes 1 byte to the length of the instruction and causes a temporary offset of up to 255 bytes *above* the address in the pointer.

Each of the Motorola 68HC11's two pointers, X and Y, support indexed addressing.

Example 4.13. The Intel 8096 also supports indexed addressing. Its *short-indexed* addressing mode temporarily adds a 1-byte number to any 2-byte word variable in the on-chip RAM. The 1-byte number is treated as a 2s-complement signed number. It is "sign extended" into its 16-bit equivalent and added to the designated

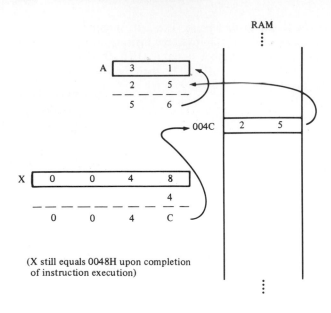

(X still equals 0048H upon completion of instruction execution)

FIGURE 4.7
The Motorola 68HC11's indexed addressing implementation of ADDA 4,X.

RAM content. Thus, short-indexed addressing produces an effective address in the range from 128 bytes below the address stored in the word variable to 127 bytes above that address. For example, the 4-byte instruction

        ADDB   TEMP,4[INDEX]

does the equivalent of the operation in Example 4.12. The ADDB instruction says to do an 8-bit addition. The 2-byte variable, INDEX, can be located in any pair of adjacent RAM locations, beginning with an even address. The 8096's addressing modes give a user lots of accumulators (e.g., TEMP in this example) and lots of index registers (e.g., INDEX in this example).

Example 4.14. The Intel 8096 also supports *long-indexed* addressing in which a 16-bit number in the instruction is temporarily added to the content of the 2-byte variable also identified by the instruction. For example, consider the following 5-byte instruction which uses a table entry in ROM to set selected bits of port 1 (leaving the remaining bits unchanged).

        ORB   P1,TABLE[OFFSET]

In this case we are able to use a 16-bit address labeled TABLE, which is in ROM* and add to this a 16-bit offset, taken from the content of a 2-byte variable in RAM called OFFSET. This effective address accesses an operand in the table. The operand is ORed with the output of port 1. The result is then written back to port 1. Note that if we have a table with only four entries, then OFFSET must still be set up to contain one of the four 16-bit numbers 0000H, 0001H, 0002H, or

---

* ROM in the Intel 8096 extends between 2080H and 3FFFH.

0003H. This may seem like overkill since a 2-byte variable, OFFSET, is needed where a 1-byte variable would do. However, it gives us a powerful method for accessing ROM tables with a variable offset.

This almost concludes our discussion of addressing modes. However, two ramifications of the Intel 8096's addressing modes offer flexibility which is valuable but is not obvious.

Example 4.15. A cursory look at the addressing modes supported by the Intel 8096 makes it appear that extended addressing is not supported. That is, it *appears* that the instruction

```
LDB     TEMP,ROM_ADDRESS
```

will not load the 1-byte RAM variable called TEMP with the content of a byte in ROM whose address is labeled ROM_ADDRESS. In actual fact, the 8096 assembler will accept this instruction and assemble it as if it had been written

```
LDB     TEMP,ROM_ADDRESS[0]
```

This works because the first two bytes of memory (i.e., addresses 0000H and 0001H) are not registers at all. Rather the designers of the 8096 fixed these two locations so that when their contents were read, 00H would be produced. Hence the above long-indexed addressing mode produces the sum of ROM_ADDRESS plus 0.

Example 4.16. The other ramification of the Intel 8096 addressing modes occurs because the stack pointer is actually located in addresses 0018H and 0019H, the first 2 bytes of RAM. Consequently, it is possible to access the stack easily. The top of the Intel 8096's stack can be read into a 2-byte RAM variable called AX (used as an accumulator) as follows:

```
LD      AX,[SP]
```

We have seen that the Intel 8096 has extensive addressing capability. It needs to be pointed out that in *any* 8096 instruction, only *one* operand (at most) can be accessed by means other than page 0 direct addressing.* With only three exceptions, an 8096 instruction must use this operand as a *source* variable. For example, the instruction

```
ADDB    BYTE1,#5
```

uses immediate addressing for the source operand whose value is 5. It adds this to the content of BYTE1 and puts the result back into the *destination* variable, which is also BYTE1.

The only 8096 instructions which can use arbitrary addressing for the *destination*, or result, of an instruction are:

---

* Refer to Fig. B.39 in Appendix B. Also, the key at the end of that figure describes the instruction syntax.

```
STB ;Store a byte
ST  ;Store a word, i.e., 2 bytes
POP ;Pop a word from the top of the stack
```

Because of this, we need the equivalent of CPU registers in the 8096 whenever both the source and the destination of an operation must be accessed using an addressing mode other than direct addressing. Within an interrupt service routine, any registers needed for this purpose can be dedicated to that task. This leads to exceedingly fast service for sources of interrupts since *no* CPU registers have to be loaded before servicing begins. In effect, each interrupt service routine has its own CPU registers, and they are already loaded with pointers, counters, and whatever when the routine is entered.

More generally, and for the sake of supporting general purpose subroutines which can operate upon a known set of CPU registers, we can define a CPU register convention for general use throughout our software development. In fact, Intel has done just exactly this for users of the Intel 8096. We will consider this convention in Sec. 4.6, in conjunction with Fig. 4.27.

4.4 INSTRUCTION SETS

In this section we will look at the entire instruction set of the Intel 8096 and of the Motorola 68HC11. We will do this to gain insight into the capabilities of typical instructions available in present-day microcontrollers.

Each of these microcontrollers has what is sometimes called an *orthogonal* instruction set, that is, an instruction set which extends all of its addressing modes to every instruction for which those addressing modes make sense. An example of an instruction-and-addressing-mode combination which would *not* make sense would be a Motorola 68HC11 instruction, STAA #05H. This says to store accumulator A to the second byte of the instruction, obliterating the immediate operand, 05H, which is there. Of course, with a program in ROM the store operation would not even take place!

Given an orthogonal instruction set, it is not really necessary to identify which addressing modes go with which instructions. Accordingly, we will describe here only the instructions themselves. For a more detailed look, see Figs. A.39 and B.39 in the appendixes.

The *data move* family of instructions are used to move data into and out of CPU registers (or in the case of the Intel 8096, any page 0 address, whether it be RAM or a register associated with an on-chip resource). Figure 4.8 shows the options available for the Intel 8096 and the Motorola 68HC11. These include pushing data onto the stack as well as popping (or pulling) it off of the stack. The Motorola 68HC11 has a few instructions for moving data *between* CPU registers (i.e., TAB and TBA, which move A to B or B to A, respectively; TPA and TAP, which move the status register to or from accumulator A; XGDX and XGDY, which *swap* one of the index registers with the 16-bit D accumulator; TSX, TSY, TXS, and TYS which move the content of the stack

	Intel 8096	Motorola 68HC11
Move a byte		
Load (arbitrary addressing for source)	LDB	LDA<1>*
Store (arbitrary addressing for destination)	STB	STA<1>
Transfer A into B or B into A		T<1><1>
Transfer A into status register		TAP
Transfer status register into A		TPA
Push accumulator onto stack		PSH<1>
Pop accumulator off stack		PUL<1>
Move a byte into a word		
Treat as an unsigned number	LDBZE	
Treat as a signed number	LDBSE	
Move a 2-byte word		
Load (arbitrary addressing for source)	LD	LD<2>
Store (arbitrary addressing for destination)	ST	ST<2>
Exchange D with X or Y		XGD<3>
Transfer from stack pointer to index register		TS<3>
Transfer from index register to stack pointer		T<3>S
Push index register onto stack		PSH<3>
Pop index register off stack		PUL<3>
Push program status word onto stack	PUSHF	
Pop program status word off stack	POPF	
Push (arbitrary addressing for source)	PUSH	
Pop (arbitrary addressing for destination)	POP	

* <1> represents accumulator A or B; <2> represents D, X, Y, or S; <3> represents X or Y.

FIGURE 4.8
Data move instructions.

pointer to or from an index register). The Intel 8096 always pushes or pops 2 bytes at a time but can identify the operand using *any* addressing mode, a powerful feature. Its PUSHF and POPF instructions push and pop the program status word, shown in Fig. 4.5. Its LDBZE and LDBSE instructions "zero extend" and "sign extend" a 1-byte variable being moved into a 2-byte variable.

The *single-operand* instructions (exclusive of the shift and rotate instructions) are listed in Fig. 4.9. These include operations on both bytes and words, that is, upon 1-byte and 2-byte variables. Motorola's instructions which operate upon a byte can be applied to either accumulator or to an arbitrary memory location. The sign-extend function enlarges the size of a 2s-complement number. If the sign bit of a 1-byte number is 0, then sign extension requires that the most-significant byte of the 2-byte number be 00H. On the other hand, if the sign bit of a 1-byte number is 1, then the most-significant byte of the sign-extended 2-byte number will be FFH.

	Intel 8096	Motorola 68HC11
Increment a byte		
In an accumulator		INC<1>*
Direct addressing	INCB	
Arbitrary addressing		INC
Increment a word		
In a CPU register		INC<2>
Direct addressing	INC	
Decrement a byte		
In an accumulator		DEC<1>
Direct addressing	DECB	
Arbitrary addressing		DEC
Decrement a word		
In a CPU register		DEC<2>
Direct addressing	DEC	
Clear a byte		
In an accumulator		CLR<1>
Direct addressing	CLRB	
Arbitrary addressing		CLR
Clear a word		
Direct addressing	CLR	
Complement bits of a byte		
In an accumulator		COM<1>
Direct addressing	NOTB	
Arbitrary addressing		COM
Complement bits of a word		
Direct addressing	NOT	
Change sign of a 1-byte 2s-complement number		
In an accumulator		NEG<1>
Direct addressing	NEGB	
Arbitrary addressing		NEG
Change sign of a 2-byte 2s-complement number		
Direct addressing	NEG	
Sign extend a 2s-complement number		
1-byte to 2-byte	EXTB	
2-byte to 4-byte	EXT	

* <1> represents accumulator A or B; <2> represents X, Y, or S.

FIGURE 4.9
Single-operand instructions.

Shift and rotate instructions are shown in Fig. 4.10. The Motorola 68HC11 shift instructions for 2-byte operands are confined to shifting the D accumulator. Each of the 68HC11 instructions shifts the operand one bit position. In contrast, the Intel 8096 instructions include a parameter to indicate how many positions to shift. For example, the instruction

       SHR   WORD1,#3

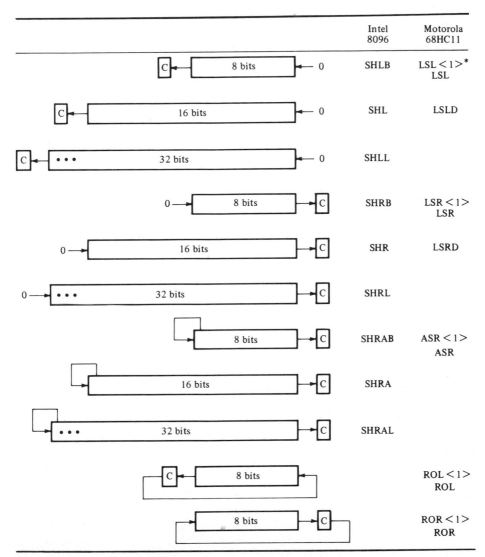

*<1> represents accumulator A or B.

FIGURE 4.10
Shift and rotate instructions.

will shift the 2-byte operand in the RAM location labeled WORD1 three places to the right. The distinction between the middle three shift operations shown in Fig. 4.10 and the bottom three shift operations relates to whether a *logic* shift or an *arithmetic* shift is called for. For arithmetic operands, each right shift of one bit position amounts to a division by two. If the operand is an unsigned number, then a 0 should be shifted in from the left (i.e., a logical shift right instruction should be used). If the operand is a 2s-complement number, then the most-significant bit should not only be shifted to the right, but also maintained in the most-significant bit position (i.e., an arithmetic shift right instruction should be used).

The Motorola 68HC11's rotate instructions are the building blocks for rotating operands of any length.

Example 4.17. Consider the code to carry out a shift-left operation on a 6-byte number stored in memory, least-significant byte first. Assume that the X index register points to the number. Then the following instruction sequence will carry out the desired operation.

```
LSL     0,X
ROL     1,X
ROL     2,X
ROL     3,X
ROL     4,X
ROL     5,X
```

Note that the carry bit serves to link the successive shift operations together.

Bit-manipulation instructions, shown in Fig. 4.11, permit the setting, clearing, and complementing of selected bits in an operand. To set selected bits in an operand, we can OR the operand with another operand having 1s in the bits to be set.

Example 4.18. Show the instruction sequence to set bits 0 and 1 on output port B of the Motorola 68HC11, leaving the other outputs unchanged.

The instruction repertoire permits us to do this in either of two ways. First we will consider a way which requires both a read and a write of the port.

```
SEI                      ;Disable interrupts
LDAA    PORTB            ;Read port B
ORAA    #00000011B       ;Force bits 0 and 1 high
STAA    PORTB            ;Store result back out to the port
CLI                      ;Reenable interrupts
```

Interrupts are disabled during this sequence because (as was pointed out in Sec. 3.7) if this sequence is interrupted, and if that interrupt service routine tries to change bit 5 (for example) on this port, then when it has done just that and returned, the rest of the sequence shown here will put bit 5 back *the way it was* before the interrupt occurred.

A second way to do this, given the Motorola 68HC11 instruction set, is to use the BSET instruction:

	Intel 8096	Motorola 68HC11
Force selected bits in a byte to 0		
In an accumulator		AND<1>*
Direct and indexed addressing		BCLR
Arbitrary addressing for one source		
operand	ANDB†	
Force selected bits in a word to 0		
Arbitrary addressing for one source		
operand	AND†	
Force selected bits in a byte to 1		
In an accumulator		OR<1>
Direct and indexed addressing		BSET
Arbitrary addressing for one source		
operand	ORB	
Force selected bits in a word to one		
Arbitrary addressing for one source		
operand	OR	
Complement selected bits in a byte		
In an accumulator		EOR<1>
Arbitrary addressing for one source		
operand	XORB	
Complement selected bits in a word		
Arbitrary addressing for one source		
operand	XOR	

* <1> represents accumulator A or B.
† Includes both two-operand and three-operand versions.

FIGURE 4.11
Bit manipulation instructions.

```
LDX   #PORTB         ;Use X as pointer to PORTB
BSET 0,X,00000011B ;Bam!
```

Now there is no reason to disable interrupts. The single BSET instruction reads the port (into a hidden register in the CPU), forces the selected bits to 1, and writes the result back out to the port. An interrupt cannot break into the middle of the execution of an instruction, so we are safe.

As was pointed out earlier, the BSET and BCLR instructions are so powerful for interacting with the 68HC11's registers that it makes sense to remap the registers to page 0 so that direct addressing can be used. If this is done, then the above sequence reduces to:

```
BSET    PORTB,00000011B   ;Set bits 0 and 1 of PORTB
```

Example 4.19. Because of the Intel 8096's multiple-operand instructions, it does not have the same need for interrupt disabling when it ORs into an output port.

Consider the following 3-byte two-operand instruction which sets bits 0 and 1 of port 1, leaving the other bits unchanged:

```
ORB     P1,#00000011B
```

To *clear* selected bits, we need to AND the operand with another operand which has 0s in the bit positions to be cleared.

> **Example 4.20.** The following Intel 8096 instruction will clear bits 2 and 3 of port 1, leaving the other bits unchanged.
>
> ```
> ANDB    P1,#11110011B
> ```

To *complement* selected bits, we need to exclusive-OR the operand with another operand which has 1s in the bit positions to be complemented.

> **Example 4.21.** The following Intel 8096 instruction will complement bits 4 and 5 of port 1, leaving the other bits unchanged. This is useful if we know, for example, that bit 4 is presently 0 and bit 5 is presently 1, and we want to change these to 1 and 0, respectively.
>
> ```
> XORB    P1,#00110000B
> ```

Arithmetic instructions require some kind of compromise between the range of operations which users want and the level of support that a given generation of technology can reasonably provide. For example, it is typical to support the addition of unsigned or signed binary numbers of arbitrary size by including add with carry instructions for adding the operands 1 or 2 bytes at a time. Note that the benefit of coding signed numbers using 2s-complement code occurs during addition and subtraction, where the circuitry is identical to that for adding and subtracting unsigned numbers. That is, the same ADD instruction works on either unsigned or signed numbers.

Motorola has carried this philosophy forward in the 68HC11 by including building block instructions for multiply and divide operations. As shown in Fig. 4.12, the 68HC11 supports the multiplication of the two 1-byte unsigned numbers (located in accumulators A and B) to produce a 2-byte result (located in accumulator D).

> **Example 4.22.** The power of building block arithmetic instructions is shown here. Consider that we are designing an instrument which uses the 68HC11 as its controller and which expects the user to key in a three-digit parameter value on a front-panel numeric keypad. We need to convert this number to the equivalent binary number. Assume that these three digits are stored as binary numbers in the three 1-byte variables:
>
> $$D100 \qquad D10 \qquad D1$$
>
> The result will be formed in the 16-bit accumulator D. The calculation we need is
>
> $$100 * D100 + 10 * D10 + D1$$

	Intel 8096	Motorola 68HC11
Add (both signed and unsigned)		
Bytes	ADDB*	ADD<1>†
Accumulators A and B; result in A		ABA
Bytes plus carry	ADDCB	ADC<1>
Words	ADD*	ADDD
Words plus carry	ADDC	
Correction for BCD addition in accumulator A		DAA
Add B into X or Y (byte into word, unsigned)		ABX
		ABY
Subtract (both signed and unsigned)		
Bytes	SUBB*	SUB<1>
Accumulator B from A; result in A		SBA
Bytes minus borrow	SUBCB	SBC<1>
Words	SUB*	SUBD
Words minus borrow	SUBC	
Multiply		
Bytes, unsigned	MULUB*	MUL
Bytes, signed	MULB*	
Words, unsigned	MULU*	
Words, signed	MUL*	
Divide		
Word/byte, unsigned	DIVUB	
Word/byte, signed	DIVB	
Double-word/word, unsigned	DIVU	
Double-word/word, signed	DIV	
N-byte/word		IDIV, FDIV

* Includes both two-operand and three-operand versions.
† <1> represents accumulator A or B.

FIGURE 4.12
Arithmetic instructions.

      or

$$10 * ((10 * D100) + D10) + D1$$

The 68HC11 sequence of instructions to carry this out is:

```
        LDAA    #10
        LDAB    D100
        MUL             ;Result is 90 or less
        ADDB    D10     ;Result is 99 or less
        LDAA    #10
        MUL             ;Result is 990 or less
        ADDB    D1      ;Result is 999 or less
        BCC     SKIP    ;If carry is set, then
        INCA            ; increment upper byte
SKIP:       .           ;Binary result is now in D
```

In like manner, the Motorola 68HC11 includes two building block instructions which are used repeatedly in a division algorithm to perform division on arbitrarily sized numbers.

The philosophy of the designers of the Intel 8096 is somewhat different from this building block approach. While they have included building block arithmetic instructions, they have also supported addition, subtraction, multiplication, and division of unsigned and signed numbers for bytes and words—with a separate, single instruction for each. This is illustrated in Fig. 4.12.

Example 4.23. To gain insight into the syntax of the Intel 8096 multiply and divide instructions, consider the following examples.

    MUL   WORD1,WORD2

This instruction multiplies the two 2-byte signed operands and puts the result into a *4-byte* location having the address WORD1, as shown in Fig. 4.13. The hard-

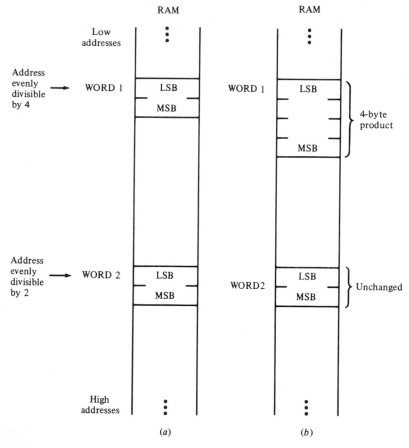

FIGURE 4.13
Intel 8096's MUL WORD1,WORD2 instruction. (*a*) Before; (*b*) after.

ware needs to be supported by having the addresses aligned as shown (with the starting address of the result evenly divisible by four).

The three-operand version of this instruction

```
MUL   WORD1,WORD2,WORD3
```

has the same alignment requirements. That is, the 4-byte result in WORD1 must have an address that is evenly divisible by four.

The more general case of operand addressing permits the operand listed last (i.e., one of the source operands) to take on any of the 8096's addressing modes, subject only to the alignment restriction of having an address which is evenly divisible by two.

Compare instructions carry out an operation similar to subtraction or bit-manipulation operations. However, they only affect the status register flags (i.e., the carry flag, etc.). That is, there is no destination operand. These instructions are shown in Fig. 4.14.

Conditional jump instructions control the flow of the execution of instructions. Many microcontrollers test *flags* and can execute a jump to an arbitrary point in a program, depending upon whether a specific flag is set (or clear). Both the Intel 8096 and the Motorola 68HC11 support such flag tests, as shown in Fig. 4.15.

Both the Intel 8096 and the Motorola 68HC11 can link a compare instruction (or a subtraction instruction) with a conditional jump instruction and execute a jump depending upon *any* specific condition desired (e.g., jump if the first operand in the compare is greater than or equal to the second operand, where both operands are signed numbers expressed in 2s-complement code). All of the variations are shown in Fig. 4.15.

Finally both of these microcontrollers can test a single bit located anywhere in RAM or in the registers associated with the on-chip facilities.

	Intel 8096	Motorola 68HC11
Subtract (without saving result)		
Bytes	CMPB*	CMP<1>†
Accumulator B from A		CBA‡
Words	CMP	CPD
		CPX
		CPY
Subtract 0 from a byte		TST<1>
		TST
AND		BIT<1>

† <1> represents accumulator A or B.
* CMPB BYTE1,BYTE2 instruction temporarily forms BYTE1-BYTE2 and sets flags accordingly.
‡ CBA instruction temporarily forms A-B and sets flags accordingly.

FIGURE 4.14
Compare instructions. Arbitrary addressing for the subtrahend is supported by both microcontrollers.

	Intel 8096	Motorola 68HC11
Jump if flag is set or clear		
if C = 1	JC	BLO
if C = 0	JNE	BHS
if Z = 1	JE	BEQ
if Z = 0	JNE	BNE
if N = 1	JLT	BMI
if N = 0	JGE	BPL
if V = 1	JV	BVS
if V = 0	JNV	BVC
if VT = 1	JVT	
if VT = 0	JNVT	
if ST = 1	JST	
if ST = 0	JNST	
Jump after a compare of unsigned numbers		
if < 0	JC	BLO
if =< 0	JNH	BLS
if = 0	JE	BEQ
if => 0	JNC	BHS
if > 0	JH	BHI
if <> 0	JNE	BNE
Jump after a compare of signed numbers		
if < 0	JLT	BLT
if =< 0	JLE	BLE
if = 0	JE	BEQ
if => 0	JGE	BGE
if > 0	JGT	BGT
if <> 0	JNE	BNE
Test a bit of a byte		
Jump if bit set	JBS	BRSET*
Jump if bit clear	JBC	BRCLR*
Loop control		
Decrement register and jump if not zero	DJNZ	

* Supported only by direct and indexed addressing; see example.

FIGURE 4.15
Conditional jump instructions.

Example 4.24. The Motorola 68HC11 instruction to test a bit of a byte located in the register area for on-chip facilities requires either that indexed addressing be used or that the registers be remapped to page 0. The test of the UART status register to see if its receive data register is full (i.e., if it has received a byte of data) and to jump to an address labeled READ_IT if it is, follows:

```
LDX    #SCSR
BRSET  0,X,00100000B,READ_IT
```

If the registers have been remapped to page 0, then this can be simplified to:

```
BRSET  SCSR,00100000B,READ_IT
```

In either case, the address, READ_IT, must be located within about ±128 addresses of this instruction since READ_IT is assembled into a single byte which is added to the program counter as a signed number.

Example 4.25. The Intel 8096 instruction to test *its* UART status register to see if a byte has been received is

```
JBS  SP_STAT,6,READ_IT
```

Note that here the 6 is not testing the same bit position as in the previous example, which was for a different microcontroller (with the bit to be tested located in a different bit position).

Example 4.26. The Intel 8096 includes a loop control instruction, DJNZ, which automates the process of executing a sequence of instructions some number of times. It requires that an arbitrary 1-byte RAM variable, called COUNT below, be initialized to the number of times the sequence is to be repeated. Then the DJNZ instruction is used as follows:

```
        LDB   COUNT,#4
LOOP:         .              ;First instruction in loop
              .
              .              ;Last instruction in loop
        DJNZ COUNT,LOOP      ;Decrement and jump if not zero
              .
```

This DJNZ instruction requires that the jump address, LOOP, be within about ±128 addresses of the DJNZ instruction since it is treated as a 1-byte signed number to be added to the program counter if the jump is taken.

Unconditional jump and *subroutine call* instructions are shown in Fig. 4.16. The short versions (SJMP and SCALL for the Intel 8096 and BRA and BSR for the Motorola 68HC11) are 2-byte instructions which can jump or call a subroutine having a nearby address. A good strategy is to use the short version and let the assembler provide the warning in those cases where the jump address is out of range (in which case the instruction can be changed to the long version). For example, if the Motorola 68HC11 subroutine call

```
        BSR      ADD_MULTIPLE_BYTE_NUMBERS
```

generates an out-of-range error, then the instruction should be changed to

```
        JSR      ADD_MULTIPLE_BYTE_NUMBERS
```

Example 4.27. The Intel 8096 includes an indirect jump instruction for which the jump address is taken from a 2-byte RAM variable (suitably aligned, starting at an even address). For example, if we label this variable JMP_ADRS, then the indirect jump instruction takes the form:

```
        BR   [JMP_ADRS]
```

	Intel 8096	Motorola 68HC11
Unconditional jump		
Within ±128 addresses		BRA
Within ±1024 addresses	SJMP	
Anywhere	LJMP	JMP
Unconditional indirect jump anywhere	BR	
Subroutine call		
Within ±128 addresses		BSR
Within ±1024 addresses	SCALL	
Anywhere	LCALL	JSR
Subroutine return	RET	RTS
Software interrupt	TRAP	SWI
Interrupt return	RET	RTI
Wait for interrupt (low-power standby mode)		WAI
Stop internal clock (lowest-power standby mode)		STOP
Software reset	RST	

FIGURE 4.16
Unconditional jumps, subroutine calls, and returns.

Example 4.28. The Motorola 68HC11 can implement indirect jump instructions and jump to subroutine instructions with the help of an instruction sequence in each case. For example, the indirect jump is:

```
LDX   JMP_ADRS ;Put content of JMP_ADRS into X
JMP   0,X ;Transfer X to the program counter
```

The jump to subroutine version of this sequence uses the JSR mnemonic in place of JMP. Note that the JMP and JSR instructions do not use indexed addressing in the normal way. That is, instead of adding the offset (0 in the case above) to the content of X and using this sum for the *address* of the operand, the JMP and JSR instructions simply transfer this sum to the program counter.

Example 4.29. Consider a *jump table* stored in ROM and shown in Fig. 4.17 for the Motorola 68HC11. We want to jump to subroutine SUB0, SUB1, SUB2, or SUB3 based upon the content of the B accumulator. To do this, we first *make sure* that B contains a number between 0 and 3. Then we execute the five-instruction sequence

```
LDX   #JMP_TBL    ;Load X with base address of table
ABX               ;Add offset twice, since there
ABX               ;are two bytes per table entry
LDX   0,X         ;Get content of table entry
JSR   0,X         ;Transfer it to program counter
```

When a subroutine is called, its return address is automatically pushed onto the stack. The subroutine return instructions shown in Fig. 4.16 simply pop this

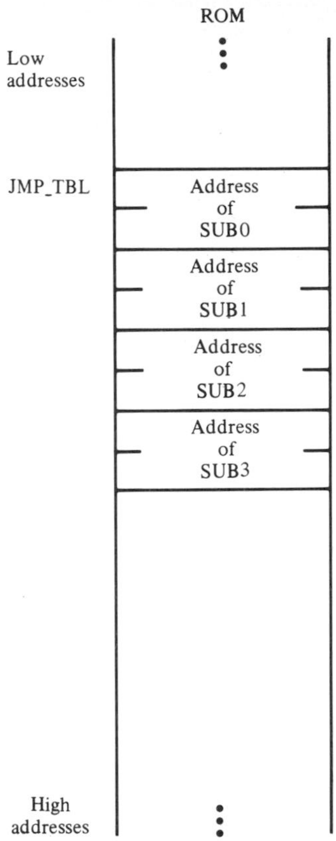

ROM

Low
addresses

JMP_TBL

| Address of SUB0 |
| Address of SUB1 |
| Address of SUB2 |
| Address of SUB3 |

High
addresses

FIGURE 4.17
A jump table.

return address off the stack and back into the program counter, returning program execution to the next instruction in line after the subroutine call.

Both the Intel 8096 and the Motorola 68HC11 include a 1-byte software interrupt or trap instruction, shown in Fig. 4.16. Execution of one of these instructions is treated by the CPU in much the same way as an external interrupt. It is not disabled by the CPU's I bit. Intel and Motorola each make use of the software interrupt instruction to implement features in their microcontroller development systems. Consequently, our use of a software interrupt will lead to a loss of some development system support capability.

The last two Motorola 68HC11 instructions listed in Fig. 4.16 give support to the 68HC11's low-power capability. We have already considered the use of the STOP instruction in Sec. 2.9. The WAI instruction can serve a similar function when it is necessary to save power and yet continue quickly when an external signal occurs. Since the internal clock is kept running, the power dissipation of the chip is not as low as when the STOP instruction has been executed.

Execution of the Intel 8096 RST software reset instruction pulls the reset pin low. Consequently, not only is the 8096 initialized, but any other chips which are reset from this same line are also initialized.

	Intel 8096	Motorola 68HC11
Status register control		
Set carry flag	SETC	SEC
Clear carry flag	CLRC	CLC
Set overflow flag		SEV
Clear overflow flag		CLV
Clear overflow trap flag	CLRVT	
Enable interrupts	EI	CLI
Disable interrupts	DI	SEI
No operation		
One-byte instruction	NOP	NOP
Two-byte instruction	SKIP	BRN
Normalize (see text)	NORML	

FIGURE 4.18
Special control instructions.

The remaining instructions for these two microcontrollers are listed in Fig. 4.18. Execution of an NOP instruction just wastes time. Sometimes during debugging it is useful to be able to change the program code, located in emulation RAM, so as not to execute a certain instruction or series of instructions. All the bytes of these instructions can be replaced by NOPs. At other times during debugging, it is convenient to be able to change the first byte of a short jump instruction so as to ensure that the jump will not take place. Replacing this first byte of the 2-byte instruction by the opcode for a SKIP or BRN instruction will achieve this.

The final instruction, NORML, listed in Fig. 4.18, gives support to arithmetic operations in the Intel 8096. The instruction

```
NORML   DOUBLEWORD1,BYTE2
```

will left shift, or normalize, the 4-byte number, labeled DOUBLEWORD1 (which must be suitably aligned at an address in RAM which is divisible by four), until its most-significant bit is 1. If, after 31 shifts, the MSB is still 0, then the Z flag is set. The number of shifts actually performed is stored in BYTE2.

To gain further insight into how instruction mnemonics and addressing modes go together for our example microcontrollers, Figs. 4.19 and 4.20 illustrate the assembly of two instructions, with all of their addressing modes, for each one. The Intel instructions of Fig. 4.19 for loading a byte have the following interpretation:

```
LDB     BYTE1,BYTE2
```

means load BYTE1 from BYTE2; that is,

```
BYTE1  <-- BYTE2
```

and the content of BYTE2 is left unchanged. Note that the required order of the operands in the assembly language source file has nothing to do with the order in which the operands appear in the assembled object file. For example, the above instruction assembles to

Address	Content	Meaning
2080	B0	Opcode
2081	25	Address of BYTE2
2082	24	Address of BYTE1

Furthermore, note that the Intel 8096 expects that full 2-byte addresses will be stored in an instruction least-significant byte first. For example, the instruction

```
LDB   BYTE1,VAL2[WORD1]
```

assembles to

Address	Content	Meaning
2090	B3	Opcode
2091	29	WORD1 (plus 1; 9th bit of opcode)
2092	34	VAL2, least-significant byte
2093	12	VAL2, most-significant byte
2094	24	BYTE1

Finally, note the object code difference between the LDB instructions using indirect addressing, with and without autoincrementing. The first bytes of these two instructions are identical (i.e., B2). The distinction between the two instructions occurs in the least-significant bit of the second byte. Because the operand represented by this second byte must necessarily have an *even* address (which the 8096 requires of all 2-byte variables), this least-significant bit is not needed by the CPU as it forms the address of this operand. Consequently, this bit was freely available to the 8096 designers to draw the distinction between two different addressing modes.

The examples of how Motorola 68HC11 instructions and addressing modes go together, shown in Fig. 4.20, illustrate one of the attractions of this microcontroller. Its instructions and addressing modes have a simplicity which is attractive to someone using the chip for the first time. However, this simplicity is deceptive. There is sufficient power in the instruction set and addressing modes to implement *any* operation as a sequence of these instructions.

4.5 ASSEMBLY LANGUAGE

An assembly language program for the microcontroller in an instrument or device is really nothing more than the listing of the *machine instructions* re-

```
          ;;;;;;;;;;  First, some definitions  ;;;;;;;;;;;;;;;;;;;;;;;;;;;;;;;;;;;

0041 =    VAL1    EQU   41H
1234 =    VAL2    EQU   1234H

0024              ORG   0024H              ;Assign variables into RAM area

0024      BYTE1:  DSB   1                  ;Byte address (reserve 1 byte)
0025      BYTE2:  DSB   1
0026      BYTE3:  DSB   1
0028      WORD1:  DSW   1                  ;Word address (reserve 2 bytes)
002A      WORD2:  DSW   1
002C      WORD3:  DSW   1

2080              ORG   2080H              ;Assign instructions into ROM area

          ;;;;;;;;;;;  Try 1-byte loads  ;;;;;;;;;;;;;;;;;;;;;;;;;;;;;;;;;;;;;

2080 B02524       LDB   BYTE1,BYTE2        ;Direct addressing      (BYTE1 := BYTE2)
2083 B14124       LDB   BYTE1,#VAL1        ;Immediate addressing
2086 B22824       LDB   BYTE1,[WORD1]      ;Indirect addressing
2089 B22924       LDB   BYTE1,[WORD1]+     ;Indirect addressing with autoincrement
208C B3284124     LDB   BYTE1,VAL1[WORD1]  ;Short-indexed addressing
2090 B3293412     LDB   BYTE1,VAL2[WORD1]  ;Long-indexed addressing
2094 24
2095 B3013412     LDB   BYTE1,VAL2[0]      ;Zero-register addressing
2099 24
```

```
209A B3013412    LDB    BYTE1,VAL2          ;Extended addressing (same as last one)
209E 24

      ;;;;;;;;;;  Try 2-byte adds    ;;;;;;;;;;;;;;;;;;;;;;;;;;;;;;;;;;;;;;;;

209F 642A28      ADD    WORD1,WORD2               ;Two operands, direct addressing
20A2 65341228    ADD    WORD1,#VAL2              ;Two operands, immediate addressing
20A6 662A28      ADD    WORD1,[WORD2]            ;Two operands, indirect addressing
20A9 662B28      ADD    WORD1,[WORD2]+           ;Two operands, indirect add., autoinc.
20AC 672A4128    ADD    WORD1,VAL1[WORD2]        ;Two operands, short-indexed addressing
20B0 672B3412    ADD    WORD1,VAL2[WORD2]        ;Two operands, long-indexed addressing
20B4 28

20B5 442C2A28    ADD    WORD1,WORD2,WORD3        ;Three operands, direct addressing
20B9 4534122A    ADD    WORD1,WORD2,#VAL2        ;Three operands, immediate addressing
20BD 28

20BE 462C2A28    ADD    WORD1,WORD2,[WORD3]      ;Three operands, indirect addressing
20C2 462D2A28    ADD    WORD1,WORD2,[WORD3]+     ;Three operands, indirect add., autoinc.
20C6 472C412A    ADD    WORD1,WORD2,VAL1[WORD3]  ;Three operands, short-indexed addressing
20CA 28

20CB 472D3412    ADD    WORD1,WORD2,VAL2[WORD3]  ;Three operands, long-indexed addressing
20CF 2A28

0000             END
```

FIGURE 4.19
Examples of Intel 8096 addressing syntax.

131

```
               ;;;;;;;;;;;  First, some definitions  ;;;;;;;;;;;;;;;;;;;;;;;;

0041 =         VAL1    EQU  41H
1234 =         VAL2    EQU  1234H

0000                   ORG  0000H       ;Assign variables into RAM area

0000           BYTE1:  RMB  1           ;Byte address (reserve one byte)
0001           BYTE2:  RMB  1
0002           BYTE3:  RMB  1
0003           WORD1:  RMB  2           ;Word address (reserve two bytes)
0005           WORD2:  RMB  2
0007           WORD3:  RMB  2

E000                   ORG  0E000H      ;Assign instructions into ROM area

               ;;;;;;;;;;;  Try one-byte loads  ;;;;;;;;;;;;;;;;;;;;;;;;;;;;

E000 9600              LDAA BYTE1       ;Direct addressing    (A := BYTE1)
E002 B61234            LDAA VAL2        ;Extended addressing
E005 8641              LDAA #VAL1       ;Immediate addressing
E007 A641              LDAA VAL1,X      ;Indexed addressing

               ;;;;;;;;;;;  Try two-byte adds  ;;;;;;;;;;;;;;;;;;;;;;;;;;;;;;

E009 D303              ADDD WORD1       ;Direct addressing    (D := D + WORD1)
E00B F31234            ADDD VAL2        ;Extended addressing
E00E C31234            ADDD #VAL2       ;Immediate addressing
E011 E341              ADDD VAL1,X      ;Indexed addressing

0000                   END
```

FIGURE 4.20
Examples of Motorola 68HC11 addressing syntax.

quired to carry out all of the tasks to be built into the operation of the instrument or device. By permitting the designer to express these instructions in
mnemonic form (e.g., as LDAA BYTE1 rather than as the hex numbers 96 and
00), an assembler provides a program with *readability*. Readability is valuable
while code is being written. It is valuable while code is being debugged. Finally,
it is valuable when code must be updated to add new features, long after the
original code was written.

Designers use assembly language code, rather than a high-level language
like Pascal, C, or PL/M, when they want to remain close to the execution of the
actual code. They do this for two good reasons:

An assembly language program runs faster than its high-level language
equivalent. By coding in assembly language rather than in a high-level
language, programmers typically improve execution speed by one or two
orders of magnitude.

An assembly language program is more ROM-efficient than its high-level language equivalent.* Thus, given the constraint of a limited amount of ROM in a given microcontroller, a designer can implement more features into the design of an instrument or device, before the ROM is filled with program code, by coding in assembly language.

The demise of assembly language programming for microcontrollers has been predicted ever since excellent high-level languages have been available. Encouragement for this demise has come as strong high-level debugging tools have made their way into microcontroller development systems. We will look at these issues in Chap. 8. For many designers who have made the transition to high-level language coding after years of assembly language coding, that *transition* has meant the ability to manage ever-larger development projects without being "bugged" to death! Nevertheless, for many designers working on many microcontroller projects, assembly language coding is still a way of life.

The assembly language program for a device typically consists of the following parts:

1. Header comments, which describe overall features of the program and how it is organized.
2. A *program hierarchy*, which gives the name of the mainline program and, indented below it, the names of the subroutines which it calls. Indented below each mainline subroutine name is a list of *its* subroutines. Each interrupt service routine has its own separate hierarchy (although in keeping with the philosophy of making interrupt service routines short, they may end up being written with no subroutines to call).

   A program hierarchy is like a table of contents for a program. When we are deep in the code, trying to remember the relationship between subroutines (i.e., who calls whom), a program hierarchy tells us immediately.
3. *Equates* which define constants used in the program. For example, the combination

```
PULSE_WIDTH    EQU   100          ;Timer-generated output
```

followed much later in the program by

```
        LDD   #PULSE_WIDTH
```

produces more readable, and therefore debuggable, code than simply

```
        LDD   #100          ;Set pulse width of output
```

Furthermore, if we want to change a constant that is used in several different places in a program, we do this with higher reliability if we can

---

* Many users of "Forth" claim that this is not necessarily true.

change an equate than if we have to hunt for all the occurrences of the constant.

4. Allocation of RAM for variables. Each variable name must be allocated a memory location. Furthermore, we need to make sure that if we assign address 0048H to a 2-byte variable called WORD1, then address 0049H will not inadvertently be assigned to another variable. Also, for a microcontroller like the Intel 8096 which *requires* that 2-byte variables be aligned with even addresses, we must make sure that we do not overlook this requirement. Fortunately, assembler directives (to be discussed shortly) can help resolve such problems.

5. The code for the mainline program. Referring to Fig. 2.28, we see a popular, commonly used structure consisting of the call of an initialization subroutine followed by a loop which contains almost nothing but subroutine calls. Each of these subroutines is executed once per loop. This structure ensures that each of these subroutines gets run periodically. We can even make calculations, and run tests, to determine the worst-case time for getting around the loop, if this is a concern.

6. The code for each of the subroutines. Each subroutine should begin with a header of comments saying what it does and how parameters are passed to and from it. It should be written to avoid any *side effects* of its being called. For example, we may establish a software convention that, in general, subroutines are to leave registers as they found them, except for the case when a register is being used to pass a parameter *back* to the calling routine. Violating such a software convention can create a lurking time bomb. If we are lucky, it will be discovered during early debugging. If we are really unfortunate, it will be ignored by the rest of our software until many months later when we add a new feature to our design and implement it by inserting a new subroutine call right after the return from this one. After much debugging, which is not due to a fault in our new subroutine, we discover the problem and finally eliminate it.

7. The code for each interrupt service routine. If the microcontroller requires polling, then the service routine will begin with a polling sequence.

8. Tables required by the program. For example, the popular CORDIC* algorithm for computing trigonometric functions is organized around a table. It needs to be stored in ROM along with the program.

9. Strings of ASCII characters required by the messages which will be presented to a user on an alphanumeric display or sent to a computer via a UART output. Assemblers include directives which permit us to enter the alphanumeric characters required. The assembler then generates the corresponding codes.

---

* Jack E. Volder, "The CORDIC Trigonometric Computing Technique," *IRE Transactions on Electronic Computers*, September 1959.

10. Interrupt and reset vectors. In contrast to the rest of our code, these vectors require exact positioning in the memory space. For the Motorola 68HC11, these vector locations are listed in Fig. 3.5. For the Intel 8096, they are listed in Fig. 3.6. At reset time, the 8096 begins execution from address 2080H, the beginning of the ROM area for user program code. (The gap between addresses 2000H to 2011H, used for vectors, and 2080H to 3FFFH, used for program code, is reserved by Intel for factory test code.)

We will see shortly that an assembler can provide much support for generating the above items into a reliable program for our application. However, its most fundamental jobs are:

The conversion of instructions expressed in mnemonic form into the *object code* (i.e., machine code) which gets stored in the microcontroller's ROM

The generation of a listing showing the relationship between the source code and the object code

We need the former to run the microcontroller. We need the latter to debug the former.

As a vehicle for considering the support which a good assembler can give us, we will consider the *cross-assembler** made by Avocet Systems, Inc., for assembling 68HC11 source code on the IBM PC computer. This is one of a popular, low-cost family of cross-assemblers for major (multiple-chip) microcomputers and (single-chip) microcontrollers. One of the reasons for their popularity is that these assemblers tend to support the *assembler directives*† defined by the manufacturer of a particular chip for use in the manufacturer's assembler. For example, Avocet uses a DS (define storage) directive in all of their assemblers for allocating memory to variables. Code written to be assembled by Motorola's assembler uses an RMB (reserve memory bytes) directive for the same purpose. Accordingly, if we have already written code using Motorola's assembler directives, we do not need to change the directives. On the other hand, if we have previously worked with an Avocet assembler for another micro, then we know Avocet's generic assembler directives, which will also work. This strategy is not unique to Avocet assemblers, of course. For example, Hewlett-Packard does the same thing in the assemblers for their top-of-the-line 64000 microprocessor development system.

Each line of a source file contains a single assembly language statement. Such a statement consists of up to four fields:

---

* A cross-assembler runs on one computer and generates machine code for another computer.
† Assembler directives are commands to the assembler. We will discuss their assorted roles shortly.

Label field. The label field contains either a *label* or a *name*. A label represents the *address* associated with the first byte of the object code generated by the line it is on. For example, an instruction label represents the address of the first byte of the instruction on that line. A label on a line which defines the constants to be stored in a table represents the address of the first byte stored in the table. A label is followed by a colon on the line where it is defined (i.e., where it appears in the label field, not the operand field). Some assemblers (e.g., the Avocet 68HC11 assembler) make colons optional.

A *name* in the label field represents a value assigned to it by the assembler in conjunction with the assembler directive on the same line. For example, the equate assembler directive

```
PULSE_WIDTH    EQU    100
```

associates the number 100 with the name PULSE_WIDTH. Names in the label field are *not* followed by a colon.

The *identifier* used for a label or a name should begin with a letter in column 1 and may consist of any number of alphanumerics as well as the underline ("_") character.* The first eight characters must not be repeated in a different identifier.

Operation field. This field usually contains an instruction mnemonic (e.g., ADDA). Alternatively, it may contain an assembler directive (e.g., EQU or RMB).

Operand field. The operation may require operands (e.g., LDAA #5. In this case #5 is the operand). If two or more operands are required, then they should be separated by commas. An operand may consist of an identifier, defined in the label field somewhere in the program. Or, an operand may consist of a number. To identify the base (i.e., binary or decimal or hex), the number may be followed by B (for binary), H (for hex), or nothing (for decimal). Hex numbers cannot begin with a letter. For example,

> E000H is not acceptable
> 0E000H is acceptable.

Operands may employ arithmetic operations upon constants *which are known at assembly time.* For example,

> TABLE+3

---

* Some assemblers, including Avocet's, permit assorted other punctuation marks to be used in identifiers.

can be evaluated by the assembler if TABLE is a label which has been defined in the program. The assembler goes to its symbol table and reads out what value it has assigned to TABLE, adds three to it, and uses this for the value of the operand. Some other typical operations which are supported by the Avocet assembler are illustrated by the examples which follow:

```
CLOCKRATE/2            ;Integral part of quotient
PERIOD*5               ;Multiplication
CLOCKRATE MOD 2        ;Remainder of CLOCKRATE/2
HIGH TABLE             ;Upper byte of address of TABLE
LOW TABLE              ;Lower byte of address
(QUEUE/8)*8            ;Forces 0s into three LSBs
QUEUE AND 11111000B    ;Same thing
```

Comment field. This field must begin with a semicolon. It is ignored by the assembler. In fact, an entire line can be a comment line if it has a semicolon in column 1. Blank lines are also treated as comment lines. Comments are critical to our understanding, and debugging, of code. They should be used liberally!

Assembler directives are commands to the assembler. Syntactically, they *look like* instruction mnemonics, being placed in the operation field. Note the placement of the EQU in the example of Fig. 4.20. However, instead of generating machine code, assembler directives

Define names, for example,

```
MAXSTACK    EQU  00FFH
```

Label variables and allocate storage to them, for example,

```
TEMP:       DS   1      ;Assign one byte to TEMP
```

Label tables and strings and assign entries to be stored in them, for example,

```
JUMP_TABLE: DW   SUB1, SUB2, SUB3
```

Set the address at which the following code is to be assigned, for example,

```
            ORG  ROM        ;Begin at address of ROM
MAINLINE:   LDS  #MAXSTACK
            .
            .
            ORG  VECTORS    ;Begin at vector address
            DW   UART       ;Vector for UART service
            .
```

Temporarily insert additional files into the source code at assembly time, for example,

```
INCLUDE 6811REGS.HDR   ;Motorola registers
```

Control the format and contents of the assembly listing, for example,

```
NOLIST     ;Do not list the following!
INCLUDE 6811REGS.HDR
LIST       ;Turn listing back on
```

Conditionally assemble portions of source code, for example,

```
IF  EXPANDED_MODE
    .      ;Assemble if EXPANDED_MODE <> 0
ELSE
    .      ;Assemble if EXPANDED_MODE = 0
ENDIF
```

Perform miscellaneous other assembler control functions, for example,

```
END    ;Terminate assembly
```

The assembler directives incorporated in the Avocet assembler for the Motorola 68HC11 microcontroller are listed in Fig. 4.21. It is not our intention to define here all of the variations on syntax permitted for each of these. Rather, the examples are included in Fig. 4.21 to give an idea of how each one is used.

The Intel 8096 places special requirements upon an assembler so as to ensure that 2-byte word variables and constants begin at even-numbered addresses. The Intel assembler for the 8096 includes several directives for this purpose:

DSW assigns RAM to word variables. Unlike DS in Fig. 4.21, the operand is a word count rather than a byte count. Also, the assigned address will be incremented by one, if necessary, to align it on a word boundary (i.e., to force it to be an even address).

DSL assigns RAM to 4-byte-long "long word" variables. The operand is a long word count. Alignment requirements will cause the assigned address to be incremented up to three times, to align on a long word boundary.

DSB assigns RAM to bytes. This directive is equivalent to DB in Fig. 4.21.

DCW stores word constants in ROM. It handles word alignment of the assigned addresses.

DCL is the analogous directive for long word constants.

DCB stores byte constants; it has no alignment requirements.

The *form* of a program listing is given shape by the assembler directives. An example is shown in Fig. 4.22.

Type of directive	Directive	Example of use
Definition of names		
Define only once	EQU	ROM EQU 0E000H
Define repeatedly	SET	AGAIN SET 3 ;For
		. ;conditional
		AGAIN SET AGAIN-1 ;assembly
Storage allocation for variables		
Operand = byte count	DS	QUEUE: DS 6 ;Reserve 6 bytes
	RMB*	BUFFER: RMB 10 ;Reserve 10 bytes
Storage of constants		
Storage of bytes	DB	TABLE1: DB 5AH, 0A5H, 31H
	FCB*	TABLE2: FCB 5AH, 0A5H, 31H
Storage of words	DW	TABLE3: DW 1234H, OFFFEH
	FDB*	TABLE4: FDB 1234H, OFFFEH
Storage of strings	DB	MESS1: DB "HELP!"
	FCC*	MESS2: FCC "HELP!"
Set location counter	ORG	ORG ROM
Insert another file	INCLUDE	INCLUDE 6811REGS.HDR
Control listing		
Title for top of page	TITLE	TITLE "GROUNDHOG PROJECT"
Second line of title	SBTTL	SBTTL "RAM ALLOCATION"
Width of lines	WIDTH	WIDTH 80 ;80 columns
Page length	PGLEN	PGLEN 66 ;11*6 lines/in
Begin a new page	PAGE	PAGE ;New
		PAGE 4 ;New page if <4
		; lines remain
Turn off listing	NOLIST	NOLIST
Turn it on again	LIST	LIST
Conditional assembly	IF..ELSE..ENDIF	(see example in text)
Miscellaneous control	END	END ;Terminate

* Motorola-defined directive supported by Avocet assembler.

FIGURE 4.21
Assembler directives.

4.6 HANDLING COMPLEXITY IN ASSEMBLY LANGUAGE CODING

The key tools we have for breaking a microcontroller development effort down into manageable tasks are the *subroutine* and the *interrupt service routine*. We can start with the fast I/O interactions which require interrupt servicing. Our job is to write interrupt service routines which run quickly, as discussed in Chap. 3. One key to speed is to pass incoming data and outgoing data through queues. Each interrupt can be dealt with immediately. Then the mainline routine can be on the other side of each queue and carry out its instructions at a more leisurely pace. We will discuss queues in more detail in the next chapter.

    The larger problem of controlling software complexity begins with the

```
            TITLE "GROUNDHOG PROJECT"
;;;;;;;;;;;;;;;;;;;;;;;;;;;;;;;;;;;;;;;;;;;;;;;;;;;;;;;;;;;;;;;;;;;;
;  Header comments ...                                             ;
;;;;;;;;;;;;;;;;;;;;;;;;;;;;;;;;;;;;;;;;;;;;;;;;;;;;;;;;;;;;;;;;;;;;
;  Program hierarchy ...                                          ;
;;;;;;;;;;;;;;;;;;;;;;;;;;;;;;;;;;;;;;;;;;;;;;;;;;;;;;;;;;;;;;;;;;;;
;  Constants used in program
            NOLIST
            INCLUDE 6811REGS.HDR ;Use already defined names
            LIST                 ;  but don't bother to list them!
RAM         EQU   0000H          ;Beginning of data area
ROM         EQU   0E000H         ;Beginning of program area
VECTORS     EQU   0FFD6H         ;Beginning of interrupt vectors
PLS_WTH     EQU   100            ;Pulse width of 100 microseconds
              .

;;;;;;;;;;;;;;;;;;;;;;;;;;;;;;;;;;;;;;;;;;;;;;;;;;;;;;;;;;;;;;;;;;;;
;  Variables used by program
            ORG   RAM            ;Beginning of data memory
TIMER:      DS    2              ;Timer input capture reg. reading
QUEUE:      DS    10             ;Queue for buffering UART input data
              .

;;;;;;;;;;;;;;;;;;;;;;;;;;;;;;;;;;;;;;;;;;;;;;;;;;;;;;;;;;;;;;;;;;;;
;  Mainline program
            ORG   ROM            ;Beginning of program memory
MAIN:       LDS   #MAXSTACK      ;Initialize stack pointer
            BSR   INIT           ;Do all of the initialization things
LOOP:       BSR   READ_QUE       ;Service queue
              .
            BRA   LOOP           ;Go back and do it again
;;;;;;;;;;;;;;;;;;;;;;;;;;;;;;;;;;;;;;;;;;;;;;;;;;;;;;;;;;;;;;;;;;;;
;  Subroutines
INIT:         .
            RTS
READ_QUE:     .
            RTS
              .

;;;;;;;;;;;;;;;;;;;;;;;;;;;;;;;;;;;;;;;;;;;;;;;;;;;;;;;;;;;;;;;;;;;;
;  Interrupt service routines
UART:         .
            RTI
SRL_IO:       .
            RTI
              .

;;;;;;;;;;;;;;;;;;;;;;;;;;;;;;;;;;;;;;;;;;;;;;;;;;;;;;;;;;;;;;;;;;;;
;  Tables and message strings
MESS1:      DB    "This is a message"
              .

;;;;;;;;;;;;;;;;;;;;;;;;;;;;;;;;;;;;;;;;;;;;;;;;;;;;;;;;;;;;;;;;;;;;
;  Interrupt vectors
            ORG   VECTORS        ;Beginning of vector storage area
            DW    UART           ;Vector for UART          (FFD6-7)
            DW    SRL_IO         ;Vector for I/O serial port (FFD8-9)
            DW    0              ;Unused                   (FFDA-B)
              .
            DW    MAIN           ;Reset vector             (FFFE-F)
            END
```

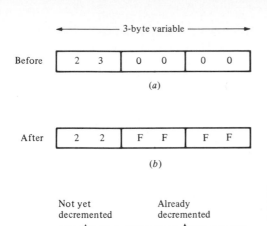

FIGURE 4.23

Erroneous interactions with a multiple-byte variable. (*a*) Content before decrementing; (b) content after decrementing; (c) value read by service routine which interrupts after lower two bytes have been decremented, and before upper byte has been.

mainline program and defining what it must do. As the large program is broken down into sequential tasks, each task can be defined as a subroutine. The header for it can be written, describing what it does, what input parameters it needs, and what output parameters it generates. Next, each subroutine can be further broken down into sequential tasks, and each task defined as a subroutine. This process is continued until all the subroutines that need to be written are both well-defined and small. Such subroutines can be written and debugged without complexity destroying us. With only a finite number of "what ifs," each one can be identified and worried about in turn. Once this work is done, we can be fairly confident that the subroutine will perform as intended.

Because we are writing code in which variables are dealt with *both* by our mainline program and our interrupt service routines, we have to be concerned about *critical regions*. This issue arose earlier in Sec. 3.7. One of the examples considered there dealt with an output port. If some pins are driven by one interrupt service routine while others are driven by another (or by the mainline program), then it is vitally important that we never read the output port, change selected bits, and write the result back out to the output port *without* disabling interrupts throughout this process.

Another example is the maintenance of a multiple-byte variable which gets *updated* by the mainline program and which gets *read* by an interrupt service routine. Consider the example shown in Fig. 4.23 which illustrates the

FIGURE 4.22
A skeleton view of a program.

case in which the mainline program decrements a 24-bit number stored in 3 bytes of memory, 1 byte at a time. Figure 4.23*a* shows the number

230000H

before it is decremented. The correct decremented result is

22FFFFH

as shown in Fig. 4.23*b*. The decrementing proceeds byte by byte. Figure 4.23*c* illustrates the moment when the two least-significant bytes have already been decremented whereas the most-significant byte has not. If an interrupt occurs at precisely this moment and reads the number, then

23FFFFH

will be read, which is very different from either 230000H or 22FFFFH—about 65,000 counts different! Again, the problem arises because an interrupt was allowed to occur at the wrong time. The solution is simple enough; disable interrupts before updating the multiple-byte variable and enable interrupts again when done. The *real* solution is to recognize the generic problem and to deal with it at that level in the following manner. Any operand

1. Which is changed by the mainline program
2. Which requires a multiple-instruction operation to make that change
3. Which can be read by an interrupt service routine

must be *protected*. The code in the mainline program which affects the operand forms a *critical region*, during which interrupts must be disabled.

The converse of this rule is also true; that is, a critical region is also formed in the mainline program where it reads (with a multiple-instruction operation) what the interrupt service routine writes.

Another potential problem arises as we break down the complexity of the mainline program into manageable tasks, i.e., a hierarchy of interrelated subroutines. We need to establish a *software convention* which will guide our use of CPU registers and what is expected of a subroutine which uses some of them. The problem which we want to resolve is illustrated in Fig. 4.24. If we have three subroutines which are called one after the other, as in Fig. 4.24*a*, then which CPU registers must have their contents preserved *between entry and exit* of each subroutine? If the DO_THIS subroutine generates a 1-byte result, then how does the subroutine return this parameter? It is common (and fine) practice to return a single-byte parameter in the accumulator (of a CPU which has one).

Now, suppose that the subroutine DO_SOMETHING_ELSE needs this parameter plus a pointer. If the pointer needs some manipulation (e.g., increment the input pointer to a queue) before it is ready to be used, then we might break this manipulation out into a short subroutine called DO_THAT. When we write DO_THAT, we will surely remember that we have to leave the content of

```
      .
      .
BSR    DO_THIS
BSR    DO_THAT
BSR    DO_SOMETHING_ELSE
      .
      .


      (a)
```

```
      .
      .
BSR    DO_THIS
BSR    DO_THAT
BSR    AND_DO_THAT
BSR    DO_SOMETHING_ELSE
      .
      .
```

FIGURE 4.24

(b)   Illustration of a problem involving the use of CPU registers.

accumulator A unchanged while we put the needed pointer in a CPU pointer register. Now the DO_SOMETHING_ELSE subroutine has the two parameters it needs to do its job.

If *remembering* is our recourse for keeping track of which CPU registers need to be maintained, then sooner or later we will find ourselves in trouble. This is illustrated in Fig. 4.24b. Six months after writing the code in Fig. 4.24a, we want to add a feature to the instrument we are designing. Our solution requires a further modification to the pointer. Rather than rewrite DO_THAT (because it is also used, as it stands, in three other places in our program), we decide to add a new subroutine, AND_DO_THAT. If in the process of modifying the CPU's pointer register, we use the accumulator as a scratch-pad register *and exit with it changed*, then DO_SOMETHING_ELSE will not work correctly. Note that AND_DO_THAT can be written, debugged, and proved to work correctly in all cases, and yet we have a problem. The problem is that AND_DO_THAT produces a *side effect*; it changes a CPU register other than the one needed to pass back a parameter. We can resolve this problem with a software convention. In fact, either of the following two will work.

Convention I. Subroutines are to leave CPU registers (other than the status register) as they found them. The only exception occurs when a CPU register is used to *return* a parameter to the calling routine.

Convention II. A calling routine must assume that any subroutine it calls may change any of the CPU registers. Consequently, it must protect the contents of any CPU registers it needs maintained (e.g., by pushing regis-

```
;;;;;;;;;;   DO_THAT SUBROUTINE   ;;;;;;;;;;;;;;;;;;;;;;;;;;;;;
;                                                             ;
;   This subroutine carries out the following operation; ...  ;
;   ......................                                    ;
;   It returns a pointer in the CPU's X index register.       ;
;   All other CPU registers are left unchanged.               ;
;                                                             ;
;;;;;;;;;;;;;;;;;;;;;;;;;;;;;;;;;;;;;;;;;;;;;;;;;;;;;;;;;;;;;;;

DO_THAT: PSHB                      ;Set aside accumulator B
         .                         ;
         .                         ;
         .                         ;B and X get used here.
         .                         ;
         .                         ;A and Y are not used (and so do
         .                         ;  not need to be set aside)
         .                         ;
         .                         ;
         PULB                      ;Restore accumulator B
         RTS                       ;  and return
```

FIGURE 4.25
Example of convention I as applied to the Motorola 68HC11.

ters onto the stack) *before* calling *any* subroutine. Then it needs to restore them after the return from the subroutine has taken place.

Example 4.30. An example of the effect of convention I upon the writing of code is shown in Fig. 4.25. This illustrates the impact of the convention upon the writing of the DO_THAT subroutine of Fig. 4.24. The CPU register structure of the Motorola 68HC11 is employed in the example. With four CPU registers (A, B, X, and Y) to be concerned about, the DO_THAT subroutine of Fig. 4.24 must return a parameter in only one, the X index register. Therefore it must preserve A, B, and Y. If, of these three, it *uses* only B, then it must set aside, and later restore, B.

Example 4.31. The effect of convention II upon this same problem is illustrated in Fig. 4.26. Here the responsibility for preserving the parameter in accumulator A is


```
        .
        .
BSR     DO_THIS
PSHA
BSR     DO_THAT
PULA
PSHA
BSR     AND_DO_THAT
PULA
BSR     DO_SOMETHING_ELSE
        .
        .
```

FIGURE 4.26
Example of convention II as applied to the Motorola 68HC11. Accumulator A is protected when calling subroutines which might corrupt its content.

carried by the calling routine. As it calls each subroutine which might affect accumulator A, it might push A onto the stack and restore it upon the return, as in Fig. 4.26.

Note that either convention solves this problem. To write code without some such convention invites recurring difficulties and potentially lurking bugs.

Each of these two approaches to maintaining register integrity during subroutine calls has its advantages. Convention I keeps the register integrity question localized within each subroutine by making the subroutine preserve the integrity of any registers it uses. It is probably easier to write bug-free code using convention I since we need look no farther than within each (hopefully short) subroutine to see which registers must be stacked at the beginning of the subroutine and unstacked again at the end. Convention I makes calls of subroutines cleaner since the calling program knows that register contents will be maintained (except for registers used to return parameters).

With convention II, the only registers which are maintained are those registers which need to be maintained. This probably leads to less code being generated and faster program execution, as examined in Problem 4.40 at the end of the chapter. However, it is more difficult to look at the code for a sequence of subroutine calls and know which ones are loading up registers for use by other subroutines. It is these registers which must be maintained during the execution of any intervening subroutines. Unless *all* registers are preserved before a subroutine is called and restored afterward, this gives rise to a potential source of bugs.

Use of the Intel 8096 gives rise to the need for another software convention. The 8096 includes no CPU registers which are dedicated by the hardware as such. We have seen that the 8096 can use arbitrary addressing for the *destination* operand of an instruction only for a very limited set of instructions (namely, stores and pops). All other instructions require that the destination operand employ page 0 direct addressing. This means that when the destination operand of an instruction is accessed with a pointer, then we need a pointer register to access it. If we used different 2-byte words in RAM every time we needed a *temporary* pointer, we would find memory to be a scarcer resource than it should really be. Accordingly, Intel recommends that the 8 bytes of RAM between addresses 001CH and 0023H be reserved as temporary storage, as shown in Fig. 4.27. These 8 bytes are used by Intel's high-level language, PL/M-96. Accordingly, as long as these bytes are not dedicated to some other purpose, we can take advantage of Intel's floating point arithmetic library that PL/M uses to operate on real (i.e, floating point) numbers. Even if we have no need for floating point arithmetic, we still ought to dedicate a few RAM addresses for temporary pointers. Intel's suggested addresses are as good a place as any.

Parameter passing to and from subroutines is another issue which warrants consistent treatment as software is written. That is, it warrants our settling upon another software convention. Given a CPU with registers (e.g., the

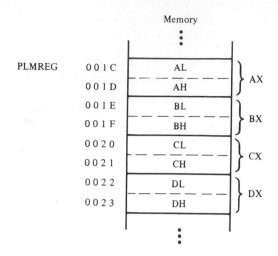

AL, AH, BL, · · · , DH   are Intel-defined labels
for 1-byte variables

AX, · · · , DX   are Intel-defined labels
for 2-byte variables

FIGURE 4.27
Intel's recommended "CPU registers."

Motorola 68HC11) or with registers designated by convention (e.g., the Intel 8096), many designers use these registers for both passing and returning parameters. When there are too many parameters to fit in the CPU registers, then some extension to this approach is needed. An interesting way to view this issue is to examine what a well-constructed compiler does for the microcontroller in question. Intel's PL/M-96 compiler *pushes all parameters* onto the stack before calling a procedure. Some procedures act upon *global** variables, in which case there is no issue of where a parameter should be returned. On the other hand, a *function* call is a specialized form of a procedure which returns a single parameter (a byte, a word, or a double word). PL/M-96's convention is to return this parameter in the address labeled PLMREG in Fig. 4.27. PL/M-96 procedures employ convention II discussed previously; that is, they assume that the addresses in Fig. 4.27 are free to be used and changed by procedures.

Example 4.32. Passing parameters to a subroutine on the Intel 8096 stack permits *all* of the addressing modes to be used in getting them there. For example,

```
PUSH  [SOME_POINTER]       ;Push WORD1
PUSH  5[ANOTHER_POINTER]   ;Push BYTE2
SCALL DO_SOMETHING_ELSE
```

---

* A variable which is global to a procedure continues to have meaning *after* the return from the procedure. A variable which is global to an entire program has meaning all the time.

might be used to call the subroutine discussed earlier, in conjunction with Fig. 4.26.

Example 4.33. Within the DO_SOMETHING_ELSE subroutine, assume that all we want to do is store BYTE2 at the address represented by WORD1. (Not too exciting, but this lets us look at the addressing involved.)

Upon entry, the stack looks like that shown in Fig. 4.28. The pointer represented by WORD1 was pushed onto the stack first (most-significant byte first, followed by the least-significant byte). Then the single-byte variable, BYTE2, was pushed onto the stack. Since pushes handle 2-byte operands only, the stack receives a garbage byte as well as the byte we care about. Finally, the subroutine address is pushed, putting it on top of the stack.

The DO_SOMETHING_ELSE subroutine below uses the CPU registers defined in Fig. 4.27 to serve as temporary variables.

```
DO_SOMETHING_ELSE:
        LDB     AL,2[SP]        ;Get BYTE2
        LD      BX,4[SP]        ;Get WORD1
        STB     AL,[BX]         ;Store BYTE2
        LD      DX,[SP]         ;Get return address
        ADD     SP,#6           ;Clean up stack
        BR      [DX]            ;Jump to return address
```

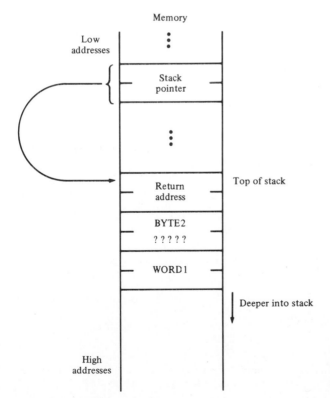

FIGURE 4.28
Intel 8096 stack, as seen by the DO_SOMETHING_ELSE subroutine.

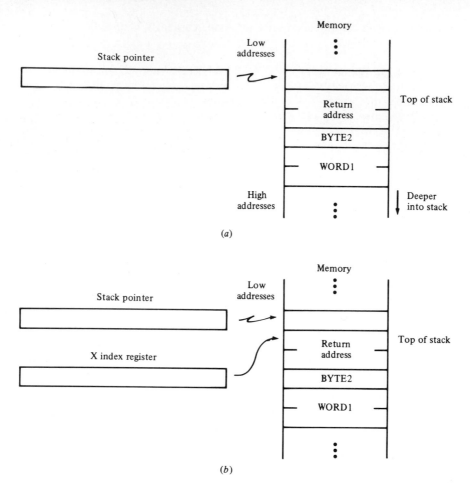

FIGURE 4.29
Motorola 68HC11 stack, as seen by the DO_SOMETHING_ELSE subroutine. (*a*) Upon entry; (*b*) after TSX instruction.

Example 4.34. In this example we will look at the same subroutine written for the Motorola 68HC11. Assume that upon entry to the DO_SOMETHING_ELSE subroutine, the stack looks like the one shown in Fig. 4.29*a*. Note that the 68HC11's stack pointer points to the next available location in memory, rather than to the top of the stack. However, a TSX instruction stores the content of the stack pointer plus one into X, making X point to the top of the stack, as in Fig. 4.29*b*. The subroutine follows:

```
DO_SOMETHING_ELSE:
            TSX                 ;X := SP+1; X points to TOS
            LDAA 2,X            ;Get BYTE2
            LDY  3,X            ;Get WORD1
            STAA 0,Y            ;Store BYTE2
```

```
LDY   0,X         ;Get return address
LDAB  #5          ;Clean up
ABX               ; stack
TXS               ;SP := X-1; restore stack ptr.
JMP   0,Y         ;Return
```

Macro capability in an assembler can help reduce our chances for erring when we have "strange" instruction sequences like those used to return from the subroutines in the last two examples. A macro is really like a nickname for an instruction sequence. We need to *define* it first, so that the assembler knows what substitution to make. Then we *invoke* it as if it were an instruction. Macro capability in an assembler usually includes the ability to employ parameters. The Avocet assembler we have looked at earlier does not include macro capability.* As an example of an assembler with good, versatile macro capability, we will look at Intel's assembler for the 8096. Figure 4.30*a* illustrates the definition of a macro called RETURN. Figure 4.30*b* shows the DO_SOMETHING_ELSE subroutine of Example 4.33, rewritten to terminate with this macro.

Some assemblers support *nested control structures.* For example, while Avocet's assemblers for 8-bit microcomputers and microcontrollers do not, their new assembler for the Motorola 68000 16-bit microcomputer does. Furthermore, many standard assemblers can be made to support nested control structures.† We will look at the features of a structured assembler using Avocet's 68000 assembler as an example.

---

* Avocet's latest generation of assemblers, for both the 68HC11 and the 8096 (as well as other microprocessors) *are* now macro-assemblers.

† The macro capability of Intel's assemblers is sufficiently powerful to support the nesting required. Hewlett-Packard's user-definable assembler for their 64000 development system can be used to modify any of their assemblers in order to support this capability.

```
RETURN    MACRO NUMBER_OF_PARAMETERS
          LD    DX,[SP]
          ADD   SP,#(2+(2*NUMBER_OF_PARAMETERS))
          BR    [DX]
          ENDM
                    (a)



DO_SOMETHING_ELSE:
          LDB   AL,2[SP]
          LD    BX,4[SP]
          STB   AL,[BX]
          RETURN 2              ;Two parameters
                    (b)
```

FIGURE 4.30
Macro capability of the Intel 8096 assembler. (*a*) Macro definition; (*b*) macro invocation.

Users of structured, high-level languages generally are quick to denigrate assembly language coding. High on their list of complaints is the criticism that algorithms must be represented with "spaghetti code"; that is, with code which is strewn with conditional and unconditional jump instructions. Spaghetti code is hard to create correctly, hard to follow when a listing is examined and therefore hard to debug. These criticisms are well placed. What is not well placed is the implicit assumption that an assembly language cannot be structured.

Many users of unstructured assemblers are quick to denigrate any features of an assembler which gets them away from the machine. They want each line of source code to produce a single machine code instruction. They get nervous when they see a structured assembler which can assemble a line like

```
IF   (A EQ #5) OR (COUNT EQ #100) THEN
     .
     .
     .
     ENDI
```

because they know that this will require the assembler to insert a variety of extra instructions to push registers on the stack at the beginning and pop them off at the end in order to make the comparisons required. Also they know that such constructs change the CPU's status flags in ways which are not always predictable. Again, their criticism is not well founded.

The Avocet assembler for the 68000 microcomputer supports tests which are nothing more than the tests built into the 68000's conditional jump instructions. These are listed in Fig. 4.31a. For example, the test

```
EQ
```

has the same meaning as that associated with the BEQ instruction, defined in Fig. 4.15 (since the 68000 supports exactly the same tests as the 68HC11). It tests whether the Z flag is set. By restricting assembler control structures to include only such tests, we indeed get the one-to-one translation from source code line to machine code instruction. We will see this shortly in an example.

Figure 4.31b shows Avocet's control structures, which can be *nested* up to eight levels deep. That is, we can have a REPEAT . . . UNTIL structure nested inside an IF . . . ENDI structure, which is in turn nested inside another IF . . . ENDI structure, and so on for up to five more levels of control structures. Given that our job is to break the complexity of our design task down into reasonably short, testable subroutines, eight levels of nesting represents overkill. On the other hand, control structures with *no* nesting capability do not do very much for us. The whole idea is to be able to express the control flow of complex algorithms in an easy-to-create easy-to-understand easy-to-debug form. The three control structures of Fig. 4.31b give us the tools we need to do this. The "early exception" BREAK IF construct of Fig. 4.31c permits the inclusion of any number of tests to get out of one of the control structures in Fig. 4.31b.

EQ NE GT GE LT LE HI HS LO LS VS VC PL MI CS CC

```
           (Refer to Fig. 4-15 for meaning)


                         (a)


IF <test> THEN          WHILE <test>           REPEAT
 .                        .                      .
 .                        .                      .
ELSE                      .                      .
 .                        .                      .
 .                        .                      .
ENDI                    ENDW                   UNTIL <test>


                         (b)



                  BREAK IF <test>


                         (c)
```

FIGURE 4.31
Some constructs of Avocet's 68000 structured assembler. (*a*) Conditional tests; (*b*) control structures.

Consider the example of Fig. 4.32. Earlier, in conjunction with Fig. 4.6, we discussed putting data into a queue. This subroutine gets data out of a queue. Note that it includes two nested IF . . . ENDI constructs. Upon entering the subroutine, two pointers are compared. If they are equal, then we are done. We set the Z flag for the benefit of the calling routine and return. On the other hand, if there is data in the queue, then the pointers will not be equal. We get the byte pointed to by QUE_OUT, the queue removal pointer. Then we increment the pointer, taking care to roll it over if it goes past the end of the RAM space allocated to the queue. Finally we clear the Z flag (to tell the calling routine that there was a byte in the queue and that it is now in accumulator A) and return.

The Avocet assembler automatically indents the control structures to accentuate the control flow. The example of Fig. 4.32 illustrates this indenting. Note that the

```
        LDX #QUEUE+4
```

instruction is indented a second level. An assembler to which you have added control structures can employ a separate text-processing program to do this indenting. It really helps!

```
;;;;;;;;;;   GETBYTE SUBROUTINE   ;;;;;;;;;;;;;;;;;;;;;;;;;;;;;;;;;;;
;                                                                   ;
;  This subroutine checks a queue for data.  If there is any        ;
;  data in the queue, the data is returned in accumulator A and     ;
;  the Z flag is cleared.  Otherwise, the Z flag is set.            ;
;                                                                   ;
;;;;;;;;;;;;;;;;;;;;;;;;;;;;;;;;;;;;;;;;;;;;;;;;;;;;;;;;;;;;;;;;;;;;;;

GETBYTE: PSHX                           ;Save X
         LDX QUE_OUT                    ;Get queue removal pointer
         CPX QUE_IN                     ;Check if queue is empty
                                        ;(set Z flag if it is empty)
         IF  NE  THEN
           LDAA 0,X                     ;Queue not empty; get data
           INX                          ;Increment removal pointer
           CPX #QUEUE+4+QLENGTH ;Check if past end of queue
           IF  EQ  THEN
             LDX #QUEUE+4               ;Past end; wrap around
           ENDI
           STX QUE_OUT                  ;Save removal pointer
                                        ;(and clear Z flag)
         ENDI
         PULX                           ;Restore X
         RTS                            ;  and return
```

FIGURE 4.32
Motorola 68HC11 subroutine for getting a byte from a queue. Written for a structured assembler.

   To illustrate the one-to-one translation between lines of structured source code and machine code instructions, this example is recast into unstructured form in Fig. 4.33. To help the visualization of this recasting, each control structure element of Fig. 4.32 is rewritten here in the comment field in capital letters. Also the indenting of Fig. 4.32 is echoed in the indenting of the comments. Note that each control structure element which includes a test translates to a conditional jump instruction having just the opposite test. Thus, referring to Fig. 4.15, we can guess that the test LE (signed number comparison for less than or equal) would translate into the instruction BGT (signed number comparison for greater). The control structure element ELSE always translates to an unconditional jump past the end of the structure. The ENDI element produces no machine code. Its purpose is to serve as a marker for the end of the control structure so that the assembler can compute the address needed by one of the other jump instructions.

4.7 SUMMARY

In this chapter we have considered two quite different microcontroller CPU register structures. The Motorola 68HC11's CPU represents an enhancement of the CPU in Motorola's earlier 6801, giving users of that earlier microcontrol-

```
;;;;;;;;;;  GETBYTE SUBROUTINE  ;;;;;;;;;;;;;;;;;;;;;;;;;;;;;;;;;;;
;                                                                 ;
;   This subroutine checks a queue for data.  If there is any     ;
;   data in the queue, the data is returned in accumulator A and  ;
;   the Z flag is cleared.  Otherwise, the Z flag is set.         ;
;                                                                 ;
;;;;;;;;;;;;;;;;;;;;;;;;;;;;;;;;;;;;;;;;;;;;;;;;;;;;;;;;;;;;;;;;;;;


GETBYTE:  PSHX                       ;Save X
          LDX QUE_OUT                ;Get queue removal pointer
          CPX QUE_IN                 ;Check if queue is empty
                                     ;(set Z flag if it is empty)
          BEQ GET_2                  ;IF  NE  THEN
          LDAA 0,X                    ;Queue not empty; get data
          INX                         ;Increment removal pointer
          CPX #QUEUE+4+QLENGTH        ;Check if past end of queue
          BNE GET_1                  ;IF  EQ  THEN
          LDX #QUEUE+4                 ;Past end; wrap around
                                     ;ENDI
GET_1:    STX QUE_OUT                ;Save removal pointer
                                     ;(and clear Z flag)
                                     ;ENDI
GET_2:    PULX                       ;Restore X
          RTS                        ;  and return
```

FIGURE 4.33
Motorola 68HC11 subroutine for getting a byte from a queue. Written without the help of a structured assembler.

ler easy access to higher capability. Code written for earlier applications using the 6801 can be used with the 68HC11. For many microcontroller users, this is no small offering.

In contrast, the Intel 8096's CPU represents a significant departure from the CPU operation in Intel's earlier microcontrollers. The designers of the 8096 have gathered together all of the experience gained from earlier work and used that experience to develop a CPU which carries out typical microcontroller tasks with speed and finesse. They have taken advantage of the limited resources of the 8096, squeezing both RAM and registers into the 256 addresses of page 0. Users employ multiple-operand instructions to deal directly with any entity, without going through CPU registers. In fact, we have seen that the designers of the 8096 have eliminated the accumulator, scratch-pad registers, and pointer registers from the CPU.

The addressing modes available for accessing operands provide one dimension for sophistication of CPU operation. At first glance, the addressing modes seem quite similar. Both microcontrollers treat page 0 addresses in a special way. Both can directly access a variable anywhere in the memory space. Both offer indexed addressing. However, the Motorola 68HC11 makes heavy use of two 8-bit accumulators and two 16-bit index registers in the CPU. In contrast, the Intel 8096 employs *any* of its page 0 RAM variables for the same purposes. Likewise, the 8096 permits indirect addressing with any of its page 0 RAM variables. The 68HC11 must combine instructions to achieve the *effect* of indirect addressing.

In the domain of instruction sets, these two microcontrollers again look quite similar at first glance. Almost every type of operation which can be carried out by one can be carried out by the other. The difference arises in *how* the operations are carried out. The 8096 provides separate versions of each operation so that one instruction will operate upon *byte* operands and another instruction will operate upon *word* operands. Separate arithmetic instructions deal directly with signed or unsigned numbers. In contrast, the 68HC11 has *building-block* instructions which typically deal with byte operands and which must be used successively to deal with larger operands.

In the microcontroller environment, RAM and ROM resources are sufficiently limited to warrant the use of assembly language, rather than a higher-level language like Pascal or C. We looked at the parts making up each assembly language instruction and the manner in which these are combined into an entire program. However, there is no substitute for learning assembly language coding by having an example of the software for a complete instrument or device which makes use of good coding practices.

The development of reliable software in a microcontroller environment is aided by keeping the following points in mind. One is the search for, and treatment of, *critical regions*. We must be ever alert for operands which are dealt with *both* by an interrupt service routine and by the instructions which it interrupts. The interrupted code must typically be protected by disabling interrupts momentarily while the operand is dealt with.

The development of reliable software can also benefit from the discipline of abiding by a convention for protecting the contents of CPU registers when a subroutine is called. One convention would have the subroutine save the content of any CPU register before the register itself is used. The other convention would have the calling program save (and later restore) any CPU register content which must be maintained even though the calling program temporarily gives up use of the CPU to a subroutine. There is less need to recommend one of these conventions over the other than there is a need to adopt one of them and stick with it.

We also looked at the support which some assemblers provide with *macro* capability and with *nested control structures*. For the casual user, one assembler may appear to be about as effective as another. For the designer who uses an assembler as a major tool in his or her professional life, these two features can make a vast difference. For example, by using macros we can solve the little intricacies in our code just once. In the next chapter, we will discuss an excellent application of macros in conjunction with queues. Nested control structures replace the spaghetti code which arises from the use of conditional branch instructions with code that has clear-cut control flow. It is common for even experienced designers to get tripped up by the sense of a conditional branch instruction (e.g., using a branch if zero instruction where a branch if *not* zero instruction is actually needed). This problem goes away for most people when replaced by

```
IF    ZS    THEN
          <execute this code if Z is set>
ENDI
```

It is also easier to read.

PROBLEMS

4.1. *CPU register structure.* Consider the CPU in a microcontroller other than the Intel 8096 or the Motorola 68HC11.

(*a*) What functions (e.g., carry function) does it support with flag bits in its status register?

(*b*) Does it have an accumulator? If so, then does the accumulator support operations on 8-bit numbers? 16-bit numbers? Anything else?

(*c*) How does it support the pointer-to-memory function?

(*d*) How big is the program counter? If it is only as big as is needed to access the on-chip program memory, then does this imply that the chip offers no way to expand program memory?

(*e*) How big is the stack pointer? How big can the stack be? Is the stack pointer initialized automatically at reset time? In which direction does the stack grow (i.e., toward lower or toward higher addresses)?

4.2. *Stack use.* We can think of the stack as if each return address were written upon a 3×5 in index card, placed on the right-hand corner of our desk. When we initialize the stack pointer, in effect we clear the corner of the desk of all cards. Each time we call a subroutine, we write the return address on a card and then put the card on the corner of the desk, on top of any other cards already there. Each time we return from a subroutine, we take the top card off the stack of cards and use it to determine which address to return to.

(*a*) Using this mechanism and expressing return addresses as labels, follow through the actions of the stack as the program of Fig. 4.3 is executed.

(*b*) Modify subroutine SUB1 so that it calls another subroutine, SUB3, just after calling SUB2. Now repeat part *a*.

(*c*) Remove SUB3. Now modify SUB2 so that sometime during its execution it calls a subroutine, SUB4. Repeat part *a*.

(*d*) Interrupts are like subroutines which can occur at any time. Repeat part *c*, considering the effect of an interrupt occurring at various times. Assume that the CPU uses the stack to store 9 bytes of CPU registers onto the stack (as the Motorola 68HC11 does) at the beginning of interrupt servicing, and restores these 9 bytes to the CPU registers upon returning from the interrupt service routine. You can represent the storage of the 9 bytes onto the stack with $4\frac{1}{2}$ cards. What is the maximum depth needed for the stack to handle this example? Assume that the stack is used *only* for the return addresses of subroutines and for the 9 bytes of CPU state information during interrupts.

4.3. *2s-complement arithmetic.* Write out a column of signed decimal numbers for the 11 integers between $+5$ and -5, with $+5$ at the top and -5 at the bottom. To the right of each of these numbers, reexpress each of these numbers as an 8-bit 2s-complement number. For example, +1 should be re-expressed as 00000001, while −1 should be reexpressed as 11111111, and −2 as 11111110.

(a) Add, as if they were unsigned binary numbers, the numbers to the right of +2 and +3. Is the result correct for addition of 2s-complement addition also?

(b) Repeat part a for +3 and −2.

(c) Repeat part a for +2 and −3.

(d) Repeat part a for −2 and −3.

(e) Now, repeat part a, but this time *subtract* +2 from +3. (How are you at the subtraction of unsigned binary numbers? If you are unsure, then check your subtraction by undoing the result you get with addition.)

(f) Repeat part e, subtracting +3 from +2.

(g) Repeat part e, subtracting −2 from +3.

(h) Repeat part e, subtracting −3 from +2.

(i) Repeat part e, subtracting +2 from −3.

(j) Repeat part e, subtracting +3 from −2.

(k) Repeat part e, subtracting −2 from −3.

(l) Repeat part e, subtracting −3 from −2.

(m) Can an adder/subtracter circuit which works for unsigned binary numbers be used for signed numbers expressed in 2s-complement code? Assuming that the result can be expressed as an 8-bit 2s-complement number, can we ignore the carry/borrow bit?

4.4. *BCD arithmetic.* The Motorola 68HC11 includes a DAA instruction for coupling with a binary addition instruction to implement BCD addition. The Intel 8096 does not have such an instruction. If we want to carry out BCD addition with the 8096, we need to implement the algorithm in software (without the help of a half-carry flag). In this problem, we will develop the algorithm.

The two operands, called BCD_A and BCD_B, can be assumed to be two-digit BCD numbers before the BCD addition (e.g., 92 is expressed as 10010010). We can mask off half of each number into temporary variables and add these. Call these half numbers BCD_A_U (upper digit) and BCD_A_L (lower digit), etc. The half carry will end up in bit 4 of the result of adding BCD_A_L and BCD_B_L (assuming that bit 0 is the least-significant bit). Express the required algorithm in words.

4.5. *Sticky bit.* The Intel 8096 sticky bit, discussed at the end of Sec. 4.2, says something about the range of the result of a right-shift operation of several places. If we take both the carry bit and the sticky bit into consideration, then express the range on the *exact* result of a multiple-position right shift operation that leaves the number 35 in a register and:

(a) C = 0 and ST = 0.

(b) C = 0 and ST = 1.

(c) C = 1 and ST = 0.

(d) C = 1 and ST = 1.

4.6. *Immediate addressing.* For the Motorola 68HC11, show the load instruction, or the load and store instruction pair, to:

(a) Load accumulator A with the number whose hexadecimal representation is 0CH.

(b) Load accumulator B with the number whose hexadecimal representation is 0FDH.

(c) Load accumulator D with the number whose hexadecimal representation is 1234H.

(d) Load the X index register with the number whose hexadecimal representation is 5678H.

(e) Load the Y index register with the number whose hexadecimal representation is 9ABCH.

(f) Load the 8-bit data direction register for port C (named DDRC) with the number whose binary representation is 00000001B.

(g) Load the 16-bit timer output compare register 1 (named TOC1) with the number whose hexadecimal representation is 4321H.

4.7. *Immediate addressing.* For the Intel 8096, show the instruction to:

(a) Load the 8-bit output port 1 (named P1) with the number whose binary representation is 11111000B.

(b) Load the 16-bit buffer (named HSO_TIME) for the timer output compare registers with the number whose hexadecimal representation is 1357H.

4.8. *Direct addressing.* For the Motorola 68HC11, show the instruction to:

(a) Load accumulator A with the content of an 8-bit variable in RAM labeled TEMP_BYTE.

(b) Load the X index register with the content of a 16-bit variable in RAM labeled QUE_IN.

(c) Store the content of the X index register in the 16-bit variable in RAM labeled QUE_IN.

4.9. *Direct and immediate addressing.* For the Intel 8096, show the instruction to:

(a) Read the UART receive data register (named SBUF), force the most-significant bit of the byte read to 0 (by ANDing the byte read with 01111111B) and write the result back out to the UART transmit data register (also named SBUF).

(b) Read the 16-bit timer 1 (named TIMER1), add to this the 16-bit number whose hexadecimal representation is 0034H, and store the result back out to the 16-bit buffer for the timer output compare registers (named HSO_TIME).

4.10. *Extended and immediate addressing.* For the Motorola 68HC11, show the instruction sequence to read the 16-bit timer counter (named TCNT), add to this the 16-bit number whose hexadecimal representation is 002FH, and store the result back out to the 16-bit output compare register 2 (named TOC2).

4.11. *Indirect with autoincrement addressing.* For the Intel 8096, show the instruction to read a 1-byte character out of a string of characters stored in ROM and pointed to by a 16-bit pointer stored in a RAM variable labeled CHAR_PTR, write the character out to the UART's transmit data register (named SBUF), and then increment the pointer so that it points to the next character in the string.

4.12. *Indexed addressing.* For the Motorola 68HC11, show the instruction sequence for the procedure of the preceding problem. The 68HC11 UART's transmit data register is named SCDR.

4.13. *Indirect with autoincrement addressing.* For the Intel 8096, show the instruction to read port 0 (named P0), store its value into a queue at an address pointed to by the content of the 16-bit variable labeled QUE_IN, and then increment the content of QUE_IN.

4.14. *Indexed addressing.* For the Motorola 68HC11, show the instruction sequence for the procedure of the preceding problem, reading from port E (named PORTE).

4.15. *Indexed addressing.* For the Intel 8096, assume that we have a table, labeled

TABLE, of 1-byte numbers stored in ROM. We want to read a single 1-byte entry from the table and write it out to port 1 (P1).

(a) Write out the fifth table entry (where the 0th entry has the address TABLE) to port 1.

(b) Write out the AXth table entry to port 1. AX is the label for a 2-byte variable in RAM.

4.16. *Indexed addressing.* For the Motorola 68HC11, repeat the last problem, writing to output port C (named PORTC), modifying the operations as follows:

(a) Write the fifth table entry to port C.

(b) Write out the Bth table entry to port C, where B represents the unsigned 1-byte variable in accumulator B. (*Note:* You will want to use the 68HC11's ABX instruction which adds the content of B to the content of X and puts the result back into X, treating all numbers as unsigned numbers.)

4.17. *Indexed addressing.* For the Motorola 68HC11, repeat Example 4.10 in the text. The 68HC11 UART's receive data register is named SCDR.

4.18. *Stack addressing.* For the Intel 8096, we might want to know what the calling routine was when the CPU is executing instructions in a subroutine that can be called from any number of places in our program. One way to do this is to look at the return address on the stack. Show the instruction(s) which will compare the subroutine return address on the top of the stack with the address labeled SUB1_1 and shown in Fig. 4.3. If these are equal, then the Z flag should be set. [*Note:* You will want to use a SUB (subtract) or CMP (compare) instruction which operates on 2-byte variables. If you need a temporary 16-bit register, assume that you have one named AX.]

4.19. *Stack addressing.* Repeat the last problem for the Motorola 68HC11. (*Note:* The TSX instruction loads the address of the top of the stack into X, so that X then points to the return address.)

4.20. *Orthogonal instruction set.* At the beginning of Sec. 4.4, we referred to the term "orthogonal" instruction set and said that both the Intel 8096 and the Motorola 68HC11 had orthogonal instruction sets. Actually, this is not quite true. For example, the 68HC11 has an ABA instruction for adding B into A, but no AAB instruction. Looking through Sec. 4.4,

(a) Are there any other 68HC11 instructions which are not orthogonal?

(b) Does the Intel 8096 have any instructions which are not orthogonal (given that only one operand in any instruction can be arbitrarily addressed)?

4.21. *Shift and rotate instructions.* The Motorola 68HC11 has no instruction analogous to the Intel 8096 instruction

```
SHRA   WORD1,#1
```

which carries out a single 16-bit arithmetic shift right operation on a 2-byte variable labeled WORD1. Show the *sequence* of 68HC11 instructions to do this. Note that the Motorola 68HC11 stores 2-byte variables in RAM *most*-significant byte first (just the reverse of what the Intel 8096 does). That is, WORD1 is the address of the most-significant byte of WORD1 while WORD1+1 is the address of the least-significant byte of WORD1.

4.22. *Shift and rotate instructions.* Repeat Problem 4.21 for the Motorola 68HC11 equivalent of the Intel 8096 instruction

```
SHL   WORD1,#2
```

4.23. *Shift and rotate instructions*. Repeat Problem 4.21 for the Motorola 68HC11 equivalent of the Intel 8096 instruction

        SHR   AX,BL

which uses the content of the 1-byte variable, BL, to determine how many right shifts to carry out on the 2-byte operand AX. To do the equivalent operation, assume that the number to be shifted is in accumulator D while the number of shifts is in the index register X (and is 16 or less).

4.24. *Shift and rotate instructions*. The Motorola 68HC11 uses its ROLA instruction to build up shift left operations on arbitrarily long numbers. This lets the carry bit serve as the bridge between bytes of the long number being shifted a byte at a time. Without actually writing the code, describe how you would carry out a shift left operation of one place on a 6-byte number using the Intel 8096. Note that the Intel 8096 likes to have multiple-byte numbers stored in memory least-significant byte first.

4.25. *Bit manipulation instructions*.   Both the Intel 8096 and the Motorola 68HC11 include excellent instructions for manipulating individual bits in a register (e.g., an output port). Show the instruction sequence needed to generate a short, positive-going output pulse (0-to-1-to-0) on bit 2 of
(*a*) Port B (named PORTB) of Motorola's 68HC11.
(*b*) Port 4 (named P4) of Intel's 8096.
In each case, no other output lines are to be affected. You may assume that the output bit has already been initialized to 0.

4.26. *Bit manipulation instructions*. Repeat the preceding problem to generate a short, negative-going output pulse on bit 3 of the same port. You may assume that the output bit has already been initialized to 1.

4.27. *Arithmetic instructions*. The support which the Intel 8096 gives to *fast* arithmetic on signed and unsigned 1- and 2-byte numbers is outstanding. For example, the execution of the instruction shown in Fig. 4.13 (which produces a signed, 32-bit product) takes only 29 cycles, or just over 7 μs! The unsigned MULU instruction takes just over 6 μs. The designers of the chip must have figured that such performance would open enough new markets for the chip to make the extra chip area required to implement these instructions worthwhile.

Consider the multiplication of two unsigned 16-bit numbers by the Motorola 68HC11. The 68HC11's MUL instruction multiplies the unsigned 8-bit operands in accumulators A and B and puts the result back into D. If we were to write a 68HC11 subroutine to take a 2-byte unsigned number pointed to by X and multiply that by the 2-byte unsigned number pointed to by X+2, and if we were to put the result back into the 4 bytes beginning at X+4, then how long would this subroutine take to run (more or less)? Note that if we express two 16-bit integers, M and N, in terms of their bytes MU, ML, NU, and NL, where

$$M = 2^8 * MU + ML \quad \text{and} \quad N = 2^8 * NU + NL$$

then the product M * N can be expressed by

$$M * N = 2^{16} * (MU * NU) + 2^8 * (MU * NL + NU * ML) + (ML * NL)$$

Assume the following execution times and list the number of adds, multiplies, loads, and stores involved.

```
MUL              5 μs
ADCA  2,X        2 μs
ADDD  2,X        3 μs
LDAA  2,X        2 μs
STAA  4,X        2 μs
STD   4,X        2.5 μs
RTS              2.5 μs
```

4.28. *Polling sequence.* A commonly recurring task within an interrupt service routine is a polling sequence. For example, when the Motorola 68HC11's UART generates an interrupt, any of four reasons for the interrupt could have occurred: (1) A byte could have been received; (2) the transmit data register could be ready to accept another byte to be sent; (3) an idle line on the receiver is detected; and (4) a "transmit complete" condition occurred. Without going into a detailed explanation of each of these, it is sufficient to know that each of these events sets a bit in the SCSR register as well as causing an interrupt with the UART vector. The bits of SCSR which are set by each of the above conditions are bits 5, 7, 4, and 6 respectively. Write a polling sequence which will test for each of these events in the (1), (2), (3), (4) order listed above and jump to an instruction labeled ACT1, . . . ,ACT4 to take appropriate action for *the first* condition with a bit set. (Each action routine can terminate by jumping back to the beginning of the polling routine. Only when the complete polling routine is executed with no bit set will we return from the interrupt.) *Hint:* Use the BRSET instruction of Fig. 4.15 and described by Example 4.24.

4.29. *Polling sequence.* When the Intel 8096's programmable timer generates an interrupt saying that an input capture event has occurred on any one of four lines, then we can go through the same kind of polling sequence as in the last problem. The status register to be read is called HSI_STATUS. We want to test bits 0, 2, 4, and 6 in that order. If a bit is set, then jump to INPUT0, INPUT1, INPUT2, or INPUT3, respectively. Show the polling sequence to do this.

4.30. *Relative addressing.* Both the Intel 8096 and the Motorola 68HC11 employ *relative* addressing for the jump address used in conditional jump instructions. This is done so that a single byte can be used to represent the address, permitting conditional jump instructions to be 2-byte instructions.

    (*a*) The relative address is an 8-bit 2s-complement number which is "sign extended" into a 16-bit 2s-complement number and then added to the program counter to form the new program counter content, when the jump is taken. If the first byte of the conditional jump instruction is at an address we will designate as ADD_JMP_INST, then what is the content of the program counter when this instruction is *executed*?

    (*b*) What is the greatest address we can reach with the conditional jump instruction (expressed as a function of ADD_JMP_INST)?

    (*c*) What is the lowest address we can reach?

4.31. *Relative addressing.* Our main building block for breaking down the complexity of a software task is the subroutine. The conditional jump instructions discussed in the last problem should *always* be able to reach to anywhere within the subroutine or else the subroutine is probably larger than it should be for optimal debugging. Consequently, the constraint of only being able to jump "near by" is more apparent than real. Nevertheless, if we have an Intel 8096 instruction

```
JC    THERE
```

which produces an EXPRESSION NOT WITHIN RANGE error message when it is assembled, then we can resolve this with an unconditional jump* to THERE, preceded by a conditional jump to *skip* the unconditional jump instruction:

```
        Jxxx    SKIP
        SJMP    THERE
SKIP:
```

(*a*) Referring to Fig. 4.15, what is the mnemonic for the required conditional test (represented by Jxxx above)?

(*b*) If this newly introduced SJMP instruction produces the same error message, then what is our next move?

(*c*) If the instruction

```
        JLT     FAR_AWAY
```

produces this error message, then what substitution should we make?

4.32. *Jump table*. Write the Intel 8096 version of Example 4.29.

4.33. *Assembler support*. Referring to Fig. 4.19, which shows the assembly of an Intel 8096 series of instructions, describe what support the assembler has given for the word alignment requirement of the 8096? *Hint:* Check the address allocated to WORD1.

4.34. *Opcode characteristics*. Referring again to Fig. 4.19, this figure shows all the 2-byte add instructions.

(*a*) Express the first byte of each instruction as a binary number.

(*b*) What determines whether the instruction uses two operands or three?

(*c*) What determines the addressing mode, and how is it coded?

4.35. *Opcode characteristics*. Referring to Fig. 4.20,

(*a*) Express the first byte of each 2-byte add instruction as a binary number.

(*b*) What determines the addressing mode, and how is it coded?

(*c*) Repeat *a* and *b* for the 1-byte load instructions. Are the addressing modes coded the same for these two (i.e., ADDD vs. LDAA) instructions?

4.36. *Assembly language*. Consider the following subset of all of the operand field operations supported by the Avocet assembler for the Motorola 68HC11:

```
        + - * / MOD HIGH LO AND OR XOR NOT
```

as well as parentheses to express precedence. Express the following functions of an 8-bit constant called MASK:

(*a*) The lower four bits of MASK (i.e., the lower "nibble")

(*b*) The upper nibble of MASK, reexpressed as a value between 0 and 15.

(*c*) A value with bit 0 being the complement of bit 0 of MASK and with the other bits equal to 1.

(*d*) A value with bit 0 being equal to 0 while the other bits are the complements of the bits of MASK.

4.37. *Assembler features*. For an assembler available to you, consider the following features:

(*a*) How do you insert entire comment lines?

(*b*) Does it accept blank lines?

---

* Refer to Fig. 4.16.

(c) Are comments on an instruction line preceded by a semicolon (or some other character)?

(d) What are the rules for creating acceptable identifiers for names and labels? A length of eight characters or less? Or any number of characters which are unique in the first eight characters? Or what? Must the first character be a letter? Can the identifier include underline characters? Periods? Anything else?

(e) Must labels in the label field be followed by a colon? May they be?

(f) What operations upon constants in the operand field are supported?

(g) How long can comments be? Do longer comments wrap around to the next line, or get truncated (or give the assembler fits)?

(h) Are names defined with EQUates? Without colons?

(i) What assembler directives are available for allocating RAM to variables? What is the syntax required?

(j) What assembler directives are available for entering tables and strings into ROM? What syntax options are available?

(k) Are ORG directives used to set the address of whatever follows? Is there any support given to differentiating between RAM and ROM (perhaps for the benefit of a linker,* if the assembler requires one)?

(l) Does the assembler support the manufacturer-defined names for the micro-computer's registers (e.g., the Motorola 68HC11 UART's SCSR register)? Are these built into the assembler? Or are they carried in a separate file which can be INCLUDEd or USEd by the assembler when it assembles our files? Or must the separate file be copied into our file in order to have the assembler know about it?

(m) Can the listing be turned off for parts of the source file? Can the listing of long tables be controlled so that only the first line of the assembled table need be listed? Can titles be inserted for the top of each page? Can the width of lines and lines per page be controlled? What is the name of the directive to begin a new page? Do directives like these get printed in the output listing?

(n) Is conditional assembly supported? Is there a SET directive which can be used to help assemble the same body of code several times, with somewhat different conditions each time? If there is, then show how you would use it to create a 100-entry table of BCD numbers from 00000000B to 10011001B.

(o) Are macros supported? If so, then are macro parameters supported? What is the syntax required in the definition of a macro? Does the assembler support the printing out of the equivalent assembly language instructions *each time* that a macro is invoked? Can this be turned off so that only the source line is shown?

(p) Does the assembler support control structures, so that the code need not be written as spaghetti code? If so, then what structures are supported and what is their syntax? How deeply can the control structures be nested?

(q) How is the assembler invoked? What options are available? For example, can the output listing be routed either to the screen, to the printer, or to a file? Can a symbol table be obtained? How about a cross-reference table saying where

---

* See Sec. 8.6.

identifiers are defined and where they are used? Can the generation of object code be turned on or off? Can the entire listing be turned on or off?

4.38. *Handling complexity.* Many people feel that if the listing of a subroutine is over a page in length, then it is too long. When looking for one or more portions of the subroutine's code to break out into further subroutines, what help can be provided by a *structured* listing, like that of Fig. 4.32 or 4.33?

4.39. *Critical regions.* Give an example of the problem which can arise if the converse of the critical region rule given in the text occurs. That is, consider the case where an interrupt service routine writes to a multiple-byte operand which the mainline program only reads, but which it must read with several instructions.

4.40. *Software convention.* The text discussed two approaches for handling CPU registers reliably between subroutines and the routines which call them. With convention I, the subroutine has the responsibility for preserving registers. With convention II, the calling routine has this responsibility.

Consider, as an example, that we have a mainline program which includes calls of two subroutines. Each of these calls two other subroutines. Each of these calls two more. And so on, up to six levels. That is, we are talking about one mainline program and

$$2(1 + 2(1 + 2(1 + 2(1 + 2(1 + 2))))) = 126$$

subroutines. Assume that each subroutine passes back a single 1-byte parameter in a CPU register and that it needs to use, and change, a 2-byte CPU pointer. Assume that the push instruction and the pop instruction are each 1-byte instructions taking 2 μs.

Finally, consider the impact of these two conventions during one loop of the mainline program, assuming that each subroutine is actually called once during this loop.

(*a*) For convention I, how many extra bytes of instructions are introduced to protect CPU registers?

(*b*) Answer *a* for convention II.

(*c*) For convention I, how many microseconds are added to the time around the loop to protect CPU registers?

(*d*) Answer *c* for convention II.

4.41. *Stack use for temporary variables.* Figures 4.28 and 4.29 illustrate the use of the stack for passing parameters to a subroutine. Sometimes a subroutine needs more *temporary* variables (i.e., variables which are created within the subroutine and which can be disposed of upon returning) than there are CPU registers to use for this purpose. One possibility is to establish a scratch area in RAM. However, this can lead to a conflict if a subroutine, which needs to use the scratch area, calls another subroutine, which also needs to use it. A foolproof way to resolve this problem is to use the stack. As a subroutine needs to set aside a variable, it simply pushes that variable onto the stack. When this has been done for all variables, the CPU can access any one of them using indexed addressing into the stack, as described in Example 4.33 for the Intel 8096 and Example 4.34 for the Motorola 68HC11.

At the end of the subroutine, the stack must be cleaned up and the return from subroutine executed. Assuming that 20 bytes of temporary variables have been created on the stack:

    (*a*) Write the Motorola 68HC11 instructions to carry out this termination of the subroutine.

    (*b*) Write the Intel 8096 instructions to carry out this termination of the subroutine.

4.42. *Macros.* Using the Intel 8096 assembler's syntax for defining macros, write a macro to handle the general case for the last problem.

    (*a*) For the Motorola 68HC11 using a parameter called NUMBYTES to represent the number of bytes used on the stack for variables.

    (*b*) For the Intel 8096 using a parameter called NUMWORDS to represent the number of words used on the stack for variables.

4.43. *Macros.* In Problem 4.41 we created a workspace on the stack by pushing initial values for the temporary variables needed onto the stack. Sometimes this is inconvenient to do because we need to *use* some of these variables before others can be initialized. In this case, as we enter the subroutine, we can *set up* a workspace on the stack, by subtracting a number from the stack pointer equal to the number of bytes of temporary storage that we need.

    Using the Intel 8096 assembler's syntax for defining macros, write a macro to set up this workspace:

    (*a*) For the Motorola 68HC11 using a parameter called NUMBYTES.

    (*b*) For the Intel 8096 using a parameter called NUMWORDS.

4.44. *Macros for a structured assembler.* The Intel assembler for their 8096 microcontroller includes a macro processing capability which can be used to implement nested control structures. Several years ago while at Georgia Tech, Stephen R. Wachtel ingeniously developed macros to support an extensive set of control structures for the Intel 8085 which worked with Intel's assembler for the 8085. Figure P4.44 shows a subset of his macros, rewritten for the 8096.

    In effect, he creates a stack for use by the macro processor as it assembles our code. Looking at the EIF macro, we see it defined with the help of two other macros, EIF_1 and EIF_2. When EIF_1 is invoked with

```
EIF_1      %LVL
```

the macro assembler evaluates the present value of LVL (e.g., 3) and passes this *value* to the EIF_1 macro. Then, within the definition of this EIF_1 macro, the line

```
TOS        SET      STK&LVL
```

concatenates this value to the letters STK, forming a previously defined variable name (e.g., STK3) whose value is then given to TOS. Study the macros listed in Fig. P4.44 to figure out more or less how they work (given the assembler's directives, syntax requirements, etc.).

    (*a*) Add a new macro called IFF_C to work with the EIF macro in a control structure IFF_C . . . EIF which will execute the included body of code if the carry bit is set.

    (*b*) Add a new macro called LSE* to complete the IFF_test . . . LSE . . . EIF control structure.

---

* Names we pick cannot conflict with the assembler's "reserved words." The Intel 8096 assembler already uses IF, ELSE, and ENDIF for conditional assembly directives. We could use ELSE_ or ELSE? or ?ELSE or _ELSE_, etc.

```
$INCLUDE (SAM96.LIB)
$LIST
```

(a)

```
$NOGEN NONCOND NOLIST
;;;;;;;;   SAM96.LIB FILE - STRUCTURED ASSEMBLER MACROS - INTEL MCS-8096 MACRO ASSEMBLER   ;;;;
;;
;;  Originally prepared by Stephen R. Wachtel for the Intel 8085.
;;  Modified by James W. King of Intel for use with the Intel MCS-8096 Macro Assembler.
;;
;;  First, initialize two variables used by the assembler's macro processor:
LVL          SET   0
;;  LVL is used as a stack pointer to help generate the stack variables STK0, STK1, etc.,
;;  which store the number of a label.  For example, at some point, STK3 might be
;;  storing the number 17, corresponding to the label LBL17.
LBLNUM       SET   1
;;  LBLNUM is used by the macro processor to generate new labels, as needed by the
;;  control structures defined in this file.  They will be named LBL1, LBL2,...
;;
;;;;;;;   IFF_E MACRO   ;;;;;;;;;;;;;;;;;;;;;;;;;;;;;;;;;;;;;;;;;;;;;;;;;;;;;;;;;;;;;;
;;
;;  This macro is used at the start of the control structure
;;                IFF_E ... EIF
;;
;;  which executes a block of code if the zero flag is set.
;;  The macro generates one line of code.  For example, it might generate
;;
;;                JNE LBL24
;;
;;  which says to jump if not equal to the 24th label generated by the macro
;;  assembler for use by these control structures.
;;  It stores the number of this label on a stack, by storing it
;;
;;            in STK0 if LVL=0
;;         or in STK1 if LBL=1
;;         or in STK2 if LBL=2
;;            etc.
```

FIGURE P4.44

Structured assembler macros for the Intel 8096 macro assembler. (*a*) Code to include at beginning of source file of program in order to invoke structured assembler; (*b*) SAM96.LIB macro file.

(continued)

```
;;   Then it increments the stack pointer LVL.
;;   Finally, it increments LBLNUM so that the next label generated will have a new number.
;;   $SAVE GEN     and      $RESTORE     turn on the listing and then put the listing state
;;   back the way it was (i.e., probably with the listing off, unless we want to list the
;;   entire macro file).
;;
IFF_E           MACRO
                IFF_E1  %LBLNUM,%LVL
                IFF_1
                ENDM

IFF_E1          MACRO   LBLNUM,LVL
$SAVE GEN
JNE LBL&LBLNUM
$RESTORE
STK&LVL         SET     LBLNUM
                ENDM

IFF_1           MACRO
LVL             SET     LVL+1
LBLNUM          SET     LBLNUM+1
                ENDM

;;
;;
;;;;;;;;;;  IFF_NE MACRO  ;;;;;;;;;;;;;;;;;;;;;;;;;;;;;;;;;;;;;;;;;;;
;;
;;   This macro is used at the start of the control structure
;;          IFF_NE ... EIF
;;
IFF_NE          MACRO
                IFF_NE1 %LBLNUM,%LVL
                IFF_1
                ENDM

IFF_NE1         MACRO   LBLNUM,LVL
$SAVE GEN
JE  LBL&%LBLNUM
```

FIGURE P4.44 (continued)

(continued)

166

```
$RESTORE    SET     LBLNUM
STK&LVL     ENDM

;;
;;
;;;;;;;;;;   EIF MACRO   ;;;;;;;;;;;;;;;;;;;;;;;;;;;;;;;;;;;;;;;;;;;;;;;;;;;;;;;
;;
;;   This macro is used at the end of any of the IFF control structures, i.e.,
;;             IFF_E ... EIF   or   IFF_NE ... EIF
;;   The macro generates a label-defining line of code, e.g.,
;;        LBL24   EQU   $
;;   which says LBL24 is to have the address of the program counter's content at this point.
;;   It first decrements the stack pointer (LVL) so that it points to the top of the stack.
;;   It does this by using LVL to select one of the STKi variables.
;;   It stores the content of this variable in TOS (top of stack) and uses this to form label.
;;
EIF         MACRO
LVL         SET     LVL-1
            EIF_1   %LVL
            EIF_2   %TOS
            ENDM

EIF_1       MACRO   LVL
TOS         SET     STK&LVL
            ENDM

EIF_2       MACRO   TOS
$SAVE GEN
LBL&TOS     EQU     $
$RESTORE
            ENDM

;;
;;;;;;;;;;   END OF MACRO FILE   ;;;;;;;;;;;;;;;;;;;;;;;;;;;;;;;;;;;;;;;;;;;;;;;
```

(b)

FIGURE P4.44 (continued)

(c) Add two new macros called REPEAT and UNTIL_E which will execute the included body of code at least once and then repeat the block of code until the Z flag is set.

(d) Add a new macro called BREAK_IF_E which causes a jump to the end of the present structure if the Z flag is set.

(e) If you have access to the actual Intel MCS-8096 macro assembler, then try to handle E, NE, etc. (for *all* of the tests which the 8096 supports) as *parameters*, redefining IFF and EIF as necessary. This will let us write

```
IFF   E
  .                ;Block of code to be executed
  .                ; if the Z flag is set.
EIF
```

(f) Again, if you have access to the Intel MCS-8096 assembler, do you have to do anything so that comments can be appended to a source code line using one of these control structures? If so, then what do you have to do?

REFERENCES

Just as the ultimate references on hardware features for a specific microcontroller are the manufacturer's user's manual and application manual, they are also the ultimate references for information on the instruction set. The appendixes at the back of this book, on the Motorola 68HC11 and the Intel 8096, include information on the number of cycles and the byte count for each instruction, for each of its addressing modes (see Figs. A.39 and B.39).

For information on a specific assembler and its features, refer to its user's manual. The assemblers used to generate the listings in this book are both cross-assemblers designed to run on an IBM PC computer.

For information on the 68HC11 cross-assembler, write

Avocet Systems Inc., P.O. Box 490, Rockport, Maine 04856 (Phone: 1-800-448-8500).

For information on the 8096 cross-assembler, write Cybernetic Micro Systems, P.O. Box 3000, San Gregorio, California 94074 (Phone: 1-415-726-3000).

As this book was being completed, Avocet introduced a new family of *macro*-assemblers to run on the IBM PC and compatibles, with versions for both the 68HC11 (AVMAC6811) and the 8096 (AVMAC96) as well as for about two dozen other microprocessors/microcontrollers. These assemblers produce relocatable output files and include a linker for producing absolute files. As will be discussed in Sec. 8.6, such a breakdown of the assembly process speeds up the reassembly process after small changes have been made in the software for a large project.

CHAPTER
5

SOFTWARE
BUILDING
BLOCKS

5.1 OVERVIEW

In this chapter, we will consider several topics which have particular bearing upon writing programs for microcontrollers. First we will look at queues, since they serve as a major way for interrupt service routines to interact with a mainline program. We will be concerned with any critical region questions which arise in the use of queues as well as any ways to make accessing them as fast as possible.

Next we will consider two ways of accessing data. Tables will be used to store constants in support of mathematical algorithms. As an example, we will look at the support which a table of binary-coded decimal numbers can give to a binary-to-BCD conversion algorithm for integers up to 100,000,000. Strings will be used to store sequences of characters to create message displays. We will look at how a queue can be used to concatenate strings together out of constant strings stored in ROM and variable strings stored in RAM. This will provide the mechanism for a mainline program to send a complex message to a device in such a way that an interrupt service routine can get the entire message without having to ask the mainline program repeatedly for more data.

We will look at program organization in a microcontroller. We will see how interrupts from a tick clock can be used to support interactions with slow electromechanical devices such as keyswitches and motors without hampering interactions with faster devices.

169

Next we will consider *state machines* and their use in solving one specific problem. We may have an interrupt service routine which must do any one of several things each time it is invoked. A state machine will keep track of the routine's state from one invocation to the next, and then vector control to the proper section of code to be executed for the given state. The mechanism we use to implement such a state machine will be relatively simple and fast, in keeping with the ways microcontrollers are used. As an example, we will look at the debouncing of the keyswitches which are used on the front panels of virtually all instruments and other stand-alone devices requiring human interaction.

Finally we will consider the *parsing* of front-panel keyswitches, that is, the sorting out of *sequences* of keyswitch entries into the actions to be taken. This is a problem which is common to virtually all instruments. It is also closely related to a problem arising in instruments which can be controlled over the IEEE-488 general purpose interface bus. When an instrument receives setup information over the bus, it does so by receiving a string of ASCII characters. The strings must be parsed into separate commands. Then action must be taken for each command. We will discuss how this problem can be translated into the equivalent problem of parsing keyswitches.

5.2   QUEUES

A *queue* is a data structure consisting of three parts, as shown in Fig. 5.1a. It consists of an insertion pointer (QUE_IN), a removal pointer (QUE_OUT), and a number of elements of data storage having a storage capacity of QLENGTH. If the queue is set up to handle the putting and getting of bytes, then QLENGTH+4 represents the number of bytes of RAM needed to support this queue structure. If the queue is set up to handle the putting and getting of 2-byte words, then (2*QLENGTH)+4 represents the number of bytes of RAM needed. These numbers assume that each pointer into the queue is represented by a 2-byte, extended address. Such a choice speeds the access into the queue (relative to the alternative of representing each pointer by an *offset* from the address labeled QUEUE). Our discussions will always assume that these two pointers are each represented by an extended address.

Figure 5.1b illustrates the initialization required by the queue structure. This initialization can be carried out at power-on time, in the initialization subroutine shown in Fig. 2.28. The queue is initialized to be empty, as indicated by both pointers pointing to the same address. As long as both pointers are equal, the routine which removes data from the queue has nothing to do—the queue is empty.

When an element of data (labeled DATA1 in Fig 5.1c) is put into the queue, the routine which puts the data into the queue puts it in the location pointed to by QUE_IN. Then it increments the QUE_IN pointer. If the element of data being stored in the queue is actually 1 *byte* of data, then we increment once; if it is actually one 2-byte *word* of data, then we must increment *twice*.

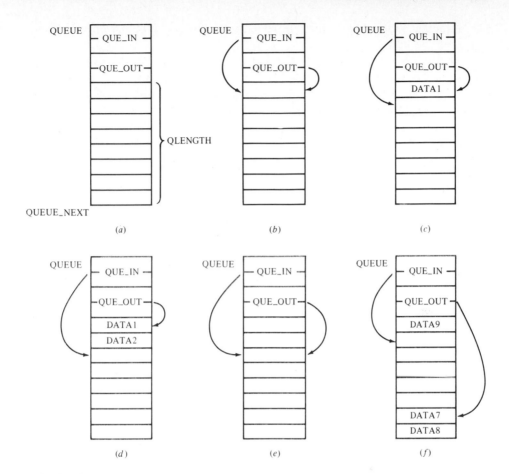

FIGURE 5.1
A queue. (*a*) Definition of terms; (*b*) initialization of queue; (*c*) queue after 1 put, 0 gets; (*d*) queue after 2 puts, 0 gets; (*e*) queue after 2 puts, 2 gets; (*f*) queue after 9 puts, 6 gets.

Also, the pointers must be rolled over to the address QUEUE+4 when they exceed the highest RAM address set aside for the queue.

The operation of the queue for more puts and gets is illustrated in Figs. 5.1*d,e,* and *f*. Note that the function of the removal pointer QUE_OUT is to chase the insertion pointer QUE_IN. Whenever the routine which gets data from the queue is ready for more, it first determines if the queue has anything in it (i.e., if the pointers are unequal). If the queue holds data, this routine gets the data pointed to by QUE_OUT and then increments QUE_OUT.

We must be sure that the put routine does not abuse a queue since it is possible to cause an *overrun* error. For example, in the queue of Fig. 5.1, which can hold eight items of data, if the put routine blindly puts a ninth item into the queue before any of these are removed, then the oldest item will be overwritten. A similar problem arises when a UART (discussed in Sec. 6.5) is ready to put a second byte of data into its receive data register before the CPU has removed the first.

In the case of the UART, the get routine can ignore the possibility of causing an overrun error by reading the receive data register without ever testing the overrun error flag. In our case, ignoring the possibility of an overrun error is satisfactory if:

> We know that an overrun error will never occur under normal circumstances, and if we do not care about the effect of an overrun error during debugging (if one might occur then), or if

> The consequences of an overrun error are outweighed by the trouble it takes to do something about it. For example, if the result of an overrun error is a garbled CRT display, and if we *think* that an overrun error will never occur but are not *sure*, then we might assume that such an error *will not* occur. We might be willing to accept the occurrence of a garbled CRT if the chance of this occurring seems to be close to zero.

We will develop PUT and GET algorithms assuming that an overrun error will never occur. Then we will develop a QHS (queue handshake) algorithm which returns a flag that serves the same function as the UART's transmit data register empty flag. That is, the flag indicates whether there is room for data. If our use of a queue calls for a handshake before the execution of a put, then we will have the mechanism ready.

At the end of the last chapter we discussed the use of macros for generating "strange" sequences of code with reliability. In this section as we deal with queues, we will see how macros can enhance the reliability of our use of a data structure which is accessed in more than one way. The significant characteristics of a queue which encourage the use of macros for their exploitation are:

> The PUT, GET, QHS, and initialization routines are each short.

> They each involve several parameters. Were we to implement these routines as subroutines, we would increase execution time and bytes of code passing parameters to the subroutines. In the case of macros, these parameters get built into the instructions which do the work needed by the routines, as we shall see.

> Each queue that we use is accessed when we define storage for it, when we initialize it, when we handshake with it, when we PUT data into it, and when we GET data from it. The assembler's macro processor can help ensure that all of these accesses to the same data structure are written correctly.

> Queues are a basic building block. By implementing their use with macros, we make them painless to use.

The macros we would like to define are shown in Fig. 5.2, as they might look when invoked for the Intel 8096 microcontroller. To make the ideas we

```
QUR:    DSQB  8                  ;Eight-byte queue for UART receiver
QUT:    DSQB  6                  ;Six-byte queue for UART transmitter
          .
          .
QIC1:   DSQW  3                  ;Three-word queue for timer's input capture register 1
          .
          .
```

(a)

```
        INITQ QUR                ;Initialize UART receiver queue
        INITQ QUT                ;Initialize UART transmitter queue
          .
          .
        INITQ QIC1               ;Initialize input capture register 1 queue
          .
          .
```

(b)

```
        PUTB  AL,QUT             ;AL --> UART transmitter queue

        GETB  SBUF,QUT           ;If queue is empty, return Z=1; otherwise return Z=0
                                 ; and UART transmitter queue output to SBUF
```

(c)

```
        PUT   HSI_TIME,QIC1  ;Timer FIFO output --> queue for input capture reg. 1
        GET   AX,QIC1        ;If queue is empty, return Z=1; otherwise return Z=0
                             ; and output of queue for input capture reg. 1 to AX
```

(d)

```
        QHSB  QUT                ;If queue (for bytes) has no more room for input data,
                                 ; return Z=1; otherwise return Z=0.

        QHS   QIC1               ;If queue (for words) has no more room for input data,
                                 ; return Z=1; otherwise return Z=0.
```

(e)

FIGURE 5.2
Macros used with queues. (*a*) Macros to allocate RAM to each queue with DSQB (define storage for queue for bytes) and DSQW (define storage for queue for words); (*b*) macros to initalize each queue; (*c*) macros to put and get bytes; (*d*) macros to put and get 2-byte words; (*e*) macros to handshake with the input to a queue to avoid an overrun error.

develop concrete, we will follow the rules associated with Intel's macro-assembler for the 8096.

First, we must allocate storage to each queue. Intel's 8096 assembler has DSB, DSW, and DSL assembler directives (defined at the end of Sec. 4.5) for allocating RAM to byte, word, and long-word variables. We will introduce what are, in effect, two new directives

```
                DSQB        ;Define storage for a queue of bytes
                DSQW        ;Define storage for a queue of words
```

for allocating RAM to queues. In Fig. 5.2*a*, the line

```
QUR:            DSQB 8      ;Eight-byte queue for UART receiver
```

says that we want to use the label QUR to identify the queue for the UART's receiver. It also says to allocate 8 bytes for the queue (and a total of 12 bytes for the entire data structure, including 2 extra bytes for each pointer).

The line

```
QIC1:           DSQW 3      ;Three-word queue for timer #1
```

says that we want to use the label QIC1 to identify a queue associated with the programmable timer's input capture register 1. In this case the directive allocates a total of 10 bytes for the entire queue data structure, since each of the three words in the queue itself requires 2 bytes. Both DSQB and DSQW must ensure that the pointers defined into each data structure are word aligned to begin at even memory addresses, as required by the 8096.

We also want a macro to initialize each of the queues, making the two pointers (called QUE_IN and QUE_OUT in Fig. 5.1*a*) point to the same address in each queue. The macro of Fig. 5.2*b*

```
            INITQ  QUT  ;Initialize UART transmitter queue
```

must be defined to do this.

The pair of macros in Fig. 5.2*c*

```
            PUTB    AL,QUT
```

and

```
            GETB    SBUF,QUT
```

put data into and get data from the QUT queue which handles bytes. The macro versions that handle 2-byte words are shown in Fig. 5.2*d*. As we shall see, the only difference between the PUTB and PUT macros is an STB instruction (to store a byte) in the former and an ST instruction (to store a word) in the latter. Finally, we want a pair of macros to support handshaking on the input to a queue of bytes, QHSB, and to a queue of words, QHS.

One thing we can do as an intermediate step in defining a macro is write the assembly language lines which should get generated by the assembler's macro processor *in a specific case*. This helps us keep track of our goal while we wrestle with the syntax required by the macro processor. The examples of Fig. 5.3 do this for each kind of macro used in Fig. 5.2. Before assembly takes place, we want the macro processor to generate each of the lines with a + sign to the left. At its most fundamental level, the macro processor simply takes each of the actual parameters employed when a macro is used (as in Fig. 5.2) and substitutes it into the text of a macro definition in place of the corresponding dummy parameter. This is illustrated most simply in Fig. 5.4*c* for the definition of the INITQ macro. We see in Fig. 5.3*c* that the actual expansion of the macro for a specific case requires the macro processor to take the actual

```
QUR:       DSQB    8              ;Allocate six aligned words of storage
+          DSW     (8/2)+2
+ QUR_NEXT EQU     $              ;Form address just past area reserved for queue
```

(a)

```
QIC1:       DSQW    3             ;Allocate five aligned words of storage
+           DSW     3+2
+ QIC1_NEXT EQU     $             ;Form address just past area reserved for queue
```

(b)

```
            INITQ   QUR
+           LD      QUR,#QUR+4    ;Load address of QUR+4 into insertion pointer
+           LD      QUR+2,QUR     ; and into removal pointer
```

(c)

```
            PUTB    AL,QUT
+           STB     AL,[QUT]+         ;Put byte into queue; increment insertion pointer
+           CMP     QUT,#QUT_NEXT     ;Roll over?
+           JNE     $+6               ;Finish if no roll over      2 bytes    6
+           LD      QUT,#QUT+4        ;Reinitialize insertion pointer    4 bytes    —
```

(d)

(continued)

FIGURE 5.3
Macro expansions for some of the macros listed in Fig. 5.2. (a) Expansion of DSQB; (b) expansion of DSQW; (c) expansion of INITQ; (d) expansion of PUTB; (e) expansion of PUTB; (f) expansion of GETB; (g) expansion of PUT; (h) expansion of GET; (i) expansion of QHSB; (i) expansion of QHS.

```
        GETB  SBUF,QUT
      + CMP   QUT,QUT+2          ;Compare pointers                      2 bytes
      + JE    $+17               ;Finish with Z flag set; empty queue   3 bytes
      + LDB   SBUF,[QUT+2]+      ;Get byte from queue; inc. rem. ptr.   4 bytes   17
      + CMP   QUT+2,#QUT_NEXT    ;Need to roll over?                    2 bytes
      + JNE   $+8                ;Finish with Z flag clear              4 bytes    8
      + LD    QUT+2,#QUT+4       ;Reinitialize removal pointer          2 bytes
      + NOT   R0                 ;Finish with Z flag clear
```

(e)

```
        PUT   HSI_TIME,QIC1
      + ST    HSI_TIME,[QIC1]+   ;Put word into queue; increment insertion pointer
      + CMP   QIC1,#QIC1_NEXT    ;Need to roll over?
      + JNE   $+6                ;Finish if no roll over                2 bytes    6
      + LD    QIC1,#QIC1+4       ;Reinitialize insertion pointer        4 bytes
```

(f)

```
        GET   AX,QIC1
      + CMP   QIC1,QIC1+2        ;Compare pointers                      2 bytes
      + JE    $+17               ;Finish with Z flag set; empty queue   3 bytes
      + LD    AX,[QIC1+2]+       ;Get word from queue; inc. rem. pts.   4 bytes   17
      + CMP   QIC1+2,#QIC1_NEXT  ;Need to roll over?                    2 bytes
      + JNE   $+8                ;Finish with Z flag clear              4 bytes    8
      + LD    QIC1+2,#QIC1+4     ;Reinitialize removal pointer          2 bytes
      + NOT   R0                 ;Finish with Z flag clear
```

(g)

(continued)

```
     QHSB   QUT              ;Z=0 when queue is ready
+    PUSH   AX               ;Make room for testing
+    ADD    AX,QUT,#1        ;Increment copy of insertion pointer
+    CMP    AX,#QUT_NEXT     ;Need to roll over?
+    JNE    $+6              ;Skip if not rolling over         2 bytes   6
+    LD     AX,#QUT+4        ;Reinitialize insertion pointer   4 bytes   —
+    CMP    AX,QUT+2         ;Set Z flag if we have reached the removal pointer
+    POP    AX               ;Restore initial content to AX
```

(h)

```
     QHS    QUT              ;Z=0 when queue is ready
+    PUSH   AX               ;Make room for testing
+    ADD    AX,QUT,#2        ;Increment copy of insertion pointer twice
+    CMP    AX,#QUT_NEXT     ;Need to roll over?
+    JNE    $+6              ;Skip if not rolling over         2 bytes   6
+    LD     AX,#QUT+4        ;Reinitialize insertion pointer   4 bytes   —
+    CMP    AX,QUT+2         ;Set Z flag if we have reached the removal pointer
+    POP    AX               ;Restore initial content to AX
```

(i)

FIGURE 5.3 (continued)

177

```
DSQB           MACRO   QNAME,NUMBER_OF_BYTES
               DSW     ((NUMBER_OF_BYTES+1)/2)+2
QNAME&_NEXT EQU        $
               ENDM
```

(a)

```
DSQW           MACRO   QNAME,NUMBER_OF_WORDS
               DSW     NUMBER_OF_WORDS+2
QNAME&_NEXT EQU        $
               ENDM
```

(b)

```
INITQ          MACRO   QNAME
               LD      QNAME,#QNAME+4
               LD      QNAME+2,QNAME
               ENDM
```

(c)

```
PUTB           MACRO   SOURCE,QNAME
               LOCAL   PUTB_1
               STB     SOURCE,[QNAME]+
               CMP     QNAME,#QNAME&_NEXT
               JNE     PUTB_1
               LD      QNAME,#QNAME+4
PUTB_1:
               ENDM
```

(d)

```
PUTB           MACRO   SOURCE,QNAME
               STB     SOURCE,[QNAME]+
               CMP     QNAME,#QNAME&_NEXT
               JNE     $+6
               LD      QNAME,#QNAME+4
               ENDM
```

(e)   (continued)

FIGURE 5.4
Macro definitions for use with queues. (a) DSQB for allocating RAM to a queue data structure
for bytes; (b) DSQW for allocating RAM to a queue data structure for words; (c) INITQ for
initializing the two pointers; (d) and (e) two ways of expressing PUTB for putting a byte into a
queue; (f) GETQ for getting a byte out from a queue; (g) QHSB for handshaking with input to
a queue for bytes; (h) QHS for handshaking with input to a queue for words.

```
GETB        MACRO   DEST,QNAME
            CMP     QNAME,QNAME+2
            JE      $+17
            LDB     DEST,[QNAME+2]+
            CMP     QNAME+2,#QNAME&_NEXT
            JNE     $+8
            LD      QNAME+2,#QNAME+4
            NOT     R0
            ENDM
```

(f)

```
QHSB        MACRO   QNAME
            PUSH    AX
            ADD     AX,QNAME,#1
            CMP     AX,#QNAME&_NEXT
            JNE     $+6
            LD      AX,#QNAME+4
            CMP     AX,QNAME+2
            POP     AX
            ENDM
```

(g)

```
QHS         MACRO   QNAME
            PUSH    AX
            ADD     AX,QNAME,#2
            CMP     AX,#QNAME&_NEXT
            JNE     $+6
            LD      AX,#QNAME+4
            CMP     AX,QNAME+2
            POP     AX
            ENDM
```

(h)

FIGURE 5.4   (*continued*)

parameter, QUR, and substitute it wherever the dummy parameter, QNAME, occurs (in the LD instructions of the macro definition of Fig. 5.4c). The effect of this macro is to initialize the two pointers so that they each point at the same address in the queue. This assumes that the QUE_IN of Fig. 5.1a is located right at the beginning of the queue structure; that is, at the address labeled QUR. QUE_OUT of Fig. 5.1a is located right after this at the address QUR+2. Finally, the queue itself begins at the address QUR+4.

A couple of problems arise in the DSQB macro definition. First, when we are allocating RAM to several queues, each one *must* begin at an even address (for the Intel 8096, because of the word alignment required for the two pointers). Because of this requirement, the definition shown in Fig. 5.4*a* actually allocates an *even* number of bytes to the queue itself. That is, if we ask for 8 bytes in the queue, we get 8. If we ask for 7 bytes in the queue, we *also* get 8. This is done so that if we use five DSQB directives in a row to set up five queues, we will not waste any bytes along the way.

A macro syntax issue is the other problem arising in the DSQB macro definition. We want to create a label which is a modification of the actual label used for the queue. Then we will use this generated label in other macro definitions. The Intel's 8096 assembler's way to do this is shown in Fig. 5.4*a* with

```
QNAME&_NEXT
```

This takes the parameter and appends _NEXT to it. The resulting assembled line, shown in Fig. 5.3*a*,

```
QUR_NEXT        EQU     $
```

uses $ to instruct the assembler to assign the present address to the label QUR_NEXT. Note that this will be the address of the *next* available byte of RAM (to be allocated by the next DSxx assembler directive).

The label we create with a DSQB directive gets used by both the PUTB and GETB macros. This is illustrated in Fig. 5.3*d* and *e*.

Another potential labeling problem arises in Fig. 5.4*d* and *e*. At the end of the macro we need to roll over the insertion pointer *if* it has been incremented past the end of the memory allocated to the queue. Otherwise, the JNE instruction skips this reinitialization of the pointer. If we use a label, as shown in Fig. 5.4*d*, then we need to tell the macro processor to avoid creating a label with this *same* name each time the macro is invoked. This is what the LOCAL directive on the first line of the macro definition does. The macro processor will create a unique label that might look like

$$??0005$$

for the fifth such label that it has had to create.

An alternative way to handle this problem is shown in Fig. 5.4*e*, where the label was avoided altogether. This keeps the symbol table generated by the assembler uncluttered by labels of the form ??xxxx. On the other hand, it requires that we count the bytes taken by each instruction *accurately*. A miscount means a jump to the wrong place! Note the byte counts located to the right of the comments in Fig. 5.3*d* to *i*. The number of bytes required for each instruction is presented in Fig. B.39 in Appendix B.

The GETB and GET macros remove a byte from the queue *if the queue is not empty*. In this case, the byte is removed and loaded into the first parameter of the macro. Then the Z flag is cleared to 0. The very last instruction of the macro is a gimmicky way of clearing the Z flag. R0 is the name of the word of the 8096's memory at addresses 0000 to 0001. As was pointed out in Example

4.15, these addresses do not access RAM memory at all. Rather, they are read as the 2-byte hexadecimal number, 0000H. The

```
NOT R0
```

instruction *tries* to complement this to FFFFH. Even though it tries to store the result back in R0, the hardware ignores this attempt. Meanwhile the Z flag has been cleared because the result of the operation is not 0000H (as far as the CPU is concerned). Our code would be less cryptic if we had defined a macro called

```
CLRZ
```

to be equivalent to

```
NOT R0
```

The queue handshake macros, QHSB and QHS, shown in Fig. 5.4*g* and *h*, *clear* the Z flag when the queue is ready to accept data (i.e., is not full). They require a temporary register which can be used to hold a copy of the incremented insertion pointer (without incrementing the pointer itself!). The PUSH and POP instructions ensure that wherever we use this macro, it will not destroy anything (i.e., the content of AX) inadvertently.

Example 5.1. The Intel 8096 has four inputs to its programmable timer for timing up to four independent events simultaneously. When the time of an event is signaled at one input, the time is stored in a FIFO (i.e., a queue built into the hardware of the programmable timer). Along with this time, a status byte is also queued up. Each FIFO entry can be thought of as holding the content of an input capture register.

We want to have the interrupt service routine sort these independent events into four software queues, called

QIC0   QIC1   QIC2   QIC3

Then in the mainline program, the subroutine which deals with input capture register 0 need only deal with QIC0, independent of the events being captured by the other three input capture registers.

First we must define the queues in the beginning of our program where we allocate RAM to variables. The code is shown in Fig. 5.5*a*. Then in our initialization subroutine we need to initialize each of these queues. This code is shown in Fig. 5.5*b*. Finally, we need to write the interrupt service routine, which is shown in Fig. 5.5*c*. After doing the normal 8096 interrupt service routine activity of reenabling interrupts for higher priority devices, the HSI_STATUS (high-speed-input status register) is polled. When bit 0 is set, the output of the programmable timer's FIFO contains a time which came from input capture register 0. Each of the four flags is checked in this way. Then an ANDB instruction checks whether removing one (or more) entry in the FIFO leaves still more entries. If so, the process is repeated. If not, then the return from interrupt is executed. Note that the ANDB instruction uses the address R0, discussed in conjunction with Fig. 5.4*f*, as a dumping place for its result. All we want to do is affect the Z flag appropriately.

```
QIC0:             DSQW  2                          ;Two-word queue for input capture register 0
QIC1:             DSQW  2                          ;Two-word queue for input capture register 1
QIC2:             DSQW  2                          ;Two-word queue for input capture register 2
QIC3:             DSQW  2                          ;Two-word queue for input capture register 3

                                (a)

                  INITQ QIC0                       ;Initialize queue for input capture register 0
                  INITQ QIC1                       ;Initialize queue for input capture register 1
                  INITQ QIC2                       ;Initialize queue for input capture register 2
                  INITQ QIC3                       ;Initialize queue for input capture register 3

                                (b)

INPUT_CAPTURE:
           PUSHF                                   ;Set aside CPU state,
           LDB   INT_MASK,#ICMASK                  ; then set up to enable higher priority devices,
           EI                                      ; then reenable interrupts
IC_0:      JBC   HSI_STATUS.0,IC_1                 ;Skip if status bit says FIFO entry is not from ICR 0
           PUT   HSI_TIME, QIC0                    ;Put time from input capture register 0 into queue 0
IC_1:      JBC   HSI_STATUS.2,IC_2                 ;Skip if status bit says FIFO entry is not from ICR 1
           PUT   HSI_TIME,QIC1                     ;Put time from input capture register 0 into queue 1
IC_2:      JBC   HSI_STATUS.4,IC_3                 ;Skip if status bit says FIFO entry is not from ICR 2
           PUT   HSI_TIME,QIC2                     ;Put time from input capture register 0 into queue 2
IC_3:      JBC   HSI_STATUS.6,IC_4                 ;Skip if status bit says FIFO entry is not from ICR 3
           PUT   HSI_TIME,QIC3                     ;Put time from input capture register 0 into queue 3
IC_4:      ANDB  R0,HSI_STATUS,#01010101B          ;Any more times in FIFO?
           JNE   IC_0                              ;If so, then poll again
           POPF                                    ;Otherwise, restore CPU state
           RET                                     ; and return

                                (c)

           GET   AX,QIC1                           ;Get queue output if it is there
           JE    DONE                              ;Not there yet; return with Z flag set
           SUB   AX,START_TIME                     ;Got it! Form difference and return it in AX
           NOT   R0                                ;Then clear Z flag
           RET                                     ;Return
DONE:

                                (d)
```

FIGURE 5.5

Code for Examples 5.1 and 5.2. (a) RAM allocation to the four queues; (b) initialization of the four queues; (c) interrupt service routine for the timer's input capture function; (d) portion of a subroutine in the mainline program which returns with the pulse width in the CPU register AX and the Z flag clear when the measurement is complete.

Example 5.2. Suppose that we have a mainline subroutine which is trying to measure a pulse width using the timer's input capture register 1. For simplicity, assume that the pulse width is shorter than the time it takes the timer to go through 2^{16} counts. The subroutine has already captured the time for the start of the pulse. Now it will check whether there is a time in the QIC1 queue. If so, it will form the pulse width in the CPU register AX, clear the Z flag, and return. Otherwise it will return with the Z flag set. In this way, the mainline program will not get hung up waiting for the pulse to be completed and can continue with other tasks. It just calls this subroutine periodically and waits for it to return with the Z flag cleared. The code to do this in a mainline subroutine is shown in Fig. 5.5*d*.

These macros permit interactions with a queue within an interrupt service routine *without* creating a critical region in the mainline program. To see this, consider the case just discussed in which the mainline program is getting data from a queue. The actual code which is executed by the GET macro in this case is shown in Fig. 5.3*g*. The one variable of interest, because it may be changed by the interrupt service routine at any time (if we leave interrupts enabled), is QIC1. This is the insertion pointer into the queue. If it is changed before the first CMP (compare) instruction, then the GET macro will see an unempty queue and get data from it. If QIC1 is changed after the first CMP instruction, then if the queue *was* empty, the GET macro will do nothing more. If the queue already had data in it, then the GET macro will get that data. It will not be bothered as the interrupt service routine puts *more* data into the queue and updates the insertion pointer. The GET macro never even looks at the insertion pointer again.

Next we will consider the case where the mainline program puts data into a queue and an interrupt service routine removes it. Does this give rise to a critical region?

Example 5.3. Consider the example of the mainline program transmitting a string of characters to an external device via the on-chip UART. What it actually does is put a byte into a queue using the exact PUTB code shown in Fig. 5.3*d*. Then the UART's interrupt service routine monitors the transmit data register empty flag. If the flag is set, then it does

```
GETB   SBUF,QUT
```

If data is available in the queue, then the GETB macro removes it and loads it into the UART's transmit data register (called SBUF). If the queue is empty, then the GETB macro never deals with SBUF. Presumably this means that the entire multiple-character message has been sent and that the UART need not interrupt again until a new message is begun.

Now, let us consider what can happen during the execution of the PUT macro *if the insertion pointer rolls over*. In this case the mainline routine carries out a multiple-instruction sequence of writes to a variable (QUT) which the interrupt service routine also monitors. If the interrupt service routine occurs *after*

```
STB    AL,[QUT]+
```

and *before*

```
LD     QUT,#QUT+4
```

then it will certainly see that the queue is not empty and will remove a byte from the queue, which is fine. What happens next depends upon how fast the interrupt service routine is. If, because of *other* sources of interrupts, the CPU does not get back to executing the mainline program for a while, and if the UART interrupts *again* during this time, then its interrupt service routine will *necessarily* detect a mismatch between the queue's two pointers and remove "data," when in fact the queue may be empty! The above sequence of events is really only a threat if two conditions are met:

> There is an interrupt service routine, ISRx, on the output side of a queue.

> The sum of the durations of *all* possible interrupt service routines is longer than the interval *between* interrupts from the source which causes the execution of ISRx above.

This is what it takes to lock out the CPU from executing the three more instructions needed to square up the queue in the mainline program. That is, *one* interrupt during this time is inconsequential, but two interrupts by the ISRx device presents a disaster. Actually, a UART sending data to a normal baud rate device (e.g., a computer) at 9600 baud will interrupt only every millisecond, so this is hardly a threat. On the other hand, one 8096 microcontroller talking to another 8096 over the serial port at the maximum baud rate of 187.5 kbaud will interrupt every 60 μs or so. It does not take many interrupt service routines piling up at the same time to put the queue in trouble in this case.

The solution is, of course, to recognize this pathological case as giving rise to a critical region and to disable interrupts during it. For most normal uses of a queue, there is no problem because the time interval between interrupts is long enough to permit the CPU to complete the PUT macro between any two interrupts of the device needing to GET data from this queue.

5.3   TABLES AND STRINGS

Tables and strings are the two ways of organizing data which are used over and over again in the development of microcontroller software. Tables are commonly used to hold any miscellaneous data which can benefit from being accessed with a known offset. Strings, on the other hand, are used to store sequences of data elements which can be of arbitrary length. They are accessed from the beginning of the string, continuing until a designated *terminator* is accessed.

An example of a table is shown in Fig. 5.6. This table includes as entries all of the BCD numbers between 00 and 99. It can be used immediately to convert from a binary number in the range from 0 to 99 to its BCD equivalent.

Example 5.4. Show the Motorola 68HC11 code to convert the binary number in accumulator B (with a value of less than 100) to its BCD equivalent, putting the result back into accumulator B.

```
LDX   #BCD_TABLE          ;Get base address of table
ABX                       ;Add offset into table
LDAB  0,X                 ;Load table entry into B
```

Example 5.5. Show the Intel 8096 code to convert any 2-byte binary number named NUMBER having a value less than 10,000 into its BCD equivalent. NUM-BER is assumed to be word aligned, as required by the 8096's operations upon words. We will assume that the CPU registers of Fig. 4.27 are available for use as scratch-pad registers.

  The unsigned divide instruction, DIVUB, divides a word by a byte and puts the quotient back in the least-significant byte position (having the lower address)

```
BCD_TABLE          0000 0000
                   0000 0001
                   0000 0010
                   0000 0011
                   0000 0100
                   0000 0101
                   0000 0110
                   0000 0111
                   0000 1000
BCD_TABLE+9        0000 1001
BCD_TABLE+10       0001 0000
                   0001 0001
                   0001 0010
                   0001 0011
                          .
                          .
                          .
                   1000 0111
                   1000 1000
BCD_TABLE+89       1000 1001
BCD_TABLE+90       1001 0000
                   1001 0001
                   1001 0010
                   1001 0011
                   1001 0100
                   1001 0101
                   1001 0110
                   1001 0111
                   1001 1000
BCD_TABLE+99       1001 1001
```

FIGURE 5.6
A 100-entry table for BCD numbers.

and the remainder back in the most-significant byte position. Indexed addressing into the table does the rest.

```
DIVUB  NUMBER,#100                ;Forms quotient and remainder
PUSH   NUMBER+1                   ;Push remainder from upper byte
CLRB   NUMBER+1                   ; and then clear the upper byte
LDB    AH,BCD_TABLE[NUMBER]       ;Form upper digits
POP    NUMBER                     ;Pop remainder to lower byte
LDB    NUMBER,BCD_TABLE[NUMBER]   ;Form lower digits
LDB    NUMBER+1,AH                ;Get back upper digits
```

Example 5.6. Show the Intel 8096 instruction sequence to convert the 4-byte binary number, DUBNUM, to BCD representation. The number can have any value less than 100,000,000. Put the result back in DUBNUM, least-significant digits first. DUBNUM is assumed to be double-word aligned (i.e., the address of DUBNUM is evenly divisible by four) and holds the binary number least-significant byte first.

The key here is another of the Intel 8096's divide instructions. DIVU is an unsigned divide of a double word by a word, leaving the quotient in the lower word position of the original double word. The remainder is left in the upper word position. The next three instructions swap these two parts.

Now we need a macro called BINBCD which does the steps in the previous example. To carry out the conversion of that problem, we would invoke it with:

```
BINBCD NUMBER
```

Then this problem can be carried out as

```
DIVU    DUBNUM,#10000    ;Form quotient and remainder
PUSH    DUBNUM           ;Swap; set aside quotient
ST      DUBNUM+2,DUBNUM  ;Move remainder
POP     DUBNUM+2         ;Now get quotient
BINBCD  DUBNUM           ;Convert the lower half
BINBCD  DUBNUM+2         ;Convert the upper half
```

These examples have illustrated the power of a table in augmenting the instructions of a microcontroller. With over 8000 bytes of ROM, both the Intel 8096 and the Motorola 68HC11 make it attractive to use tables liberally to speed up what might otherwise be slow algorithmic processes.

Sometimes the entries in a table need to be larger than a single byte. What does this mean for accessing them?

Example 5.7. Suppose that we have a table called WORD_TABLE, in which each entry is a 2-byte word. Show the Motorola 68HC11 instruction sequence to load accumulator D with the Bth entry, where B represents the content of accumulator B.

```
LDX #WORD_TABLE    ;Get base address of table
ABX                ;Add offset into table
ABX                ; twice
LDD 0,X            ;Load table entry into D
```

For a table with 3-byte entries, we would use *three* ABX instructions.

Example 5.8. Using the Intel 8096 and the same WORD_TABLE as in the last problem, use the offset in the 2-byte word variable, OFFSET, to access the table. Put the table entry back into OFFSET.

```
SHL OFFSET,#1                        ;Double offset
LD  OFFSET,WORD_TABLE[OFFSET] ;Get table entry
```

Strings are needed whenever we have a variety of messages to display on a CRT. Or whenever we need to transmit a sequence of control codes to a device serially to set up its mode of operation. Or whenever . . . . The applications are endless.

A string consists of a number of bytes of data followed by a defined *terminator*. An *ASCII string* is the common example of a string. It consists of the ASCII codes for letters, numbers, punctuation marks, and control codes, shown in Fig. 5.7. One of these codes is reserved for use as a terminator. For example, we might treat the EOT (end of transmission) ASCII character, with the hexadecimal representation 04, as a terminator. Then in reading an ASCII string representing a canned message out of ROM, we continue until 04 is accessed. A string *beginning* with the terminator is called a *null string*.

A *pointer* to the beginning of the string is our key to its access. Consider the following example.

Example 5.9. Suppose that we make an instrument which either can be operated in a stand-alone mode with a seven-segment numeric display, or else can be connected to a CRT terminal to provide more extensive output. Depending upon the mode of operation of the instrument, we want to send out any one of 10 ASCII strings. They can contain characters to do such things as clear the screen, position the cursor, write header messages with reverse video turned on for emphasis, etc. Show how the Motorola 68HC11 would send one of these strings, which begins at an address labeled MESSAGE_5, out to a device using the on-chip UART. Assume that the UART has already been initialized but that it is standing idle.

To make sense out of this job, we must first consider the Motorola 68HC11 UART's registers, shown in Fig. 5.8. Of all of its capabilities, all we need be concerned with here are the three shaded entities in that figure.*

> SCDR. When written to, this is the *transmit data register*. The character written to it will be transmitted with the serial protocol established by the bits previously written to the two control registers (SCCR1 and SCCR2) and the baud rate register (BAUD).

> TDRE. When set by the UART, this *transmit data register empty* bit in the status register (SCSR) signifies that any byte previously written to the transmit data register has been transferred to the transmit data *shift* register, starting the serial output. The transmit data register is empty, ready for another byte (if there is another byte to be sent). This TDRE bit is cleared by the two-step process:

---

* For a more complete description of the 68HC11's UART, refer to Sec. A.8.

ASCII Char.	Dec	Binary	Oct	Hex
NUL	0	00000000	000	00
SOH	1	00000001	001	01
STX	2	00000010	002	02
ETX	3	00000011	003	03
EOT	4	00000100	004	04
ENQ	5	00000101	005	05
ACK	6	00000110	006	06
BEL	7	00000111	007	07
BS	8	00001000	010	08
HT	9	00001001	011	09
LF	10	00001010	012	0A
VT	11	00001011	013	0B
FF	12	00001100	014	0C
CR	13	00001101	015	0D
SO	14	00001110	016	0E
SI	15	00001111	017	0F
DLE	16	00010000	020	10
DC1	17	00010001	021	11
DC2	18	00010010	022	12
DC3	19	00010011	023	13
DC4	20	00010100	024	14
NAK	21	00010101	025	15
SYNC	22	00010110	026	16
ETB	23	00010111	027	17
CAN	24	00011000	030	18
EM	25	00011001	031	19
SUB	26	00011010	032	1A
ESC	27	00011011	033	1B
FS	28	00011100	034	1C
GS	29	00011101	035	1D
RS	30	00011110	036	1E
US	31	00011111	037	1F

ASCII Char.	Dec	Binary	Oct	Hex
space	32	00100000	040	20
!	33	00100001	041	21
"	34	00100010	042	22
#	35	00100011	043	23
$	36	00100100	044	24
%	37	00100101	045	25
&	38	00100110	046	26
'	39	00100111	047	27
(	40	00101000	050	28
)	41	00101001	051	29
*	42	00101010	052	2A
+	43	00101011	053	2B
,	44	00101100	054	2C
-	45	00101101	055	2D
.	46	00101110	056	2E
/	47	00101111	057	2F
0	48	00110000	060	30
1	49	00110001	061	31
2	50	00110010	062	32
3	51	00110011	063	33
4	52	00110100	064	34
5	53	00110101	065	35
6	54	00110110	066	36
7	55	00110111	067	37
8	56	00111000	070	38
9	57	00111001	071	39
:	58	00111010	072	3A
;	59	00111011	073	3B
<	60	00111100	074	3C
=	61	00111101	075	3D
>	62	00111110	076	3E
?	63	00111111	077	3F

ASCII Char.	Dec	Binary	Oct	Hex
@	64	01000000	100	40
A	65	01000001	101	41
B	66	01000010	102	42
C	67	01000011	103	43
D	68	01000100	104	44
E	69	01000101	105	45
F	70	01000110	106	46
G	71	01000111	107	47
H	72	01001000	110	48
I	73	01001001	111	49
J	74	01001010	112	4A
K	75	01001011	113	4B
L	76	01001100	114	4C
M	77	01001101	115	4D
N	78	01001110	116	4E
O	79	01001111	117	4F
P	80	01010000	120	50
Q	81	01010001	121	51
R	82	01010010	122	52
S	83	01010011	123	53
T	84	01010100	124	54
U	85	01010101	125	55
V	86	01010110	126	56
W	87	01010111	127	57
X	88	01011000	130	58
Y	89	01011001	131	59
Z	90	01011010	132	5A
[	91	01011011	133	5B
\	92	01011100	134	5C
]	93	01011101	135	5D
^	94	01011110	136	5E
_	95	01011111	137	5F

ASCII Char.	Dec	Binary	Oct	Hex
`	96	01100000	140	60
a	97	01100001	141	61
b	98	01100010	142	62
c	99	01100011	143	63
d	100	01100100	144	64
e	101	01100101	145	65
f	102	01100110	146	66
g	103	01100111	147	67
h	104	01101000	150	68
i	105	01101001	151	69
j	106	01101010	152	6A
k	107	01101011	153	6B
l	108	01101100	154	6C
m	109	01101101	155	6D
n	110	01101110	156	6E
o	111	01101111	157	6F
p	112	01110000	160	70
q	113	01110001	161	71
r	114	01110010	162	72
s	115	01110011	163	73
t	116	01110100	164	74
u	117	01110101	165	75
v	118	01110110	166	76
w	119	01110111	167	77
x	120	01111000	170	78
y	121	01111001	171	79
z	122	01111010	172	7A
{	123	01111011	173	7B
|	124	01111100	174	7C
}	125	01111101	175	7D
~	126	01111110	176	7E
DEL	127	01111111	177	7F

FIGURE 5.7 Table of ASCII codes.

1. Read the SCSR status register (while TDRE=1).
2. Write a byte of data to the SCDR transmit data register (sending another byte off chip).

TIE. This *transmit interrupt enable* bit in the UART's control register 2 (SCCR2) is set and cleared under program control. If it is set, then when the TDRE bit is also set, the interrupt line from the UART to the CPU will be pulled low and will cause an interrupt (if interrupts are enabled in the CPU).

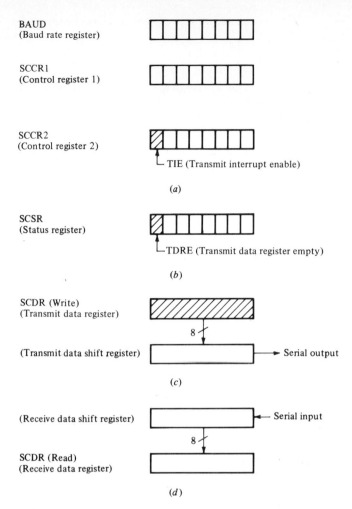

BAUD
(Baud rate register)

SCCR1
(Control register 1)

SCCR2
(Control register 2)

└ TIE (Transmit interrupt enable)

(*a*)

SCSR
(Status register)

└TDRE (Transmit data register empty)

(*b*)

SCDR (Write)
(Transmit data register)

8

(Transmit data shift register)   → Serial output

(*c*)

(Receive data shift register)   ← Serial input

8

SCDR (Read)
(Receive data register)

(*d*)

FIGURE 5.8
Motorola 68HC11 UART's registers. (*a*) Control registers; (*b*) status register; (*c*) transmit data register; (*d*) receive data register.

If it is clear, then this disables the UART from pulling the interrupt line low even though the TDRE bit may well be set.

To send an ASCII string under interrupt control, we need an interrupt service routine which uses a pointer to send the next character in the string. This pointer is stored in RAM when not being used by the 68HC11's CPU. The Motorola assembler directive to allocate 2 bytes of RAM with the label TX_PTR is shown in Fig. 5.9*a*. When the terminator (defined in Fig. 5.9*b*) is accessed, the interrupt service routine sends nothing but clears the TIE bit. This interrupt service routine is shown in Fig. 5.9*c*. To start this sequence, the CPU must load the interrupt service routine's pointer with the starting address of the ASCII string. Then it must set the TIE bit. This is illustrated in Fig. 5.9*d*.

```
TX_PTR: RMB    2                    ;Pointer for UART's transmitter

                                    (a)


EOT     EQU  04H                    ;End of transmission ASCII code

                                    (b)


UART:   BRCLR SCSR,10000000B,U_RX   ;Branch if not transmit interrupt

U_TX:                               ;Serve transmitter
        LDX   TX_PTR                ;Get pointer
        LDAA  0,X                   ;Get character pointed to
        INX                         ;Increment pointer
        STX   TX_PTR                ; and save it for next interrupt
        CMPA  #EOT                  ;End of string?
        BEQ   U_TX_1                ;If so, turn off transmitter
        STAA  SCDR                  ;Otherwise send byte (clearing TDRE)
        RTI                         ; and return from interrupt

U_TX_1: BCLR  SCCR2,10000000B       ;Clear TIE bit to disable interrupts
        RTI                         ; and return from interrupt

U_RX:       .                       ;Handle UART's receiver
            .
        RTI
                                    (c)


        LDD   #MESSAGE_5            ;Get pointer to ASCII string
        STD   TX_PTR                ; and store it in transmitter's pointer
        BSET  SCCR2,10000000B       ;Set TIE bit to initiate first interrupt

                                    (d)
```

FIGURE 5.9
Motorola 68HC11 code for Example 5.9 to transmit a single string. (a) RAM allocation for a
pointer dedicated for use by UART's transmit function; (b) an EQUate for inclusion in the
definitions section of program; (c) interrupt service routine; (d) mainline program sequence of
instructions to initiate transmission.

Note in these routines that the BSET and BCLR instructions (which set
and clear individual bits in an addressed byte) cannot employ extended addressing. And yet we need to set or clear a bit in the SCCR2 UART control
register. Likewise, the BRCLR instruction which tests SCSR cannot employ
extended addressing but requires either indexed addressing or direct addressing. It is for just this reason that the discussion at the beginning of Appendix A
on the Motorola 68HC11 recommends mapping the registers to page 0. When
this is done, the hex address of SCCR2 becomes 002DH and the direct addressing mode can be used with the BSET and BCLR instructions. The hex address
of SCSR becomes 002EH and is again accessible with direct addressing.

Example 5.10. The last example was for practice. In this example we will develop the real thing. This time the UART's transmitter will function identically as in the previous example *until* it sees the terminator at the end of a string. Then, instead of turning off interrupts, the transmitter will determine if the mainline program wants another string transmitted. The mainline program will queue up a *pointer* to each string to be transmitted. When it has sent an entire string, the interrupt service routine will look at the output of this queue to see if it holds a pointer to another string to be transmitted. If the queue is not empty, it transmits the first character of the next string. Only when the queue is empty does the interrupt service routine shut itself down by clearing the TIE bit, which disables further interrupts.

To implement this, we need to allocate RAM for a queue for the UART's transmitter. However, instead of its being a queue for the bytes which will be transmitted, it is a queue for *pointers* to the *strings* which will be transmitted. This RAM allocation is shown in Fig. 5.10*a*.

The modification of the interrupt service routine is almost trivial. It consists of the four lines beginning at the U_TX_1 label in Fig. 5.10*c*.

The mainline program no longer interacts directly with the transmitter's pointer. Now it puts pointers into the transmitter's queue of pointers, as in Fig. 5.10*d*. Note that if the mainline program has to transmit more than four ASCII strings, we must be concerned about overrunning the queue (since the queue is defined in Fig. 5.10*a* to hold a maximum of four pointers).

Example 5.11. Show the mainline program to display the following message on a CRT driven by the Motorola 68HC11's UART:

OUTPUT VOLTAGE = 23.5 VOLTS

We can break this up into three strings, assuming that the number, 23.5, is a measured result. Two strings can be stored as fixed ASCII strings in ROM, with labels ROMMESS_1 and ROMMESS_2, as in Fig. 5.11*a*. The variable string must be formed in a RAM buffer, called RAMMESS and defined in Fig. 5.11*b*. Then assume that somehow we have formed the number 23.5 and have used a subroutine called CONVERT to convert this number to three ASCII digits, pushing them onto the stack, one by one, least-significant digit first. The code in Fig. 5.11*c* shows each of these digits being popped off the stack and put into the RAM buffer, along with the ASCII code for a decimal point and the string terminator. Thus the RAM buffer will contain five ASCII codes:

D D . D T

where D represents a digit, . represents a period, and T represents the string terminator.

Pointers to these three strings are put into the UART transmitter's queue, and the transmitter's interrupts are enabled so that the interrupt service routine will begin by picking up the first character in the first string.

5.4 PROGRAM ORGANIZATION

In this section we will consider a simple scheme for organizing the software for an instrument which is controlled by front panel keyswitches. Me-

```
TX_PTR:  RMB   2                           ;Pointer for UART's transmitter
QUT:     DSQW  4                           ;Four-word queue for pointers for UART transmitter

                                           (a)


EOT      EQU   04H                         ;End of transmission ASCII code
                                           (b)


UART:    BRCLR SCSR,10000000B,U_RX         ;Branch if not transmit interrupt

U_TX:    LDX   TX_PTR                       ;Serve transmitter
         LDAA  0,X                          ;Get pointer
         INX                                ;Get character pointed to
         STX   TX_PTR                       ;Increment pointer
         CMPA  #EOT                         ; and save it for next interrupt
         BEQ   U_TX_1                       ;End of string?
         STAA  SCDR                         ;If so, check queue
         RTI                                ;Otherwise send byte (clearing TDRE)
                                            ; and return from interrupt

U_TX_1:  GET   D,QUT                        ;Get pointer from QUT queue and put it into D
         BEQ   U_TX_2                       ;If queue is empty, turn off interrupts and quit
         STD   TX_PTR                       ;Otherwise copy pointer into transmitter's pointer
         BRA   U_TX                         ; and begin again
```

```
U_TX_2:  BCLR SCCR2,10000000B      ;Clear TIE bit to disable interrupts
         RTI                       ; and return from interrupt


U_RX:    .                         ;Handle UART's receiver
         .
         .
         RTI

                                   (c)


         LDD  #MESSAGE_5           ;Get pointer to ASCII string
         PUT  D,QUT                ; and put it into transmitter's queue
         BSET SCCR2,10000000B      ;Set TIE bit to initiate first interrupt
         LDD  #ANOTHER_MESSAGE     ;Next ASCII string to be sent
         PUT  D,QUT
         .                         ;More of same (but do not overrun queue)

                                   (d)
```

FIGURE 5.10
Motorola 68HC11 code for Example 5.10 to transmit any number of strings concatenated together with the help of a queue. (a) RAM allocation for a pointer and a queue dedicated for use by UART's transmit function; (b) an EQUate for inclusion in the definitions section of program; (c) interrupt service routine; (d) mainline program sequence of instructions to initiate the transmission of concatenated ASCII strings.

```
ROMMESS_1: FCC  "OUTPUT VOLTAGE = "  ;First part of message
           FCB  04H                   ;Terminator
ROMMESS_2: FCC  " VOLTS"              ;Last part of message
           FCB  04H                   ;Terminator
```

(a)

```
RAMMESS:   RMB  11                     ;Reserve memory for variable messages
```

(b)

```
BSR  CONVERT              ;Pushes three ASCII characters on stack
PULA                     ;Get tens digit
STAA RAMMESS             ;First byte of RAMMESS
PULA                     ;Get units digit
STAA RAMMESS+1           ;Second byte of RAMMESS
LDAA #"."                ;Get ASCII code for a period
STAA RAMMESS+2           ;Third byte of RAMMESS
PULA RAMMESS             ;Get tenths digit
STAA RAMMESS+3           ;Fourth byte of RAMMESS
LDAA #EOT                ;Form terminator
STAA RAMMESS+4           :Fifth byte of RAMMESS

LDD  #ROMMESS_1          ;Get pointer to first part of message
PUT  D,QUT               ; and put it into transmitter's queue
BSET SCCR2,10000000B     ;Set TIE bit to initiate first interrupt
LDD  #RAMMESS            ;Get pointer to variable part of message
PUT  D,QUT               ; and put it into transmitter's queue
LDD  #ROMMESS_2          ;Get pointer to last part of message
PUT  D,QUT               ; and put it into transmitter's queue
 .                       ;Done! Go do other things
 .
```

(c)

FIGURE 5.11
Motorola 68HC11 code for Example 5.11 to send concatenated strings to a terminal. (a) Fixed messages stored in ROM; (b) allocation of RAM to store variable part of a message; (c) code to form variable ASCII string and then to initiate the transmission of the three-part message.

chanical keyswitches (and other electromechanical devices) are relatively slow, by microcontroller standards. They have keybounce associated with them whereby a pressed keyswitch will usually exhibit closure, open, closure, open, . . . ,etc., before settling to the steady closed state. With a typical keybounce time of less than 10 ms, a keyswitch can be interrogated once every 10 ms with the assurance that if it was in the middle of keybounce the last time it was interrogated, then it will have settled out by this time.

Other electromechanical devices have response times in the same range. For example, a stepper motor* which can be stepped at 100 steps per second

---

* Discussed in Chap. 7.

can handle a new command to take a step after each 10-ms interval. Again, this is orders of magnitude slower than the time it takes a microcontroller to execute instructions.

In light of this disparity between the time needed to do electromechanical things and the time needed to do microcontroller things, we can make good use of a *tick clock* to help with the timing of electromechanical activities. By this we mean a real-time clock* or programmable timer which generates an interrupt at equal intervals, perhaps every 10 ms. Its TICK_CLOCK interrupt service routine can control the timing for many electromechanical devices. As soon as it has reset the interrupt in the programmable timer to occur again in another 10 ms, it can reenable interrupts for *all* devices.† Then it can continue to handle its tick-clock jobs at a more leisurely pace than is usual for an interrupt service routine. It need only complete its activities sometime before the next tick occurs.

This system organization is depicted in Fig. 5.12. It shows an interrupt service routine which does two electromechanical jobs. It interrogates keyswitches with an algorithm called NEWKEY?.‡ If a keyswitch is newly pressed, then NEWKEY? will return a flag to indicate this. If the flag represents a yes answer, then the CPU puts the keycode for the key pressed into a queue called QKEYCODE. The mainline program shown in Fig. 5.12 includes a block for parsing keycodes.§ This means that it reads the output of the queue, interpreting *sequences* of keyswitch inputs and then taking whatever action is required.

> **Example 5.12.** If we use a stepper motor as an actuator, and if we want to have it step at a 100 step per second rate when it steps, then the organization shown in Fig. 5.12 will serve us well. We could define a 2-byte, signed variable, NUMSTEPS, which is used by the STEP subroutine shown there. If the PARSE_KEYS subroutine wants the stepper motor to take 35 steps counterclockwise, then it simply executes the command
>
>   LD   NUMSTEPS,#(−35)
>
> Then the job of STEP, each tick time when it is called, is to look at NUMSTEPS. If its content is 0, then it does nothing. If its content is positive, then it decrements NUMSTEPS and takes one clockwise step. If its content is negative, then it increments NUMSTEPS (toward 0) and takes one counterclockwise step.

For tasks which need to be timed at an even slower pace, we can organize around a tick-time counter. Its job is to make events happen during every *n*th tick time.

---

* See Sec. 3.4.

† Including itself. This eliminates any need to shuffle priorities from what they were, making it simple to implement with both the 68HC11 and the 8096.

‡ Discussed in Sec. 5.5.

§ Discussed in Sec. 5.6.

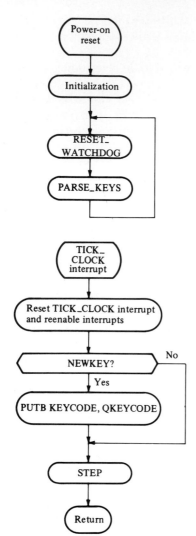

FIGURE 5.12
Use of a tick-clock interrupt service routine.

Example 5.13. Suppose we want to use a stepper motor which has a heavier inertial load than the motor discussed earlier. Because of this load, we want to limit the stepping rate to 33⅓ steps per second (i.e., to a step every third tick time). Define a 1-byte variable, called COUNT_THREE_TICKS, which is incremented every tick time by a subroutine called THIRD_TICK?. It counts

···· 0 1 2 0 1 2 0 1 2 ····

Furthermore, THIRD_TICK? returns a flag to indicate when the counter is reset to the 0 state. Given this operation, the resulting flowchart is shown in Fig. 5.13.

As these examples have shown, a tick-clock interrupt service routine provides an easy way to have slow events take place when they should. By reenabling

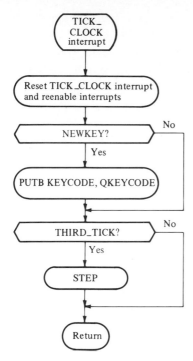

FIGURE 5.13
Use of a tick counter to control events which occur at an integral multiple of the tick time.

interrupts during its execution, it does not get in the way of any other interrupt service routines.

5.5 STATE MACHINES

A recurring problem arises in the development of microcontroller software. We want to use an interrupt service routine to execute an algorithm which extends over many interrupts. To do this, we need a *state machine*; that is, a sequencing algorithm. We will develop this algorithm using as an example the NEWKEY? subroutine alluded to in the last section.

What follows is a *powerful* approach for handling state machine complexity in an organized manner. It separates the tests and the actions to be taken from the sequencing mechanism. In so doing, it provides a means for implementing state machines which can be built, debugged, and upgraded with high reliability.

Before we get ankle-deep in the intricacies of the example which we will use as a vehicle for developing the approach, it is worthwhile to consider the features of the approach:

> It permits the *complexity* of an algorithm to be broken apart systematically. We can concentrate on each part, *independent* of the other parts.

Each test is implemented as a subroutine. As such, it can have arbitrary complexity and it can be independently tested.

Each action is also implemented as a subroutine. Again, this means that actions can have arbitrary complexity and can be independently tested.

The *sequencing* from state to state, making tests and carrying out required actions, is implemented in a short table-driven routine. The instructions for this routine are independent of the state machine being implemented.

The actual table-driven routine which we shall examine assumes that a state machine has at most

Eight states

Eight different tests

Eight different actions

These constraints arise because of the use of a *state table* to represent the algorithm. We will use a state table having 16-bit entries, each of which will be divided into five fields of three bits each (with one bit left over). By changing to a state table having 32-bit entries, we *could* deal with state machines having up to

Sixty-four states

Sixty-four different tests

Sixty-four different actions

We would find that the routine which uses this new table is no more complex than that which we develop here. However, it should be pointed out that the large majority of applications for state machines in microcontrollers fit the constraints which we will use here. A little state machine complexity lets us do a lot!

Example 5.14. Consider the array of keyswitches in Fig. 5.14, shown connected to port 1 of an Intel 8096 microcontroller. While they are shown arrayed in a square, they actually represent up to 16 momentary-action keyswitches laid out in an arbitrary arrangement on the front panel of a device or instrument. If more keyswitches are needed, the scheme would be changed very little. We would add a few more lines from another port to form as large a square array of keyswitches as is needed. For example, with four additional lines on another port, we can handle up to 36 keyswitches.

The scheme we use to examine the keyswitches assumes that we have a bidirectional port, so that at one time we may drive some lines as outputs while at other times we will read these same lines as inputs. If a bidirectional port were not available to us, then we would have to change a couple of the subroutines which deal with the keyswitches, but the state machine (i.e., the sequencing mechanism) would remain unchanged.

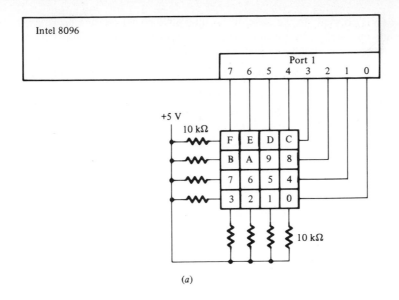

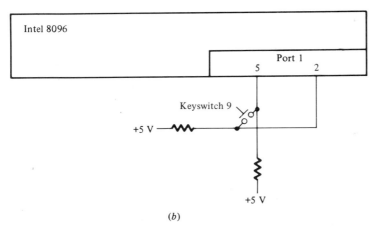

FIGURE 5.14
Keyswitch connections to microcontroller. (*a*) Overall circuit; (*b*) connection to keyswitch 9.

As shown in Fig. 5.14, each switch does nothing more than close a contact between a column wire and a row wire. The 10-kΩ pullup resistors ensure that each line on port 1 will see a 1 if it is set up as an input and if no keyswitch is pressed.

To test to see if no keyswitch is pressed, we can set up bits 7 . . . 4 as outputs, bits 3 . . . 0 as inputs, and then drive bits 7 . . . 4 low. Now if we read the port, we should see

00001111

if no keyswitch is pressed. In the case of the Intel 8096 port 1, setting up which lines are inputs and which are outputs is easy. Just write 1s to any lines which should be inputs, as was discussed in Sec. 2.6. That is, write

$$00001111$$

to the port to drive bits 7 . . . 4 low and to have lines 3 . . . 0 be inputs. Note that what we write to a port and what we read back from the same port are not necessarily the same when some lines are serving as inputs.

Implementing an arbitrary state machine using our *table-driven* approach begins with a *state diagram* which shows:

Each state (represented by a rectangle)

The test called for, corresponding to each state (represented by a Yes/No box)

The action to be taken, given a specific state and the result of the corresponding test (represented by an oval)

Example 5.15. In this example, we will consider the state machine diagram which we will use for debouncing keyswitches, shown in Fig. 5.15. We are going to end up with a table-driven subroutine called NEWKEY? which uses a variable called KEYSTATE to go from state to state in this diagram. That is, each time we call NEWKEY?, that subroutine looks at the value of KEYSTATE and decides what to do. For example, if KEYSTATE = 1 upon entry to NEWKEY?, then it executes a test subroutine which determines whether the same keyswitch which was pressed during the last call of NEWKEY? has been released. If so, it changes KEYSTATE to 0 and then clears the carry bit (which serves as a flag bit back to the calling routine). If not, it changes KEYSTATE to 2 and sets the carry bit. This setting of the carry bit informs the calling routine that a keyswitch has been pressed and that its keycode value is stored in a variable called KEYCODE. Figure 5.14*a* shows the hexadecimal keycode value which will be returned in response to the pressing of each of the 16 keys.

Notice, in Fig. 5.15, that this is a four-state machine. Even with so few states, we are able to get rather sophisticated performance. Before a keyswitch is treated as being pushed, we have to see it pressed *twice* in succession. Likewise, to see the release of a keyswitch, we have to see it released *twice* in succession. It is state machines like this which can actually make "a silk purse out of a sow's ear!" Some low-cost membrane switches look erratic while being pressed. They may look closed *most* of the time, but if you roll your finger around on the switch (not release it, just roll it around), it will release momentarily. Without the help of a good algorithm to handle the keyswitches, such a keyswitch can *look* like it has been pressed two or more times when, in fact, it has only been pressed once—by someone who had slow, rolling fingers. The change in the algorithm to handle this requires the addition of more states after state 3 which kick back to state 2 if a keyswitch is seen as still pressed. For example, to call a keyswitch released, we can require that it be seen as released during four successive calls of NEWKEY?.

Example 5.16. The translation of the state machine diagram of Fig. 5.15 into a table-driven algorithm begins with the allocation of RAM variables, as in Fig.

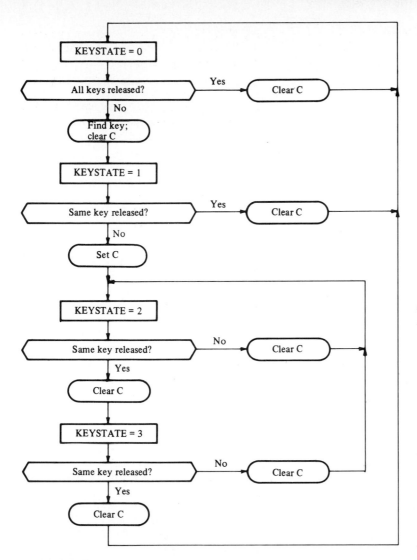

FIGURE 5.15
State machine diagram for keyswitch debouncing.

5.16*a*. The variable KEYSTATE is a 2-byte variable even though it takes on only one of four values. We do this so that in an implementation for the Intel 8096, we can use KEYSTATE as a variable index (into KEYSTATETABLE, defined in Fig. 5.16*d* and *e*). In this simple example, we have only two tests. They have the names listed in Fig. 5.16*b*. The Intel 8096 assembler permits question marks in labels, which helps us identify the sense of each test (i.e., what Yes and No mean). Our definition of state machine operation requires that each test subroutine return the carry bit set if the answer to the test question is Yes.

Notice in Fig. 5.16*b* that each test has been assigned a number (0 and 1 in

```
KEYSTATE:  DSW  1                              ;NEWKEY?'s state variable
KEYCODE:   DSW  1                              ;Identifies which key was pressed
                                 (a)
KEYTEST:   DW   ALL_KEYS_RELEASED?  ;Test 0 — C=1 if all keys are released
           DW   SAME_KEY_RELEASED?  ;Test 1 — C=1 if same key is released
                                 (b)
KEYACTION: DW   CLR_C                  ;Action 0 — Returns C=0
           DW   SET_C                  ;Action 1 — Returns C=1
           DW   FIND_KEY_AND_CLR_C  ;Action 2 — Loads KEYCODE; returns C=0
                                 (c)
```

Present state	Test	Yes answer (C = 1)		No answer (C = 0)	
		Action	Next state	Action	Next state
0	0	0	0	2	1
1	1	0	0	1	2
2	1	0	3	0	2
3	1	0	0	0	2

(d)

```
KEYSTATETABLE:  DW   00210Q          ;Entry for present state = 0
                DW   00121Q          ;Entry for present state = 1
                DW   03021Q          ;Entry for present state = 2
                DW   00021Q          ;Entry for present state = 3
                           (e)
```

FIGURE 5.16
RAM allocation and ROM tables for NEWKEY?, the subroutine which services keyswitches.
(a) Allocation of RAM for two variables; (b) jump table of addresses of subroutines which
execute tests; (c) jump table of addresses of action subroutines; (d) KEYSTATETABLE entries; (e) KEYSTATETABLE.

our case). What is shown in this figure is a *table* of the addresses where each of the
test subroutines begin. This will be used by the NEWKEY? instructions as a jump
table to get to the right test. Consequently, the numbering of the tests corresponds
to their entry number in this table (i.e., 0, 1, 2, 3, 4, . . .). This lets us make a
compact KEYTEST table of addresses which we can index into easily.

In a similar vein, the KEYACTION table holds the addresses of all of our
action subroutines. Again, the action subroutines are numbered 0, 1, 2, 3, 4, . . .
according to their entry number in this table.

Now we are in a position to translate the state machine diagram of Fig. 5.15
into the KEYSTATETABLE of Fig. 5.16d and e. Again, consider that we enter
NEWKEY? with KEYSTATE = 1. We see from Fig. 5.15 that the test to run is
SAME KEY RELEASED? This is test 1, as listed in the KEYTEST jump table of
Fig. 5.16b. It is entered into the table of Fig. 5.16d under "Test," on the "Present
State = 1" row. The other entries on this same row of Fig. 5.16d are filled in by

picking up the "Next state" numbers for Yes and No answers to the test from Fig. 5.15. The "Action" numbers are filled in by translating the actions called for in Fig. 5.15 for Yes and No answers to the test into the action numbers listed in the KEYACTION jump table of Fig. 5.16c.

The actual KEYSTATETABLE of Fig. 5.16e is formed by assigning a 16-bit word to each entry. Each entry is divided into five 3-bit fields, one field for each column of Fig. 5.16d to the right of the vertical line. The test field is listed last, to simplify the use of the table. By coding each entry in *octal* code, we can simply write the numbers from the table in Fig. 5.16d into the DW (define word) assembler directives in Fig. 5.16e. The assembler knows to interpret these entries as octal numbers because of the Q appended to the end of each entry.

Example 5.17. Carrying on with the keyswitch state machine, we are now ready to consider the NEWKEY? subroutine itself, shown in Fig. 5.17. This is a subroutine which sits in the tick-clock interrupt service routine and which is called once every 10 ms or so. The PARSE_KEYS subroutine in the mainline program of Fig. 5.12 is continually asking whether there is any keyswitch input so that it can make the appropriate response when there is. Each time NEWKEY? is called with no keyswitch pressed, the subroutine is entered with KEYSTATE = 0. It runs the ALL_KEYS_RELEASED? subroutine and finds that indeed all keyswitches are released. It exits with KEYSTATE still equal to 0 and a cleared carry flag. The tick-clock routine tests the carry flag and realizes that it has nothing to put into the queue for PARSE_KEYS in the mainline program.

The NEWKEY? subroutine, written for the Intel 8096, uses the same Convention II of Sec. 4.6 that Intel recommends (and which their PL/M compiler uses). That is, it assumes that its caller expects the general purpose registers of Fig. 4.27 (defined by convention, not by hardware) to be used and possibly changed. Consequently, NEWKEY? makes no attempt to set aside AX and BX at the beginning of the routine and restore them at the end. Rather, it protects *itself* before calling the selected test subroutine by pushing AX before that call and popping AX back again afterward, as we shall see.

NEWKEY? begins by copying KEYSTATE into AX, doubling it, and using it to form an index into KEYSTATETABLE. Doubling is necessary since the table has two addresses assigned to each 2-byte entry. NEWKEY? loads this table entry into AX. Then it forms

$$2 * \langle \text{test number} \rangle$$

in BX to access the KEYTEST jump table. It does this by masking off all bits other than bits 2 to 0 (refer to Fig. 5.17b) and shifting them to bit positions 3 to 1, filling in all the other bits with 0s. Indexed addressing then loads the *address* of the required test subroutine into BX. The KEYSTATETABLE entry in AX is next shifted right three places (throwing away the test bits) and saved on the stack.

We want to execute a subroutine call to the test routine whose number now resides in BX. Unfortunately, the 8096 does not have an indirect subroutine call instruction. However, we can make one (we could even make a macro out of it!) by first pushing the return address (labeled NEWKEY1) onto the stack and then using the 8096's indirect jump instruction, JP[ ].

We pushed AX onto the stack and will restore it after the return from the

```
;;;;;;;;;  NEWKEY? SUBROUTINE  ;;;;;;;;;;;;;;;;;;;;;;;;;;;;;;;;;;;;;;
;
; This subroutine uses a state machine to service front-panel keyswitches. It must be called no more
; frequently than the keybounce time of the keyswitches. It debounces the switches.
; It returns C=1 and a valid KEYCODE value once per key press.
; It returns C=0 otherwise.
; It uses the AX and BX general purpose registers and assumes that the calling routine has saved
; their content (if their content needed to be saved).
;;;;;;;;;;;;;;;;;;;;;;;;;;;;;;;;;;;;;;;;;;;;;;;;;;;;;;;;;;;;;;;;;;;;;

NEWKEY?:    LD   AX,KEYSTATE                ;Get KEYSTATE and double it to serve as an offset
            SHL  AX,#1                      ; into a table of words
            LD   AX,KEYSTATETABLE[AX]       ;Get state table entry

            AND  BX,AX,#0000007Q           ;Get test number, to call test subroutine
            SHL  BX,#1                      ;Mask off all but test number in bit positions 2..0
            LD   BX,KEYTEST[BX]             ; and shift it to bit positions 3..1 (doubling it)
            SHR  AX,#3                      ;Load BX with address of subroutine
            PUSH AX                         ;We are done with the test field; dump it,
                                            ; preserving remainder of table entry
            PUSH #NEWKEY1                   ;Now create an indirect subroutine call
            JP   [BX]                       ;Push return address onto stack
                                            ; and jump indirect to subroutine which executes test

NEWKEY1:    POP  AX                         ;Restore AX
            JNC  NEWKEYNO                   ;If C=0, test was answered No
NEWKEYYES:  SHR  AX,#6                      ;If C=1, test was answered Yes
                                            ; so move Yes table entries to bit positions 5..0
```

```
NEWKEYNO:  AND  BX,AX,#00007Q     ;Update KEYSTATE with next state
           ST   BX,KEYSTATE       ;Mask off all but next state number
                                  ;Store next state in KEYSTATE

           AND  BX,AX,#000070Q    ;Execute action subroutine
           SHR  BX,#2             ;Mask off all but action number
           LD   BX,KEYACTION[BX]  ;Shift right only two places to double its value
           JP   [BX]              ;Load BX with address of subroutine
                                  ;Jump indirect to action subroutine
                                  ; whose return from subroutine will return to caller
                                  ; of NEWKEY?
```

(a)

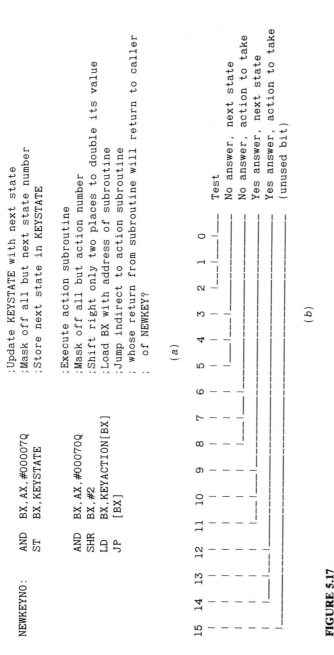

```
15  14  13  12  11  10  9   8   7   6   5   4   3   2   1   0
|___|___|___|___|___|___|___|___|___|___|___|___|___|___|___|___|
                                                              Test
                                                         No answer, next state
                                                    No answer, action to take
                                               Yes answer, next state
                                          Yes answer, action to take
                                     (unused bit)
```

(b)

FIGURE 5.17
NEWKEY? subroutine. An Intel 8096 table-driven state machine to service keyswitches. (a) The subroutine; (b) meanings of bit positions in table entry taken from KEYSTATETABLE.

205

test subroutine, in order to protect its content from change by the test subroutine. Since the POP instruction leaves the flags alone, the flag passed back from the test subroutine in the carry bit will still be intact.

NEWKEY? responds to the carry flag returned by the test subroutine by picking which set of 6 bits in the state table entry to work with during the rest of the routine. If the test response were No ($C = 0$), then bits 5 to 0 of AX hold the values for the next state and for the action to be taken. On the other hand, if the test response were Yes ($C = 1$), then bits 11 through 6 of AX hold these values. In this case a shift of six positions to the right permits the rest of the subroutine to proceed with the correct values in *either* case by working with bits 5 to 0.

The next state is extracted from this 6-bit field and stored back into KEY-STATE. The action subroutine is called in much the same way as the test subroutine was called. However, instead of pushing a return address onto the stack before doing an indirect jump to the action subroutine, this time we execute the indirect jump by itself. Then when the action subroutine has been completed, instead of returning to NEWKEY?, it will return to the routine which *called* NEWKEY?.

At this point, we have developed the entire table-driven state machine algorithm. It can be used in arbitrary applications. While the version shown in Fig. 5.17 is dedicated to keyswitch debouncing, that entire algorithm could be written as a *macro* and then recast for use with a different state table, different test subroutines, and different action subroutines. Alternatively, it could be written as a subroutine, perhaps passing the addresses of the three tables involved to the subroutine on the stack.

To continue with our specific example of keyswitch debouncing, consider the KEYCODETABLE shown in Fig. 5.18. This table is used in two ways. It is used by the SAME_KEY_RELEASED? subroutine to sense whether the specific keyswitch identified by KEYCODE is pressed or released. It is also used by the FIND_KEY_AND_CLR_C subroutine when a keyswitch is newly pressed to identify which keyswitch it is.

> **Example 5.18.** For each keyswitch of Fig. 5.14a, the KEYCODETABLE of Fig. 5.18 has a corresponding entry for which the row and column bit positions hold 0s. All of the remaining bit positions hold 1s. For example, keyswitch 9 is accessed via bit 2 to select a row and bit 5 to select a column, as shown in Fig. 5.14b. Appended to the end of the table is an entry of all 1s which will be used to support a graceful recovery when we look for which keyswitch is pressed and do not find a match with any of the table entries, for whatever reason.

Next we come to a discussion of the two test subroutines, ALL_KEYS_RE-LEASED? and SAME_KEY_RELEASED?. They are listed in Fig. 5.19.

> **Example 5.19.** The ALL_KEYS_RELEASED? subroutine has a quick job to do, with a little four-instruction subroutine. It first drives the column lines low (refer to Fig. 5.14a again) on the keyswitch array. At the same time, it writes 1s to the row lines which, for port 1 of the 8096 microcontroller, sets these lines up as

```
KEYCODETABLE:    DB   11101110B        ;KEYCODE=0000H
                 DB   11011110B        ;KEYCODE=0001H
                 DB   10111110B        ;KEYCODE=0002H
                 DB   01111110B        ;KEYCODE=0003H
                 DB   11101101B        ;KEYCODE=0004H
                 DB   11011101B        ;KEYCODE=0005H
                 DB   10111101B        ;KEYCODE=0006H
                 DB   01111101B        ;KEYCODE=0007H
                 DB   11101011B        ;KEYCODE=0008H
                 DB   11011011B        ;KEYCODE=0009H
                 DB   10111011B        ;KEYCODE=000AH
                 DB   01111011B        ;KEYCODE=000BH
                 DB   11100111B        ;KEYCODE=000CH
                 DB   11010111B        ;KEYCODE=000DH
                 DB   10110111B        ;KEYCODE=000EH
                 DB   01110111B        ;KEYCODE=000FH
KCT_END:         DB   11111111B        ;"No match" table entry
```

FIGURE 5.18
KEYCODETABLE used by both SAME_KEY_RELEASED? subroutine and
FIND_KEY_AND_CLR_C subroutine.

inputs. Now, if no keyswitch is pressed, then reading back port 1 will produce all
1s in the row bits because of the pullup resistors to the left of the keyswitch array
in Fig. 5.14*a*. In this case, the CPU will read back

00001111

On the other hand, if one of the keyswitches is pressed, then one of the row lines
will be pulled low by the 0 on one of the column lines. For example, if keyswitch 9
is pressed, then the CPU will read back

00001011

since bit 2 has been pulled low by the output from pin 5 (which pulls low much
harder than the pullup resistor pulls high). The instruction

```
        CMPB   AL,#00001111B
```

subtracts

00001111

from one of these two numbers. If the subtraction is

00001111–00001111

then no borrow is generated. In this case, the Intel 8096 *sets* the carry bit. That is,
C = 1 when *no* keyswitch is pressed (i.e., if *all* keyswitches are released). The
subtraction resulting when keyswitch 9 is pressed is

00001011–00001111

In this case a borrow *is* generated and the carry bit is cleared to zero, C = 0. In
this simple fashion, we can tell very quickly if a keyswitch is pressed.

```
;;;;;;;;;;   ALL_KEYS_RELEASED? SUBROUTINE  ;;;;;;;;;;;;;;;;;;;;;;;;;;;;;;;;;;;;;;;
;
;  This subroutine checks the keyswitches of Fig. 5.14
;  It returns C=1 if all keys are released. It returns C=0 if any key is pressed.
;  It uses the AL general purpose register and assumes that the calling routine
;  has saved its content (if the content needed to be saved).
;
;;;;;;;;;;;;;;;;;;;;;;;;;;;;;;;;;;;;;;;;;;;;;;;;;;;;;;;;;;;;;;;;;;;;;;;;;;;;;;;;;;;;

ALL_KEYS_RELEASED?:
          LDB  P1,#00001111B ;Drive all column lines low
          LDB  AL,P1         ;Read back from port 0000XXXX
          CMPB AL,#00001111B ;Generate borrow, and make C=0, if any key is pressed;
                             ; otherwise make C=1 if all keys are released
          RET                ;Return
```

(a)

```
;;;;;;;;;;;;   SAME_KEY_RELEASED? SUBROUTINE   ;;;;;;;;;;;;;;;;;;;;;;;;;;;;;;;;;;;;;;;;;
;
;  This subroutine checks the single switch identified by KEYCODE.
;  It returns C=1 if that key has been released.
;  It returns C=0 if that key is still pressed.
;  It uses the AX (i.e., AH and AL) general purpose register and assumes that the
;  calling routine has saved its content (if the content needed to be saved).
;
;;;;;;;;;;;;;;;;;;;;;;;;;;;;;;;;;;;;;;;;;;;;;;;;;;;;;;;;;;;;;;;;;;;;;;;;;;;;;;;;;;;;;;

SAME_KEY_RELEASED?:
          LDB  AH,KEYCODETABLE[KEYCODE]   ;Get table entry, to serve as a mask
          LDB  AL,AH         ;Make copy to fiddle with in AL
          ORB  AL,#00001111B ;Drive all lines of P1 high, except for
          LDB  P1,AL         ; the selected column bit

          NOTB AH            ;Form a mask with only
          ANDB AH,#00001111B ; selected row bit set
          ANDB AL,P1,AH      ;Read P1 and mask all bits other than selected row bit
          CMPB AL,AH         ;Generate borrow and make C=0 if key is pressed;
                             ; otherwise make C=1 if selected key is released
          RET                ;Return
```

(b)

FIGURE 5.19
KEYTEST subroutines. (a) ALL_KEYS_RELEASED? subroutine; (b) SAME_KEY_RE-
LEASED? subroutine.

Example 5.20. The SAME_KEY_RELEASED? subroutine carries out a test on port 1 using the KEYCODETABLE of Fig. 5.18. First the entry in the table pointed to by KEYCODE is obtained. This is written out to port 1 after being modified so that the column bits reflect the table entry while the row bits are written with 1s. The single column holding the keyswitch to be tested will be driven low. Any switches in that column which are pressed will drive their corresponding row lines low. If keyswitch 9 is pressed and if the column driven by bit 5 of port 1 is being tested, then bit 2 of port 1 will be pulled low. On the other hand, if keyswitch 9 is not pressed, then the pressing of no other (single) keyswitch in the entire array will lead to bit 2 being pulled low. Consequently, we now only need to test the value read on bit 2 (if we are testing keyswitch 9).

To do this, first a mask is formed in AH which will be

$$00000100$$

Then port 1 is read and ANDed with this mask to form

$$00000000$$

if keyswitch 9 is pressed, or

$$00000100$$

if keyswitch 9 is not pressed. Finally the mask is subtracted from this result. If keyswitch 9 is pressed, then this subtraction,

$$00000000-00000100$$

produces a borrow, which (in the 8096) clears the carry flag, $C = 0$. If keyswitch 9 is released, then $C = 1$ is produced.

We complete this keyswitch handler with the three action routines whose addresses are kept in the KEYACTION jump table of Fig. 5.16c. Two of these, CLR_C and SET_C, are two liners, and are listed in Fig. 5.20a and b. They exist because our state machine structure imposes the requirement that *all* actions, no matter how simple, must be relegated to action subroutines. A discussion of the interesting third action subroutine follows.

Example 5.21. The FIND_KEY_AND_CLR_C action subroutine, listed in Fig. 5.20c, first of all translates the pressing of a single keyswitch into a number which will match that keyswitch's entry in KEYCODETABLE. For example, the subroutine translates the pressing of keyswitch 9 into the number

$$11011011$$

in the AL register. This is done in the first five instructions of the subroutine, as follows. First

$$00001111$$

is written to the port, driving all column lines low and establishing the row lines as inputs to the port. Note that keyswitch 9 being pressed now makes its row line go low. The second instruction thus reads back

$$00001011$$

The third and fourth instructions write

$$11111011$$

to port 1, establishing the column lines as inputs. The fifth instruction reads

$$11011011$$

back from the port because keyswitch 9 connects row 2 (being driven low by the port output on bit 2) to column 5.

The next part of the subroutine compares this number with each entry in KEYCODETABLE, stopping if no match has been found. Since the pointer used to access the table is the actual ROM address of a table entry, the end of the subroutine converts this to a number between 0 and 15 if one of the keyswitches was found to be pressed and 16 otherwise. The subroutine also clears the carry bit since the action routines are the last thing done by NEWKEY? before it returns to the TICK_CLOCK interrupt service routine with a flag in the carry bit. The NEWKEY? state machine is designed so that it *never* says a keyswitch is pressed until it has first been identified by FIND_KEY_AND_CLR_C and verified 10 ms or so later by SAME_KEY_RELEASED? Consequently, the FIND_KEY_AND-_CLR_C subroutine *never* returns the carry bit set, even though it usually finds a pressed key.

Before we leave this keyswitch handler, we should look at several remaining issues:

How does it handle keybounce?

What does it do if more than one keyswitch is pressed at the same time?

Example 5.22. To consider the effect of keybounce, refer back to the state machine diagram of Fig. 5.15. We see that when a keyswitch is first pressed, while in state 0, the FIND_KEY_AND_CLR_C subroutine is called right then. If the keyswitch is exhibiting keybounce, this action subroutine may indeed not find a keyswitch pressed, putting the number 0010H into KEYCODE (i.e., the offset to the 11111111B entry at the end of KEYCODETABLE). In spite of this, the state machine progresses to state 1. Ten milliseconds later, when NEWKEY? is called again, it will ask whether the same key is still pressed or has been released. To do this, the SAME_KEY_RELEASED? subroutine uses the value

$$11111111$$

taken from the table to drive a selected row bit low. With 1s in all the row bit positions, *no* row line will get driven low. Accordingly, when the port is read again, this same all-1s value will be read. The algorithm will proceed to set the carry bit, indicating that the "selected" key has been released.

Looking back at the state machine diagram of Fig. 5.15, we see that keybounce can have the effect of making the state machine "hiccup" to state 1. Ten milliseconds later, it returns to state 0. Ten milliseconds later it gets the right value for the keycode, going to state 1 again. Finally 10 ms later, it sets the carry flag and the mainline program takes appropriate action. Consequently, the keyswitch might not be noted until 30 ms after it is pressed. For the front panel of an instrument, this is still much faster than needed, even when used by a fast-fingered operator.

```
;;;;;;;;;;   CLR_C SUBROUTINE   ;;;;;;;;;;;;;;;;;;;;;;;;;;;;;;;;;;;;;;;;;;;;;;
;
; This subroutine clears the carry flag to pass back to the caller of NEWKEY?.
;;;;;;;;;;;;;;;;;;;;;;;;;;;;;;;;;;;;;;;;;;;;;;;;;;;;;;;;;;;;;;;;;;;;;;;;;;;;;;;
CLR_C:      CLRC
            RET
```

(a)

```
;;;;;;;;;;   SET_C SUBROUTINE   ;;;;;;;;;;;;;;;;;;;;;;;;;;;;;;;;;;;;;;;;;;;;;;;
;
; This subroutine sets the carry flag to pass back to the caller of NEWKEY?.
; Caller will expect the code for the key pressed to be in KEYCODE.  It will be.
;;;;;;;;;;;;;;;;;;;;;;;;;;;;;;;;;;;;;;;;;;;;;;;;;;;;;;;;;;;;;;;;;;;;;;;;;;;;;;;
SET_C       SETC
            RET
```

(b)

```
;;;;;;;;;;   FIND_KEY_AND_CLR_C SUBROUTINE   ;;;;;;;;;;;;;;;;;;;;;;;;;;;;;;;;;;
;
; This subroutine tests the keyswitches.
; If it finds one pressed, it identifies it with a number stored in KEYCODE.
; If it does not find one pressed, it loads KEYCODE with a pointer to 11111111B.
; This will let the algorithm straighten itself out, even though the
; ALL_KEYS_RELEASED? subroutine said a key was pressed.
; The subroutine returns with the carry bit cleared.
; It uses the AX and BX general purpose registers and assumes that the calling
; routine has saved their content (if their content needed to be saved).
;;;;;;;;;;;;;;;;;;;;;;;;;;;;;  ;;;;;;;;;;;;;;;;;;;;;;;;;;;;;;;;;;;;;;;;;;;;;;;
FIND_KEY_AND_CLR_C:
            LDB   P1,#00001111B    ;Drive all columns low
            LDB   AL,P1            ;Read back from port 0000XXXX
            ORB   AL,#11110000B    ;Set columns to be inputs
            LDB   P1,AL            ;Write out all 1s to port except for selected row
            LDB   AL,P1            ;This will have 0s in selected row and column only

            LD    BX,#KEYCODETABLE ;Use BX to index into KEYCODETABLE of Fig. 5.18
FIND_KEY1:  CMPB  AL,[BX]+         ;Compare with each possibility
            JE    FIND_KEY2        ;Found it!
            CMP   BX,#KCT_END      ;Tried all possibilities?
            JNE   FIND_KEY1        ;No, so try the next one

            LD    KEYCODE,#KCT_END-KEYCODETABLE   ;No match; point to 11111111B
            SJMP  FIND_KEY3        ;Clear carry bit and return

FIND_KEY2:  SUB   KEYCODE,BX,#(KEYCODETABLE+1) ;Form number of keycode
FIND_KEY3:  CLRC                   ;Clear carry bit
            RET                    ; and return
```

(c)

FIGURE 5.20
KEYACTION subroutines. (a) CLR_C subroutine; (b) SET_C subroutine; (c) FIND_KEY-
_AND_CLR_C subroutine.

Example 5.23. To discover how this algorithm responds when a second keyswitch is pressed before a first keyswitch has been released, consider Fig. 5.14 again. If keyswitch 9 is pressed and noted by NEWKEY?, which puts 0009H into KEYCODE, then when another keyswitch is pressed, it will not be noted as long as keyswitch 9 is still pressed. When keyswitch 9 is released, the state machine reverts to state 0 and the new keyswitch is then picked up. This is what is called *two-key-rollover* capability. It provides an excellent way to handle instrument front-panel keyswitches. For an instrument with keyswitches mounted so close together that a slightly off-target finger presses one keyswitch and barely presses another one also, even if the second keyswitch's contacts close, the closure will probably occur after that for the first keyswitch and will be released before the first keyswitch is released. In this event the grazed keyswitch goes unnoticed by the instrument, just as we would want it to be.

Before leaving this section on state machines, we might do well to look back upon what we have done. The table-driven state machine approach has the benefit of breaking a complex algorithm down into manageable parts. It all begins with the definition of the algorithm with a state machine diagram, analogous to that of Fig. 5.15. After this, the algorithmic development process becomes compartmentalized. We can worry about the pieces, one at a time, with calm assurance that when the pieces are developed and debugged, the whole algorithm will work.

This compartmentalization does not come without a price. If we are hard pressed for CPU time to accomplish all that needs to be done, then there are faster ways to organize all our tasks.. For example, most of the time the keyswitch handler determines that no keyswitch is pressed. However, it does this every 10 ms or so. We could determine the percentage of CPU time taken by the keyswitch handler when no keyswitch is pressed. If this small percentage of time is more than we want to spend *every* 10 ms, then we can modify the table-driven approach and speed it up considerably.

Example 5.24. Without modifying NEWKEY? one iota, we can *use* it in such a way that its response is speeded up when no keyswitch is pressed. Now, the TICK_CLOCK interrupt service routine calls a new subroutine which we will name FASTNEWKEY?. This subroutine, shown in Fig. 5.21, *immediately* calls the ALL_KEYS_RELEASED? subroutine. If a keyswitch is pressed, or if we have not finished servicing a previously pressed keyswitch (i.e., if KEYSTATE <> 0), then a jump to NEWKEY? is taken. (The return from the NEWKEY? subroutine is back to the TICK_CLOCK interrupt service routine.) Otherwise, the carry bit is cleared (passing back the proper parameter) followed by a return to the TICK_CLOCK interrupt service routine.

Note that we have not lost anything that we have done to create the table-driven state machine approach to the problem. We get the benefit of compartmentalization while writing and debugging the code. And yet with virtually no extra effort, we can end up with a fast version if we need it.

The emphasis of this section has been upon the use of a state machine algorithm to serve as a keyswitch handler. While it makes the real-time clock

```
;;;;;;;;;;   FASTNEWKEY? SUBROUTINE   ;;;;;;;;;;;;;;;;;;;;;;;;;;;;;;;;;;;;;;;;;;;
;
; This subroutine speeds up keyswitch servicing when no key is pressed.
; When it finds a pressed key, it reverts to the more thorough treatment given
; the keyswitches by NEWKEY?.
;
;;;;;;;;;;;;;;;;;;;;;;;;;;;;;;;;;;;;;;;;;;;;;;;;;;;;;;;;;;;;;;;;;;;;;;;;;;;;;;;;;;

FASTNEWKEY?: SCALL  ALL_KEYS_RELEASED? ;C=1 if all keys are released
             JNC    NEWKEY?             ;Do the full thing; a key has been pressed
             CMP    KEYSTATE,#0         ;No key is pressed; finish up a previous key
             JNE    NEWKEY?             ;A previous key's handling needs finishing
             CLRC                       ;Clear the carry bit and return
             RET
```

FIGURE 5.21
FASTNEWKEY?, a speeded-up version of NEWKEY?.

interrupt service routine run long (relative to how long we would prefer an interrupt service routine to take), this is not really a problem. Recall that the real-time clock interrupt service routine is executed only every 10 ms or so. Even if this state machine algorithm were to take 100 to 200 μs to execute, that would still leave 98 to 99 percent of the CPU time available for other tasks.

We normally must worry about the duration of one interrupt service routine as it affects the latency of other interrupt service routines. In the case of the real-time interrupt service routine this is not a problem because (as we discussed earlier) it can start out by clearing the real-time interrupt flag and then reenabling interrupts to the CPU. At that point it is not ready to cause another interrupt, but any other source of an interrupt which was enabled before will again be able to obtain service. Consequently, the real-time interrupt tasks can be made to hold the lowest of interrupt priorities and will be serviced only when no other interrupt source needs service. On the other hand, the real-time interrupt tasks hold precedent over the mainline program. They will get handled every tick time of 10 ms or so, perhaps with some delay, but nevertheless handled.

A hardware alternative to this software approach to handling up to 20 keyswitches is shown in Fig. 5.22a. It uses a CMOS chip designed specifically for this purpose. The chip continuously scans the keyswitches. When a key is pressed, the data out lines emit the keycode and the data available line emits an active-high strobe pulse. The rising edge of the strobe pulse occurs just before* the keycode becomes valid on the data out lines. The falling edge of the strobe pulse occurs after the key is released. This sequence of events is illustrated in Fig. 5.22b. The keycode output is internally latched and remains valid, even after the present key is released, until a new key is pressed. Since the keycode will remain valid for at least tens of milliseconds, this is a good application for a low-priority interrupt input.

---

* Worst case, 150 ns.

The rising edge of the data available signal should be used to trigger an interrupt. This will produce a response from the microcontroller when a key is first pressed rather than when it is released. However, since the keycode is not valid until something up to 150 ns after this edge occurs, it is important to avoid using an input port which *latches* its input in response to the signal which triggers the interrupt.

The 20-key encoder chip of Fig. 5.22 includes its own oscillator. The two capacitors shown control the oscillator frequency and the keybounce time. The values shown produce reasonable response characteristics.

This chip implements the same two-key-rollover capability exhibited by the software algorithm discussed previously. That is, if a second key is pressed before a first key has been released, then the chip ignores the second key. The second key is handled only if it is still pressed *after* the first key has been released.

If keyswitches are handled using this hardware approach, then the interrupt service routine is very short. In this case, keyswitch handling need not have anything to do with the real-time interrupt service routine. Only when a key is pressed is there any need for service, and the rising edge of the strobe pulse signals this need.

This is an example of a tradeoff between hardware and software. While the hardware solution eliminates hassling with keyswitches, most designers prefer the software solution. *Any* reduction in hardware tends toward a more reliable, and more easily maintainable, design. On the other hand, there are applications where other considerations intrude. For example, the personal computer shown in Fig. 1.6 employs an entire microcontroller chip to handle the keyboard. In this case, the ability to connect the keyboard to the system unit with a simple four-wire cable (sending keycodes serially) is an overriding consideration.

5.6 KEYSWITCH PARSING

For instruments and devices which include front-panel keyswitches, a prime task of the mainline program is the *parsing* of these keyswitches. By parsing we mean taking appropriate action each time a keyswitch is pressed; the action is *based upon* which keyswitches have been pressed just before the current one. For example, if an instrument front panel includes among its keyswitches a numeric keypad for entry of set-up information, then the action that would *normally* be taken when a numeric keyswitch is pressed is updating the multi-digit number entered by the numeric keyswitch pressings prior to the current entry. However, the instrument may offer a selection of set-up options which are identified by pressing a special keyswitch followed by a numeric keyswitch. In this case, the numeric keyswitch pressing has an entirely different meaning from that of normal data entry.

> **Example 5.25.** Consider the 20-channel "scanner" shown in Fig. 5.23. This instrument is designed to be used with other instruments which are tied together on

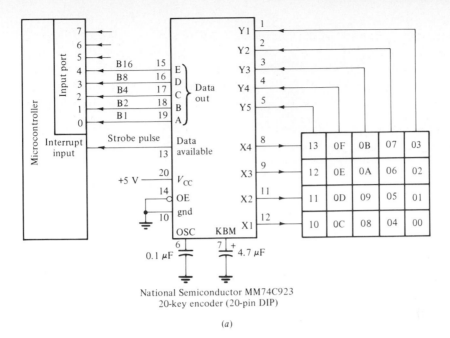

National Semiconductor MM74C923
20-key encoder (20-pin DIP)

(a)

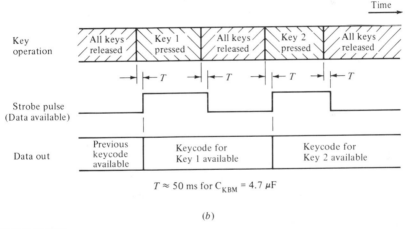

$T \approx 50$ ms for $C_{KBM} = 4.7 \ \mu F$

(b)

FIGURE 5.22
Hardware keyswitch debouncing. (a) Circuit; (b) timing.

the IEEE-488 general-purpose instrument bus to make automatic measurements. This scanner might be used to connect any of 20 test points in a device being tested to the input of a digital multimeter. Then a programmable controller can send a message over the IEEE-488 bus telling the scanner to close the relay contacts which connect test point 5 to the multimeter's input. In this mode of operation, the front-panel keyswitches are not used at all. The instrument is set up and run by the programmable controller over the bus.

FIGURE 5.23
A 20-channel scanner for auto-
mated measurements. (*Keithley
Instruments, Inc.*)

Before the control program has been written, it is helpful to be able to carry
out any of the scanner's bus-controlled functions from the front panel. However,
one philosophy for front-panel design would keep it clear of extra keys for rarely
used functions. The scanner of Fig. 5.23 does this. Its normal multiplexing function can be immediately accessed with a CHANNEL keyswitch followed by a
numeric keyswitch (to select a channel) followed by an OPEN or CLOSE keyswitch (to open or close a relay connecting that input channel to the output). Its
hidden functions are accessed by pressing the PRGM (program) keyswitch followed by a numeric keyswitch. For example, the scanner uses battery backup to
maintain several set-up parameters including the instrument address on the IEEE-
488 bus. This bus address is changed to address 14 by pressing

```
PRGM   3   1   4   ENTER
```

For another example, a cold-start reset, which resets battery backup parameters
to factory-set values, is executed with

```
PRGM   9   9
```

The parsing of keyswitches is related to another problem which arises in the
design of instruments which accept ASCII command strings from a programmable controller over the IEEE-488 bus. If each command from a controller
has its counterpart front-panel command, then we might make whatever translation is necessary so that the two sources of input sequences can be handled by
the same parsing algorithm.

Parsing of keyswitch sequences (or of IEEE-488 command strings) is
highly syntax dependent. The simplest syntax occurs in an instrument which
has a single keyswitch for identifying each parameter to be set up, plus a
numeric keypad for entering parameter values. For example, the function generator of Fig. 5.24 is set up with its front-panel keyswitches to generate a 200-
KHz 1.5-V peak-to-peak square wave output by pressing:

```
SQUARE
FREQ   2   0   0   EEX   3
AMPL   1   .   5   0   0
ON
```

Parsing such a keyswitch sequence consists of two parts:

1. A numeric buffer for the numeric parameter is *initialized* to 0 at reset time. It is updated each time a *numeric keyswitch** is pressed. It is *used* and then reinitialized to 0 each time a *function keyswitch†* is pressed.
2. Each press of a function keyswitch *terminates* the entry of the previous function. In the example above, pressing AMPL terminates the entry of the FREQ function. It leads to the call of an action routine for the *previous* function. If that function keyswitch required a numeric parameter, then the routine checks the numeric buffer content to make sure that it is within range. Then it uses the number as appropriate for setting up that function. Finally, a check is made as to whether the action of the *new* function keyswitch requires a numeric parameter. If not, then it is carried out also. That is, each press of a function keyswitch leads to the execution of *two* action routines. The action routine for the last function keyswitch pressed is executed *with* a numeric parameter (if one was entered). Then the numeric parameter is zeroed and the action routine for the latest function keyswitch pressed is executed *without* a numeric parameter. Of course, if this second action routine requires a numeric parameter, then its action is deferred until *another* function key is pressed.

As keyswitches are pressed, the first task of the parser is to sort out the role of each keyswitch, and determine whether it is serving to define a function or is helping to update a numeric parameter. For the function generator of Fig. 5.24, the keys themselves make this distinction. For the scanner of Fig. 5.23, this is true for almost all keyswitches (e.g., CHANNEL, CLOSE, and OPEN). However, the PRGM keyswitch is linked to the single numeric keyswitch which follows it to define a hidden function of the instrument (like carrying out a cold-start reset).

Once the parser has sorted out keyswitch entries into the identification of a function versus the updating of a numeric parameter, the function which has been identified can be translated into a function number of

$$0 \text{ or } 1 \text{ or } 2 \text{ or } 3 \text{ or } 4 \cdots$$

and used with a jump table to get to the appropriate action routine.

Example 5.26. Keyswitch entries for the function generator of Fig. 5.24 make this first job of the parser trivial. The keycodes themselves can make the distinction. For example, the 12 numeric keyswitches (0, . . . , 9, +/-, EEX) might be assigned the hexadecimal keycode values, 00 through 0B. Then when a keyswitch is pressed it is checked to see if it falls within the range 00 to 0B. If so, then it is used

---

* For the function generator of Fig. 5.24, numeric keys include 0, . . . 9,+/−, and EEX.

† For example, SQUARE, FREQ, AMPL, or ON.

FIGURE 5.24
A function generator which uses a single key to identify each function of the instrument.
(*Tektronix Inc.*)

to update the numeric parameter. Otherwise 0B is subtracted from the keycode and this value used to define a *function code*. This function code is the number used to access the jump table of action routine addresses.

Example 5.27. Consider now the scanner of Fig. 5.23. Keyswitch entry for this instrument is handled in almost the same way *unless* the PRGM keyswitch is pressed. When PRGM is pressed, the single numeric keyswitch which follows it is used by the parser's *preprocessor* to define a function code. Thus, numeric keyswitch entries are normally used to update a numeric parameter being entered. However, after PRGM is pressed, the next numeric keyswitch is trapped and used to identify a function code.

Example 5.28. As a final example consider the way in which the function generator of Fig. 5.24 parses character strings sent to it over the IEEE-488 bus. The same 200-KHz 1.5-V peak-to-peak square wave discussed previously can be initiated when a programmable controller sends the following string of characters over the bus:

```
SQUARE;FREQ 200E+3;AMPL 1.5;OUT ON
```

In this case, the parser's preprocessor has more of a job to do. It must identify character strings like SQUARE and OUT ON and translate them into function codes. It also treats semicolons as separators between functions. After these considerations have been handled by the preprocessor, the parser's job is identical to that for entry of front-panel keyswitch sequences.

5.7 SUMMARY

We have discussed the implementation of queues with some carefully defined macros. Queues are important to microcontrollers because they can serve as the link between interrupt service routines, for either input or output, and the mainline program. For input data, they permit an interrupt service routine to accept data quickly while postponing the handling of the data by the mainline program to some later time, when time is available. For output data, a queue keeps a backlog of data available so that when a device is ready for more data from the microcontroller, the data is immediately available. In this way queues provide the means to keep an interrupt service routine short, reducing the latency time seen by other interrupt sources.

While our interactions with queues might have been implemented with subroutines, the use of macros maintains *consistency* between the definition of the queue data structure, the initialization of the queue, the putting of data into the queue, and the getting of data from the queue. Each access of a queue requires only its name.

Tables represent another data structure which is important for microcontroller work. Each entry in a table can be accessed with a pointer formed by summing a fixed base address and a variable offset. We saw examples of algorithms which could be simplified and speeded up with the use of a table. With 8192 bytes of internal ROM, both the 68HC11 and the 8096 have enough ROM to encourage the use of table-driven algorithms.

Strings represent a third data structure encountered frequently when using a microcontroller. Characters in a string are accessed in sequential order, beginning with a pointer to the first character in the string and ending with a terminating character. We saw how strings could be used to support the output of successive ASCII characters via a serial port. We also saw how a queue could be used to concatenate several strings.

For interacting with devices which are slow by microcontroller standards, we saw how the program for an instrument or device could make use of a tick clock, implemented with interrupts from a real-time clock. Events which are to take place every 10 ms, or multiples of 10 ms, can be executed in the interrupt service routine for a real-time clock which interrupts every 10 ms. Examples arose in the software debouncing of keyswitches and in the stepping of a stepper motor. In contrast with other interrupt service routines, we found that the real-time clock interrupt service routine has the good feature that its execution time does not affect the latency of other interrupt sources. This is a result of reenabling interrupts almost immediately after the interrupt service routine is entered (after clearing the flag which caused the real-time interrupt to occur). We took advantage of this to execute a somewhat long keyswitch debounce algorithm with no worry that it might run as long as 100 to 200 μs. This still occupied only 1 to 2 percent of each tick time, occurring every 10 ms. Consequently, this approach leaves 98 to 99 percent of the CPU time available for other tasks . . . with interrupts enabled to give these other tasks access to CPU time. Whatever tick time is left over is given to the mainline program.

With interrupts playing such an important role in microcontroller applications, we found that what an interrupt service routine is supposed to do in response to an interrupt is often a function of what it has done during previous interrupts. State machines present a systematic approach for handling this linkage between present and past events. The description of a particular algorithm is represented by a state machine diagram, as exemplified by that of Fig. 5.15 for keyswitch debouncing. We looked at a mechanism for implementing state machines so that the complexity of the state machine diagram could be put into a table and into some test subroutines and action subroutines. In this way, the algorithm is compartmentalized. When changes to the algorithm arise, it is a straightforward matter to make the changes without destroying the correctly working parts of the algorithm.

We concluded the chapter with a discussion of keyswitch parsing. This is a specialized application of a state machine. It arises when the proper response to the pressing of a keyswitch is dependent upon the sequence of keyswitch pressings which preceded it. We looked at how two instruments organized the parsing of their keyswitches quite differently. We also examined the relationship between keyswitch parsing and the parsing of strings of characters sent by a controller to an instrument over the IEEE-488 bus. With the help of some preprocessing, the response of the instrument to the strings received over the IEEE-488 bus can be translated to sequences of keyswitch pressings. By being aware of this commonality, we are armed to handle what looks like two distinct processing problems with the same algorithm.

PROBLEMS

5.1. *Macros versus subroutines.* In Sec. 5.2 we developed Intel 8096 macros to define and initialize a queue, to put data into it, to get data from it, and to handshake with the input so as to avoid overrun errors. In this problem we will look at the alternative of implementing PUTB as a subroutine instead of as a macro.

A major reason for implementing an algorithm as a subroutine rather than as a macro is to save bytes of program memory when the algorithm is used again and again. However, a subroutine requires that we load registers to pass parameters to it whereas a macro can have built-in parameters.

(*a*) Write a PUTB subroutine to support putting a byte of data from the AL register into a queue if the insertion pointer has already been loaded into BX and the queue length (or its equivalent, represented in whatever manner you wish) has already been loaded into CX. How many bytes did this require?

(b) Write instructions loading these registers appropriately before calling the PUTB subroutine written in part a. How many bytes did this require?

(c) Write instructions restoring the incremented pointer back to memory. How many bytes did this require?

(d) What is the total number of bytes needed to handle a single application of a PUTB subroutine?

(e) What is the total number of bytes needed to handle 10 applications of a PUTB subroutine?

(f) What is the total number of bytes needed to handle 10 applications of the PUTB macro developed in the text?

5.2. *Queues.* Define a PUTB macro which does the same thing as the macro defined in the text but write it for the Motorola 68HC11 microcontroller. It should be able to put the 1-byte content of an arbitrary page 0 RAM location into any queue which is also located on page 0. Assume that you have a macro assembler which employs the same macro-defining syntax as that used in the text. Be sure to push and pop any CPU registers which you use. Compare the number of bytes generated by this macro with the number of bytes generated by the version written for the Intel 8096. Also compare the execution time for these two versions, using the cycle time per instruction information available in Fig. A.39 of Appendix A. Each cycle takes 0.5 μs.

5.3. *Queues.* Repeat Problem 5.2 for:

(a) DSQB, define storage for a queue of bytes, macro.

(b) DSQW, define storage for a queue of words, macro.

(c) INITQ, initialize queue, macro.

(d) PUT, put a word, macro.

(e) GETB, get a byte, macro.

(f) GET, get a word, macro.

(g) QHSB, queue handshake for bytes, macro.

(h) QHS, queue handshake for words, macro.

5.4. *Macros.* Write the macro definition for each of the following new instructions for the Intel 8096. Try to minimize the number of bytes of code which will be generated when these macros are invoked. Determine this number.

(a) CLRZ, a macro to clear the Z flag (assume that it is OK to change any other flags).

(b) SETZ, a macro to set the Z flag.

5.5. *Queues.* For the INPUT_CAPTURE interrupt service routine of Fig. 5.5c, determine its execution time in each case below. The Intel 8096 takes 5.25 μs to go from the interrupted program before it executes the first instruction of the interrupt service routine. Refer to Fig. B.39 of Appendix B for cycle time per instruction information for the Intel 8096. Each cycle takes 0.25 μs.

(a) For a single FIFO entry from input capture register 0.

(b) For a single FIFO entry from input capture register 3.

5.6. *Queues.* For the Motorola 68HC11,

(a) Write an interrupt service routine for the programmable timer's input capture register 1 (called TIC1) which will put the 2-byte value obtained into a queue called QTIC1. This interrupt source has its own vector (as do virtually *all* of the 68HC11's interrupt sources). It also has an input capture 1 flag which must be cleared in the interrupt service routine by writing 00000100B to TFLG1.

(b) How does this use of a queue help the mainline program to make timing measurements? Can it measure shorter intervals between successive edges than it could if it measured them directly? Or is the mainline program helped by being freed to do other activities than just the input timing measurements? Or what?

5.7. *Queues.* The GET macro defined in the text manipulates a flag for the benefit of the user of the macro. What would happen if this flag were ignored? That is, what would happen if we simply do a GET whenever we are ready to handle another byte from a queue?

5.8. *Queues.* Example 5.3 in the text looked at the potential for disaster when a mainline program and an interrupt service routine communicate via a queue and when the mainline program *does not* disable interrupts during its interactions (i.e., does not handle this as a critical region). We will explore this problem for the Motorola 68HC11 which must do in several instructions what the Intel 8096 can do in just one.

(a) When the mainline program does a PUTB macro, it must push A and X onto the stack, load into A the byte to be PUT, load the insertion pointer into X, store the byte to be PUT into the queue using indexed addressing, increment the pointer, roll it over if need be, store the incremented pointer back in memory, and restore X and A. Does any of this define a critical region? Explain.

(b) Repeat the analysis of part *a* for the case where the mainline program executes a GETB macro.

5.9. *BCD table.* If you have not already done Problem 4.37n to create the table of BCD numbers in Fig. 5.6, then do it now. You will use the conditional assembly facility of an assembler to keep from having to write 100 lines of assembler directives. What percentage of the Intel 8096's ROM is taken up by this table?

5.10. *BCD-to-binary table.* In this problem we will explore BCD-to-binary conversion using a table with the Intel 8096. For the macros which you write, make sure to push the content of any CPU registers used (e.g., AX) onto the stack initially and restore their content at the end.

(a) Use the conditional assembly facility of an assembler to create a table called BINARY. The 0th entry of this table will be the binary number 00000000. The 10011001th (i.e., 99H) entry of this table will be formed with

        DB   99

The assembler will treat this as a decimal number and convert it to 01100011. This table needs to reserve the 154 addresses between these two values, even if you only fill 100 of them with table entries.

(b) Given a 1-byte (two-digit) BCD number in a RAM location labeled NUM1, convert NUM1 to binary using this table.

(c) Given a 2-byte (four-digit) BCD number in a RAM location labeled NUM2, figure out how to convert NUM2 to binary using this table, doing whatever else you need to do. Assume that the least-significant byte has the address NUM2 (and the most-significant byte has the address NUM2+1).

(d) Write a macro definition called BCDTOBIN which does the operation in part *c*.

(e) Convert a 4-byte (eight-digit) BCD number in a RAM location labeled NUM4 to binary. Use the macro developed in part *d*, if it can help.

5.11. *Strings and macros.* This is a problem to try only if you have Intel's 8096 macro assembler available to you. With another macro assembler, you might try to do the same operation, but the format of the macro, when it is invoked, will probably be different. Using the Intel 8096's assembler directives for defining macros, create a macro called STRING which creates concatenated strings in ROM memory out of the parameters listed after it. Use the same EOT (04H) terminator discussed in the text. For example, the invocation of the macro with

```
IEEE488_COMMANDS: STRING ‹SQUARE,FREQ,AMPL,OUT ON,OUT OFF›
```

should store the ASCII codes for

$$S, \ Q, \ U, \ A, \ R, \ E, \ EOT, \ F, \ R, \ E, \ Q, \ EOT, , , , \ F, \ EOT$$

To do this note that

```
DCB 'SQUARE'
```

causes six ASCII characters to be stored. Also note that

```
DCB '&PARAM&'
```

will concatenate beginning and ending single quotes around a parameter. Also note that the indefinite repeat macro directive

```
IRP  PARAM, ‹SQUARE,FREQ,....,OUT OFF›
DCB  '&PARAM&'
ENDM
```

causes the body of a macro definition to be repeated once for each parameter in a list. The parameters in the list are separated by commas. Finally note that macro definitions can be nested. You will probably have to experiment with the assembler before you get this going.

5.12. *Strings.* Assume that the ASCII codes of the last problem reside in ROM memory, beginning at the address labeled IEEE488_COMMANDS. Write an Intel 8096 instruction sequence which will compare the string which is in RAM at an address labeled INPUT_STRING with each string, in turn, in IEEE488_COMMANDS. If INPUT_STRING matches with the first string (i.e., SQUARE), then load AL with the number 1. Continue looking for a match until one is found and load AL with the number of the string. Assume that a match *will* be found.

5.13. *Strings and macros.* Write the solution to the previous problem as a macro. However, modify it to terminate, whether or not a match has been found, when it gets to a null string. If a match has not been found, then return 0 in AL. This will let us put a null string (i.e., two terminators back-to-back) at the end of the list and have the search stop there.

5.14. *Strings and macros.* For the Motorola 68HC11
(*a*) Do Problem 5.12.
(*b*) Do Problem 5.13.

5.15. *Strings.* Example 5.10 in the text developed an interrupt service routine for driving a UART's transmitter. Successive ASCII characters are taken from strings pointed to by pointers passed to the interrupt service routine in a queue.
(*a*) How can the mainline program know whether the last message which it set up to be sent has indeed been entirely sent?
(*b*) How can the mainline program terminate an incompletely sent message prematurely?
(*c*) How can the mainline program know whether there is room in the queue for

the beginning of another message, even though the last message is still being sent?

(*d*) If the interrupt service routine is just finishing up one multiple-pointer message when the mainline program wants to initiate another one, can the mainline program set the TIE bit (even though it has been set) and *ensure* that the TIE bit will not be turned off by the interrupt service routine and be left turned off? Explain.

5.16. *Strings.* For the Intel 8096's UART,

(*a*) Write the interrupt service routine analogous to the one in Fig. 5.10.

(*b*) Write the code to display the three-string message of Example 5.11 (analogous to the code of Fig. 5.11).

5.17. *Program organization.* For the Motorola 68HC11, write the TICK_CLOCK interrupt service routine corresponding to Fig. 5.12. Assume that NEWKEY? and STEP are subroutines which have already been developed. STEP needs nothing but a subroutine call (since it looks at a global variable to decide what to do). NEWKEY? also works with a global variable, so it has no parameters which we need to pass to it. It returns the carry bit set to serve as a YES answer.

(*a*) Use the 68HC11's real-time interrupt (which has its own vector). This interrupt source can be assumed to have been set up to interrupt every 8.19 ms (i.e., the 8-MHz crystal clock period of 0.125 μs times 2^{16}). Before reenabling interrupts for *all* sources (with a CLI instruction), it is necessary to clear the real-time interrupt's flag so that it will no longer pull low the interrupt line which goes to the CPU. This flag is cleared by writing 01000000B to TFLG2.

(*b*) Use the 68HC11's output compare 5 interrupt (which has its own vector). Before reenabling interrupts, the timer must be reinitialized to cause its next interrupt in another 10 ms. Do this by reading the 16-bit register, TOC5, adding 5000 to it, and putting the result back into TOC5. This assumes that the programmable timer's prescaler has been set up to divide the 68HC11's internal clock by four (so that the timer's free-running counter is counted every 2 μs). Also the output compare 5 interrupt flag must be cleared by writing 00001000B to TFLG1.

5.18. *Watchdog timer.* Figure 5.12 shows a mainline program which includes a subroutine to reset the watchdog timer. If the watchdog timer is set up to time out after 16 ms, then the implication of this organization is that the PARSE_KEYS subroutine must never take longer than 16 ms to execute. If PARSE_KEYS is designed to be called to handle each new keycode passed to it in a queue from the tick-clock interrupt service routine, then this means that the action taken in response to any keyswitch sequence must never take as long as 16 ms. If this constraint is difficult to overcome, then there are two alternatives. For one, PARSE_KEYS can be designed as a state machine using the techniques of Sec. 5.5 so that its execution is completed when it has done some subtask of handling a keyswitch sequence, rather than the entire task. Upon exiting, the watchdog timer is reset and then PARSE_KEYS is reentered and it continues on its way. Its *state* information keeps track of where it is in handling the actions associated with a particular keyswitch sequence. Only when PARSE_KEYS is entered, and when its state information implies that the response to the last keyswitch sequence has been completed, does it go to the queue to get another keycode. This approach divorces the duration of PARSE_KEYS from the time it takes to execute the response to a specific keyswitch sequence.

An alternative approach would put the RESET_WATCHDOG subroutine in the tick-clock interrupt service routine. One advantage of this approach is that it will ensure the recovery from a malfunction which somehow turns off tick-clock interrupts. On the other hand, it does *not* check on a malfunction which abuses the stack. By putting RESET_WATCHDOG in the mainline program, it will only be called when all of the subroutine returns and interrupt service routine returns successfully unfold *all* of the return addresses from the stack.

(*a*) With RESET_WATCHDOG in the mainline program, describe an error which can occur because the program counter is loaded with a wrong address during a jump instruction (i.e., where the CPU missed a bit while fetching an instruction from ROM) or because a subroutine return address was pulled off the stack with one or more bits in error (perhaps because a RAM chip dropped a bit of the return address which was stored in it) which will lead to the CPU never executing RESET_WATCHDOG again.

(*b*) With RESET_WATCHDOG in the tick-clock interrupt service routine, describe how you might have it check on the mainline program once every *second* (i.e., once every 60 ticks) to see that PARSE_KEYS has returned to the mainline program more often than every second. That is, what would you insert into the mainline program to support such a test? What would you insert in the tick-clock interrupt service routine to support such a test?

5.19. *Flag testing.* The Intel 8096 packs eight interrupt flags into a single 8-bit register, IOS1. Furthermore, any access (i.e., write *or* read) of this register clears six of the 8 bits. Because of this, we need an IOS1_TEST subroutine which will do the following:

Have a mask passed to it in the AL register which identifies a single bit of IOS1 to be tested (e.g., using 01000000B to test bit 6)

OR the IOS1 register into a single-byte RAM variable called IOS1_COPY (using an ORB instruction)

Test the selected bit in IOS1_COPY

Clear the selected bit in IOS1_COPY

Return a flag (e.g., the Z flag or the C flag), indicating whether the selected bit was set *either* in IOS1 or in IOS1_COPY.

Write this subroutine and include a header of comments indicating what it does, what bit is being used as a flag, and the sense of the flag (e.g., the Z flag will be set if the selected bit was set).

5.20. *Polling routine.* The Intel 8096 has a single vector to handle four software timers. They are four output compare registers, each of which can generate an output compare interrupt (but which has no output pin associated with it which is affected at the time the output compare occurs).

(*a*) Write the interrupt service routine for handling these software timers which immediately reenables interrupts for all sources except this one by executing

```
LD INT_MASK,#11011111B
EI
```

Then the interrupt service routine should use the IOS1_TEST subroutine of the last problem to initiate a subroutine call to ST0 if bit 0 has been set, to ST1 if bit 1 has been set, to ST2 if bit 2 has been set, and to ST3 if bit 3 has been set. Upon returning from any of these four subroutines, repeat the test until *no* flag is found set. Then return from the interrupt.

(b) Write the ST3 (software timer 3) subroutine to serve as a tick clock by reenabling interrupts for *all* sources, including software timer interrupts. Reset the timer to interrupt again in 10 ms by reading a 2-byte variable, OC3_COPY, adding 5000 to it, and writing the result back to OC3_COPY. Then the timer's output compare register is loaded by

Disabling interrupts (beginning of a critical region)

Writing 3BH to HSO_COMMAND

Writing OC3_COPY to HSO_TIME

Enabling interrupts (ending of critical region)

Finally, do the other jobs shown in Fig. 5.12 for the TICK_CLOCK interrupt service routine.

5.21. *Keyswitches.* The 16-keyswitch array shown in Fig. 5.14 can be expanded to a larger array by using more lines from another port. Before we do this, it might be helpful to know whether we need bidirectional lines for both the column lines *and* the row lines.

(a) Does ALL_KEYS_RELEASED? (Fig. 5.19a) use row lines as inputs? As outputs? Does it use column lines as inputs? As outputs?

(b) Answer part *a* for SAME_KEY_RELEASED? (Fig. 5.19b).

(c) Answer part *a* for FIND_KEY_AND_CLR_C (Fig. 5.20c).

(d) Do we need column lines to be inputs? Outputs?

(e) Do we need row lines to be inputs? Outputs?

(f) For up to 25 keyswitches, we might use bits 6 and 7 of port 2 of the 8096. These are bidirectional lines also. Port 2 is convenient to use because it has only two other outputs. One of these is the UART output. The other is a pulse-width modulation output (which can be used to drive a dc motor). Neither of these outputs is affected by writes to port 2. Consequently, we can do our keyswitch writes to port 2 as if none of the other lines were being used. This usage will not mess up any other outputs.

Redraw Fig. 5.14a showing row and column connections for a 5 × 5 array of keyswitches. Label them with successive hexadecimal numbers from 00 through 18.

(g) Rewrite ALL_KEYS_RELEASED? (Fig. 5.19a) for testing this array of up to 25 keyswitches.

(h) Rewrite SAME_KEY_RELEASED? (Fig. 5.19b) for testing this array of up to 25 keyswitches.

(i) Rewrite FIND_KEY_AND_CLR_C (Fig. 5.20c) for testing this array of up to 25 keyswitches.

(j) Rewrite KEYCODETABLE (Fig. 5.18) for testing this array of up to 25 keyswitches.

(k) Is there anything else that needs to be changed?

5.22. *State machine.* Consider the modification to the state machine depicted by Fig. 5.15 so that a key is checked not twice but four times to see if it has *really* been released. This is the "silk purse out of a sow's ear" modification mentioned in the text.

(a) Draw the new state machine diagram.

(*b*) Rewrite the KEYSTATETABLE of Fig. 5.16*e* to reflect this change.

(*c*) Is there anything else which needs to be modified to make this change?

5.23. *State machine.* Write an Intel 8096 macro called STATE_MACHINE which can be used to write a subroutine like NEWKEY?. It has passed to it

> The name of a subroutine to be defined, without the question mark at the end (e.g., NEWKEY)
>
> The name of a state table (e.g., KEYSTATETABLE)
>
> The name of a 2-byte state variable (e.g., KEYSTATE)
>
> The name of a test jump table (e.g., KEYTEST)
>
> The name of an action jump table (e.g., KEYACTION)

Note that the Intel 8096 macro assembler lets us modify parameter names to form label names. For example, if we define the above macro with

```
STATE_MACHINE  MACRO    NAME,STABLE,STVAR,TEST,ACTION
```

and define the subroutine name with

```
NAME&?:
```

then when we invoke it with

```
STATE_MACHINE NEWKEY,KEYSTATETABLE,KEYSTATE,····
```

the subroutine will become labeled with

```
NEWKEY?:
```

5.24. *Keyswitch handler.* For the Motorola 68HC11, assume that port C has been set up for interrogating keyswitches, using the circuit of Fig. 5.14. Access port C by reading from or writing to PORTC. Its data direction register (DDRC) must have 0s in any bit positions corresponding to port lines which are to be driven externally as inputs. If this is not done, then contention will occur on such lines, possibly shortening the life of the 68HC11's drive circuitry for those lines.

(*a*) Write the NEWKEY? subroutine corresponding to Fig. 5.17*a*.

(*b*) Write the ALL_KEYS_RELEASED? subroutine corresponding to Fig. 5.19*a*. Note that the 68HC11 *sets* the carry bit when a borrow occurs during a subtract, or compare, instruction.

(*c*) Write the SAME_KEY_RELEASED? subroutine corresponding to Fig. 5.19*b*. If it helps to have KEYCODE defined as a 1-byte variable, then make this change.

(*d*) Write the FIND_KEY_AND_CLR_C subroutine corresponding to Fig. 5.20*c*.

(*e*) Write the FASTNEWKEY? subroutine corresponding to Fig. 5.21.

5.25. *Keyswitch handler.* We have developed the subroutines which deal with the keyswitch array of Fig. 5.14 assuming that port 1 is a bidirectional port. If we were using another microcontroller, or if we needed to use the Intel 8096's port 1 in some other way, then we might choose to use the row lines solely as inputs and the column lines solely as outputs. In this case we can eliminate the four pullup resistors on the column lines.

(*a*) Will the ALL_KEYS_RELEASED? subroutine be affected by this change? If so, then modify it to work in this new situation.

(*b*) Will the SAME_KEY_RELEASED? subroutine be affected by this change? If so, then modify it to work in this new situation.

(c) The FIND_KEY_AND_CLR_C subroutine will be affected by this change. Its first five instructions form a pattern in the AL register which will be the same as one of the rows of KEYCODETABLE if a single keyswitch is pressed. These are the only instructions in the subroutine which are affected, *providing* we can form this pattern in some way when a keyswitch is pressed. Modify this routine so as to drive each of the column lines low, one at a time, and then see if this causes a row line to be pulled low. If so, then form the pattern required by the rest of the subroutine. Can you build this testing of columns into a loop of instructions which is executed four times, or until a pressed keyswitch has been found, whichever is less?

(d) Rewrite the FIND_KEY_AND_CLR_C subroutine by using an entirely different approach. This time set up a loop which will be executed at most 16 times, once for each keyswitch. Each time around the loop call the SAME-_KEY_RELEASED? subroutine with a different value of KEYCODE. Stop when a keyswitch is detected as being pressed and return KEYCODE and C = 0. If no keyswitch is detected as being pressed, then return KEYCODE = 10H and C = 0.

(e) Compare the worst-case execution times of the subroutines of parts c and d.

(f) How does the algorithm of part d handle multiple keyswitches being pressed? When a first keyswitch is detected and found, it will hang up until that keyswitch is released. Before this release happens, assume that two more keyswitches are pressed and are still pressed when the first keyswitch is released. Consider the case where the two new keyswitches are in the same row. In the same column. In different columns and rows.

5.26. *Keyswitch parsing.* The scanner instrument shown in Fig. 5.23 handles the pressing of the PRGM keyswitch followed by the pressing of any numeric keyswitch between 0 and 8 *as if* a separate function keyswitch had been pressed. If the PRGM keyswitch is followed by a 9, then one more digit is looked for to define 10 more functions. In this way, the scanner can make its 25 keyswitches work *as if* they were 44 keyswitches. This has the advantage of hiding the complexity of 19 functions from the novice user (by not cluttering the front panel with keys for them) and yet offering these functions to the knowledgeable user.

Make a little state machine diagram (analogous to that of Fig. 5.15) for a subroutine called FULLKEYS, which will read the output of the tick-clock interrupt service routine's queue, QKEYCODE, and which will generate an output, called FULLKEYCODE, which is a number between 0 and 43 corresponding to the derived keycode discussed here. Assume that the original numeric keyswitches generate keycodes between 00 and 09 and that the PRGM keyswitch generates a keycode of 0A, while all of the other function keyswitches generate keycodes between 0B and 18. The new derived keycodes will range from 19 to 2B. FULLKEYS should set a flag each time that it comes up with a new value in FULLKEYCODE.

5.27. *Keyswitch parsing.* Following on the tail of the last problem, make a state machine diagram for a subroutine called SORTKEYS which accepts FULLKEY-CODE input and which sorts according to whether the value read represents a numeric keycode (between 00 and 09) or a function keycode (between 0B and 2B). Each numeric keycode should be converted to its ASCII equivalent and appended to the end of a string, called NUMSTRING, for subsequent handling. Set a flag when a function keycode is read, for the benefit of the subroutine which follows this one and which will use a jump table to take the appropriate action. Also go to a state which will clear the numeric string (to a null string) when SORTKEYS is next called.

5.28. *Temporary variables*. In the last several problems we have been casually introducing new variables as if unlimited RAM were available. Some of these variables are used internally in a state machine and must be maintained between executions of a subroutine as global variables. Other variables are used temporarily to pass parameters between subroutines. This is a good time to use a CPU register instead of permanently dedicating a RAM location to such use. Other variables may only be needed as temporary variables *within* a subroutine. Consider the variables needed in the solution to Problems 5.26 and 5.27. Which ones are global, requiring the allocation of RAM specifically to them? How are the others best handled?

5.29. *Keyswitch parsing*. Implement the FULLKEYS subroutine of Problem 5.26 using your favorite microcontroller. Be sure to include a header of comments saying what it does, how it handles parameters passed to and from it, and any RAM variables that it needs.

5.30. *Keyswitch parsing*. Implement the SORTKEYS subroutine of Problem 5.27. Include an appropriate header of comments.

5.31. *IEEE-488 bus command parsing*. Consider the example of a character string sent over the IEEE-488 bus to the function generator of Fig. 5.24, described in Example 5.28 in the text.

(*a*) Make a state machine diagram for a subroutine called SORTBUS. The subroutine is to read the output of a queue called QBUS and generate two strings called BUSFNSTRING and BUSNUMSTRING. Each time that it reads a semicolon, it is to set a flag as a signal to the subroutine following it that it has sorted a complete command into BUSFNSTRING. If that command includes a parameter, then it will be in BUSNUMSTRING.

(*b*) Implement the SORTBUS subroutine using your favorite microcontroller.

5.32. *String to function conversion*. When SORTBUS, in the last problem, sets its flag saying that it has gathered together a complete function from the bus, the subroutine which follows it, called STRINGTOFUNCTION is to use BUSFNSTRING and compare it with each substring in IEEE488_COMMANDS, defined in Problems 5.11 and 5.12. Its output is a function number which can be used to take appropriate action. Define and write STRINGTOFUNCTION appropriately, using your favorite microcontroller. The output of this subroutine should include the setting of a flag *only if* the input command was valid. For example, do not set a flag if the string in BUSFNSTRING matched with *none* of the strings in IEEE488_COMMANDS.

REFERENCES

Probably the best reference for the kinds of topics discussed in this chapter is a listing of the assembly language code for an instrument. Many issues arise which must be handled in a consistent manner in the development of the program for an instrument. By scanning such a listing, you can quickly gain valuable insights, even as you ignore the intricate details (and peculiarities) of the program.

CHAPTER

6

MICROCONTROLLER EXPANSION METHODS

6.1 OVERVIEW

A major concern when considering the use of a microcontroller chip for a specific application is whether it has sufficient internal resources to meet the needs of the application. If it does not, then the microcontroller might still be used, for any of several reasons:

We might select it because the on-chip real-time control resources make it the fastest controller around.

Our application might emphasize the minimization of size, weight, or power dissipation. The microcontroller's on-chip resources probably lead to a reduction in the overall system chip count relative to what would be obtained using a general-purpose microprocessor CPU chip. Minimizing chip count leads to the minimization of size and weight. The use of a CMOS microcontroller fosters the minimization of power dissipation.

We may be far along in a design which we thought would fit in the single-chip microcontroller, but which actually will not, while still achieving the performance goals we have set for the instrument or device. Operating the same microcontroller in an expanded mode to add resources may be the simplest revision to our plans.

Consider this latter point further. If we *think* that the resources of a specific microcontroller chip will be sufficient to meet the needs of a project and if we initiate the design process using this microcontroller chip, then what will be the consequences if, in fact, we run out of a needed resource? That is, what are the consequences of learning, six months into a project, that we need more RAM than a microcontroller has available? Or more ROM? Or more ports? Or a second UART? Will we have to walk away from six months of work and start all over with a different microcontroller chip having more capability, or with a multiple-chip microcomputer configuration? If this latter is the penalty for mis-judging our needs at the outset, then the risk may be too high to chance. On the other hand, if the penalty is the same as for a multiple-chip microcomputer (i.e., add more chips), then there is not much risk at all.

Different microcontroller *families* of chips have handled this expansion problem in rather different ways. We will explore some of the possibilities in the next section.

In the following section we will consider the expansion of the Motorola 68HC11 and the Intel 8096, and the manner in which each chip brings out the internal address bus, data bus, and bus control lines. We will see how to connect assorted peripheral chips to these bus structures.

The next section will be devoted to the timing considerations which must be met by peripheral chips (e.g., RAM and EPROM) to ensure their reliable operation with a microcontroller chip. We will learn how to translate microcontroller timing requirements and circuit topology considerations into timing specifications for peripheral chips.

A UART serial port represents one way to expand the capabilities of a microcontroller. It has the advantage of using only two pins on the chip to achieve bidirectional communication with another chip, perhaps another microcontroller. By using several lines of an output port together with a couple of "glue" parts,* we will see how to expand this capability for interactions with several other chips. Because a UART serial port presents a standard and popular interface, knowledge of UART use will permit us to include such an interface in our designs.

Sometimes the processing power of a single microcontroller chip is insufficient for the job at hand. In such a case, one alternative is to split the required tasks between *two* microcontrollers. The two microcontrollers can then be coupled together. We will look at the means for doing this using the on-chip UART and at the implications of this coupling. We will also consider the connection possibilities for more than two microcontrollers.

An I/O serial port is similar to a UART serial port in that it lets us expand the resources of a microcontroller using very few lines. Since it is basically a shift register interface (with clock, data input, and data output lines), it presents a versatile interface. In this chapter we will see that it permits the expansion of I/O lines via shift registers. For an application in which the microcontroller has

---

* That is, multiplexers and decoders.

sufficient on-chip resources other than I/O lines, expansion through an I/O serial port permits the microcontroller to be operated in the single-chip mode while still being able to access expanded I/O ports easily and within microseconds. In the next chapter we will see examples of a variety of peripheral devices which are accessed by means of a built-in I/O serial port interface. Because of this flexibility, we will explore several ways in which a microcontroller's I/O serial port can be expanded.

The microcontroller resource which plays the major role in real-time control is the programmable timer. Both Intel and Motorola have expanded the capability of the programmable timer in the chips discussed in this book, relative to the capability of earlier generation chips. Nevertheless, there are occasions when not enough timer inputs or outputs are available. For those times, we will see that the circuitry needed for expansion is similar to that needed for the expansion of serial ports.

For applications requiring real-time control that is even faster than that achievable with a microcontroller, we will consider a peripheral chip which is, in essence, a programmable state machine. It runs much faster than the microcontroller. Its functioning can be shaped to fit the application. In effect, it provides an avenue for expanding the real-time processing power of the microcontroller.

6.2   ALTERNATIVE FAMILY MEMBER CHIPS

One of the most widely used microcontroller chips ever produced is the Intel 8048 microcontroller, introduced in 1976. Since its introduction, it has been cloned into a dozen or so variations, all with essentially the same register structure and instruction set, and all able to be expanded to include more ROM, RAM, and I/O ports. The variations include the amount of on-chip ROM, EPROM, and RAM. Package size represents another variation, with the original 40-pin DIP being cloned into lower-cost 28-pin and 20-pin versions of the microcontroller. One variation even includes a two-channel A/D converter.

From a designer's point of view, the pressure from having to judge at the project's outset the amount of ROM and RAM needed is greatly reduced by the availability of the three members of the 8048 family listed in Fig. 6.1. Each of the three has the same register structure and the same instruction set. Each has the same 40-pin package, with identical "pinout"; that is, each of the 40 pins is defined in exactly the same way. A designer who thinks that the 8048 will serve as the controller for a specific project is comforted by the availability of the 8049 and 8050. If the 1024 bytes of ROM become insufficient, or if the 64 bytes of RAM become insufficient, then the 8049 can be used in place of the 8048. The designer can proceed to make printed circuit boards early in the project, knowing that if the choice of microcontroller has to be upgraded from the 8048 to the 8049 or even to the 8050, the board that the microcontroller plugs into will be

Microcomputer	ROM	RAM	
8048	1024 bytes	64 bytes	
8049	2048 bytes	128 bytes	Identical pinout, register structure, and instruction set
8050	4096 bytes	256 bytes	

Each is packaged in a 40-pin DIP.
Each has 27 I/O lines.
Each employs one 8-bit timer-counter.

FIGURE 6.1
Intel 8048 family members for ROM/RAM expansion.

identical. Similarly, the pressure is off for software development. A change to one of the upgraded chips will not entail the rewriting of a single line of code. All of the resources of the simpler chip are present in the upgraded chip, augmented by more ROM and RAM.

Intel also supports I/O port expansion for the 8048 family with a special 24-pin I/O port expander chip, the 8243. This chip ties up five pins of the 8048 in return for 16 I/O pins, configured as four 4-bit ports, as shown in Fig. 6.2*a*. By using 8048 output port lines to decode among multiple 8243 chips, the I/O can be expanded to an unlimited amount. This is illustrated in Fig. 6.2*b*, where a single additional line from port 2 is used to select either of two chips. Before executing the instruction which is designed to interact with a single 8243 I/O expander chip, it is necessary to write a 0 or a 1 to bit 4 of port 2 to select between the two 8243 chips. This process of writing to a port to select between chips is called *bank switching*.

The Intel 8048 also supports the expansion of ROM and RAM to a limited extent (up to 4096 bytes of ROM and up to 256 bytes of RAM). The limitation is posed by the size of the program counter (12 bits) in the case of program memory. Data memory is limited by a hardware and instruction combination. The instruction to access an external address uses an *8-bit* register as a pointer to the external address. Again, an output port can be used to support the bank switching of even more ROM and RAM. However, a bank switch is an awkward operation for program memory since it is equivalent to a jump from address N in one ROM to address $N+1$ in another ROM as the 1-byte write to port instruction is executed. Bank switching of RAM means that software must be written with an awareness, at all times, of *which* external RAM bank is being accessed every time an access to external RAM is made.

Because of such considerations, many a user of the Intel 8048 has been ready to graduate to a newer microcontroller chip. The general approach taken to support unencumbered expansion of ROM and RAM in newer microcontroller chips has been to bring out the internal bus, using it to access additional memory chips.

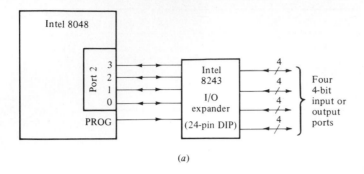

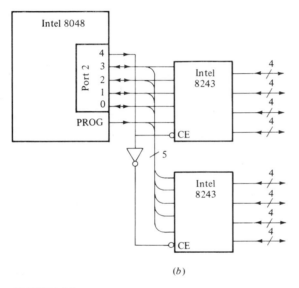

FIGURE 6.2
Intel 8048 I/O port expansion. (*a*) Use of dedicated port 2 pins plus PROG pin for built-in port expansion; (*b*) bank switching of expander chips.

6.3   BRINGING OUT THE INTERNAL BUS

Both the Intel 8096 and the Motorola 68HC11 handle expansion in a straightforward manner. They relinquish two ports, plus a few more lines for control, and permit any of the 65,536 addresses which are not being used by internal resources to be used to access external resources. The instructions for accessing external resources are identical to those for accessing internal resources.* The memory map for the Intel 8096 is shown in Fig. 6.3. As discussed in Sec. B.2 of Appendix B, the chip can be set up with either an 8-bit or a 16-bit external data bus. It then recognizes accesses to off-chip addresses and generates the proper

---

* Of course, the Intel 8096 cannot use direct (page 0) addressing for the off-chip addresses.

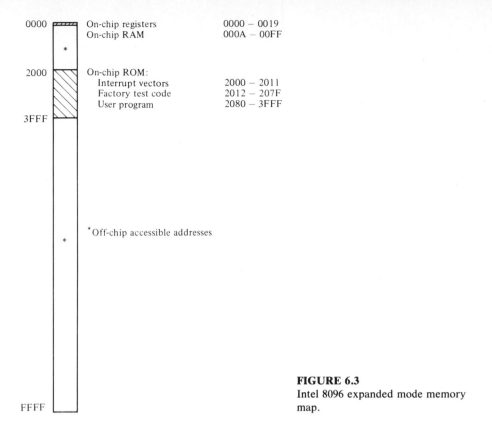

0000	On-chip registers 0000 – 0019
	On-chip RAM 000A – 00FF
*	
2000	On-chip ROM:
	Interrupt vectors 2000 – 2011
	Factory test code 2012 – 207F
	User program 2080 – 3FFF
3FFF	
*	*Off-chip accessible addresses
FFFF	

FIGURE 6.3
Intel 8096 expanded mode memory map.

signals for a multiplexed address bus/data bus structure built around the lines which would have been ports 3 and 4 in the single-chip mode.

The memory map for the Motorola 68HC11 operating in its expanded mode is shown in Fig. 6.4. To operate in this mode, two pins on the chip, MODA and MODB, must be tied to +5 V. This map shows the on-chip registers as having been remapped to hex addresses 0000 to 003F from their default location of 1000 to 103F. As is pointed out in Sec. A.2 of Appendix A, both the on-chip registers and the on-chip RAM can be remapped. The remapping shown in Fig. 6.4 lets all registers and RAM be accessed using direct (page 0) addressing. On the other hand, it throws away the 64 bytes of RAM, which *would* have occupied addresses 0000 to 003F if the registers had not been mapped to those addresses. Another remapping alternative would put on-chip RAM at addresses 1000 to 10FF. Such a remapping provides more RAM, but it requires either extended addressing or indexed addressing to access it.

Example 6.1. Consider the expansion of the Motorola 68HC11's I/O ports. A circuit to serve this purpose is shown in Fig. 6.5. Port B becomes the upper half of the address bus. Port C becomes a multiplexed bus, serving as the lower half of

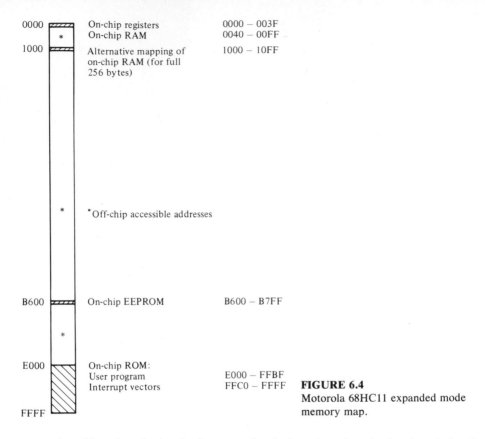

0000	On-chip registers	0000 – 003F
	On-chip RAM	0040 – 00FF
1000	Alternative mapping of on-chip RAM (for full 256 bytes)	1000 – 10FF
*	*Off-chip accessible addresses	
B600	On-chip EEPROM	B600 – B7FF
E000	On-chip ROM:	
	User program	E000 – FFBF
	Interrupt vectors	FFC0 – FFFF
FFFF		

FIGURE 6.4
Motorola 68HC11 expanded mode memory map.

the address bus during the first part of a clock cycle and as the data bus during the latter part of a clock cycle, as shown in Fig. 6.6. The STRB handshake line becomes a R/W line (high during a read cycle, low during a write cycle). The STRA handshake line becomes an AS (address strobe) signal, used to latch the lower half of the address when this is needed later in the clock cycle.

Motorola has built a 68HC24 port replacement unit chip to support a virtually exact reconstruction of ports B and C as well as the STRA and STRB handshaking lines. This chip provides excellent support for anyone building an emulator for the single-chip 68HC11. An emulator needs access to the internal bus. It needs to augment the internal RAM with more RAM for program storage during development. And it needs ports which operate like the single-chip ports.

If exactly reconstructed ports B and C are not needed, or if even more ports are needed, then Fig. 6.5 illustrates how these can be built with octal flip-flops for output ports and with three-state buffers for input ports. Two of these 20-pin "skinny DIPs" can implement two ports and take less room on a board than the 40-pin 600-mil-wide (i.e., 0.6-in-wide) 68HC24 chip. The 74HC138 decoder chip on the left does a partial decode on the addresses from 0000 to 3FFF, assuming that nothing else will be enabled in this range (other than on-chip RAM and on-chip registers). Note that a write to any address of the form

00XX X001 XXXX XXXXB

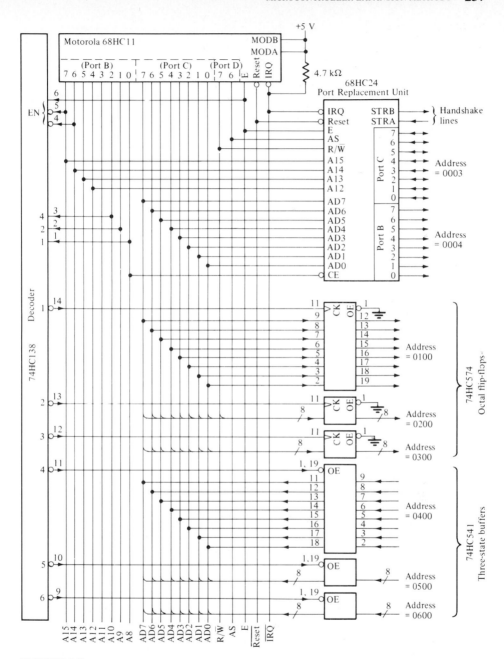

FIGURE 6.5
Motorola 68HC11 I/O port expansion.

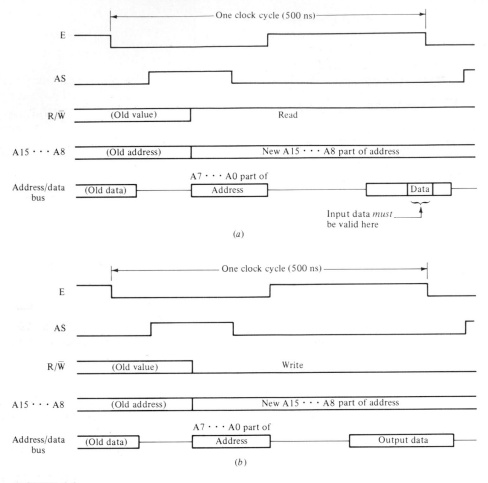

FIGURE 6.6
Motorola 68HC11 expanded mode timing. (*a*) Read cycle; (*b*) write cycle.

will cause a write to the octal flip-flop labeled Address=0100. This occurs because any address of the form

$$00XX\ XXXX\ XXXX\ XXXXB$$

will enable the 74HC138 decoder chip during the second half of the clock period, when E is high. This decoder decodes address lines A10, A9, and A8, which are

$$A10 = 0 \qquad A9 = 0 \qquad A8 = 1$$

for the address range being considered. Hence, the decoder's "1" active-low output is driven low in this case, clocking the octal flip-flop.

In similar fashion we can see that the three-state buffer labeled Address=0400 is actually enabled for any address of the form

$$00XX\ X100\ XXXX\ XXXXB$$

since in this case address lines A10, A9, and A8 are

$$A10 = 1 \qquad A9 = 0 \qquad A8 = 0$$

and since this three-state buffer is enabled by the "4" active-low decoder output. Thus this buffer serves as an input port by driving the data bus with the state of its eight inputs during the second half of the clock period when a read from address 0400 is executed.

Partial decoding of the address space of any multiple-chip microcomputer configuration is commonly done. However, it does require care. Note that the 74HC138 decoder's 0 output must not be used since reads and writes to any on-chip register or on-chip RAM address uses an address of the form

$$\underline{0000} \quad \underline{0000} \quad XXXX \quad XXXXB$$

If a 74HC574 output port were clocked with the decoder's "0" output, then it would be clocked by every read or write of an internal register or RAM address. That is, its desired output would be inadvertently and regularly corrupted.

Also note that the read/write line has not been used here, thereby simplifying the decoding. However, we must be careful not to write inadvertently to addresses 0400, 0500, or 0600 since such a write will lead to contention on the data bus between the 68HC11 and the selected buffer chip.

This type of I/O port expansion has the advantage of being fast and simple to use—as fast and as simple as a write to normal on-chip ports. It has the disadvantage that two ports are lost (i.e., ports B and C) in the process of creating the external bus. For example, if we want to obtain two more ports than are on the single-chip 68HC11, then we actually need to add four ports (implemented with four 20-pin DIPs plus a decoder). In contrast, if we use the I/O serial port* to implement these two extra ports, then we will find that we need the addition of only two 20-pin DIPs. But the serial interactions make for port accesses which are an order of magnitude slower than those obtained here.

Example 6.2. Consider, now, I/O port expansion for the Intel 8096. A circuit for this purpose is shown in Fig. 6.7. The circuitry is simplified to the extent of saving one octal latch chip by using an 8-bit external data bus instead of a 16-bit external data bus. (With an 8-bit external data bus, the upper half of the address bus remains available for the duration of the read or write cycle.) It makes use of the timing in Fig. B.4 of Appendix B. As discussed in conjunction with that figure, the active-low ADV line goes low only during accesses of *external* devices. Its use to enable the 74LS138 chip ensures the distinction between enabling internal and external resources.

The operation of this circuit is almost identical to that for the Motorola 68HC11 just discussed since the 8096 multiplexes the lower half of the address bus with the 8-bit data bus. Early in a clock cycle the eight lines of port 3 (which is no

---

* See Sec. 6.6.

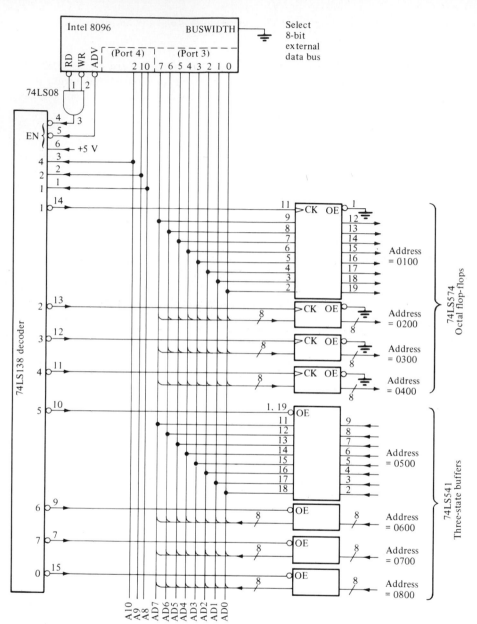

FIGURE 6.7
Intel 8096 I/O port expansion.

longer available as a port) emit the eight lower bits of the address. Later in the clock cycle, these lines become an 8-bit data bus.

The circuit of Fig. 6.7 looks only at address lines A10, A9, and A8 to do a partial address decode. Thus, the top output port, shown with Address=0100, will be updated by any write of the form

XXXX X001 XXXX XXXXB

As a consequence, the partial decoding of Fig. 6.7 works fine for port expansion as long as further devices (e.g., extra RAM or EPROM) are not added.

The 74LS08 AND gate shown in Fig. 6.7 serves as an active-low OR gate so that when *either* an active-low read or write pulse occurs, the 74LS138 will be enabled. An alternative would be to eliminate the 74LS08 and use *two* 74LS138 decoders, one for reads, the other for writes.

If there were a variety of devices to be enabled onto the bus (e.g, RAM and EPROM as well as I/O ports), then a more versatile decoding circuit would employ a programmed PAL. A PAL is the decoder of choice for most designers in this situation. It simply implements arbitrary boolean functions between its inputs and outputs. But it must be programmed with a PAL programming unit.

Note that the circuit of Fig. 6.7 uses 74LSxxx parts whereas the circuit of Fig. 6.5 uses 74HCxxx parts. In the latter case, the use of HC parts combines CMOS circuitry with a CMOS microcontroller, minimizing power and otherwise giving essentially the same characteristics. While HC parts could be used with the NMOS 8096 microcontroller, the outputs from the 8096 to the CMOS gates would require the addition of pullup resistors to ensure the pulling up of all outputs to +5 V instead of the specified +2.4 V. In contrast, +2.4 V is within specification as a logic 1 level when driving a 74LSxxx TTL part. Consequently, it can be interfaced directly, without extra resistors.

Expansion of program memory requires that we handle address lines not only to select a chip but also to select addresses within a chip.

Example 6.3. Program memory expansion for the Intel 8096 is illustrated in Fig. 6.8. Again, a partial decoding is shown, assuming that the EPROM is the only added device on the bus. The ADV line not only times the latching of the lower half of the address into the 74LS573 latch chip, it also enables the EPROM output onto the bus only for *external* reads.

By using an 8-bit external data bus, two simplifications occur. First, only a single latch chip and a single EPROM chip are required. Second, the EPROM programming is simplified because the content of consecutive addresses in the 8096's address space are mapped into consecutive addresses in the EPROM. That is, the program which is supposed to reside at addresses 4000 to 7FFF of the 8096's address space is programmed into (consecutive) addresses 0000 to 3FFF of the EPROM's memory.

The disadvantage of using an 8-bit external data bus in the circuit of Fig. 6.8 is a reduction in memory access speed. Fetching 2 bytes takes two memory cycles instead of the one which occurs with every fetch over a 16-bit external data bus. The alternative of using the 8096's 16-bit external data bus is illustrated in Fig. 6.9. That circuit shows how two 2764 EPROMS can provide the same amount of

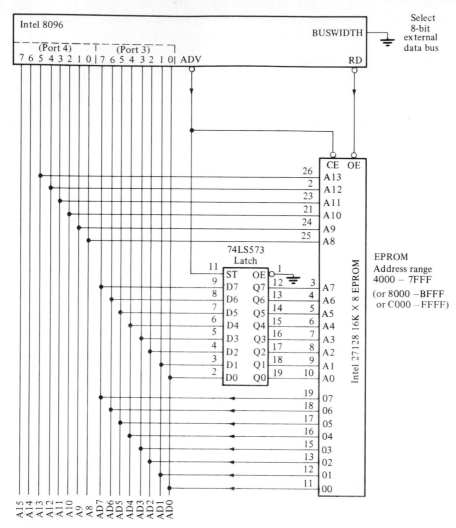

FIGURE 6.8
Intel 8096 EPROM memory expansion with 8-bit external data bus.

memory as the 27128 EPROM of Fig. 6.8 but with faster access. Note that an extra latch is needed and that each EPROM provides data to half of the bus. The object code file of our program must be divided into even bytes and odd bytes for programming into the two EPROMs. This capability is supported by the software associated with some EPROM programmers.

The expansion of microcontroller RAM is implemented most easily with *static* RAM. Furthermore, *CMOS* static RAM offers the opportunity for long-term battery backup of data. In contrast, *dynamic* RAM (DRAM) must be

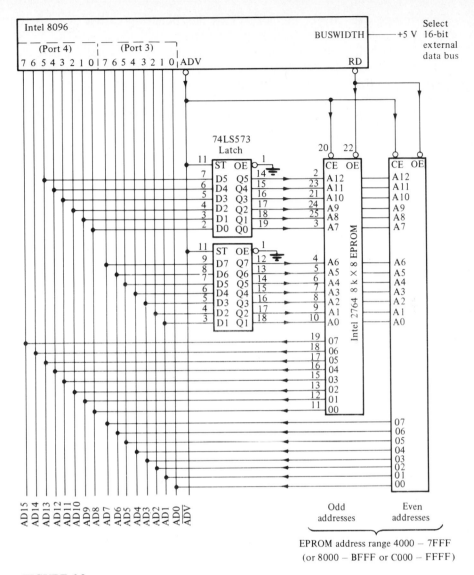

FIGURE 6.9
Intel 8096 EPROM memory expansion with 16-bit external data bus.

periodically refreshed, essentially by carrying out repeated reads of data. Because dynamic RAM permits the storage of more bits per unit of board area than does static RAM, it is widely used for computers needing more memory than can be acquired with only a few chips. In such a circumstance, the dynamic RAM is refreshed by a separate refresh controller chip so that this function requires no CPU time.

The storage capacity of memory chips has been increasing ever since semiconductor memory chips were first introduced. With the first 32K × 8 static RAM chips being introduced in 1985, we have reached the point where all excess memory space of a microcontroller, operating in the expanded mode, can be filled with RAM using only two memory chips!

Example 6.4. Figure 6.10 shows the addition of 24 kbytes of RAM to a Motorola 68HC11 microcontroller. This circuit is quite similar to that of Fig. 6.5 for I/O port expansion for the 68HC11. For I/O port expansion, we ignored the lower half of the address. Here we need it. Since it appears on the multiplexed bus early in the

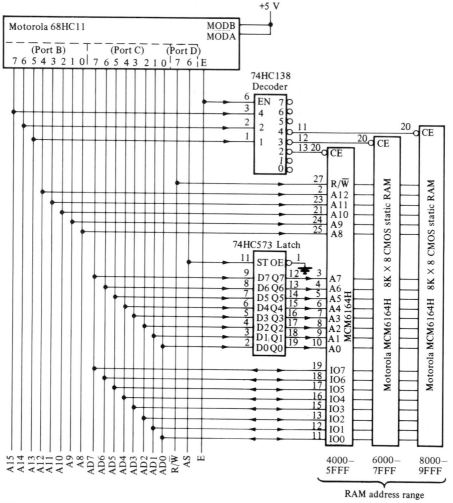

FIGURE 6.10
Motorola 68HC11 static RAM expansion using Motorola RAM chips.

clock cycle and then goes away, it must be latched if it is needed later in the clock cycle. The 74HC573 transparent latch uses the AS (address strobe) signal of Fig. 6.6 to capture this lower half of the address. Consequently, all three RAM chips receive their full 13-bit address for over one-half of the clock period. The read/write line also comes out early in the clock cycle and lingers through to the beginning of the next clock cycle. The decoder is not enabled until the second half of the clock period, when E goes high. This occurs when all of the other signals are stable, thus ensuring that all writes to RAM will be glitch-free.* It also occurs after the 68HC11 has stopped emitting the lower byte of the address, when the address strobe is no longer active, thus ensuring that no contention will occur between the 68HC11 driving the multiplexed bus with this lower byte of the address and a RAM chip which drives the multiplexed bus with data being read from it.

An alternative static RAM expansion circuit for the Motorola 68HC11 is shown in Fig. 6.11a. This circuit uses the generic (6264) RAM chip which has active-low WE (write enable), OE (output enable), and CE (chip enable) inputs. These control inputs are more standard than the R/W (read/write) input and CE (chip enable) input to the Motorola 6164 RAM chip of Fig. 6.10. However, they are not as easy to decode. The circuit of Fig. 6.11a illustrates the use of a programmed PAL for decoding between this mismatch of control signals. The boolean equations implemented by the PAL are shown in Fig. 6.11b.

Example 6.5. A static RAM expansion circuit for the Intel 8096 is shown in Fig. 6.12. This circuit uses the full 16-bit data bus. Consequently, the full 16-bit address must be latched before these lines become a 16-bit data bus later in the clock cycle.

Each RAM chip is enabled by two chip enable inputs, one active-high and the other active-low. Since they are connected to address lines A15 and A14, two of the chips will be enabled by any address of the form

$$01XX\ XXXX\ XXXX\ XXXXB$$

while the other two chips will be enabled by any address of the form

$$10XX\ XXXX\ XXXX\ XXXXB$$

The first addresses range between 0400 to 7FFF while the second addresses range between 8000 to BFFF.

A new wrinkle in this circuit is the need for two write signals. An STB (store byte) instruction to an even address needs to strobe the lower byte (only) of the 16-bit data bus into RAM. The 8096 manages this by strobing WRL and not WRH. In similar fashion a STB instruction to an odd address leads to the strobing of WRH and not WRL. On the other hand, an ST (store word) instruction will necessarily be made to an even address because of the word-alignment requirement of the 8096. When an ST instruction is executed, the 8096 writes to both the even-address RAM and the odd-address RAM by strobing *both* WRL and WRH.

---

* If address lines are still settling out when a RAM chip gets its write signal, then a wrong address can be inadvertently changed along with the right address.

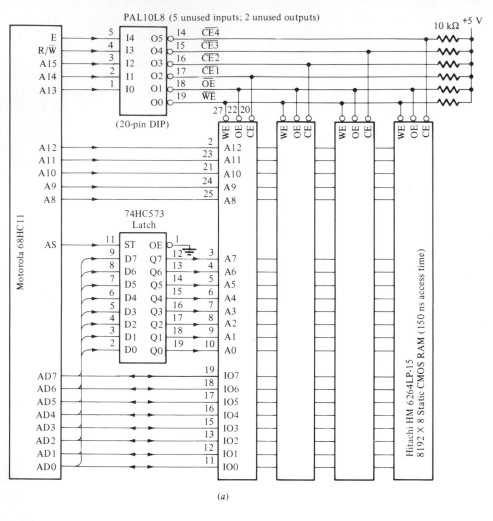

(a)

$$WE = E \cdot \overline{(R/\overline{W})}$$

$$OE = E \cdot (R/\overline{W})$$

$$CE1 = \overline{A15} \cdot \overline{A14} \cdot A13 \qquad (2000 - 3FFF)$$

$$CE2 = \overline{A15} \cdot A14 \cdot \overline{A13} \qquad (4000 - 5FFF)$$

$$CE3 = \overline{A15} \cdot A14 \cdot A13 \qquad (6000 - 7FFF)$$

$$CE4 = A15 \cdot \overline{A14} \cdot \overline{A13} \qquad (8000 - 9FFF)$$

(b)

FIGURE 6.11

Motorola 68HC11 static RAM expansion using generic RAM chips. (a) Circuit; (b) PAL equations.

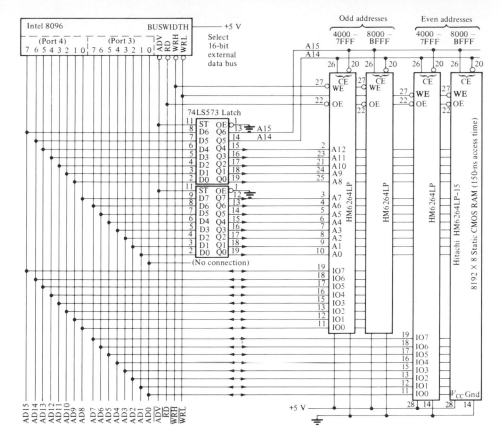

FIGURE 6.12
Intel 8096 static RAM expansion (32K).

Reading from RAM in the circuit of Fig. 6.12 does not entail the same problem. Whether the 8096 is trying to read a byte or a word, *both* the even-address and the odd-address RAM chips are read, putting their content onto the 16-bit data bus. The 8096 ignores one of these bytes when it is reading only a byte, not a word.

The circuits of Figs. 6.10, 6.11, and 6.12 make use of 8K × 8 static RAM chips. While these may seem like overkill for the amount of RAM needed in an application, the phenomenal truth is that 8K × 8 static RAMs became "commodity," or "jelly-bean," parts in 1985, priced well below $10, even in small quantities.

6.4 TIMING CONSIDERATIONS

We really have no business *using* a RAM chip (or an EPROM chip, for that matter) without checking whether it is fast enough for the application. The

"-15" listed in the Hitachi part number (HM6264LP-15) in Figs. 6.11 and 6.12 signifies a RAM chip having an *access time* of 150 ns. This is their *slowest* (and therefore cheapest) version of the chip, but as we shall see, it is fast enough for this application.

When checking the timing requirements for devices attached to the external bus of a microcontroller operating in the expanded mode, we can treat the *read* timing and the *write* timing as two separate requirements. First we will consider the timing requirements for reliably *reading* from a RAM chip (or from a ROM chip, for that matter). Three timing requirements must be met:

1. The *access time* of the RAM chip must be less than that required by the circuit. Access time of a RAM chip is the time which begins when a new address is applied to a RAM and ends when the RAM emits the content of that address (assuming that the chip and the output have long since been enabled).

2. The *output enable to output time* of a RAM chip must be less than that required by the circuit. This is the time beginning when the input which controls the RAM's three-state output is driven active and ending when the content of the selected RAM address appears on the output (assuming that the address input and the chip enable input have been present for a long time). In Fig. 6.12, this is the time for a RAM chip to respond when pin 22 is driven low.

3. The *chip enable to output time* of the RAM chip must be less than that required by the circuit. This time begins when the RAM's chip-enable input (or inputs) is driven active and ends when the content of the selected RAM address appears on the output (assuming that the address input and the output enable input have been present for a long time). In Fig. 6.12, this is the time for a RAM chip to respond after both pin 26 is driven high and pin 20 is driven low. The chip enable to output time is generally longer than the output enable to output time because chip enable inputs are often used to power down much of the internal circuitry of the RAM chip, dropping its power dissipation significantly.

> **Example 6.6.** The Intel 8096 timing specifications for reading from external devices (like RAM chips) are depicted in the top half of Fig. 6.13. The bottom half of this figure examines how these values play out, given the Hitachi 6264 RAM chip and the specific circuit of Fig. 6.12. Timing specifications for this RAM chip are listed in Fig. 6.14.
>
> The ADV signal in Fig. 6.12 strobes the address from the 8096 into transparent latches. Using transparent latches here produces a major advantage over using edge-triggered flip-flops. While ADV is high, then *as soon as* the address appears on the output of the 8096, it will propagate through the latches and appear on their outputs within the D-to-Q propagation delay of the 74LS573 latches (i.e., within 18 ns). If edge-triggered flip-flops were used, we would have to trigger off the falling edge of ADV. Because of this, we would lose the time between when the address is assured of being valid and when ADV drops low.

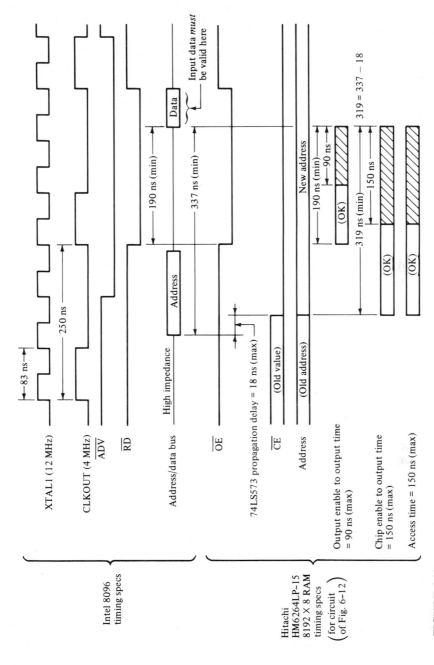

FIGURE 6.13
Intel 8096 timing for reading from RAM.

249

Package: 28-pin DIP

Operating power supply current (chip enabled):
Current   =   40 mA (typical)
                                           =   80 mA (max)

Standby power supply current (chip disabled):
Current   =   2 μA (typical)
                                           = 100 μA (max)

Read specifications:
  Access time   = 150 ns (max)
  Output enable to output time   =   70 ns (max)
  Chip enable to output time   = 150 ns (max)

Write specifications:
  Write pulse width   =   90 ns (min)
  Address set-up time before write   =   0 ns
  Address valid to end of write   = 100 ns (min)
  Address hold time after write   =   0 ns
  Data valid to end of write   =   60 ns (min)
  Data hold time after write   =   0 ns

FIGURE 6.14
Specifications for Hitachi HM6264LP-15 8192×8 static CMOS RAM.

The bottom half of Fig. 6.13 shows the signals which are actually applied to the RAM chip. Looking at the circuit of Fig. 6.12, we see that a RAM chip does not actually see the address put out by the 8096 until it has had a chance to propagate through the 74LS573 latches (18 ns, maximum). Consequently, the RAM access time must be less than $337 - 18 = 319$ ns. Since the Hitachi RAM has an access time specification of 150 ns (maximum), this is OK.

The RAM chip enables are driven by address signals traversing the same latch circuitry. Accordingly, the RAM chip enable to output time must be less than this same 319 ns. This is again met satisfactorily by the Hitachi RAM, which requires no more than 150 ns.

Finally, we can check on the output enable to output time. As shown in Figs. 6.12 and 6.13, this begins when the 8096's active-low RD (read) line goes low and ends when data must first be valid. Figure 6.13 shows this time to be 190 ns; it is again met satisfactorily by the RAM chip, which requires no more than 90 ns.

Next, we will consider the timing requirements for reliably *writing* to a RAM chip. We will be concerned with six timing parameters: a pulse width, three setup times, and two hold times.

1. The *write pulse width* which the microcontroller generates must be sufficiently wide to satisfy the minimum write pulse width requirement of the RAM chip.
2. The *address setup time before write* time is measured from address valid to

the *beginning* of the write pulse. The microcontroller must allow sufficient time to ensure that the write will not change the wrong address, accessed while the address lines are still settling out from the last clock cycle.

3. The *address valid to end of write* time is measured from address valid to the *end* of the write pulse. The microcontroller must allow sufficient time to ensure that the selected address will be reliably written into.

4. The *data valid to end of write* time is measured from data valid to the *end* of the write pulse. The microcontroller must allow sufficient time to ensure that the correct data will be written into the selected address.

5. The *address hold time after write* is measured from the *end* of the write pulse until the address changes. This is the time during which the RAM chip requires that the address be held since it is still using the address for the write operation. The microcontroller must maintain the address for longer than this amount of time.

6. The *data hold time after write* is measured from the *end* of the write pulse until the data changes. Again, this is the time during which the RAM chip requires that the data be held since it is still using the data for the write operation. The microcontroller must maintain the data for longer than this amount of time.

Example 6.7. Consider once again the Intel 8096 RAM expansion circuit of Fig. 6.12, the 8096 write timing given at the top of Fig. 6.15, and the Hitachi 6264 write specifications of Fig. 6.14. For the 8096 and the RAM to work together reliably, the six timing requirements must be met. We will examine each in turn.

1. The write pulse width which the RAM sees is equal to the width of the 8096's write pulse. Consequently, the minimum write pulse width requirement for the RAM chip, listed in Fig. 6.14 as 90 ns, is easily met by the 8096's 147-ns minimum pulse width.

2. Consider next the address setup time before write requirement. Figure 6.15 shows that the RAM's 0-ns specification for this parameter is easily met by the 162 ns provided by the 8096 circuit.

3. For the RAM's address valid to end of write requirement, Fig. 6.15 shows that the RAM's requirement of 100 ns is again easily met by the 309 ns provided by the 8096 circuit.

4. To consider the data valid to end of write requirement, look back at Fig. 6.12. The data will appear at the input to the RAM as soon as it is emitted by the 8096 since there is no buffering of the data lines. The RAM's write pulse is identically equal to the write pulse emitted by the 8096. This leads to the value listed in Fig. 6.15 for this parameter of 190 ns. Since the RAM only requires 60 ns, this is satisfactory.

5. For the address hold time after write requirement, Fig. 6.12 shows that the 74LS573 latches will hold the address until ADV goes high at the end of the write cycle, plus their (minimum) propagation delay. The minimum propagation delay of a 74LS573 is an unspecified parameter, but it is certainly greater

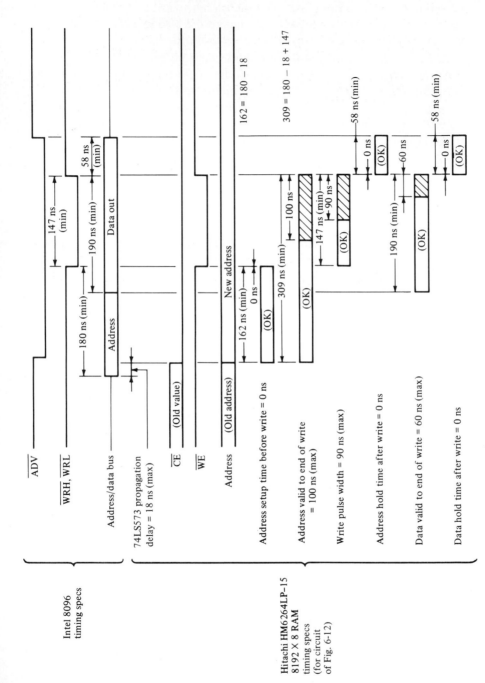

FIGURE 6.15
Intel 8096 timing for writing to RAM.

252

than 0 ns. Anyway, since the RAM chip has a zero hold time specification and since the address is held for at least 58 ns longer than this, we are OK.

6. The data hold time after write requirement of 0 ns is easily met, since the data is actually held for at least 58 ns after the completion of the write pulse.

We will conclude this section with the timing specifications for the Motorola 68HC11. They are shown in Fig. 6.16.

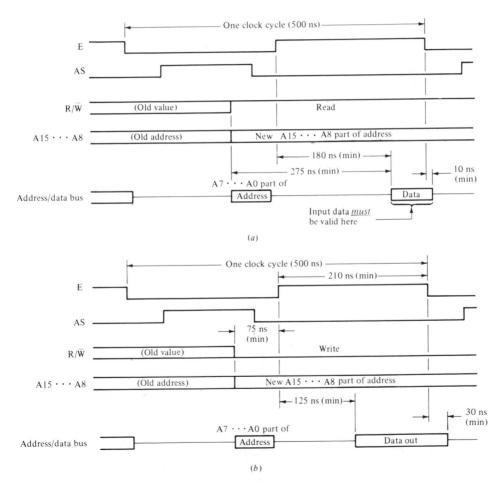

FIGURE 6.16
Motorola 68HC11 timing. (*a*) Read timing; (*b*) write timing.

6.5   UART SERIAL PORTS

A *UART* is a widely used device for converting bytes of data into a serial data stream and vice versa. The discussion of strings in Sec. 5.3 of the last chapter dealt briefly with UART use. We considered the common application of a UART to transmit and receive strings of data. In Secs. A.8 and B.9 of the appendixes, we deal with the details involved in the use of the UARTs in the Motorola 68HC11 and the Intel 8096. In this section, we discuss UART use and see how it can be used to expand the capabilities of a microcontroller.

Serial data can be transmitted between two UARTs imbedded in two devices. Each device requires only three wires for this form of communication: one wire for transmission, one for reception, and one wire for ground. This simplicity of cabling encourages asynchronous serial data transmission between digital devices such as computers and terminals. The need to dedicate only two pins of a device for data exchanges with another device makes this form of communication popular for connecting microcontrollers to each other and to other devices. Consequently, it is not uncommon for a microcontroller to include an on-chip UART.

When a character of data is transmitted serially, the transmitting device and the receiving device must be in agreement on the *communications protocol*. Figure 6.17a shows a popular protocol. When data is not being transmitted (that is, when the line is idle), the line is high. When a character is sent, it is *framed* as follows:

The line goes low for one "bit time" to create a "start" bit

Eight data bits are transmitted, least-significant bit first, with each bit lasting for one bit time

The line goes high for one bit time to create a "stop" bit

For the receiving device to receive this data reliably, it must understand not only that the data is framed in this way, but also it must know the duration of a

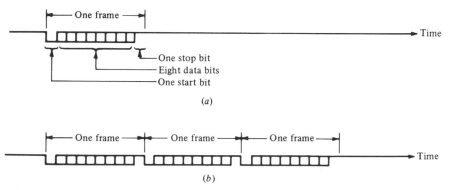

(a)

(b)

FIGURE 6.17
A popular serial communications protocol. (a) Transmission of 1 byte, framed by a start bit and a stop bit; (b) transmission of three successive bytes.

bit time. This is commonly specified as a *baud rate*. For example, a baud rate of 9600 baud specifies that the bit time is $\frac{1}{9600}$ s.

Frame synchronization requires the receiving device to see the line drop from high to low at the beginning of the frame and to synchronize upon this falling edge. This is the reason that idle bits are high and start bits are low. It is also the reason that each frame terminates with a stop bit which is high. When a succession of bytes are transmitted, as in Fig. 6.17b, the receiver synchronizes upon the first falling edge following a long string of idle bits and then resynchronizes upon the next falling edge after the frame is completed. That is, the synchronization circuitry in the receiver waits 9.5 bit times after the falling edge at the start of the frame and then starts looking for the next falling edge to resynchronize upon (which occurs at 10 bit times for an immediately following byte of data or which occurs sometime later than this if some idle bits occur between frames).

Within the UART there are the following registers, shown in Fig. 6.18:

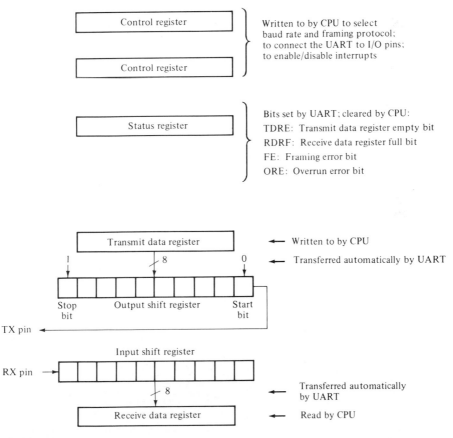

FIGURE 6.18
UART registers.

Transmit data register. The CPU writes each byte to be transmitted to this register.

Output shift register. After a byte has been entirely shifted out of this register, the UART automatically transfers the content of the transmit data register to this register and begins shifting it out. It also sets the TDRE bit in the status register, signifying that the transmit data register is ready to accept another byte from the CPU.

Input shift register. Serial data from another device is received here. When the full byte has been received, the UART automatically transfers it to the receive data register. It also sets the RDRF bit in the status register to tell the CPU that the receive data register is full. If, when the UART is ready to make this transfer of the received byte of data, it discovers that a previously received byte of data has not yet been removed, then it sets the *overrun error* bit in the status register. If the UART reads a 0 when it is expecting a stop bit, it sets the *framing error* bit in the status register.

Receive data register. When this register is read by the CPU, the RDRF bit in the status register is automatically cleared.

Control register(s). These registers provide the means for the CPU to set up the UART's communications protocol. They also enable interrupts from the UART to the CPU so that the CPU can know when it needs to send or receive another character. They also enable the circuitry which connects the UART to actual pins on the chip.

Status register. This register holds the status bits described above.

To send a sequence of bytes out the UART's serial port, the CPU monitors the TDRE bit and when it is set, the CPU writes a byte to the transmit data register (which also clears the TDRE bit). Then it goes back to monitoring the TDRE bit, waiting until that bit is set so that another byte can be sent.

Alternatively, the UART can be set up to *interrupt* the CPU each time it is ready for another byte. This way, the CPU is free to do other things, knowing that it will be informed when the time comes for it to take more action with the UART. Even at the relatively fast baud rate of 9600 baud, interactions with the UART only occur every millisecond (actually 10 bit times $* 1/9600$ s $= 1.042$ ms), so the interrupt approach provides *a lot* of time to do other things.

Receiving data involves the same sorts of interactions. When the RDRF bit is set (perhaps sending an interrupt to the CPU), the CPU knows that a byte has been received. It first checks the framing error bit and the overrun error bit in the status register to make sure that nothing has gone awry. Then it reads the byte from the receive data register, which automatically clears the RDRF bit.

Note that after the RDRF bit is set, the UART can be ignored by the CPU for almost 10 bit times without an overrun error occurring. This is true because the next byte of data received needs 10 bit times to shift into the input shift register before it is ready to be automatically transferred to the receive data register. Again, at 9600 baud this means that the CPU has about a millisecond to respond after the RDRF bit has been set and before it is in trouble. Having a separate receive data register (instead of reading the input shift register directly) is called *double buffering*. Some UARTs actually go so far as to add more registers of buffering, organized as a *FIFO* (first-in–first-out memory) between the input shift register and the receive data register, so that even more time can elapse without incurring an overrun error. However, this would be unusual for the UART on a microcontroller chip, where chip "real estate" can be used more profitably in other ways.

  To illustrate the variety of communications protocols which are supported by microcontrollers, Figs. 6.19 and 6.20 show those supported by the Motorola 68HC11 and the Intel 8096. The use of a programmable ninth data bit, in Figs. 6.19*b* and 6.20*d*, supports serial communications between a master microcontroller and any number of slave microcontrollers, as depicted in Fig. 6.21. Each of the slave microcontrollers can be set up so that if it receives a frame which has the ninth data bit *set*, then it will generate an interrupt to its CPU; otherwise, a frame received without this bit set will be ignored by the UART.

  When the master microcontroller wants to send a sequence of bytes to slave 1, it first transmits the number 1, together with the ninth bit set. All of the slaves interrupt their CPUs. The interrupt service routine in each microcontroller checks the received number to see if it is "its" number. All but slave 1 see

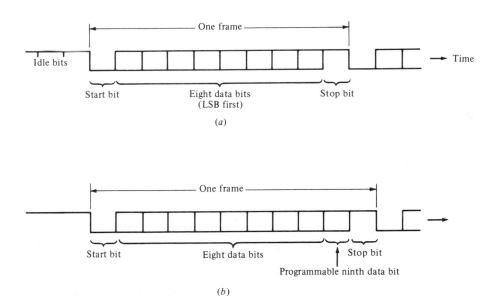

FIGURE 6.19
Motorola 68HC11 UART protocols. (*a*) Ten-bit frame with one start bit, eight data bits, one stop bit; (*b*) 11-bit frame with one start bit, eight data bits, one programmable ninth data bit, one stop bit.

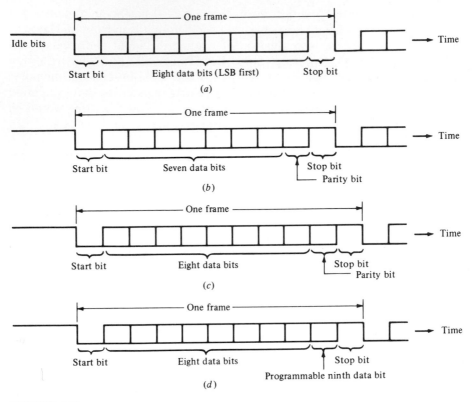

FIGURE 6.20
Intel 8096 UART protocols. (*a*) Ten-bit frame with one start bit, eight data bits, one stop bit; (*b*) 10-bit frame with one start bit, seven data bits, one parity bit, one stop bit; (*c*) 11-bit frame with one start bit, eight data bits, one parity bit, one stop bit; (*d*) 11-bit frame with one start bit, eight data bits, one programmable ninth data bit, one stop bit.

that it is not. They return from their interrupt service routines and will not be interrupted by any of the bytes which follow, which are directed by the master to slave 1 and which all have the ninth bit *cleared*. On the other hand, slave 1 sees that the master wants to send it data. *It changes its mode of operation* so as to be interrupted by *all* received bytes, regardless of the state of the ninth bit. Then it handles all of the remaining bytes sent to it. The master microcontroller sends the ninth data bit by setting or clearing a bit in the control register before writing to the transmit data register. The ninth data bit is received by the addressed slave (slave 1 in our case) as a bit in the status register. That slave's interrupt service routine must test this bit each time a frame is received so that it can know when the master wants to address another slave. If it receives a frame with this bit set, then it checks if the byte sent represents its address (of 1). If not, then it "puts itself to sleep" by setting its mode so as not to be interrupted until the next time the master sends a frame with the ninth data bit set.

In this master-slave configuration of multiple microcontrollers, the slaves

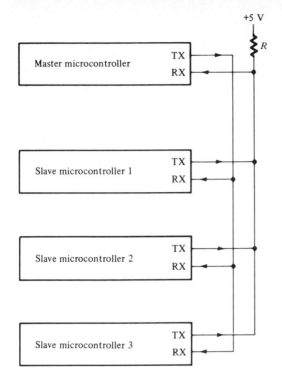

FIGURE 6.21
Master-slave interconnection of microcontrollers.

are never permitted to transmit data until told to do so by the master. In this way *contention* on the shared line is avoided. That is, the line connecting the UART transmitter output from each slave to the UART receiver input to the master will never be driven by more than one slave at a time. The circuit of Fig. 6.21 assumes that each TX output is an open collector or three-state output so that the line can be shared in this way. If this is not true, then an external gate can combine these TX outputs from the slaves into the RX input to the master.*

The function of a microcontroller is to have all the resources it needs on the chip. In the case of the UART, this includes the *baud rate generator*. Of course, to interconnect multiple microcontrollers as in Fig. 6.21, *any* baud rate will work so long as all of the microcontrollers employ the same crystal frequency and all are set up to divide the internal clock by the same number to obtain the baud rate. In fact, we might want to use the maximum baud rate available. The values for the Intel 8096 and the Motorola 68HC11 are listed near the top of Fig. 6.22 for their standard crystal frequencies (of 12 and 8 MHz, respectively).

If the crystal used to generate the internal clock of the microcontroller is not carefully chosen, then it will be impossible to divide it down and obtain standard baud rates exactly. The Intel 8096 and Motorola 68HC11 have both been designed to get close to standard baud rates even while using their stan-

---

* For an example, see Fig. B.33 of Appendix B.

		Intel 8096	Motorola 68HC11
Crystal frequency		12 MHz	8 MHz
Maximum baud rate		187,500 baud	125,000 baud
Accuracy of "nominal" baud rates	19200 baud	2.40%	1.73%
	9600 baud	2.40%	0.16%
	4800 baud	0.16%	0.16%
	2400 baud	0.16%	0.16%
	1200 baud	0.16%	0.16%
	600 baud	0.16%	0.16%
	300 baud	0.00%	0.16%

FIGURE 6.22
Baud rate generator characteristics.

dard crystal frequencies. Figure 6.22 shows how close the actual baud rates can be set to approximate a selection of standard baud rates. For example, the Motorola 68HC11 divides the 8-MHz crystal clock rate by $4 * 13 * 16 = 832$ to get an actual baud rate of 9615.4 baud, when what is desired is 9600 baud. Thus the actual baud rate is $100*15.4/9600 = 0.16$ percent high.

Refer again to Fig. 6.17 to consider the effect of a slightly inaccurate baud rate. Consider the case where the transmitting UART is transmitting at 9615.4 MHz while the receiving UART in another device is receiving this same data with its own baud rate generator set up for exactly 9600 baud. The receiver is going to synchronize on the leading edge of the start bit (to within one-sixteenth of a bit time) and then count over to the middle of the first data bit to read that bit. However, that middle will be off by 0.16 percent—a very slight amount. The inaccuracy plays out to a maximum value when the receiver tries to read the stop bit 9.5 bit times after synchronization. At that point the inaccuracy of 0.16 percent translates into 0.16 percent of 9.5 bit times, or 0.015 bit time. It can actually be off by any amount under 0.5 bit time (or better, under 0.4 bit time) and still read the stop bit, which extends from $9.5 - 0.5 = 9.0$ bit times to $9.5 + 0.5 = 10.0$ bit times after synchronization. Consequently, this inaccuracy of 0.16 percent means that the times when the received data is being read are nearly perfect! Even an error of 2.40 percent translates into 0.23 bit time, which should serve satisfactorily.

The interface of a UART to a personal computer or a terminal or some other device which employs an RS-232C interface requires some special considerations. We have to match the baud rate and the communications protocol (e.g., one start bit, eight data bits, and one stop bit) of that device. Furthermore, the RS-232C standard requires that voltage levels be translated from the 0 V to +5 V world at the pins of the microcontroller to levels of above +3.0 V or below −3.0 V on the cable between the two devices. One circuit for achieving this voltage translation is shown in Fig. 6.23a. However, this circuit requires that we have available +12-V and −12-V power supplies. If we have

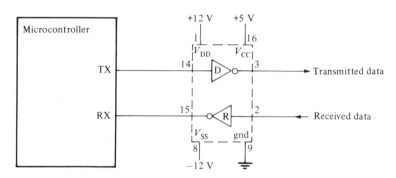

Motorola MC145406
RS-232C drivers/receivers
(three drivers, three receivers)

(a)

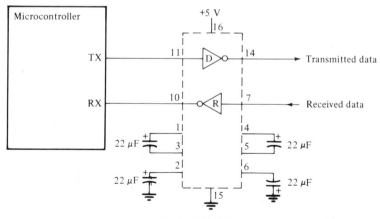

Maxim MAX232
RS-232C drivers/receivers
(two drivers, two receivers)

(b)

FIGURE 6.23
RS-232C interface circuits. (a) Given three power supply voltages; (b) given only a supply voltage of +5 V.

only +5-V power available, then the RS-232C interface chip shown in Fig. 6.23*b* includes two dc-to-dc converters as well as two drivers and two receivers, all in a 16-pin DIP.

Both the Intel 8096 and the Motorola 68HC11 include a single UART. What can we do to meet the needs of an application requiring *more than one* UART? The first thing we can do is realize that if the application involves a serial port which receives serial data but never transmits it, then we have a leftover serial transmitter which can be used for another purpose.* For example, a printer buffer with serial input port and serial output port would appear to need two serial ports. And in fact it does need two complete UARTs if it employs XON-XOFF flow control. A flow control protocol provides the means for a printer buffer to tell a computer to stop sending data because the buffer is almost full. Under XON-XOFF flow control, the printer buffer sends an ASCII control code *back* to the computer which tells the computer to pause in its sending of characters. When the buffer has sufficient room again, the printer buffer sends a different ASCII control code† back to the computer which tells the computer to resume its sending of characters. While a printer buffer needs a flow control protocol of some sort, it might use a hardware handshake protocol instead of XON-XOFF, if the application permits.

More generally, filling a requirement for more than one UART is usually met by operating the microcontroller in the expanded mode and adding a separate UART chip. This is exactly what the printer buffer shown in Fig. 1.8 does. Because it needs to have the versatility of being able to handle any of the normal flow control protocols, it employs a Zilog 8031 asynchronous serial communications controller. This 40-pin DIP includes *two* independent UARTs, each with its own baud rate generator. Since the microcontroller is operated in the expanded mode anyway (to access up to 2 Mbyte of RAM for its buffer function), adding an extra chip is not a big decision.

In those few applications where several UARTs are needed but no more than one is used at any one time, the single UART on a microcontroller chip can be *multiplexed*. Examples of such applications arise where the microcontroller drives several output devices serially. A serial interface is commonly used for a CRT display. We consider the use of a built-in CRT display in Sec. 7.4. Another common way to produce a minimum-cost device is to require the user to supply the display in the form of a CRT terminal. The device then just needs to provide an RS-232C interface to the terminal. A printer with a serial interface represents another such output device.

The feature shared by these output devices which is important here is that they do not talk back, except perhaps to handshake. Consequently, they do not generate inputs at times which could be missed by the microcontroller while it

---

* Provided the serial communications protocol and the baud rate can be the same.
† The two ASCII control codes are called XON and XOFF. They are normally the hexadecimal numbers 11H and 13H, respectively.

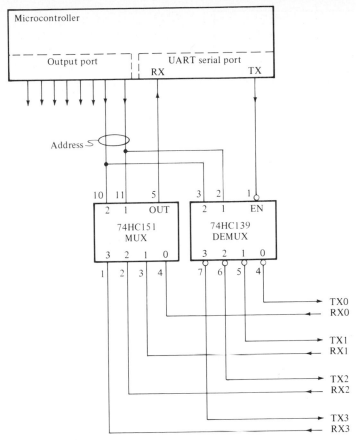

FIGURE 6.24
Multiplexing the on-chip UART.

has reconnected its UART to another device. In such a circumstance, the circuit of Fig. 6.24 serves well.

In other applications we would worry about the introduction of glitches onto the outputs of the 74HC139 or the output of the 74HC151 when the 2-bit address is changed. However, since virtually all UARTs are designed to be immune to glitches, that is not a big issue. In fact, the 74HC139 is inherently glitchless in this application (because its active-low enable input will be high when the 2-bit address is changed, forcing the selected output high, the same level to which the unselected outputs are held).

6.6 I/O SERIAL PORTS

For many applications, the most limited resource of a microcontroller chip is the number of I/O lines available. Some microcontrollers, including both the

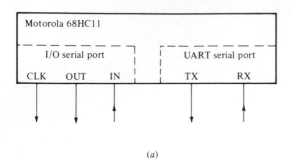

(a)

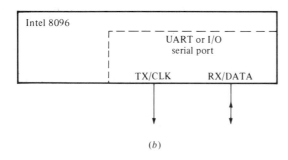

(b)

FIGURE 6.25
Serial port capabilities. (a) Motorola
68HC11; (b) Intel 8096.

Intel 8096 and the Motorola 68HC11, support I/O expansion with a few pins configured to be an I/O serial port and the internal hardware to service them quickly. This is illustrated in Fig. 6.25. The Motorola 68HC11 has the distinct advantage of having separate pins for its UART serial port and its I/O serial port. That is, these two facilities can be operated independently of each other. In contrast, users of the Intel 8096 have to give up use of the UART serial port during I/O serial port transfers and vice versa. In addition, the 8096's I/O serial port combines the data in and data out lines into one bidirectional data line, thereby complicating the interconnections to external devices somewhat.

An I/O serial port can be connected to shift registers to serve the purposes of I/O expansion. In its simplest application, when the CPU writes to an 8-bit register which controls the I/O serial port, the 8 bits are shifted out of the microcontroller along with a pulse train consisting of eight pulses. The timing involved is shown in Fig. 6.26 for both the Motorola 68HC11's and the Intel 8096's I/O serial port. A simple circuit for output expansion using this capability is shown in Fig. 6.27. In addition to the serial clock and the serial data output lines, it also requires one output line from a port.

To change these three output ports, the CPU might first update 3 bytes of RAM which hold a copy of what is presently on the ports. This permits a few bits to be changed while holding the remainder unchanged. Then a subroutine is called which writes the byte directed to the bottom shift register to the serial output register. When a status bit indicates that the serial port is ready for the next byte, it is written out, followed shortly thereafter by the third byte. This

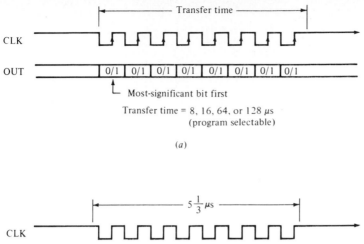

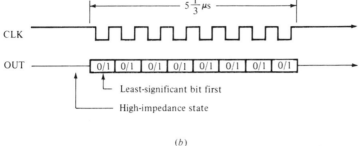

FIGURE 6.26
Output timing for I/O serial port. (*a*) Motorola 68HC11 (this is one of four options; see Fig. A.28 in Appendix A); (*b*) Intel 8096.

will shift all 24 bits out to the shift registers. The 74HC595 part includes an 8-bit buffer register between the shift register's parallel output and the eight parallel output lines of the chip. Consequently, none of the 24 outputs change during the shifting process. Now, with all 24 bits in place, the CPU writes a 0 and then a 1 to the bit of the output port labeled Transfer in Fig. 6.27. This generates a rising edge to the Latch Clock input of each 74HC595, which causes the internal transfer to take place between its shift register and its buffer register driving the output lines.

The circuit of Fig. 6.27 shows the pullup resistors needed if an NMOS Intel 8096 is interfaced to the CMOS 74HC595 parts. Alternatively, 74LS595 shift registers could be used with *either* microcontroller and the resistors avoided. However, the CMOS parts have the advantage of dissipating much less power than the LSTTL parts.

An alternative circuit that is faster than the one in Fig. 6.27 is shown in Fig. 6.28. A 2-bit address is first written to the output port shown (carefully, to avoid changing the other lines of the port). The single byte which is then transmitted by the I/O serial port will thus be steered to the selected shift register. When the 8-bit transfer is complete, a 0 and then a 1 is written to the

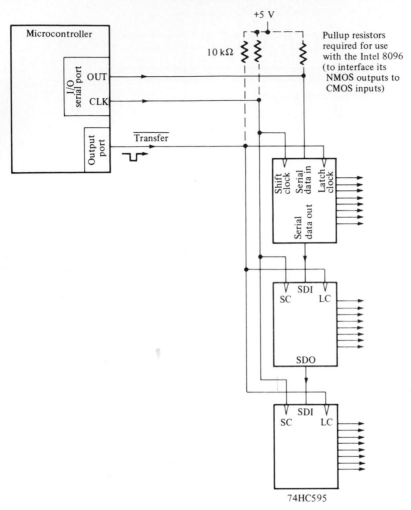

FIGURE 6.27
Output expansion via I/O serial port.

output port bit which drives the active-low Strobe line. Again, this is steered to the selected shift register.

The circuit of Fig. 6.28 is really a universal output expander for an I/O serial port. In the next chapter we survey a variety of output devices having an I/O serial port input. The interface to every one of these output devices is shown in Fig. 6.28. The output device simply takes the place of one of the shift registers of Fig. 6.28.

Input port expansion can be handled in much the same way. Figure 6.29 shows the input timing for the I/O serial port of both the Motorola 68HC11 and the Intel 8096. The circuit of Fig. 6.30 shows three 74HC165 parallel-in serial-

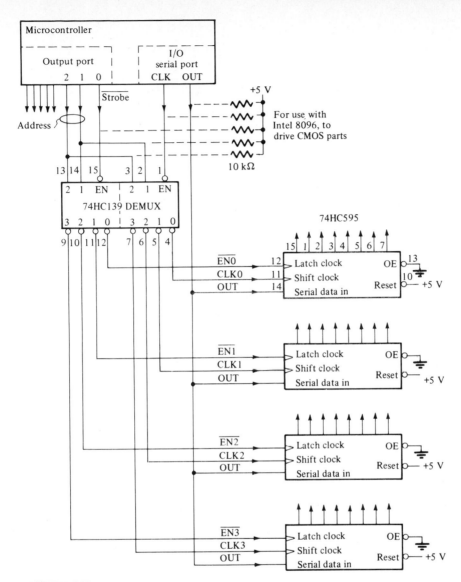

FIGURE 6.28
Alternative output expansion via I/O serial port.

out shift registers configured to support this application. When the CPU is ready to read the three ports, it might call a subroutine which first writes a 0 followed by a 1 to the line of the output port labeled Load. This parallel-loads the 24 inputs into the three shift registers. Then the CPU initiates the serial transfer of 3 bytes into the I/O serial port, waiting after each transfer until a status bit indicates that the 8-bit transfer is complete. Again, it might store these 3 bytes into 3 bytes of RAM and then return from the subroutine written

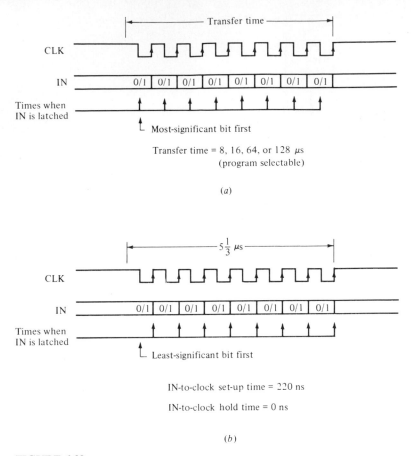

FIGURE 6.29
Input timing for I/O serial port. (*a*) Motorola 68HC11 (this is one of four options; see Fig. A.28 in Appendix A); (*b*) Intel 8096.

to carry out this transfer. The calling routine can now read any of these 3 bytes of RAM just as if it were reading any of the three input ports created with the three 74HC165 parts.

There is one timing concern implied by Fig. 6.29*b* which calls for our consideration. Note that the serial input data is latched by the Intel 8096 on the rising edges of the serial clock. Also note, in Fig. 6.30, that the shift registers are clocked by this same rising edge. Consequently, the serial input data *changes* at the very moment that it is being read. The designers of the 8096 were careful and designed the serial input port to have *zero hold time*. That is, the serial input data need not be held for *any* nanoseconds after the rising edge of the serial clock in order for reliable data transfer to occur. Since the serial input data actually changes a flip-flop propagation delay *after* the serial clock's rising edge, the transfer will indeed be reliably carried out.

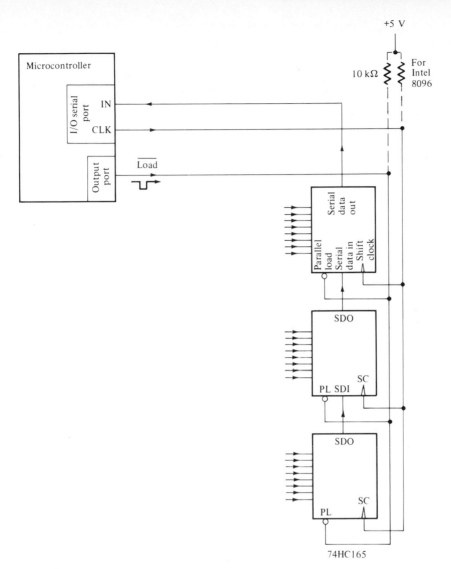

FIGURE 6.30
Input expansion via I/O serial port.

While the Intel 8096 permits this expansion capability, it does so with limited flexibility since the serial data input line and the serial data output line are, in fact, one and the same. Consequently, if we want to expand both input ports and output ports, we need to add some extra gating to accomplish this expansion. In particular, the circuit of Fig. 6.30 raises an immediate concern because the serial data output of the top 74HC165 will cause contention when the data line of the I/O serial port is configured to be an output. In addition, recall from Fig. 6.25 that the Intel 8096 also shares these two lines with the UART function discussed in the last section. Consequently, if an application requires *both* capabilities, then extra gating must be added to disable the UART function while the I/O serial port function is being used and vice versa.

Example 6.8. Expansion of the Intel 8096 serial port is illustrated in Fig. 6.31. Note that the input multiplexer has a three-state buffer to drive the bidirectional data line.

The enabling of the 74HC138 decoder is a little tricky. When a UART is being used, we must write

00001xx0

to port 1, where the xx bits select one of the four UART ports. This assumes that bits 5, 6, and 7 of the port are used as inputs (as indicated in the Note in Fig. 6.31) so that we do not affect these bits by writing to them. The 1 which is written to bit 3 is ANDed with the TX output of the I/O serial port to generate one of the active-low enable inputs to the 74HC138 demultiplexer. The other active-low enable input to the 74HC138 is enabled by the 0 we wrote to bit 0 of port 1. Consequently, the 74HC138 will be enabled each time the UART emits a 0. The selected output of the 74HC138 will follow this signal. The deselected outputs of the 74HC138 will all idle high.

When one of the four expanded I/O serial ports of Fig. 6.31 is being used, we must first write

000y0xx1

to port 1. The value of bit 4, labeled y above, is determined by whether we are going to carry out an input transfer or an output transfer. The two bits labeled xx above select the I/O port. Then, depending upon the nature of the device connected to the I/O serial port, we must deal with bit 0 of port 1 appropriately. For example, if a device requires an active-low enable signal during the entire serial transfer, then we will clear bit 0 before the serial transfer is begun and set it again upon completion of the transfer. If the device requires an active-high enable signal during the entire serial transfer, then we will do the same thing. However, in this case we will add an inverter between the active-low output of the 74HC138 decoder and the active-high enable input to the device.

If the device connected to the I/O serial port is a 74LS595 shift register used to create an output port, then the 74LS595's latch clock must be strobed *after* the transfer has been completed. To do this, we wait until the transfer has been completed before clearing bit 0 of port 1. Then we immediately set bit 0 again. The 74LS138 output will thereby serve as the latch clock for the selected 74LS595 chip.

The Intel 8096 I/O serial port is given either an input command or an output command so that it can use the single DATA line for either input or output. It then generates either the input timing of Fig. 6.29b or the output timing of Fig. 6.26b.

Note that the I/O serial port's clock output is shown going directly to all of the expanded I/O serial ports without being gated. This will work fine with the 74LS595 shift registers discussed earlier. The serial data sent to one shift register will be shifted into all of the shift registers. Only the shift register whose latch clock input is strobed will transfer this data to its output pins. The output pins on the other 74LS595 chips remain oblivious to what has been going on. For connec-

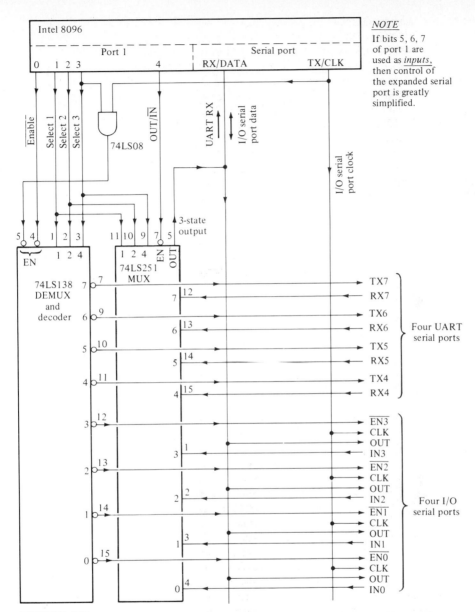

FIGURE 6.31
Intel 8096 serial port expansion.

tion to some more sophisticated peripheral which cannot afford to be clocked when it is not selected, the decoder approach of Fig. 6.28 can be used. Alternatively, a two-input OR gate will solve the problem, as in Fig. 6.32.

Example 6.9. For the Motorola 68HC11, a circuit analogous to that of Fig. 6.31 is shown in Fig. 6.33. Note that the interactions with the UART serial port are completely independent of the interactions with the I/O serial port, permitting

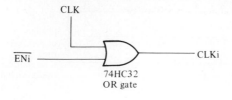

(a)

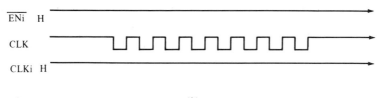

(b)

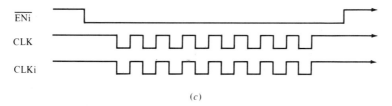

(c)

FIGURE 6.32
Use of an OR gate to generate a gated clock signal. (a) Circuit; (b) timing diagram with gated clock output disabled; (c) timing diagram with gated clock output enabled.

these two to be used simultaneously. Also note that the expansion circuit requires only three chips, not four, since a single 74HC139 decoder/demultiplexer chip serves in two capacities.

Example 6.10. As a final example of expanding internal microcontroller resources with the help of the I/O serial port, we will consider the addition of a small amount of nonvolatile memory to the Intel 8096. The 32 bytes of EEPROM in the Xicor X2444 chip shown in Fig. 6.34a are organized into 16 words of 16 bits each. In fact, the EEPROM is isolated from the serial port by a 16 × 16 RAM. Using the I/O serial port we can

Write a 16-bit word into a specified RAM address
Read a 16-bit word from a specified RAM address

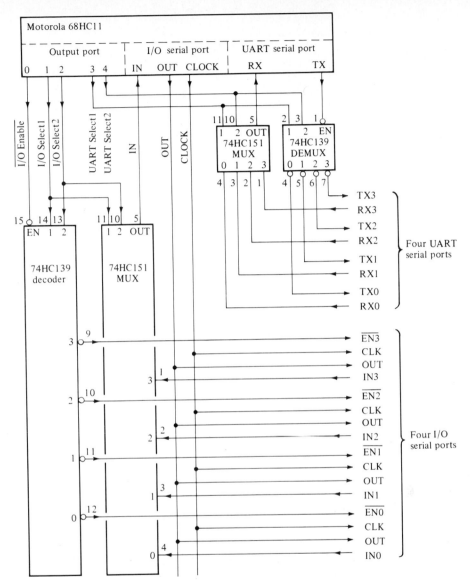

FIGURE 6.33
Motorola 68HC11 serial port expansion.

Command the chip to store the entire RAM into EEPROM

Command the chip to recall the entire RAM from EEPROM

Command the chip to enable or disable all writes to RAM and all stores to EEPROM

The ability to enable and disable writes and stores is the key to maintaining data integrity when power is lost. An inadvertent power outage may lead to the CPU flailing on its address and data bus. However, if writes and stores are normally left disabled, then it is extremely unlikely that this flailing will send just the right serial command to the Xicor chip to enable writes and stores. It is even more improbable that such an unlikely event will be followed by a write command or by a store command.

The Xicor X2444 chip requires an active-high chip enable signal. This signal must be disabled between commands to the X2444, giving it a way to resynchronize with each new command. The circuit of Fig. 6.34a derives this signal by inverting the active-low EN1 output of the 8096 serial port expansion circuit of Fig. 6.31. This EN1 line can be changed by setting or clearing bit 0 of port 1 in Fig. 6.31.

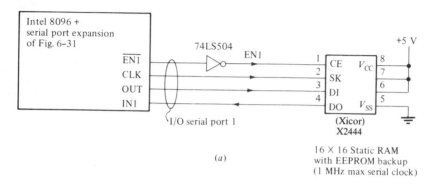

(a)

16 X 16 Static RAM with EEPROM backup (1 MHz max serial clock)

Command name	Hex coding	Function
WRDS	01	Disables writes and stores
WREN	21	Enables writes and stores
STO	81	Store RAM data into EEPROM
RCL	A1	Recall RAM data from EEPROM
WRTMP	C1	Template for forming write command
RDTMP	E1	Template for forming read command

(b)

Write command = WRTMP + 2* address

Example: C7 is hex coding for command to write to address 3
(C1 + 2 X 3 = C7)

Read command = RDTMP + 2 * address

(c)

FIGURE 6.34
Xicor X2444 16×16 serial static RAM with EEPROM backup. (a) Interface circuit; (b) commands; (c) first byte of 3-byte write and read commands.

Each of the Xicor X2444 commands has a hex coding, shown in Fig. 6.34*b*. If these codes are defined in a program with

```
WRDS    EQU    01H              ;EEPROM disable command
WREN    EQU    21H              ;EEPROM enable command
STO     EQU    81H              ;EEPROM RAM to EEPROM command
  .
  .
```

then we can use them to help build the commands which must be sent to the chip. The timing for these commands is shown in Fig. 6.35. To execute the WREN command (to enable writes and stores), we must generate a sequence such as:

```
LDB     P1,#00010011B    ;Select but don't enable
ANDB    P1,#11111110B    ;Now enable X2444
LDB     SBUF,#WREN
```

When the I/O serial port is ready for the next transfer, we can write the content of the AX register to RAM address 5 in the EEPROM chip with three writes to the I/O serial port. The first write is:

```
TOGGLE                   ;Macro to toggle EN1 (1-0-1)
LDB     SBUF,#WRTMP+(2*5)
```

This uses the formula of Fig. 6.34*c* to create the first byte of the EEPROM chip's write command. When the serial port has finished sending this, we can send the least-significant byte of AX, called AL, with

```
LDB     SBUF,AL
```

Then, after waiting for the serial port to finish executing the last command, the most-significant byte is sent with

```
LDB     SBUF,AH
```

The other commands are carried out in much the same way.

The command to read a word from the Xicor 2444's RAM back to the AX register is the most intricate of the commands. After toggling the Xicor X2444's chip enable pin and after sending the first byte, we must change the output to the serial expansion circuit of Fig. 6.31 so that the I/O serial port is redirected from output to input. This is done with

```
ANDB    P1,#11101111B    ;Select serial input
```

Now a serial input command is executed.* When the serial transfer is complete, we can load the received byte into AL with

```
LDB     AL,SBUF          ;Get least-significant byte
```

Now we initiate another transfer. When it is complete,

```
LDB     AH,SBUF
```

Thus far in this section we have discussed several alternative expansion circuits for a microcontroller's I/O serial port. We have dealt with devices

---

* Refer to Sec. B.9 of Appendix B.

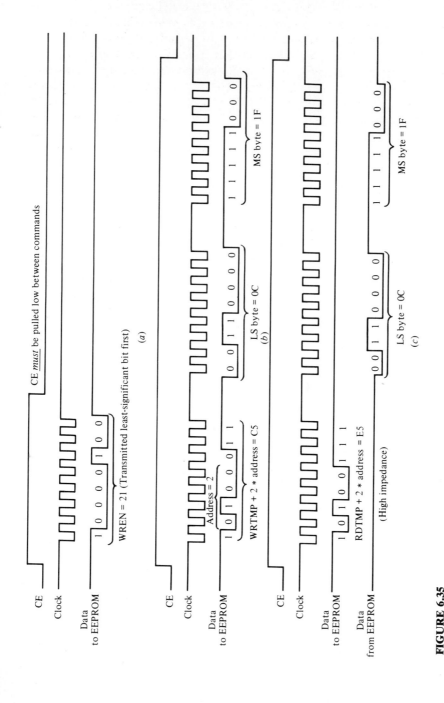

FIGURE 6.35
Timing diagrams for Xicor X2444 16×16 serial static RAM with EEPROM backup. (*a*) WREN command; (*b*) write of 1F0C to address 2; (*c*) read of 1F0C from address 2.

connected to the expansion circuit as if the I/O serial port were idle at the moment when a device requires service. More realistically, the I/O serial port's interrupt service routine is the appropriate place to sort out which device gets serviced whenever the I/O serial port is ready to make another transfer.

As an example of the interactions involved, assume that the I/O serial port is used solely for output expansion and that the circuit of Fig. 6.28 is employed for this purpose. Assume that 74HC139 outputs 0, 1, and 2 are connected to the 74HC595s shown and are to be serviced with that priority (i.e., output 0 with the highest priority and output 2 with the lowest of the three). To illustrate the interactions involved with a more complicated I/O device, assume that output 3 is connected to a liquid-crystal display circuit of the kind to be discussed in Sec. 7.3. This display can be assumed to have the lowest priority of all. When it needs to be updated, the updating process is permitted to take as long as 0.1 s or so (before the LCD display will exhibit a momentary blink of erratic data).

Associated with each of these outputs is a *buffer,* located in RAM. For example, if EXT0 represents the output of the 74HC595 connected to output 0, then its buffer might be a 1-byte RAM variable called EXT0_BUF. In like manner, EXT1_BUF and EXT2_BUF might be the RAM variable buffers for the other two 74HC595 outputs. As will be seen in Fig. 7.14, the LCD might have a RAM buffer called LCD_BUF consisting of *seventeen* bytes.

The link between code calling for an I/O transfer and the I/O serial port's interrupt service routine is a *state variable*, perhaps called IOSP_STATE. The bits of IOSP_STATE might be assigned as follows:

Bit 0 is set when EXT0 is to be updated.

Bit 1 is set when EXT1 is to be updated.

Bit 2 is set when EXT2 is to be updated.

Bit 3 is set when the LCD is to be updated.

Whenever the I/O serial port completes a transfer, it generates an interrupt. Within its interrupt service routine, it might first drive the active-low Strobe line of Fig. 6.28 high. This will complete the *last* transfer. Then ISOP_STATE is examined. If bit 0 is set, then this bit is cleared, the I/O serial port's interrupt flag is cleared, the I/O serial port expansion circuit of Fig. 6.28 is set up to select EXT0, and the I/O serial port itself is set up to transfer data in the format required by EXT0 (i.e., with the format of Fig. 6.26). Then the content of EXT0_BUF is written to the I/O serial port, and the interrupt service routine is exited. This scheme uses the I/O serial port's interrupt service routine to transfer exactly 1 byte. By returning from the interrupt service routine following the setup of each transfer, it opens the door to the servicing of any higher priority interrupt sources between I/O serial port transfers. For a microcontroller like the 8096 which can reenable interrupts *within* the interrupt service routine, an alternative approach would be to wait for the flag to be set which indicates that the transfer has been completed and then jump back to the beginning of the interrupt service routine.

The above procedure outlines what happens when the highest priority device, EXT0, requires updating. On the other hand, if bit 0 of IOSP_STATE is not set, then the interrupt service routine polls bit 1 to see if EXT1 requires updating. If not, then it polls bit 2 to see if EXT2 requires updating. If not, then it polls bit 3 to see if the LCD is ready to receive one more of the 17 bytes it gets during its updating. If none of the above conditions is true, then after the active-low Strobe line of Fig. 6.28 is driven high, I/O serial port interrupts are disabled, but the I/O serial port's *interrupt flag* is left set. In this way, the subsequent reenabling of I/O serial port interrupts will cause an immediate interrupt, leading to immediate service.

We have seen the role of the I/O serial port's interrupt service routine in the updating process. The other half of any transfer takes place as illustrated by the following example. Suppose that the least-significant bit of EXT1 is used to drive a single light-emitting diode (LED). Perhaps we turn on the LED by setting that bit to 1. To do this without changing any of the other bits of EXT1, we can use an OR instruction to force bit 0 of EXT1_BUF to a 1. Alternatively, if the microcontroller has a bit set instruction, then it can be used. Bit 1 of IOSP_STATE is then set to notify the I/O serial port's interrupt service routine of the need to update EXT1. Finally, I/O serial port interrupts are enabled. If they are *already* enabled, then this will be like trying to turn on a switch which is already on; no harm is done. On the other hand, if I/O serial port interrupts had been disabled, because of no other I/O activity going on at that moment, then this enabling of interrupts will lead to the immediate invocation of the interrupt service routine.

We conclude this section on I/O serial port expansion by pointing out that the I/O serial port can be used to interconnect multiple microcontrollers. In the last section we saw how the UART serial port could achieve this function. While a discussion of this capability will not be dealt with here, it is included in Sec. A.7 for the Motorola 68HC11. It should be pointed out that using the I/O serial port for this role permits a speed enhancement of over an order of magnitude compared to using the 68HC11's UART serial port for this purpose.

6.7   PROGRAMMABLE TIMERS

In this chapter, we have been considering how to expand the various capabilities of a microcontroller. In fact, the programmable timer is so crucially important for so many applications that the designers of both the Intel 8096 and the Motorola 68HC11 have provided expanded programmable timers in each of these chips. That is, where the programmable timer on earlier microcontrollers might have been able to generate one output or measure one input, these microcontrollers can deal with a handful of inputs and outputs simultaneously. Nevertheless, there are times when even more of a good thing becomes valuable.

As long as we can partition timed outputs into groups such that no two outputs of any one group must be active simultaneously, we can easily expand

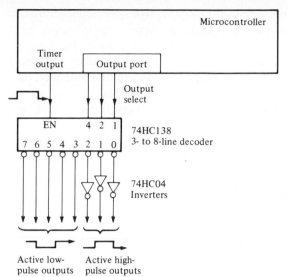

FIGURE 6.36
Generation of nonoverlapping accurately timed outputs.

the number of timed outputs. The circuit is shown in Fig. 6.36. We simply select an output channel by writing to the output port shown. Then the timer output will be steered to the selected output line.

For expanding the programmable timer's inputs, we proceed in an analogous fashion with the circuit of Fig. 6.37. Again, only one of the input lines can be monitored at a time.

6.8   FUSE PROGRAMMABLE CONTROLLERS

The final capability of a microcontroller which we want to augment is the *speed* of its interactions with an external process. In Chap. 3 we saw how a program-

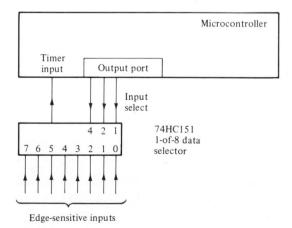

FIGURE 6.37
Circuit for making timing measurements on up to eight inputs, one at a time.

mable timer could be used to generate outputs which have speed characteristics that are hard or impossible to match solely under program control. In the next chapter we will see how specialty chips can be used to carry out activities (e.g., control of a CRT display) that are too fast for a microcontroller to handle directly. In this section we look at how fast interactions can be controlled by a single-chip, fast (up to 20 MHz), programmable state machine.

Single-chip, programmable state machines are available from several vendors:

> PAL (programmable array logic) chips are available from Monolithic Memories (and assorted second sources) which include D-type flip-flops. These include fusible links for implementing random logic inputs to the flip-flops. PALs are good for implementing strange sequential circuits. Since the random logic can be a function of both inputs and flip-flop outputs, it is also possible to use them to implement simple state machines.

> FPLS (field programmable logic sequencer) chips from Signetics include JK flip-flops and are somewhat more versatile than PALs for implementing sequential circuits. Again, they include fusible links which can be programmed to implement strange sequential functions. While they are not optimized for implementing state machines, nevertheless they can be used for simple sequencing operations.

> FPC (fuse programmable controller) chips from Advanced Micro Devices are designed specifically for implementing state machines. Furthermore, they employ what AMD calls a shadow register, which provides a superb link to a microcontroller chip. We will see that an FPC requires only *a single output* from the microcontroller, in addition to an I/O serial port.

AMD's 29PL141 fuse programmable controller is packaged in a 28-pin DIP. It includes a 64-word by 32 bit-per-word PROM which can be programmed so as to execute a user state machine algorithm. Its features include:

> Fourteen outputs, for controlling fast circuitry

> Six inputs, for sensing conditions in the fast circuitry and deciding what to do next

> Twenty-nine instructions, for determining the next state of the state machine in any of 29 different ways

> A return address register, called SREG, permitting subroutine calls

> A counter register, called CREG, permitting a sequence of states to be repeated n times

> Instructions linking SREG and CREG so that subroutines can be nested to a depth of two, or else counting loops can be nested to a depth of two

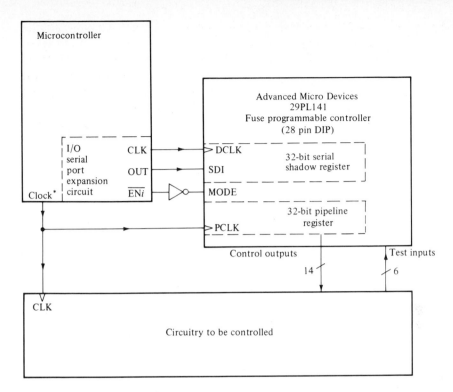

* For 68HC11, this is "E" (2 MHz)
  For 8096, this is "CLKOUT" (4 MHz)

FIGURE 6.38
Connection of AMD 29PL141 fuse programmable controller to a microcontroller, using the microcontroller's clock as the FPC clock.

The 14 outputs of the FPC come from 14 of the 32 bits of the internal PROM after being buffered in a 32-bit pipeline register. Accordingly, with only 64 words in the internal PROM, the chip cannot generate more than 64 different combinations on the 14 output lines. On the other hand, the ability to execute subroutines and loops implies that there are considerably more than just 64 states in the state machine. All in all, this chip represents a breakthrough in bringing fast control to circuitry which needs it from a microcontroller which cannot keep up with the required speed.

The circuit interconnection between a microcontroller and the FPC is illustrated in Fig. 6.38.* It is the shadow register which makes interactions between the two so simple. This is a 32-bit shift register whose clock, called

---

* Not shown is an active-low reset pin. By tying it to the microcontroller's active-low reset pin, the FPC will power up into a known state.

To initialize the content of the FPC's pipeline register:

1. With MODE low, execute four serial outputs (32 bits) from the microcontroller to the FPC's shadow register.
2. Raise MODE and let PCLK transfer the shadow register to the state flip-flops. If several PCLK edges occur while MODE is high, that is fine. However, the *edges* of MODE must meet setup and hold time requirements relative to PCLK.
3. Lower MODE.

To read the content of the FPC's pipeline register:

1. Turn off PCLK. This would call for extra gating in our circuit.
2. Raise MODE.
3. Clock the DCLK input one or more times. This will transfer the FPC's 32-bit pipeline register to the shadow register (without affecting the pipeline register). This operation can be carried out by transmitting anything to the FPC via the I/O serial port.
4. Lower MODE.
5. PCLK can now be turned on again.
6. Execute four serial input commands, to transfer the 32-bit shadow register back to the microcontroller.

FIGURE 6.39
Microcontroller interactions with the AMD 29PL141 fuse programmable controller (FPC).

DCLK, is operated by the microcontroller's I/O serial clock *independent* of the clocking of the FPC's state machine function (with a clock called PCLK). The shadow register can be loaded by four successive 8-bit outputs from the microcontroller via its I/O serial port. The FPC's MODE input is then used to control the transfer of this information into the 32-bit pipeline register. In this way, the microcontroller can get control of the FPC.

The exact sequence of steps necessary to carry out this transfer is listed in Fig. 6.39. Note the constraint upon when the MODE input can change. If the microcontroller clock is used to clock the FPC's PCLK input, then this synchronization problem is taken care of automatically. On the other hand, if a faster PCLK is desired, then it is necessary to synchronize the MODE input to this faster clock, as shown in Fig. 6.40.

The pipeline register in the FPC does *not* directly control the state of the FPC. The state information is tied up in three registers:

PC, the 6-bit program counter

SREG, the 6-bit subroutine return address register

CREG, the 6-bit counter register

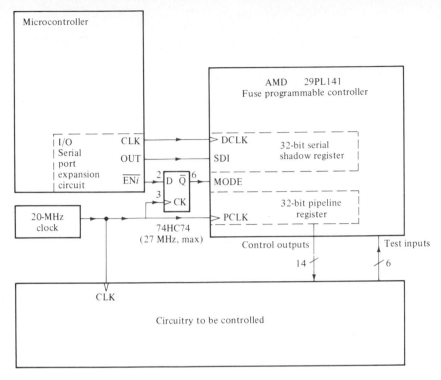

FIGURE 6.40
Connection of AMD 29PL141 fuse programmable controller to a microcontroller using an independent clock for the FPC clock.

However, the pipeline register is normally loaded with the 32-bit content of one word of the PROM. When we force the transfer of the shadow register into the pipeline register, we can select the *next* instruction to be executed as well as affecting the *present* outputs. This is illustrated in Fig. 6.41, which shows the role of the pipeline register bits. It also shows the relationship of the pipeline register to the shadow register and the fact that the serial data is transferred least-significant bit first. The pipeline register bits are defined in the following fields:

> Bits 15 . . . 0 generate the outputs of the chip. Bits 5 . . . 0 go directly to output pins, while bits 15 . . . 8 pass through three-state buffers. Bit 31 of the pipeline register should have a 1 in it to enable these outputs. Bits 7 and 6 are not available as outputs when the shadow register feature is used. The shadow register's DCLK and MODE inputs usurp the two pins

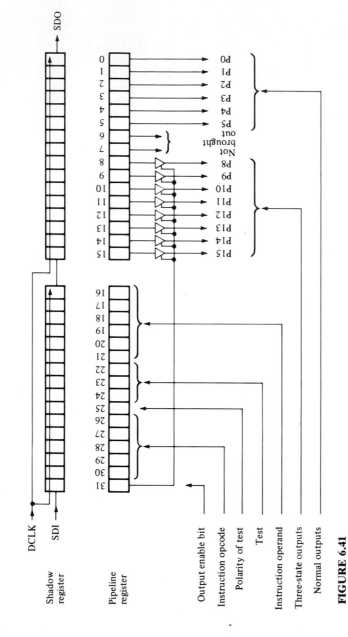

FIGURE 6.41
Function of bits of pipeline register and relationship to shadow register.

284

Assembly language mnemonic for test	Encoding of test in pipeline register bits			Parameter tested
	24	23	22	
T0	0	0	0	External pin T0 (pin 25)
T1	0	0	1	External pin T1 (pin 24)
T2	0	1	0	External pin T2 (pin 23)
T3	0	1	1	External pin T3 (pin 22)
T4	1	0	0	External pin T4 (pin 21)
T5	1	0	1	External pin T5 (pin 20)
CC	1	1	0	(Unavailable test when using the serial shadow register mode.)
EQ	1	1	1	Internal flag (see text)

FIGURE 6.42
Tests implemented in the AMD 29PL141 FPC chip.

which otherwise would have been connected to bits 7 and 6 of the pipeline register.

Bits 30 . . . 26 form the opcode for the next instruction to be executed.

Bits 24 . . . 22 select the test to be associated with the next instruction. Almost all instructions have a test built into them. These bits select one of the choices listed in Fig. 6.42. Six of them provide the inputs to the state machine from the circuit being controlled by the FPC chip. A seventh test looks at a pin which *would* have been available for testing if we were not using the shadow register feature of the chip. The eighth test, of an internal EQ flag bit, is used to test the equivalent of a sum-of-products boolean expression on the six input pins to the FPC chip. Each product is formed by an individual instruction. The EQ flag implements the sum. The designers of the chip were wise to include such a capability after ascertaining desirable features for a fast state machine.

Bit 25 is used to set the *sense* of the test selected above. For example, there is a conditional jump instruction which can be used to jump on the basis of the state of one of the external inputs, say T3. If bit 25 = 0, then the jump will be taken if T3 = 1. If bit 25 = 1, then the jump will be taken if T3 = 0.

Bits 21 . . . 16 form the operand of an instruction. For a jump instruction, these bits represent the 6-bit address to be loaded into the FPC's program counter. The program counter, in turn, holds the address input to the 64-word PROM which holds the 32-bit words, one of which gets loaded into the pipeline register during each PCLK clock period. For one of the mask instructions, these bits are ANDed with the six inputs T5 . . . T0 to form the operand for the instruction.

To initialize the state of the FPC, we need to send it an unconditional jump instruction to the 6-bit PROM address for the instruction sequence which we want executed. As shown in Fig. 6.43, the actual bytes we send depend upon whether the I/O serial port of the microcontroller handles data least-significant bit first or most-significant bit first. For example, to begin an instruction sequence at address 3 of the PROM, with only outputs P0, P2, and P8 high, the Intel 8096 (which transmits least-significant bit first) must send the following sequence of bytes out to the FPC's shadow register, for loading into the pipeline register.

$$
\begin{array}{lll}
00000101\mathrm{B} & \text{or} & 05\mathrm{H} \\
00000001\mathrm{B} & \text{or} & 01\mathrm{H} \\
00000011\mathrm{B} & \text{or} & 03\mathrm{H} \\
11100100\mathrm{B} & \text{or} & 0\mathrm{E}4\mathrm{H}
\end{array}
$$

In contrast, the Motorola 68HC11 (which transmits *most*-significant bit first) must send

$$
\begin{array}{lll}
10100000\mathrm{B} & \text{or} & 0\mathrm{A}0\mathrm{H} \\
10000000\mathrm{B} & \text{or} & 80\mathrm{H} \\
11000000\mathrm{B} & \text{or} & 0\mathrm{C}0\mathrm{H} \\
00100111\mathrm{B} & \text{or} & 27\mathrm{H}
\end{array}
$$

To gain insight into the actual use of this FPC chip as a state machine, we need to look at its instruction set. AMD supports state machine development with an assembler for the 29PL141 which runs on the IBM PC. Accordingly, we will represent each of its instructions with the assembly language syntax shown in Fig. 6.44. The upper case words represent required syntax for an instruction. The lower case words represent parameters which require our substitution of suitable values. For example, the instruction

```
IF(test) THEN GOTO PL(label);
```

might become

```
IF(T3) THEN GOTO PL(NEXT);
```

where T3 is one of the six testable input pins and NEXT represents a label that we might define in our assembly language program. The term PL in this instruction indicates that the operand comes from the operand field of the instruction in the PipeLine register.

Some of the FPC instructions employ an operand expressed as

```
TM(mask)
```

This operand is formed by ANDing the 6-bit operand taken from the instruction in the pipeline register with the six testable inputs

```
T5 T4 T3 T2 T1 T0
```

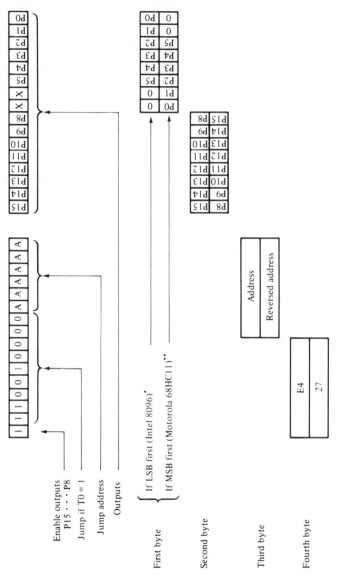

FIGURE 6.43

State initialization sequence of bytes to be sent to the FPC.

Enable outputs
P15 · · · P8

Jump if T0 = 1

Jump address

Outputs

First byte {
If LSB first (Intel 8096)*
If MSB first (Motorola 68HC11)**

Second byte

Third byte

Fourth byte

* Intel 8096 transmits and receives I/O serial port data *least*-significant bit first.
** Motorola 68HC11 transmits and receives I/O serial port data *most*-significant bit first.

Address

Reversed address

E4

27

Op code	Assembly language syntax	CREG content	TRUE			FALSE		
			PC	SREG	CREG	PC	SREG	CREG
0B	IF(CREG=0) THEN GOTO PL(label);	<>0 =0	PC+1 label	— —	— —		— —	— —
0F	IF(test) THEN GOTO TM(mask);		T*M	—	—	PC+1	—	—
19	IF(test) THEN GOTO PL(label);		label	—	—	PC+1	—	—
18	IF(test) THEN GOTO PL(label) ELSE GOTO (SREG);		label	—	—	SREG	—	—
1C	IF(test) THEN CALL PL(label);		label	PC+1	—	PC+1	—	—
1D	IF(test) THEN CALL PL(label), NESTED;		label	PC+1	SREG	PC+1	—	—
1E	IF(test) THEN CALL TM(mask);		T*M	PC+1	—	PC+1	—	—
1F	IF(test) THEN CALL TM(mask), NESTED;		T*M	PC+1	SREG	PC+1	—	—
04	IF(test) THEN LOAD PL(value);		PC+1	—	value	PC+1	—	—
05	IF(test) THEN LOAD PL(value), NESTED;		PC+1	CREG	value	PC+1	—	—
06	IF(test) THEN LOAD TM(mask);		PC+1	CREG	T*M	PC+1	—	—
07	IF(test) THEN LOAD TM(mask), NESTED;		PC+1	CREG	T*M	PC+1	—	—
15	IF(test) THEN PUSH;		PC+1	PC+1	—	PC+1	—	—
17	IF(test) THEN PUSH, NESTED;		PC+1	PC+1	SREG	PC+1	—	—
14	IF(test) THEN PUSH, LOAD PL(value);		PC+1	PC+1	value	PC+1	—	—
16	IF(test) THEN PUSH, LOAD TM(mask);		PC+1	PC+1	T*M	PC+1	—	—

Opcode	Instruction	Cond						
02	IF(test) THEN RET;		SREG	—	—	PC+1	—	—
03	IF(test) THEN RET, NESTED;		SREG	CREG	—	PC+1	—	—
00	IF(test) THEN RET, LOAD PL(value);		SREG	—	value	PC+1	—	—
01	IF(test) THEN RET NESTED, LOAD PL(value);		SREG	CREG	value	PC+1	—	—
09	IF(test) THEN DEC;		PC+1	—	decr.	PC+1	—	—
0C	WHILE(CREG<>0) WAIT ELSE LOAD PL(value);	<>0	PC	—	decr.	—	—	—
		=0	PC+1	—	value	—	—	—
0E	WHILE(CREG<>0) WAIT ELSE LOAD TM(mask);	<>0	PC	—	decr.	—	—	—
		=0	PC+1	—	T*M	—	—	—
1B	WHILE(CREG<>0) IF(test) THEN GOTO PL(label) ELSE WAIT;	<>0	label	—	—	PC	—	decr.
		=0	PC+1	—	—	PC+1	—	—
1A	WHILE(NOT test) WAIT ELSE GOTO PL(label);	<>0	label	—	—	PC	—	—
08	WHILE(CREG<>0) LOOP TO PL(label);	<>0	label	—	decr.	—	—	decr.
		=0	PC+1	—	—	PC+1	—	—
0A	WHILE(CREG<>0) LOOP TO PL(label) ELSE NEST;	<>0	label	—	decr.	—	—	—
		=0	PC+1	—	SREG	—	—	—
0D	CONTINUE;		PC+1	—	—	PC+1	—	—
10-13	CMP TM(mask) TO PL(data);		PC+1	—	—	PC+1	—	—

FIGURE 6.44

AMD 29PL141 fuse programmable controller instruction table.

Note that some instructions load this value into the counter register CREG while others load it into the program counter PC (permitting a *fast* response to any one of several conditions via a multiple branch in a single instruction).

These instructions are most easily understood by looking to the right of the vertical line in Fig. 6.44. Three columns show the effect of each instruction upon the three 6-bit registers, PC and SREG and CREG, when the condition tested is true. The other three columns show the effect of the instruction upon these three registers when the condition tested is false. For example,

```
IF(T2) THEN PUSH;
```

increments the program counter regardless of the value of the T2 input. However, if T2 = 1 (i.e., TRUE), then this new value of the program counter is pushed onto a single-level stack consisting of just the SREG register. On the other hand,

```
IF(T2) THEN PUSH, NESTED;
```

does the same thing, with the addition of pushing the new value of the program counter onto a *two*-level stack consisting of the SREG and CREG registers. These two variations show the versatility of the chip, given only three registers. If we want to nest return addresses two deep we can. Alternatively, if we want to use CREG as a counter register, we can still call a subroutine. But in this case we cannot call a subroutine which calls another subroutine.

The last instruction listed in Fig. 6.44 is used to form sum of products expressions on the testable inputs.

> **Example 6.11.** Assume that we want to take some action if either T1 and T2 both equal 1 or if T1 and T3 both equal 1. The action to be taken is represented by a sequence of instructions labeled ACTION. First, the instruction
>
> ```
> CMP   TM(000110#B) TO PL(000110#B)
> ```
>
> will set the EQ flag (described in conjunction with Fig. 6.42) if
>
> ```
> (T5 T4 T3 T2 T1 T0) ANDed with (0 0 0 1 1 0)
> ```
>
> equals the value
>
> ```
>              0 0 0 1 1 0
> ```
>
> If these are *not* equal, then the EQ flag is *unaffected*. Now the instruction
>
> ```
> CMP   TM(001010#B) TO PL(001010#B)
> ```
>
> checks the second product term, again setting the EQ flag if this second product term is true, and leaving it alone otherwise. Finally the instruction
>
> ```
> IF(EQ) THEN GOTO PL(ACTION);
> ```
>
> will produce a jump to the instruction labeled ACTION if the sum-of-products expression is true. This instruction will also clear the EQ flag (as will *any* instruction which tests the EQ flag).

To tie all of the ideas of this section together, we will consider as an example the waveform recorder of Fig. 6.45. This circuit uses a flash A/D converter to convert an analog input, between -1 V and $+1$ V, to its 8-bit digital equivalent. The FIFO in this figure is a hardware queue, buffering a burst of 512 8-bit samples, collected at an extremely fast data rate, and feeding them into the microcontroller at a more leisurely rate. The state machine is called for because we want to sample the waveform faster than the microcontroller can do it.

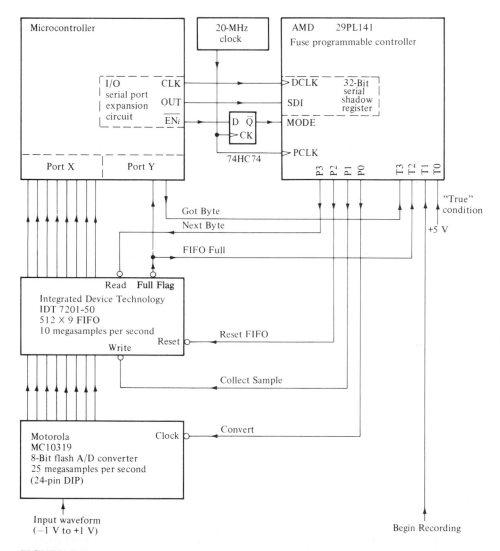

FIGURE 6.45
Waveform recorder.

Example 6.12. Consider first the sequence of events required of the waveform recorder circuit of Fig. 6.45. We will assume that the microcontroller sets up the collection of data by initializing the fuse programmable controller with a sampling period, and telling it to begin collecting samples upon the next rising edge of the Begin Recording signal. So, the sequence of events is:

1. Microcontroller initializes fuse programmable controller with sampling period.
2. FPC resets FIFO.
3. FPC waits for Begin Recording signal to equal 0.
4. FPC waits for Begin Recording signal to equal 1, defining the rising edge of this signal.
5. FPC triggers a conversion of the A/D converter.
6. FPC collects the converted sample into the FIFO.
7. FPC repeats this until the FIFO is full, at which point it halts, waiting for the microcontroller to grab the oldest sample, sitting on the output of the FIFO.
8. Microcontroller periodically queries the FIFO full signal. When it sees that the FIFO is full, it reads the FIFO output. Then it raises and lowers the Got Byte signal to the FPC.
9. FPC sees Got Byte go high. It strobes the (active-low) Next Byte line once, clearing out the oldest sample from the output of the FIFO. Then it waits for Got Byte to go low before going back and waiting for the microcontroller to read the next byte and raise Got Byte again.
10. Microcontroller counts the bytes which it reads from the FIFO. It stops after reading the 512th sample.

We are now in a position to consider the assembly language program for this waveform recorder, shown in Fig. 6.46. It generates a listing output, shown in Fig. 6.47, and a fuse map output (in a form which can be handled by a PROM programmer), shown in Fig. 6.48. As with any assembler, we need to get the syntax ground rules straight before we are able to make sense out of the file. Some of these follow:

> Comments may be inserted anywhere. They are delineated by quotation marks.
>
> The assembler ignores case. Either upper-case characters, lower-case characters, or a mixture may be used.
>
> AMD's assembler is written to support a variety of parts. The second line
>
> ```
> device (am29pl141)
> ```
>
> is used to denote that the assembler is to assemble this source file for the am29pl141 fuse programmable controller.
>
> A define statement can be used to give names to the test inputs we use. Whereas the assembler knows about test input names such as t0 and t1, error-free code becomes easier to write if we can rename the test inputs with our variable names. These definitions use an equal

sign in much the same way that we have used an EQU in the assemblers which we have discussed heretofore.

We can also use the define statement to give names to the output combinations which we want to generate. For example, when we want to reset the FIFO, Fig. 6.45 shows that we need to drive the FPC output P2 low. At the same time, we must drive P0, P1, and P3 high or we will get some nasty side effects when we go to reset the FIFO. Accordingly, we define

```
reset = 1011#b
```

Note that this assembler uses #b to denote that this is a binary number. Alternatively, we can use #h for hexadecimal numbers. No suffix at all is needed for denoting decimal numbers.

This assembler uses a semicolon to terminate the entire define statement. It also uses a semicolon to terminate each instruction. It uses a colon to terminate a label for an instruction. It even cares about that little period after the final end statement.

Each instruction has as many as four parts:

An optional label, followed by a colon.

The output to be generated, followed by a comma.

The instruction, with the syntax of Fig. 6.44 (including the semicolon at the end).

An optional comment, delineated by quotes.

Example 6.13. For this state machine program, consider how parameters are passed to the FPC by the microcontroller. With three registers in the FPC (PC, SREG, and CREG), we have up to three parameters to worry about. Earlier in this chapter we dealt with the initialization of the program counter, PC. We loaded a jump instruction into the shadow register and then let the fast PCLK clock transfer it to the pipeline register. Note that the third line from the end of the file in Fig. 6.46 lists the instruction which we would like to load in this way. While this instruction is really not needed in the PROM, its inclusion in the PROM will not *hurt* anything, and we will be able to look at the assembled output and read the bytes which must be sent by the microcontroller in order to have this instruction executed.

For this application, the microcontroller actually wants to load up both CREG and SREG with a parameter which controls the sampling interval of the input waveform. CREG will be used as a counter, getting counted down to 0 at a 20-MHz rate. When it reaches 0, it is automatically reloaded from SREG, to set up the timing for the next sampling interval.

The first two instructions of the program (which starts with the instruction labeled Start) are subroutine calls to accomplish this loading of CREG and SREG. At the same time, we want to come out of this loading process with the program counter in a known state. Loading a jump instruction from the shadow register

```
                                    "WAVEFORM RECORDER"

device (am29pl141)
ssr = 1;                            "Specifies use of serial shadow register mode"
default = 0;                        "This leaves unused fuses intact"
define      "Define names for test conditions"
            true            = t0    "Input which will always equal 1"
            record          = t1    "Begin recording waveform"
            fifo_not_full   = t2    "512 samples are still being collected"
            got_byte        = t3    "Handshake signal from microcontroller"
            "Define names for outputs"
            do_nothing      = 1111#b    "Do not drive any of the active-low outputs low"
            reset           = 1011#b    "Reset the FIFO"
            convert         = 1110#b    "Strobe the flash A/D converter"
            collect         = 1101#b    "Write converter output into FIFO"
            next_byte       = 0111#b;   "Put next byte collected on output of FIFO"

begin
"Initialize state of FPC's three registers."
start:      reset,      if(true) then call pl(get);     "Reset FIFO; 'get' gets parameter"
            do_nothing, if(true) then call pl(fix);     "'fix' copies CREG to SREG"

"Now, with count parameter in both CREG and SREG, begin data collection as soon as the external
 'begin recording' signal goes from 0 to 1."
level0:     do_nothing, if (record) then goto pl(level0);     "Wait for 'Begin Recording' = 0"
level1:     do_nothing, if (not record) then goto pl(level1);  "Wait for 'Begin Recording' = 1"

"Collect samples until FIFO is full."
wait2:      convert,    while(creg<>0) loop to pl(wait2) else nest;  "Count down CREG; reload when zero"
            collect,    if(fifo_not_full) then goto pl(wait2);       "Strobe FIFO's write; FIFO full?"

"512 samples have been collected; now dump them to microcontroller.
```

```
Let microcontroller count bytes to tell when done."
dump:    do_nothing,   if (not got_byte) then goto pl(dump);       "Wait for flag from microcontroller"
         next_byte,    continue;                                    "Strobe FIFO output once"
         do_nothing,   while (got_byte) wait else goto pl(dump);    "Wait for flag to drop"
"————————— SUBROUTINES USED TO GET SAMPLING INTERVAL PARAMETER INTO CREG AND SREG —————————"

get:     do_nothing,  if(true) then goto pl(get);         "Wait to get parameter"
"Wait for microcontroller to load pipeline register with the following instruction:
         do_nothing,   if(true) then ret, load pl(value);
where 'value' is a number between 1 and 63 which determines the sampling interval."

"The following subroutine copies CREG into SREG and then returns to the address which was in SREG."
fix:     do_nothing,  if(true) then ret, nested;
"—————————— INSTRUCTIONS WHICH MICROCONTROLLER MUST SEND ——————————"

"Since there is extra PROM available, assemble the instructions which the microcontroller must
send so that we can look at the assembler output and get the bytes to be sent."
         do_nothing,   if(true) then goto pl(start);       "Unconditional jump to start"
         do_nothing,   if(true) then ret, load pl(4);      "Let sample rate value = 4"

"—————————— RESET ——————————"

"The following statement is assembled to address 63 and is executed at reset time to halt FPC."
63#d:    do_nothing,  if(true) then goto pl(63#d);
end.
```

FIGURE 6.46
FPC assembly language code for a waveform recorder.

```
PROM Contents are :
             OE   OPCODE   POL TEST    DATA          OUTPUT
word    0  [ 1 |  11100  | 0 |  000 |  001001 |  0000000000001011 ]
word    1  [ 1 |  11100  | 0 |  000 |  001010 |  0000000000001111 ]
word    2  [ 1 |  11001  | 0 |  001 |  000010 |  0000000000001111 ]
word    3  [ 1 |  11001  | 1 |  001 |  000011 |  0000000000001111 ]
word    4  [ 1 |  01010  | 1 |  111 |  000100 |  0000000000001110 ]
word    5  [ 1 |  11001  | 0 |  010 |  000100 |  0000000000001101 ]
word    6  [ 1 |  11001  | 1 |  011 |  000110 |  0000000000001111 ]
word    7  [ 1 |  01101  | 1 |  111 |  111111 |  0000000000000111 ]
word    8  [ 1 |  11010  | 0 |  011 |  000110 |  0000000000001111 ]
word    9  [ 1 |  11001  | 0 |  000 |  001001 |  0000000000001111 ]
word   10  [ 1 |  00011  | 0 |  000 |  111111 |  0000000000001111 ]
word   11  [ 1 |  11001  | 0 |  000 |  000000 |  0000000000001111 ]
word   12  [ 1 |  00000  | 0 |  000 |  000100 |  0000000000001111 ]
word   63  [ 1 |  11001  | 0 |  000 |  111111 |  0000000000001111 ]
```

FIGURE 6.47
Assembled output of source file in Fig. 6.46.

into the pipeline register accomplishes this since it forces the jump address into the program counter each time the FPC chip is clocked (every 50 μs). The fact that MODE (the signal from the microcontroller which tells the FPC to use its clocks for this purpose) is high for many PCLK clock pulses does not matter. During each PCLK clock pulse, the program counter is reloaded with the jump address.

Looking at Fig. 6.44, we want an instruction which loads a value into CREG from the pipeline register and which retains control of the program counter (i.e. which does *not* simply increment it) as the FPC is clocked many times with this same instruction in the pipeline register. Only one instruction meets these two

```
F0*
L0000  0 00011 1 111 110110 1111111111110100 *
L0032  0 00011 1 111 110101 1111111111110000 *
L0064  0 00110 1 110 111101 1111111111110000 *
L0096  0 00110 0 110 111100 1111111111110000 *
L0128  0 10101 0 000 111011 1111111111110001 *
L0160  0 00110 1 101 111011 1111111111110010 *
L0192  0 00110 0 100 111001 1111111111110000 *
L0224  0 10010 0 000 000000 1111111111111000 *
L0256  0 00101 1 100 111001 1111111111110000 *
L0288  0 00110 1 111 110110 1111111111110000 *
L0320  0 11100 1 111 000000 1111111111110000 *
L0352  0 00110 1 111 111111 1111111111110000 *
L0384  0 11111 1 111 111011 1111111111110000 *
L2016  0 00110 1 111 000000 1111111111110000 *
C2149*
```

FIGURE 6.48
Fuse map corresponding to source file of Fig. 6.46.

requirements. It is

```
IF(test)  THEN RET, LOAD PL(value);
```

A casual look at the instruction just below this in the table of Fig. 6.44 seems to indicate that it meets these two requirements also. However, after three PCLK clocks with MODE high, the parameter that we wanted to load into CREG will actually be loaded into CREG, SREG, and PC. That is, if we try to load the number 6 into CREG, we will also *jump* to address 6 (after this number has been loaded into the program register and when MODE drops low again).

    To use the subroutine return instruction listed above, we want to be *in* a subroutine when it is executed. This is the reason for the instruction labeled Start. It puts the address of the next instruction (address = 1) into the stack register, SREG. The Get subroutine, which is called by this first instruction, does nothing but wait, jumping to itself repeatedly, as illustrated in Fig. 6.49. When the micro-controller loads the parameter with the subroutine return instruction listed above, the FPC breaks out of this loop and returns to address 1 with the parameter in CREG. The instruction at address 1 does another subroutine call so that we can execute the return nested instruction, which will copy CREG into SREG, even as it returns by copying the old content of SREG into the program counter.

Example 6.14. Next consider how the FPC program detects a rising edge on the Begin Recording input. The next instruction, labeled Level0, ensures that the FPC detects at least one 0 input at the Begin Recording input before going on. Then the instruction labeled Level1 waits until the Begin Recording signal goes high. In this way, we sense a rising edge.

Example 6.15. We now get to the two instructions, labeled Wait2, which form the heart of this application. The first instruction sends a convert output to the flash A/D converter. The A/D converter latches the input as it is at the time that Convert goes high. This sampling obviates the need for a sample-and-hold circuit. The second instruction strobes this sample into the FIFO. Then it jumps back to the timing loop instruction. The two-instruction loop shown here has not taken into account the timing requirements of the flash A/D converter (i.e., how long must its strobe input be low? be high?), the FIFO (i.e., have set-up time require-ments been met? How long does it take to emit the Full Flag signal?), or the FPC (i.e., does the Full Flag signal get to the FPC in time to be acted upon?). We will consider some of these issues in the end-of-chapter problems.

    Note that these two instructions implement a loop within a loop. If the parameter loaded is 0, then we will execute the convert instruction just once before going on to the collect instruction. This represents the fastest possible sampling rate, of one sample every two cycles, or 10 megasamples per second. (Again, this may be faster than the timing considerations actually allow.) If the parameter loaded is 1, then we will collect a sample every three cycles, giving a sampling period of 150 ns. In general, the sampling period is given by

```
Sampling period = (parameter + 2) * 50 ns
```

The algorithm of Fig. 6.46 ends by controlling the transfer of bytes from the FIFO to the microcontroller, after the FIFO has been filled. The microcontroller reads a byte. Then it tells the FPC that it is ready for the next one. Since the FPC can

Opcode	Program counter			Instruction	New values		
					PC	SREG	CREG
1C	0	start:	reset,	if(true) then call pl(get);	9	1	x
19	9	get:	do_nothing,	if(true) then goto pl(get);	9	1	x
	.						
	.						
	.						
19	9	get:	do_nothing,	if(true) then goto pl(get);	9	1	x
00	Forced by pipeline register		do_nothing,	if(true) then ret, load pl(4);	1	1	$\underline{4}$
00	Forced by pipeline register		do_nothing,	if(true) then ret, load pl(4);	1	1	$\underline{4}$
	.						
	.						
00	Forced by pipeline register		do_nothing,	if(true) then ret, load pl(4);	1	1	$\underline{4}$
1C	1	fix:	do_nothing,	if(true) then call pl(fix);	10	2	$\underline{4}$
03	10		do_nothing,	if(true) then ret, nested;	2	$\underline{4}$	$\underline{4}$
19	2	level0:	...				

FIGURE 6.49
Instruction sequence to load parameter (e.g., the value 4) into CREG and SREG.

respond faster than the microcontroller, it need not tell the microcontroller when the new byte is available. Instead, it just puts out the new byte *before* the microcontroller can possibly be ready for it. The FPC can do this since it is clocked every 50 ns and since it is dedicating itself single-mindedly to this task.

Before completing this section, we will look, one more time, at the instructions of the FPC. While the table of Fig. 6.44 holds a lot of subtle application information, we will try to make this more explicit. The waveform recorder example should be a help in this direction.

Example 6.16. We have seen that the call and ret instructions work together to implement a subroutine structure. Furthermore, the nested versions include CREG in the stack, letting us nest subroutines to a depth of two. In contrast, the unnested versions employ only SREG for a stack, protecting CREG so that it can be used as a counter.

Example 6.17. The push and ret instructions work together to implement a loop structure. The push instruction marks the *next* instruction as the beginning of the loop by pushing its address onto the stack (i.e., SREG). The ret instruction ends the loop by getting us back to the beginning of the loop. It loads the program counter with the content of the SREG. For example, the following structure

```
            <output>, if(true) then push;
begin_loop:                     .

end_loop: <output>, if(test) then ret;
```

will cause endless looping until the test at the end of the loop is not met (or until a test within the loop causes a jump out of the loop).

Example 6.18. To generate precisely timed outputs, we can use the counter to pause while an output is being generated for a specified number of clock periods. For example, to generate a 1-μs pulse output, called Pulse, we can use

```
            do_nothing, if(true) then load pl(19);
            pulse,      while(creg<>0) wait else load pl(0);
```

This assumes a 20-MHz clock.

Example 6.19. To execute a loop three times, consider the following:

```
            <output>,    if(true) then load pl(2);
begin_loop:    .

end_loop: <output>,    while (creg<>0) loop to pl(begin_loop);
```

This will execute the loop the first time with CREG = 2, the second time with CREG = 1, and the third time with CREG = 0.

6.9   SUMMARY

In this chapter we began with a consideration of the variety of reasons *why* we might need to expand the resources of a microcontroller. The remainder of the chapter considered *how* to achieve this expansion for each of the resources on the chip. We began by looking at an early microcontroller family. Up to a point, expansion was easy. After that, a designer using that microcontroller experienced intense pain.

In contrast, modern-day microcontrollers are generally designed for expansion by throwing away several ports and using the relinquished pins to bring out the internal address bus, data bus, and bus control lines. We saw how this is done for both the Intel 8096 and the Motorola 68HC11. We saw how this could be used to expand RAM and ROM as well as I/O ports. Then we looked at the nitty-gritty timing issues which must be met to *ensure* that these added chips will work with the microcontroller at the speeds determined by the microcontroller's clock.

The remainder of the chapter considered the expansion of microcontroller resources *whether or not* the internal bus is brought out. We explored how a *UART serial port* operates. This can provide an outstanding I/O option that the user of an instrument or device might see. For example, a simple, low-cost device might include a liquid-crystal display for showing status and output information. It might *also* include a UART serial port which can optionally be connected to a CRT terminal for a much more extensive display of status and output information. Alternatively, the UART serial port might be used to connect the microcontroller to another facility *within* the instrument. For example, it might connect to the drive circuitry for a built-in CRT display. We will discuss this application in the next chapter. Finally, for an application which overpowers a single microcontroller, one option would be to divide the tasks between several microcontrollers and then to use the UART serial port to interconnect the several microcontrollers. We saw how this played out, with master and slave microcontrollers interacting serially.

In a microcontroller environment, the *pins* of the microcontroller chip become a precious resource. While a UART serial port uses just two pins to communicate with other devices, it requires a matching UART in each of these other devices. A more generally useful serial expansion method is to provide a shift register interface to external devices. The *I/O serial port* does just this. We saw how to get extra I/O lines easily, in a manner which helped to minimize overall chip count. We also saw how to use this capability to interface to an EEPROM chip, thus adding nonvolatile memory to a microcontroller like the Intel 8096 which has none. In the next chapter we will see a variety of I/O devices which can be connected to a microcontroller via this I/O serial port. To support such applications, we considered how to expand a single I/O serial port into several I/O serial ports.

Since programmable timers represent such a fundamental tool for fitting a microcontroller to its real-time control tasks, designers of present-day microcontrollers have expanded programmable timer capability considerably. Nevertheless, applications arise which require even more capability. For example, an antiskid braking system in an automobile needs to monitor timing inputs from all four wheels. If the microcontroller's programmable timer does not have four inputs available, then this does not mean that a different microcontroller is needed. Rather, the input timing capability need only be expanded. We saw how to do this as well as how to expand output timing capability.

We closed the chapter with an overview of a chip which can serve us as a fast state machine. This is useful for applications which require real-time control at speeds unattainable even from a microcontroller. AMD's *fuse programmable controller* offers significant flexibility for implementing a state machine with a single chip. In so doing, we avoid all of the chips which would result if the state machine were implemented with PROMs, counters, data selectors, decoders, and other ''glue'' logic chips. Unfortunately, we discovered that the problem of using a chip like AMD's entailed struggling with the peculiarities of yet another complex chip. While it would implement a desired state machine with blinding speed, it did require exceedingly careful attention, as we mapped the things we wanted to do to the instructions available for doing them.

PROBLEMS

6.1. *Expansion.* Discuss how a specific microcontroller, other than the two discussed in this book, supports expansion of ROM or EPROM? Of RAM? Of I/O ports?

6.2. *Motorola 68HC11 expansion.* The circuit of Fig. 6.5 shows a single chip, a 74HC138 decoder, used to obtain a simple decoding of extra ports. Note that this decoding does not use the 68HC11's R/W (read/write) line. What are the possibilities for *contention* on the data bus if we have an error in our program in which we

(*a*) Read from an output port address like 0100H?

(*b*) Write to an input port address like 0400H?

6.3. *Motorola 68HC11 expansion.* A partial decoding like that of Fig. 6.5 ties up a lot more addresses than needed, just to address the ports involved. This is an excellent way to simplify circuitry if we can afford to waste the unused addresses (which is usually the case with a microcontroller). On the other hand, if we want to use almost the entire address space as a gigantic RAM buffer, we can use a 74HC133 13-input NAND gate (instead of address lines A15 and A14) to enable the 74HC138 decoder chip of Fig. 6.5. This same NAND gate output can be used to disable another decoder chip used for enabling RAM chips over the rest of the memory address space. We will not have to invert address lines going to the NAND gate if we locate the I/O ports just below the on-chip ROM, at the eight addresses DFF8 to DFFF. The eight outputs of the decoder can enable a mix of up to eight 74HC574 octal flip-flops and 74HC541 three-state buffers.

(*a*) Show the connections to the 74HC138 decoder chip to implement this decoding.

(*b*) Label the address decoded by each output.

(*c*) Check that no read of the ROM in the range E000 to FFFF will *also* enable an input port or an output port. You should be able to use one of the two active-low enable inputs to the 74HC138 to prevent this.

(*d*) If the condition checked in part *c* is violated, then what happens as we fetch an instruction from a ROM address which also enables an input port?

(*e*) Repeat part *d* for a ROM address which also enables an output port.

6.4. *Expanded I/O port TTL compatibility.* The Motorola 74HCxxx family of parts has been designed to combine 74LSxxx speed (more or less) with the best features of CMOS gates (e.g., *negligible* quiescent power dissipation). Its outputs drive more or less what a 74LSxxx gate will drive. However, if *inputs* are being driven from TTL logic, then special care must be taken since a normal 74HCxxx gate requires that a high input be pulled higher than 3.5 V, whereas a 74LSxxx output is only guaranteed to pull up to 2.7 V.

(*a*) One solution is to use a 74LSxxx part instead of a 74HCxxx part to implement an input port. Referring to a data book for 74LSxxx parts, find the quiescent power dissipation of an octal three-state buffer.

(*b*) Another solution is to add a pullup resistor to any 74HCxxx input which must have TTL compatibility. Then when the TTL gate drives an input high, the pullup resistor will pull that input the rest of the way up to +5 V. For reasonable loading, we might use a 10-kΩ pullup resistor. Given this value, if the line from the TTL output to the 74HCxxx input exhibits a stray capacitance to ground of 15 pF, then how long will this *RC* circuit take to charge from 0 V to 2.3 V (which is the typical threshold voltage of a 74HCxxx gate; that is, the voltage below which an input is treated as a 0 and above which an input is treated as a 1)? What is the effect of this time upon the effective propagation delay of the 74HCxxx gate?

(*c*) A third solution to this problem is to use a 74HCTxxx version of the three-state buffer, if one is available. These parts substitute a TTL-compatible input stage in place of the normal 74HCxxx input stage. However, not all family members are available in the 74HCTxxx version. For example, Motorola does not (yet) offer a 74HCT541 version of the 74HC541 shown in Fig. 6.5. On the other hand, the 74HCT244 is an octal buffer which could be used as a TTL-compatible input port in the circuit of Fig. 6.5. It does not have as convenient a pinout as that of the '541 (which has all of its inputs on one side of the chip and all of its outputs on the other side). Check a Motorola high-speed CMOS logic data book to see what 74HCTxxx parts are available.*

6.5. *Motorola 68HC11 page 0 use.* The designers of the 68HC11 have included a feature on the chip which permits all the on-chip registers to be located on page 0 so that they can be accessed using direct addressing. The RAM, which defaults to page 0 addresses, can be moved so that it begins at any user-specified address of the form X000. Alternatively, it can be left to share page 0 with the on-chip registers (just as is *necessarily* done by the Intel 8096). If this is done, then the 64 addresses from 0000 to 003F will access the on-chip registers and the corresponding RAM locations as these same addresses become inaccessible. This still leaves 192 bytes of RAM available for our use, from 0040 to 00FF.

The key to this use is the 68HC11's INIT register, located initially at address 103D. The upper nibble of INIT specifies the upper hex digit to be used in

---

* Or a second source vendor for the 74HCTxxx family of parts.

accesses of RAM. The lower nibble of INIT does the same for the on-chip regis-ters. For example, if INIT contains the hexadecimal number CD, then RAM will be located at addresses C000 to C0FF while the on-chip registers will be located at addresses D000 to D03F. The on-chip registers and the RAM can also be superim-posed upon the ROM addresses, with the former taking precedence over ROM during any access to a shared address.

To protect users from quirky behavior of the chip, the INIT register can be written to *only once*. Furthermore, that write must occur during the first 64 clock cycles after reset. Thereafter, the INIT register becomes a read-only register.

If the INIT register is loaded with FF, and if the chip is used in the expanded mode, then redraw the memory map for the chip corresponding to Fig. 6.4.

6.6. *Motorola 68HC11 internal ROM/EEPROM use.* When the Motorola 68HC11 is used in the expanded mode, the program does not necessarily have to come from the internal ROM. On the other hand, there *must* be either internal or external memory at addresses FFC0 to FFFF to handle the reset vector and all of the interrupt vectors. The chip includes a CONFIG register (located initially at ad-dress 103F) which is implemented in EEPROM (in spite of the register address) and which can be changed only by doing an EEPROM write to the CONFIG address. Writing a 0 to bit 0 of CONFIG disables the 512-byte internal EEPROM. Writing a 0 to bit 1 of CONFIG disables the 8-kbyte internal ROM. Furthermore, writing a 0 to bit 2 of CONFIG *enables* the watchdog timer feature of the chip. The virgin, fresh-from-Motorola, state of each of these bits is 1.

A printer buffer is a device to which a computer can send a file quickly (using either a serial port or a parallel port from the computer) and which will then handle the relatively slow interactions with the printer at the printer's rate. It permits any computer (which cannot handle printing as a background task) to be finished with the print job quickly and to get back to work on other things. In this problem, consider constructing a printer buffer using the 68HC11 and using exter-nal RAM to fill as much of the 64K memory space as possible. We might disable the internal EEPROM and the internal ROM, map the internal registers and RAM to page F0XX, and then fill out the remaining addresses in the range F000 to FFFF with an external EPROM or ROM.

(*a*) Show the memory map of the 68HC11 for this use.

(*b*) Using little more than a 74HC133 13-input NAND gate, show the connections to enable an EPROM when the NAND gate output is 0 (i.e., low) and to enable RAM chips when the output is high. The EPROM can be enabled for internal RAM and on-chip register addresses with no problem *as long as* it is disabled during writes (i.e., when the 68HC11's R/W line is low). When the CPU reads from an address which enables both EPROM and an internal resource (e.g., internal RAM), it uses an internal three-state buffer to disable the external data bus from driving the internal data bus. Consequently, it reads the internal resource without contention.

(*c*) Does it matter whether the NAND gate in part *b* enables the addresses F000 to F0XX occupied by internal registers and RAM? Answer this question in light of this output *also* being used as an active-high enable for external RAM. Consider both the requirement for reading and writing to each resource cor-rectly and also the need to avoid contention. Make sure that your circuit of part *b* satisfies the requirements examined here.

(d) Does it matter if the circuit decoding developed above enables external RAM when we do a write to an internal register? Explain. Does this have any bearing upon your circuit in part b?

6.7. *Motorola 68HC11 read timing.* Consider the read timing diagram shown in Fig. 6.6a for expanded mode operation of the 68HC11.

(a) When during a read cycle will contention on the multiplexed bus occur *if* the clock signal E is *not* used to enable external devices onto the bus? Draw the timing diagram as it occurs in this case. Assume a worst-case propagation delay of 0 ns for the external device (i.e., worst-case for contention).

(b) Contention can also occur if the output *disable* time for an external device is too slow, even though the external gating is designed to disable external devices when E goes low at the end of the clock period. Redraw the (idealized) timing diagram of Fig. 6.6a, marking the maximum time which a device has to get off the multiplexed bus. Does the propagation delay of chip-enable gating (e.g., the 74HC138 of Fig. 6.5) affect this? If so, then reflect this in your redrawn timing diagram.

(c) Arm yourself with timing specifications for the Motorola 68HC11 (Fig. 6.16), the 74HC138 decoder, and the 74HC541 (or 74HC244) input port. Check the worst-case timing for the potential contention problem of part b. If you need a minimum propagation delay specification where only a maximum value is given, use the rule-of-thumb minimum value of one-third the maximum value.

6.8. *Motorola 68HC11 write timing.* Consider one of the 74HC574 output ports of Fig. 6.5.

(a) Draw a timing diagram for this port, showing when it is stable with its old value, when its output changes, and when it is stable at its new value. Without knowing specific timing parameters, label the times involved (e.g., prop. delay of 74HC138 decoder).

(b) Note that this output port is a positive-edge-triggered flip-flop. What impact does this have upon the timing diagram?

(c) If a transparent latch (e.g., a 74HC573) were substituted for the positive-edge-triggered flip-flop, then what polarity should its strobe input have to work with the circuit of Fig. 6.5? (If the polarity is incorrect, it could be fixed with an inverter.)

(d) Draw the timing diagram for this case, given the transparent latch (with correct input strobe polarity) of part c.

(e) The circuitry connected to an output port may not be able to stand *glitches*. That is, it may require that no momentary 0 output occur when a 1 is written to an output bit which already has a 1 on it. Likewise, it may require that no momentary 1 output occur when a 0 is written to an output bit which already has a 0 on it. The octal positive-edge-triggered flip-flop satisfies this requirement. Does the transparent latch also? Explain.

(f) The edge-triggered flip-flop output port of Fig. 6.5 requires that the data on the data bus be held for some amount of hold time after the falling edge of E occurs. If the 68HC11 specification says that it will hold output data for a minimum of 30 ns after the falling edge of E, and if the 74HC138's maximum enable-to-output propagation delay is 30 ns, and if the output port's hold time is 5 ns maximum, then do we have guaranteed reliable operation? Explain.

(g) The hold time situation of part f is actually alleviated because of stray capacitance on the multiplexed bus lines. If a driver pulls one of these lines high and then releases it, the line will remain high until some other device drives it low.

In light of this, what is the critical 68HC11 specification to consider, with regard to the hold time problem of part f?

6.9. *Intel 8096 I/O port expansion.* The circuit of Fig. 6.7 uses a 74LS08 AND gate to form an active-low read or write pulse. With any decoding circuitry, we should be concerned about its propensity for generating glitches (i.e., momentary 0-1-0 spikes or momentary 1-0-1 spikes).

(a) Will the output of this AND gate be glitchless? Explain.

(b) Using either a TTL data book or a 74HCxxx data book, find the logic diagram for the '138 gate. Given the input signals to the gate of Fig. 6.7, will the outputs of the 74LS138 chip be glitchless? Explain.

6.10. *Intel 8096 expansion bus timing.* The timing diagrams of Figs. B.3 and B.4 in Appendix B illustrate that the Intel 8096 uses two of its internal cycles to access an external device. This makes accesses to external resources slower than for internal resources. Comparing the 4-MHz 8096 read timing of Fig. B.3 with the 2-MHz 68HC11 read timing of Fig. 6.6a, roughly how do the access times compare? This is the time which starts when the chip emits an address and ends when the content of that address has to be returned to the CPU on the external data bus. To answer this question, use the idealized waveforms shown in the two figures.

6.11. *Intel 8096 I/O expansion.* The circuit of Fig. 6.7 adds I/O ports using an 8-bit data bus. Using ideas from Fig. 6.12, modify this circuit so as to obtain two *16-bit* output ports having addresses 0100 to 0101 and 0200 to 0201 and two *16-bit* input ports having addresses 0300 to 0301 and 0400 to 0401.

6.12. *Motorola 68HC11 EPROM use.* Before going to a mask-programmed ROM for an application program for the 68HC11, we might want to operate the chip in the expanded mode with an Intel 2764 (8K × 8) EPROM mapped to addresses E000 to FFFF. As discussed in Problem 6.6, we can disable the internal ROM to support this. Using decoding circuitry that is as simple as possible (i.e., using as few standard gate packages as possible), draw the circuit. Assume that no other external devices will be on the bus. Also assume that the program is flawless in the sense that it will never do a write to an EPROM address. Consequently, the decoding circuitry need not look at the R/W line.

6.13. *Motorola 68HC11 RAM expansion.* In this problem we will explore RAM expansion using a single Hitachi HM62256 (32K × 8) static RAM chip. It has 15 address lines, eight bidirectional I/O lines, an active-low output-enable input (for reading), and active-low write-enable (for writing) and an active-low chip-enable (which must be driven low for *either* reading *or* writing).

(a) Draw the circuit analogous to Fig. 6.10. Enable the RAM for all addresses from 0000 to 7FFF.

(b) As pointed out in Problem 6.6, part b, the 68HC11 has no problem with contention when an external device and an internal device are both enabled for the same address. What is an example of an address where this will happen? What will the external RAM do for a write to this address? What will the external RAM do for a read from this address?

(c) If we are really greedy for RAM, how can we use the INIT register discussed in Problem 6.5 to increase the RAM in the system by 256 bytes? Can we make all of the RAM *contiguous* (i.e., have no gaps in its address space)? If so, then how? If not, then why not?

6.14. *Intel 8096 RAM expansion.* Consider the circuit of Fig. 6.12. This circuit uses the active-low WRL signal to help enable the even numbered addresses and the active-low WRH signal to help enable the odd numbered addresses. List the state (i.e., high or low) of the WRL and WRH lines during the write which takes place during each of the following instructions:

(a) STB AL, 4000H

(b) STB AL, 4001H

(c) ST AX, 4000H

6.15. *Intel 8096 read timing considerations.* Given the read timing specifications of Fig. 6.13 for the Intel 8096, check the timing for reading from an Intel 2764 EPROM using the circuit of Fig. 6.9. Do this by drawing the timing diagram, analogous to the bottom half of Fig. 6.13. Assume the following maximum propagation delay:

```
74LS573        18 ns
```

Consider the following speed-selected versions of the EPROM: 2764-20, 2764-25, 2764-30, and 2764-45. Their worst-case timing characteristics are given below.

	-20	-25	-30	-45
Access time, ns	200	250	300	450
Chip enable to output time	200	250	300	450
Output enable to output time	75	100	120	150

Show your timing diagram for the slowest version of the 2764 which will still work with an 8096 running with a 12-MHz crystal and with the circuit of Fig. 6.9.

6.16. *Motorola 68HC11 read timing considerations.* In this problem you are to reconsider the EPROM expansion circuit which you developed in Problem 6.12 together with the timing specifications given in Fig. 6.16. Show your timing diagram for the slowest version of the 2764 EPROM (whose timing specifications are given in the previous problem) which will still work with a 68HC11 running with an 8-MHz crystal.

6.17. *Motorola 68HC11 read timing considerations.* In this problem you are to reconsider the RAM expansion circuit of Fig. 6.10 together with the timing specifications given in Fig. 6.16. Assume the following maximum propagation delays:

```
74HC573        30 ns
74HC138        30 ns (enable to output)
               40 ns (address to output)
```

(a) Determine the longest RAM access time which will still ensure reliable reading of data from the RAM.

(b) Determine the longest RAM chip enable to output time which will still ensure reliable reading of data from the RAM.

6.18. *Motorola 68HC11 read/write timing considerations.* In this problem we will consider Fig. 6.11. Assume maximum propagation delays of 35 ns for the PAL and 30 ns for the 74HC573.

Using the timing specifications for the Motorola 68HC11 shown in Fig. 6.16, do each of the following:

(a) Check the read timing with a timing diagram analogous to that given at the bottom of Fig. 6.13. Be sure to point out any timing requirements which will *not* be met.

(b) Check the write timing with a timing diagram analogous to that given at the bottom of Fig. 6.15. Be sure to point out any timing requirements which will *not* be met.

6.19. *UART.* For a specific microcontroller which includes an on-chip UART:

(*a*) What alternative protocols does it support (e.g., one start bit, eight data bits, one stop bit)?

(*b*) What baud rates does it support? Do these baud rates require a special crystal frequency, or is the normal, maximum crystal frequency for the chip satisfactory?

(*c*) Does the UART status register include a bit to announce the detection of a framing error? An overrun error?

6.20. *UART.*

(*a*) When data is being transmitted under interrupt control at 300 baud, then how often do the interrupts occur? Assume that the protocol is one start bit, eight data bits, and one stop bit.

(*b*) How long, after the interrupt occurs, does the CPU have to respond to the interrupt before an error occurs?

(*c*) If data is being *received* at 300 baud with this same protocol, how long does the CPU have to respond to the interrupt for a received byte before an error occurs?

6.21. *UART.* The Intel 8096 obtains the baud rates listed in Fig. 6.22 by first dividing the internal clock rate of 3 MHz by an arbitrary 16-bit number. The result is divided by 16 to obtain the desired baud rate. (The extra divide by 16 results because the UART actually needs a clock input which is 16 times the desired baud rate so that it can look at received data 16 times per bit time. It uses this 16 × clock to count from the start edge to the *middle* of the first data bit.)

(*a*) What is the divisor used to obtain approximately 9600 baud?

(*b*) What is the actual baud rate obtained?

(*c*) Do these two values lead to the accuracy listed in Fig. 6.22?

(*d*) If the divisor were changed by 1 to get to the other side of 9600 baud, what would the accuracy be?

6.22. *UART.* Consider the effect of the 16 × clock required by the UART discussed in the previous problem.

(*a*) Using graph paper with four or five lines per inch, draw 40 cycles of this 16 × clock.

(*b*) Now show the start edge (of received data) occurring *just microseconds before* the rising edge of the 16 × clock. Show the resulting start bit, the first data bit (with a value of one), and the second data bit (with a value of 0).

(*c*) How many rising edges of the 16 × clock occur after the start edge of the received data and up to (and including) the edge which gets closest to the middle of the first data bit?

(*d*) Now show the start edge (of received data) occurring *just microseconds after* the rising edge of the 16 × clock. Again, show the resulting start bit, the first data bit (with a value of one), and the second data bit (with a value of 0).

(*e*) How many rising edges of the 16 × clock occur after the start edge of the received data and up to (and including) the edge which gets closest to the middle of the first data bit?

(*f*) If you were the designer of the UART, how many rising edges of the 16 × clock would you count to get to the middle of the first data bit?

(*g*) If a frame consists of 11 bits, the last of which is a stop bit, then how close does the actual receive baud rate have to be to the actual transmit baud rate (of another UART) for the stop bit to be sampled correctly, *taking into account* the effect of the 16 × clock?

6.23. *Multiple microcontrollers.* Consider an off-the-shelf instrument of some complexity, for which you have information describing its features. Describe how you might partition the jobs of the instrument between two microcontrollers, one a master and one a slave. Can you afford to tie up the UART serial port on each one for nothing but interprocessor communication? If not, then which one, or ones, must be expanded?

6.24. *I/O serial port.* Consider a specific microcontroller, other than the two considered in this book, which includes the pins and hardware to support shift register input/output. (Note that *any* microcontroller can use three lines of a general-purpose I/O port for this purpose, by "bit twiddling" on the port. The advantage of a built-in facility is speed and simplicity. One instruction shifts out an entire byte.)

(*a*) Using a sheet of graph paper with four or five lines to the inch, show each possible serial port timing, analogous to Figs. 6.26 and 6.29. Be sure to indicate what the state of the output line is both before and after the transfer, as well as the times *when* it changes during the transfer.

(*b*) Can one of these be used to support output transfers to the circuit of Fig. 6.27? If not, then can it be made to work by inverting any of the output lines?

(*c*) Repeat part *b* for the circuit of Fig. 2.30.

(*d*) What is the *hold-time* requirement for serial input data? Reconsider your answers in part *c* in light of this requirement.

(*e*) In this part, we want to expand both input and output of the I/O serial port. We also want to be able to access any one port *directly*, analogous to Fig. 6.33. Will the circuit of that figure work as shown in our case here? If not, then neatly draw up a working circuit.

6.25. *I/O serial port.* In this problem we want to consider the circuit of Fig. 6.28 and write a subroutine called OUTPUT which will update one of the four expanded output ports. We might first write to the microcontroller's output port (generically called PORT) the address which selects one of the four expanded output ports (with the Strobe line high). Then we can call an OUTPUT subroutine which does everything to update the port. Check the I/O serial port to wait until it is done with any transfer begun earlier. Then reinitialize it to transfer data at its maximum rate. Return before checking that the transfer has been completed. Specify how the data for the port is to be passed to OUTPUT.

(*a*) Do this for the Motorola 68HC11, referring to Appendix A for information on setting up and using its serial peripheral interface. Using the instruction timing data of Fig. A.39, determine how long it takes to execute OUTPUT.

(*b*) Using the OUTPUT subroutine developed in part *a*, show the code to write 00H to one expanded port and FF to another. Determine how long it takes to do *the entire operation* associated with the write to one port.

(*c*) Repeat parts *a* and *b* for the Intel 8096. Refer to Appendix B for information on setting up and using its serial port.

6.26. *EEPROM for the Intel 8096.* If none of the serial port capability of the 8096 is needed in an application, then we can use this otherwise unused pair of lines to communicate with a Xicor X2444 chip and thereby obtain some nonvolatile memory. The X2444 chip supports this by permitting the DI and the DO pins to be tied together. The 8096 also supports this by putting the I/O serial port DATA line into the high impedance state when it is not outputting data. Compare the timing diagrams of Figs. 6.26*b* and 6.29*b* for the 8096's I/O serial port with that shown in Fig. 6.35 for the X2444 chip. In this application, what do we have to do (*a*) To send a command to the X2444 chip?

    (*b*) To read the content of one address of the X2444 chip?

    (*c*) To write to one address of the X2444 chip?

6.27. *Fuse programmable controller.* It can be useful to interrogate the AMD 29PL141 fuse programmable controller to see if it has finished some process. One way to do this is to use the FPC's shadow register to read the pipeline register and to transfer this information back to the microcontroller. An alternative is to have the FPC use one of its outputs as a signal back to the microcontroller to indicate the same thing. The serial scheme has the advantage of being able to indicate any of several states of the FPC without tying up any extra microcontroller inputs.

    The circuit of Fig. 6.40 supports transfers in only one direction. In this problem, we will consider a circuit for transferring data *in either direction* between the microcontroller and the shadow register of the FPC. The circuit, shown in Fig. P6.27, uses a Texas Instruments 74LS320 crystal oscillator which includes not only the oscillator but also a D-type flip-flop and some buffered outputs. With the addition of the OR gate (used as an active-low input, active-low output AND gate), the PCLK signal is turned off whenever MODE = 1. This is exactly what we need in order to interrogate the FPC's shadow register. Given the ability to turn PCLK off, we can now carry out the required sequence listed at the bottom of Fig. 6.39 of the text. Of course, this process stops the running of the state machine while MODE = 1.

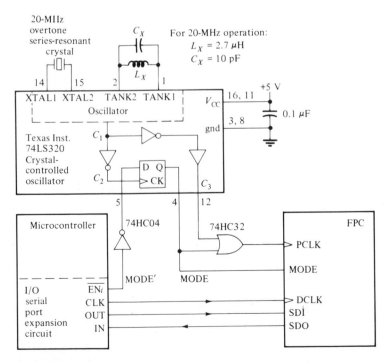

FIGURE P6.27

Clock-gating circuit.

If PCLK is thought of as a waveform with active-low pulses, then the function of the circuit is to inhibit those pulses when MODE = 1. The trick is to gate PCLK without getting glitches as MODE changes. (A glitch occurring on PCLK just as MODE is changing may produce a strange change in the state of the FPC's pipeline register.) Construct a timing diagram showing five cycles of C1, the oscillator output, beginning with a falling edge and ending with a falling edge. Now show C2 and C3, each of which is an inverted version of C1 but slightly delayed from C1. Show each delay as ⅛ of a clock period, so that you can see what is happening. So C2 should be shown delayed ⅛ of a clock period while C3 should be shown delayed ¼ of a clock period. Note that the rising edges of C2 define successive clock periods for the MODE′ input since this signal is looked at by the D-type flip-flop only during these rising edges. Mark vertical lines through these rising edges of C2 to indicate the successive clock periods. Now show MODE′ changing from 0 to 1 during the second of the five clock cycles and changing back to 0 during the fourth clock cycle. Next show the synchronized MODE signal, with its changes delayed ⅛ clock period from the rising edges of C2. Finally, show the PCLK signal generated by the OR gate of Fig. P6.27.

(a) Does this circuit indeed gate PCLK with MODE?

(b) Does the propagation delay from C1 to C2 relative to the propagation delay from C1 to C3 make any difference as far as the generation of glitches is concerned? If so, then which propagation delay must be longer in order to avoid glitches?

6.28. *Waveform recorder.* The waveform recorder of Fig. 6.45 uses some blazingly fast parts in order to be able to sample a waveform at rates up to 10 MHz. Nevertheless, it is questionable whether the circuit shown will actually work at 10 MHz, with the program of Fig. 6.46.

(a) Note that the program of Fig. 6.46 drives the Collect Sample output low at the beginning of a 50-ns clock period and tries to test the FIFO Full flag during the same clock period. The FIFO specifications state that this propagation delay will be less than 45 ns. What requirement does this lead to for the FPC, for reliable operation?

(b) Assuming that the FPC is not fast enough to do this, then change the FPC program so that it will work (with the FPC still being clocked at 20 MHz). You will (probably) have to slow down the maximum sample rate. Can you do this so that the minimum sampling period is 150 ns instead of 100 ns?

6.29. *Waveform recorder.* The circuit of Fig. 6.45 uses a flash A/D converter to achieve a fast 8-bit A/D conversion. For a circuit which can be triggered to generate a waveform, the approach used in a sampling oscilloscope leads to a lower cost. In this approach, a pulse, which we might call TRIGGER, is emitted by the FPC to trigger the waveform generation. Then one clock period later (i.e., 50 ns later), a sample-and-hold circuit* captures the circuit output, in response to the FPC's raising of an output called HOLD. The FPC then signals the microcontroller with a CONVERT output (while continuing to keep HOLD high), telling the microcontroller to convert the output of the sample-and-hold circuit with its on-chip A/D converter (if it has one). When the conversion is complete, the microcontroller

---

* For example, the Analog Devices HTC-0300 ultra high speed hybrid track-and-hold amplifier, packaged in a 24-pin DIP.

sends a DONE signal back to the FPC. The FPC drops the HOLD signal so that the sample and hold can begin tracking the input again. Then it starts the process all over again. It emits the TRIGGER signal and raises the HOLD signal during the *second* clock period after TRIGGER. In this way a waveform of any duration can be sampled at 50-ns intervals over an arbitrarily long duration. The microcontroller can count the samples and turn the process off when it has gotten all the samples it wants by reloading the FPC's pipeline register into an idle state.

(a) Develop the FPC program, in a form analogous to that of Fig. 6.46, for this application. Do this so as to sample the waveform at 50-ns intervals.

(b) How many samples will your solution in part a collect before running out of steam? Is it just 64 or so, as determined by the size of CREG?

(c) Can you rearrange the algorithm so as to generate the timing for more than 64 samples with the FPC?

6.30. *Waveform recorder*. Rework the last problem so as to generate the timing to collect just one sample. The time between the TRIGGER output and the HOLD output will be passed to the FPC by the microcontroller by loading up SREG and CREG. This time can be assumed to be larger than 64 * 50 ns (since the algorithm of the last problem can be used to collect the first 64 samples).

Using this algorithm, the microcontroller can pass the parameter to the FPC to collect each sample. In this way it can collect samples over a much longer period, but by means of a slower process since it has to set up the FPC to generate the timing for each sample. On the other hand, it is no longer necessary to collect samples spaced 50 ns apart. A larger time between samples just requires the microcontroller to modify its algorithm for generating the parameter which it sends to the FPC.

(a) Develop the FPC program, analogous to that of Fig. 6.46, which will use the content of SREG and CREG to generate the timing for collecting a sample. Your algorithm should use SREG and CREG so as to be able to specify any time interval up to about 63 * 63 * 50 ns.

(b) Now that you have an algorithm, write an equation which relates the initial content of SREG and CREG to the time interval produced.

(c) Given the equation in part b, can you use a divide instruction, such as the 8096's, which produces both an integer quotient and a remainder, to solve for values for SREG and CREG, given a time interval? If so, then describe the algorithm.

REFERENCE

An outstanding source of information about integrated circuits and what is available is *IC MASTER*, published by Hearst Business Communications, Inc., 645 Stewart Avenue, Garden City, NY 11530. It can be entered with a part number, to find out what the part is. It can be entered with a category, like A/D converters, to find the IC versions of this component sorted by number of bits and by speed. It gives the alternative sources for any IC which has them. It lists local distributors and local representatives for all IC manufacturers. It even includes data sheets for those parts which manufacturers have been willing to advertise in this way.

CHAPTER
7

I/O
HARDWARE
ALTERNATIVES

7.1 OVERVIEW

We begin this chapter with a discussion of alternative methods for displaying data, which might be used on the front panel of an instrument. Integrated circuit manufacturers have used the advances in technology, which permit more and more logic to be put onto a silicon chip, to implement some startling drivers for display devices. In particular, we are interested in driving displays without using many I/O lines from a microcontroller chip. We will find that device drivers are available which can be driven from the microcontroller's serial ports.

For the displays themselves, we will look at three alternatives. First, we consider LED (light-emitting diode) displays in two forms: LED annunciators (i.e., on-off lights) to indicate status information and seven-segment LED displays to present just a few digits of information.

Second, we consider liquid crystal displays (LCDs), which are commonly used on instrument front panels to display seven-segment numeric information together with annunciators in the form of words (e.g., Hz, kHz, MHz, FREQUENCY, PERIOD, DC VOLTS, AC VOLTS) or graphic symbols (e.g., displaying a bell when an alarm function is enabled). They can also be used to display 16-segment alphanumeric information or dot-matrix alphanumeric information. Instrument front panels have been cleaned up with the help of custom LCDs. Their manufacturing process supports the development of custom parts without introducing undue delay or cost into the development of an instrument.

Finally, we will look at CRT (cathode-ray tube) displays, for applications where extensive alphanumeric information is needed. While multiplexed, dot-matrix LCD and plasma discharge displays are making inroads into such applications, the CRT still provides a crisp, relatively simple display for such applications. Furthermore, what used to be a fairly complex drive circuitry problem has been drastically simplified by the introduction of chips like National Semiconductor's NS455 Terminal Management Processor.

Next, we will look at several approaches to actuation. D/A conversion permits us to obtain an analog voltage which can be used to drive dc motors and other devices which require a *proportional* output rather than an on-off output. Pulse-width modulation is a form of D/A conversion which can be used to obtain *high-resolution* control of any variable which does not change fast. For a microcontroller which is designed to support it well (e.g., the Motorola 68HC11 for up to four outputs), pulse-width modulation provides a simple way to get a resolution of one part in 2^{16} without tying up *any* CPU time (other than to get it started) and tying up only one pin for each PWM output.

To handle faster changing proportional outputs, or more outputs, we consider D/A converters which latch a binary output and produce a voltage proportional to it. We are particularly interested in D/A converters which employ serial input to minimize the interconnections to a microcontroller.

Stepper motor actuation permits a microcontroller to control mechanical positioning with speed, accuracy, and simplicity. Stepper motors are available to handle a broad range of torques, speeds, and step sizes, making them extremely versatile positioners. While they are normally configured to cause the rotation of a shaft, we also consider a linear stepping actuator. This permits a microcontroller to control linear position directly.

We consider three approaches for sensing mechanical position. Slotted optical switches serve to define a single position. They can be used to define the end limits for a positioning system driven by a stepper motor or a dc motor. They can also be used to define a zero-reference point for a positioning system. Incremental shaft encoders quantize the angular position of a shaft. As the shaft rotates, the incremental encoder notes each quantum change in position, including the direction of change. A microcontroller can accumulate these changes into position information. If an incremental encoder is to be used to indicate absolute position, rather than just changes in position, then it requires the addition of a mechanism to indicate a zero-reference position. Also, until the incremental encoder has been driven past this zero-reference position initially, the encoder and microcontroller combination cannot begin to produce absolute position information. Absolute shaft encoders resolve this problem by producing position information directly with each reading. However, if the encoder is moving while it is being read, then care must be taken to ensure a correct reading of position. We will see that a combination of encoder design and software decoding can resolve this problem.

Actuation generally requires a translation of the drive capabilities of a microcontroller output to the drive requirements imposed by the actuator. We will consider some IC drivers for interfacing to devices which require higher current drive capability or higher voltage switching capability than can be handled directly by the output of a microcontroller. We will also consider other devices which optically isolate an output circuit from the sensitive circuitry of a microcontroller, permitting the microcontroller to switch ac power without getting "fried" in the process.

7.2   LED DISPLAYS

LED displays are commonly used in one of three forms.

As annunciators which can display a single bit of information

As seven-segment displays for numeric information

As dot-matrix displays for a limited amount of alphanumeric information

Some miniature LED lamps for use as annunciators are shown in Fig. 7.1a. On an instrument front panel, these might be mounted next to a word (e.g., uncalibrated) to indicate some condition of the instrument. A drive circuit for them is shown in Fig. 7.1b. It makes use of the I/O serial port expansion circuit of Sec. 6.6. The circuitry takes advantage of high-efficiency LEDs to produce a bright red output, even with an output current that is lower than that which can be supplied by the 74HC595 shift register outputs.*

The LED light bars shown in Fig. 7.2 can be mounted behind a silk-screened transparent front panel to light up a silk-screened word. In this way, status information is presented to a user more directly than it is by a lamp mounted *next* to the word of interest to the user. Furthermore, with red, yellow, and green light bars available, the nature of the status information can be encoded by the color of the word displayed, giving a user an intuitive color cue.

For the display of numeric information, seven-segment LED displays, such as those shown in Fig. 7.3, have long been popular. The availability of the simple drive circuitry for a microcontroller shown in Fig. 7.4 makes these displays easier than ever to use. Motorola's 14499 chip is designed to display four digits. It accepts input data for the four digits serially. It also accepts four more bits for the four decimal points on the seven-segment displays (or for four separate annunciators, wired up in place of the decimal point connections). These 20 bits of data can be imbedded into a string of 24 bits obtained by outputting 3 bytes via the I/O serial port. It is only necessary to drive the 14499 enable input low (i.e., active) before transferring the data. When this line is raised, after the 24 shifts, the data which the 14499 has accepted into its shift register is transferred to an internal latch. From there, the 14499 circuit multiplexes the displays with it.

---

* Six milliamperes, maximum.

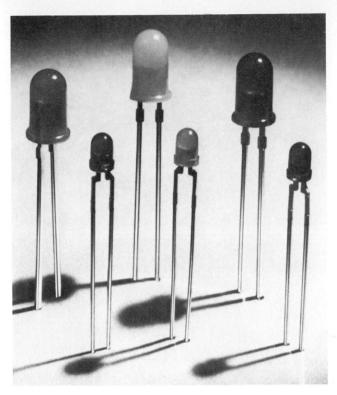

(a)

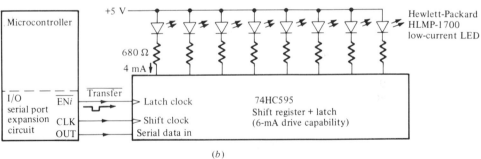

(b)

Hewlett-Packard low-current LED lamps:

Red: HLMP-1700
Yellow: HLMP-1719

Luminous intensity = 1.0 mcd min at 2 mA
                   = 1.8 mcd typical at 2 mA

Voltage drop   = 1.8 V typical at 2 mA
                  = 2.2 V max at 2 mA

Current   = 7 mA max

(c)

FIGURE 7.1
LED annunciator lamps. (*Hewlett-Packard Co.*) (*a*) Typical units; (*b*) drive circuit; (*c*) characteristics.

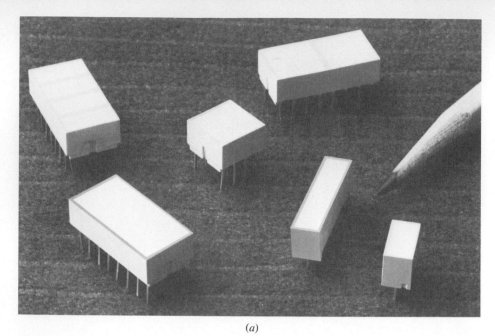

(a)

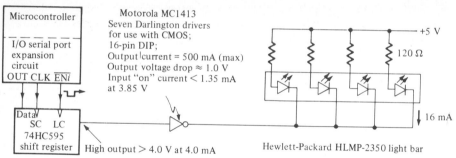

Microcontroller

I/O serial port
expansion
circuit

OUT CLK ENi

Data
  SC   LC
74HC595
shift register

High output > 4.0 V at 4.0 mA

Motorola MC1413
Seven Darlington drivers
for use with CMOS;
16-pin DIP;
Output current = 500 mA (max)
Output voltage drop ≈ 1.0 V
Input "on" current < 1.35 mA
at 3.85 V

+5 V

120 Ω

16 mA

Hewlett-Packard HLMP-2350 light bar

(b)

Size	Color	Hewlett-Packard part number	Description
0.150 × 0.350 in	Red Yellow Green	HLMP-2300 HLMP-2400 HLMP-2500	Light bar contains two LEDs
0.150 × 0.750 in	Red Yellow Green	HLMP-2350 HLMP-2450 HLMP-2550	Light bar contains four LEDs

HLMP-2300 luminous intensity = 6 mcd min at 20 mA
                            = 23 mcd typical at 20 mA
HLMP-2350 luminous intensity = 13 mcd min at 20 mA
                            = 45 mcd typical at 20 mA
Voltage drop (per LED)   = 2.0 V typical at 20 mA
                            = 2.6 V max at 20 mA
Current (per LED)   = 25 mA max

(c)

FIGURE 7.2
LED light bar annunciators. (*Hewlett-Packard Co.*) (*a*) Typical units; (*b*) drive circuit; (*c*) characteristics.

FIGURE 7.3
Seven-segment LED displays.
(*Hewlett-Packard Co.*)

Multiplexing seven-segment LED displays provides some benefits besides drastically reducing the number of lines to the driver chip. Given the same average current through one of the LED segments, the LED will appear somewhat brighter* when multiplexed than when the current is dc. On the other hand, the *peak* currents which arise in a multiplexed LED display circuit require special consideration. The 14499 chip is specified to drive peak segment currents no higher than 50 mA. When divided by four, this limits the average current to 12.5 mA, well below the 20 mA specified for many LED displays. To get more brightness, a multiplexed display can take advantage of *high-efficiency* displays, which produce a brighter output with less current than a normal efficiency display.

Figure 7.5*a* and *b* shows the order in which data must be sent to the 14499 chip, beginning with the decimal point data, followed by the most-significant digit (most-significant bit first), etc. If decimal points (or separate annunciators wired in place of the decimal point LEDs) are not used, then the 14499 can be loaded with just *two* 8-bit serial transfers from the microcontroller. The sixteenth clock pulse will leave the least-significant bit of the least-significant digit in bit 20 (the first bit) of the 14499's shift register—just where it should be. The shifting of each byte most-significant bit first is exactly what the Motorola 68HC11 does, and just the opposite of what the Intel 8096 does. However, if the microcontroller used in an application has ROM to spare (as is likely, for many applications of the Intel 8096 with its 8K ROM), then 256 bytes of ROM can be used to create a table for switching the order of the bits in a byte by means of a simple table lookup.

If more than four digits of display are needed, then several 14499s can be used. A separate I/O serial port can be used to drive each one using the circuit of Fig. 6.28.

---

* The Hewlett-Packard data sheet for the HDSP-7503 indicates about a 30 percent increase in brightness for the multiplexed display circuit of Fig. 7.4 relative to a nonmultiplexed circuit having the same segment current.

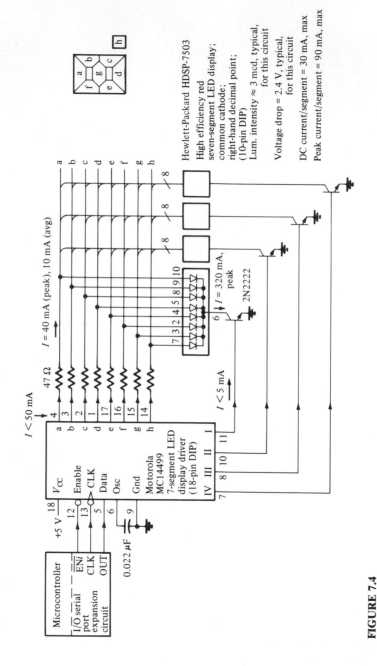

FIGURE 7.4

Multiplexed four-digit seven-segment LED display.

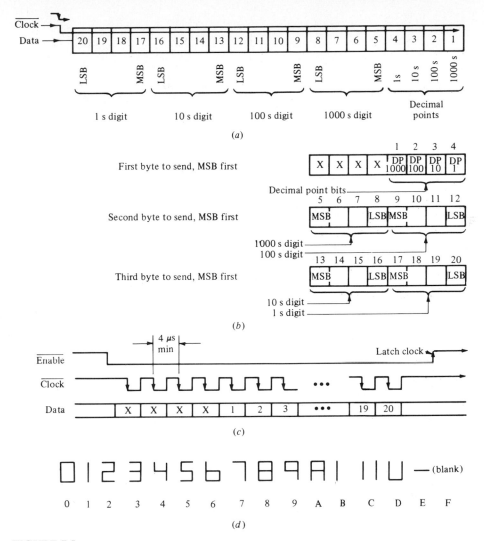

FIGURE 7.5
Timing and coding for display of Fig. 7.4. (*a*) MC14499 shift register bits and what the corresponding latch bits drive; (*b*) sequence of bytes for microcontroller to send most-significant bit first; (*c*) timing diagram; (*d*) segment decoding.

    The Motorola MC14449 seven-segment LED display driver was designed some years ago using old CMOS technology. This is not a trivial point; it means that data cannot be shifted to it reliably at the data rates normally used with the I/O serial ports of either the Motorola 68HC11 or the Intel 8096. In fact, since the pulse width of the clock output of the Intel 8096 I/O serial port cannot be varied from its nominal value of ⅓ μs, a one-shot (e.g., a 74LS123) is needed to stretch the active-low pulses. The circuit is shown in Fig. 7.6. Note that the leading (i.e., falling) edge of CLK*, which clocks the data into the 14499 chip,

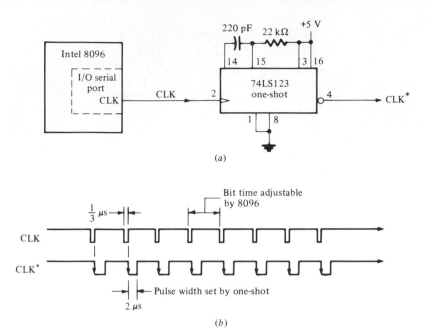

(a)

(b)

FIGURE 7.6
Pulse-stretching circuit. (a) Circuit; (b) timing diagram.

occurs at essentially the same time as the trailing (i.e., rising) edge of CLK. This is required so that the *data* on the I/O serial port output will be aligned with these falling edges of CLK*. Refer to Fig. B.30 in Appendix B.

7.3   LCD DISPLAYS

Liquid crystal displays are finding broad use in instrument design. As illustrated in Fig. 7.7, the display can easily handle the display of numeric variables (i.e., 100.000000 and −140.0), units (i.e., MZ and DM), and annunciators (i.e., FREQ and AMPTD).

A liquid crystal display element consists of a liquid crystal material which is imbedded between a frontplane conductor and a backplane conductor. When an ac voltage of sufficient root-mean-square (RMS) value is impressed across the two conductors, the liquid crystal changes its reflectivity (for the usual, front-lit version) or its transmittivity (for a panel which uses backlighting). A *direct drive* liquid crystal display employs a separate driver for each segment of the entire display. For a display with six digits and 16 annunciators, a direct drive LCD requires 6 * 7 + 16 = 58 drivers. Each driver is commonly built into an LCD driver chip and consists of the CMOS exclusive-OR gate and oscillator circuit shown in Fig. 7.8. While the current involved (and therefore the power dissipation) is negligible, a direct drive circuit has the problem of handling the sheer number of signals involved. Direct drive displays tend to employ several

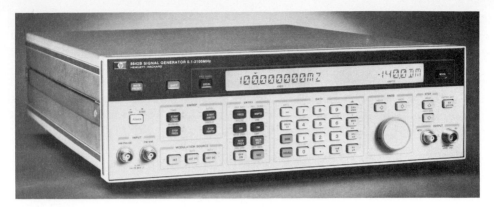

FIGURE 7.7
A signal generator which makes versatile use of a liquid crystal display. (*Hewlett-Packard Co.*)

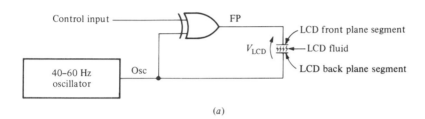

(*a*)

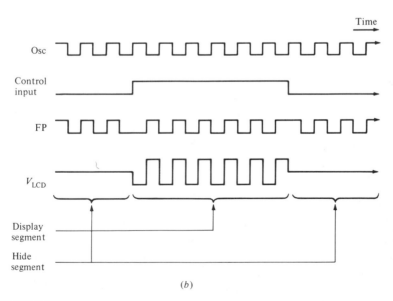

(*b*)

FIGURE 7.8
Direct drive of an LCD segment. (*a*) Circuit; (*b*) timing diagram.

driver chips, each having many pins. On the other hand, the direct drive approach puts the least demands upon the liquid crystal fluid since it is subjected to either a 0-V signal or a V_{CC} square wave. Consequently, the fluid must be selected to respond to a voltage whose RMS value is V_{CC}.

A liquid crystal display imposes the requirement that the drive voltage *must not* have a dc component. As a practical matter, any dc component present must be less than 50 mV. If the oscillator shown in Fig. 7.8*a* has a CMOS output stage, and if both this output stage and the exclusive-OR gate share the same supply voltage, then this condition is easily met. If it is not met, the liquid crystal fluid deteriorates irreversibly and, consequently, the display deteriorates.

Advances in liquid crystal technology make it possible to *multiplex* the display elements. One incentive to do this is to simplify the artwork for the frontplane and the backplane conductors of the display. Instead of having to bring out separate conductors from every segment, the *backplane* segments can be daisy-chained together into a few horizontal rows. The *frontplane* segments are connected to form many vertical columns. The conductors needed for connection to the drive circuitry are relatively few in number and are easily brought out to the edges of the display. From the edges of the display, the connection of each conductor is easily made to the corresponding conductor on the printed circuit board holding the drive circuitry. This is illustrated in Fig. 7.9, which shows an energy management unit together with the frontplane and backplane artwork used in designing its display.

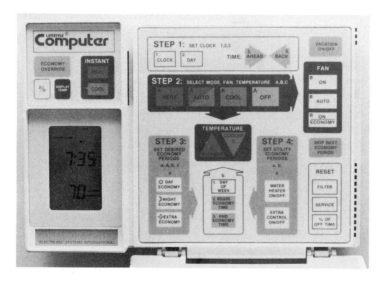

(*a*)

FIGURE 7.9
An energy management unit which has a liquid crystal display driven by the Motorola MC145000 and MC145001 drivers discussed in this section. (*Electronic Systems International*) (*a*) The unit; (*b*) the liquid crystal display, shown with all segments turned on; (*c*) the LCD frontplane and backplane artwork.

(b)

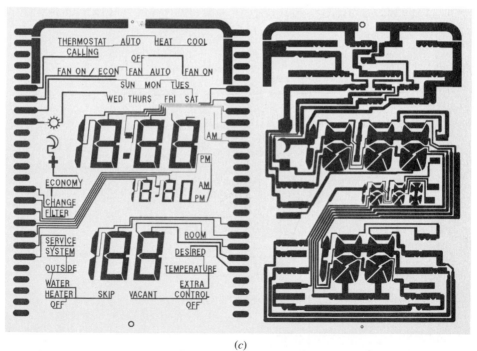

(c)

FIGURE 7.9 (*continued*)

Several multiplex formats have become popular during the past few years. Intersil is a company which makes widely used LCD drivers. Their ICM7230 series of LCD drivers employ ⅓ multiplexing. This means that the backplane is divided into three daisy-chained conductors. The Motorola LCD drivers which are discussed here employ ¼ multiplexing, as illustrated in Fig. 7.10. While this figure shows just one column of segments, the segments for the entire display would be interconnected on the backplane so as to be driven by just four backplane drivers: BP1, BP2, BP3, and BP4. The frontplane segments would be divided up into groups of four segments, with each group being driven by a separate driver. Consequently, driving a liquid crystal display which employs ¼ multiplexing means driving four backplane conductors and $N/4$ frontplane conductors, where N is the number of segments in the entire display. The advantage of multiplexing lies in the reduction in the number of frontplane drivers (and the number of separate wires which must be handled) from the value of N required by a direct drive display to the value of $N/4$ required by a ¼-multiplexed display.

An advantage of using ¼ multiplexing for a display to be used with a microcontroller is that each character can be efficiently represented with a multiple of four segments. For example, a numeric display made up of seven-segment displays (plus decimal points between each pair of digits) can be developed as shown in Fig. 7.11, with two frontplane drivers required for each digit. When the microcontroller sends display data to the Motorola LCD driver chips

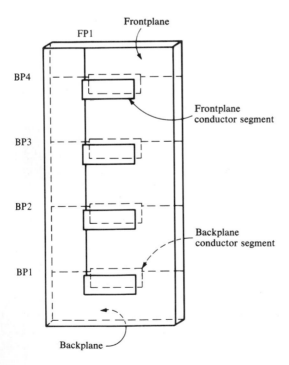

FIGURE 7.10
Frontplane-backplane multiplexing of a liquid crystal display.

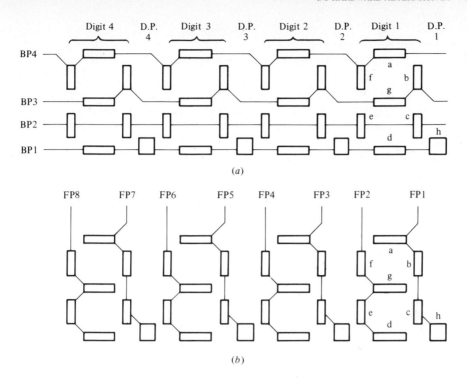

FIGURE 7.11
Typical 2 × 4 multiplex format for seven-segment numeric display. (*a*) Backplane; (*b*) front-plane; (*c*) segment truth table.

(which we shall discuss), it sends 8 bits for each digit, encoded to turn on the segments corresponding to the digit to be displayed.

An alphanumeric display using the popular starburst configuration of segments shown in Fig. 7.12 represents each character with 16 segments. A ¼-multiplex scheme means that each character requires four frontplane drivers. The microcontroller sends 16 bits (2 bytes) of display data to the Motorola LCD driver chips for each character to be displayed. Any decimal points and annunciator words or symbols are gathered together into frontplane groups of four and driven with extra frontplane drivers.

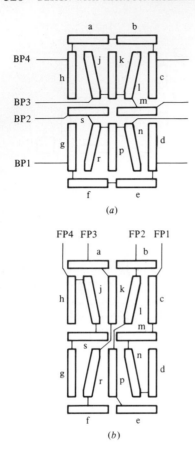

(a)

(b)

	FP4	FP3	FP2	FP1
BP4	h	a	b	c
BP3	j	k	l	m
BP2	s	r	p	n
BP1	g	f	e	d

(c)

FIGURE 7.12
Typical starburst alphanumeric display. (a) Backplane; (b) frontplane; (c) segment truth table.

At the end of this section, we consider *how* the drivers drive the front-plane and backplane lines so as to turn on one segment while leaving others turned off. We will see exactly what requirements the manufacturer of a liquid crystal display must meet for successful multiplexing. These requirements are least stringent for a direct drive display, more stringent for a ⅓-multiplexed display, and still more stringent for a ¼-multiplexed display. They play out for the manufacturer as a need to use one fluid for direct drive displays and another for multiplexed displays. They play out for the instrument designer as a possible need for temperature compensation for the display (by controlling the supply voltage to the driver chips). They play out for the user of the instrument as a decreased angle of viewing (from straight-on viewing) for a multiplexed display as compared with a direct drive display. In spite of all these caveats, a $\frac{1}{4}$-multiplexed LCD is an excellent choice for most instrument applications.

The Motorola MC145000 and MC145001 serial input multiplexed LCD driver chips are shown driving a liquid crystal display in Fig. 7.13. The Motorola designers gave careful thought to the integration process. As can be seen in the figure, the circuitry has only two adjustments. One resistor is used to set the frequency of an internal oscillator so that the operating frequency required by the liquid crystal fluid can be optimized easily. A potentiometer (or a temperature compensation circuit) can be used to adjust the supply voltage to the chip for optimum contrast of the display. Even the pinout of the chips has been optimized for easy connection to the liquid crystal display itself. A display which requires more frontplane drivers than provided by the circuit of Fig. 7.13 gets them from the addition of more MC145001 18-pin DIPs, connected to the right of those shown in Fig. 7.13.

The interface between a microcontroller having an I/O serial port and the LCD driver chips could hardly be simpler. The data from the I/O serial port is clocked into a long shift register, extending over several chips (via the Data Out to Data In connections between chips). Data is clocked into the shift register on the falling edge of the clock pulses, with only a 50-ns setup time, a zero hold time, and a maximum clock rate of over 7 MHz. These times and clock rates are easily met by the I/O serial ports for the two microcontrollers we have been considering in this book, even at their highest transfer rate. Furthermore, the two inputs which connect to the microcontroller are buffered so that their voltages can exceed the supply voltage for the chips (which is normally not acceptable for a CMOS chip). Since the specifications for the chips call for a supply voltage anywhere between +3 and +6 V (to optimize LCD performance), this was an important consideration for the designers.

The third connection between the microcontroller and the LCD display controller chips is an interrupt signal, used to synchronize the data transfers. This is an active-high pulse. The microcontroller needs to transfer data while it is not high. It can transfer data (when data has been changed) in response to the trailing edge (i.e., the falling edge) of this "frame-sync" pulse.

Example 7.1. Consider the interactions which must take place when the microcontroller wants to update the liquid crystal display.
1. It updates an array in RAM by adding in any new values to those values which are presently being displayed and which do not need to be changed. This array is called LCD_BUF in Fig. 7.14. For numeric data, this updating of the array involves a table lookup to convert each digit into its seven-segment code representation. For alphanumeric data, the updating involves a table lookup for ASCII to starburst code representation. For decimal points and annunciators, the updating involves setting or clearing selected bits in the array.
2. It initializes the two-byte variable LCD_POINTER, shown in Fig. 7.14, with the address of LCD_BUF. This pointer will be used by the I/O serial port's

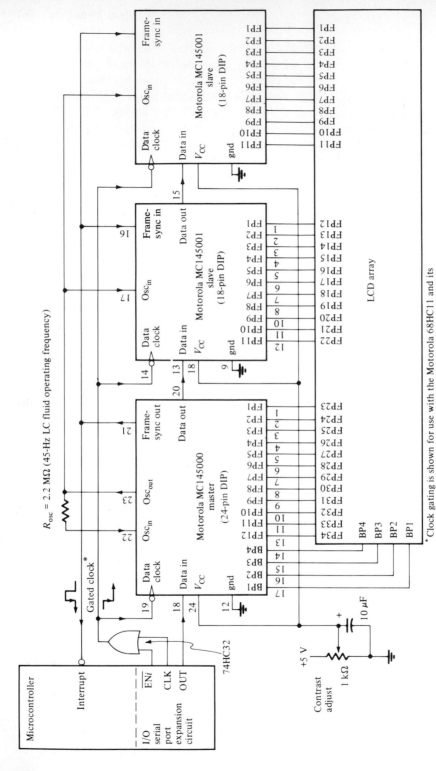

FIGURE 7.13
Multiplexed LCD display circuit.

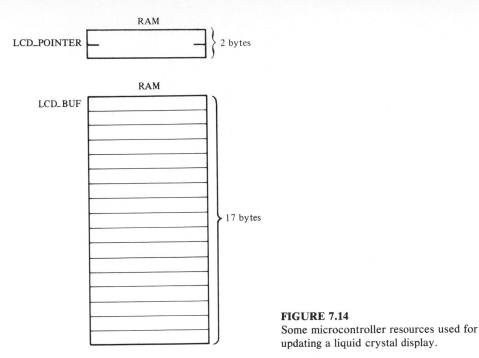

RAM

LCD_POINTER   } 2 bytes

RAM

LCD_BUF

17 bytes

FIGURE 7.14
Some microcontroller resources used for updating a liquid crystal display.

interrupt service routine to indicate which byte to output next and when the last byte of LCD_BUF has been transferred.

3. It clears the flag for the (presently disabled) interrupt input which is driven by the frame-sync output of the LCD master driver chip.

4. It enables interrupts from this interrupt source.

When the interrupt occurs (up to $1/11$th s later, as we can see in a moment), the interrupt service routine does the following:

1. It disables further interrupts from itself.

2. It sets a bit in the state variable, IOSP_STATE, as discussed at the very end of Sec. 6.6. Then it sets the bit which enables interrupts to the CPU from the I/O serial port. Whether or not this bit is already set, this will leave it set so that the I/O serial port will cause an interrupt to occur either immediately, or whenever it completes a presently occurring transfer. These two actions serve to communicate with the I/O serial port's interrupt service routine.

As discussed in Sec. 6.6, the I/O serial port may well be used to provide serial output data to other devices in addition to the LCD display. Each time that it completes a transfer, it interrupts the CPU. The interrupt service routine polls bits in the IOSP_STATE variable to determine which device is to be served next. If it is the LCD display, then the I/O serial port is set up to emit data which is stable on the *falling* edges of the clock (e.g., using the timing of Fig. A.28*b* for the Motorola 68HC11). Then LCD_POINTER is used as a pointer into LCD_BUF. The selected byte is written to the I/O serial port and LCD_POINTER is incremented. When the last byte of LCD_BUF has been sent, then the bit of IOSP_STATE for the LCD display is cleared.

A block diagram for the Motorola MC145000 master LCD driver is shown in Fig. 7.15. Note that the internal circuitry runs from an autonomous clock circuit. With a 2.2-MΩ oscillator resistor, the latch which drives the display is updated from the shift register 11 times a second. A buffered version of this frame-sync signal is brought out and used to synchronize the slave LCD drivers as well as the microcontroller's serial data transfers.

Note that the backplane and frontplane drive circuits require that the V_{CC} supply voltage be split into three parts. The voltage splitter circuit is one of the critical parts of the chip. Motorola specifies that the average dc offset voltage across any segment will be less than 50 mV. This is really a requirement on the voltage splitter in this chip *relative to* the voltage splitter in the slave chips. The Motorola designers have done their part to maintain the life of the LCD panels

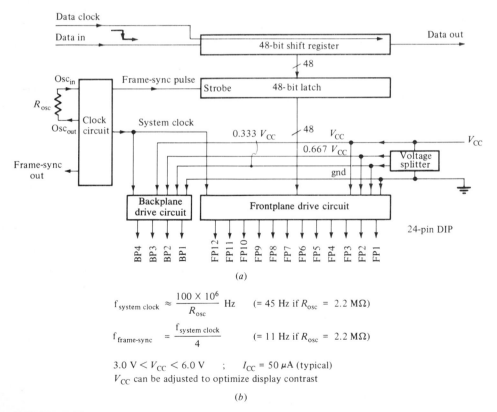

(a)

$$f_{\text{system clock}} \approx \frac{100 \times 10^6}{R_{\text{osc}}} \text{ Hz} \qquad (= 45 \text{ Hz if } R_{\text{osc}} = 2.2 \text{ M}\Omega)$$

$$f_{\text{frame-sync}} = \frac{f_{\text{system clock}}}{4} \qquad (= 11 \text{ Hz if } R_{\text{osc}} = 2.2 \text{ M}\Omega)$$

$$3.0 \text{ V} < V_{CC} < 6.0 \text{ V} \quad ; \quad I_{CC} = 50 \text{ }\mu\text{A (typical)}$$
V_{CC} can be adjusted to optimize display contrast

(b)

FIGURE 7.15
Motorola MC145000 serial input, multiplexed LCD driver (master). (a) Block diagram; (b) a few specifications.

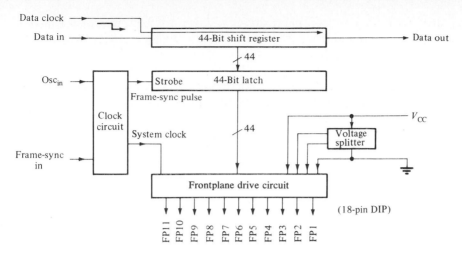

FIGURE 7.16
Motorola MC145001 serial input, multiplexed LCD driver (slave).

used with their chips by holding this average value below the value which can do harm to the LCD fluid. We see, shortly, how these voltages are used to make up the frontplane and backplane driver waveforms.

A block diagram for the MC145001 slave unit is shown in Fig. 7.16. It is a simpler circuit, in a smaller package, since it does not have an on-chip oscillator and since it only derives frontplane driver outputs. It has one less frontplane driver than the MC145001 master unit does. This is a consequence of using an 18-pin package for the circuit.

As designers using these LCD driver chips, the key data we need is the relationship between each bit position in the shift register and the LCD segment which it controls. This is presented in Fig. 7.17. The very first bit shifted out of the microcontroller will, after 48 shifts, be in the 1 position shown. If this bit position holds a 1, then the LCD segment at the intersection of the BP4 back-

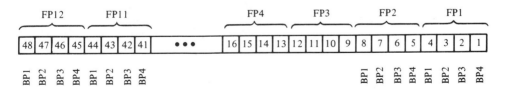

The "1" bit of the 48-bit shift register is the first bit shifted in.
The "48" bit is the last (i.e., most recent) bit shifted in.

A "0" shifted in turns the corresponding LCD segment off.
A "1" shifted in turns the corresponding LCD segment on.

Example: A "1" in bit position 6 turns on the FP2-BP3 segment.

FIGURE 7.17
Relationship between shift register bit position and the LCD segment which it affects.

plane driver and the FP1 frontplane driver will be turned on. Look back now at Fig. 7.13 which shows one master and two slaves driving an LCD array requiring up to 34 frontplane drivers. In this case, the first bit shifted out of the microcontroller will, after $34 \times 4 = 136$ shifts, be in the FP1 position for the display (i.e., the FP1 position of the rightmost slave). We are now in a position to figure out how to code digits and alphanumerics.

Example 7.2. Considering the following configuration of parts, create a table called LCD_DIG which can be used to translate digits to the coding needed for the display.

A Motorola 68HC11 (whose I/O serial port transfers data *most*-significant bit first).

An LCD array which connects the segments as shown in Fig. 7.11. There is no standard for this used by LCD manufacturers. Their treatment of segment interconnections necessarily will vary depending upon whether direct drive, $1/3$ multiplexing, or $1/4$ multiplexing is to be used. Even for just $1/4$ multiplexing, there is no standard. However, while creating this table is one more thing which we have to do as part of a design, it also means that there are no *wrong* configurations of segments. We create a table which puts the bits where they belong.

The LCD array will be positioned as shown in Fig. 7.13. That is, BP4 will be the top backplane driver and FP1 will be the rightmost frontplane driver. Note that this is not necessarily a given. The display could be flipped over, depending upon whether the driver chips are mounted above or below the display. Again, this discussion only bears upon the entries which we put into the LCD_DIG table and upon the order in which we write the bytes of the LCD_BUF array of Fig. 7.14 out to the display.

The derivation of the LCD_DIG table is shown in Fig. 7.18. Figure 7.18*a* shows how the bits of the byte sent by the 68HC11 must be arranged. Since the 68HC11 sends the most-significant bit first, the leftmost bit of this byte will end up in the rightmost position of the byte in the LCD drivers' shift register. It is labeled with FP1 and BP4 to correspond to the rightmost bit position of Fig. 7.17. Figure 7.18*b* just translates the table of Fig. 7.11*c* into these bit positions. Figure 7.18*c* shows the coding for each of the digits together with the byte needed for the table entry. Finally, Fig. 7.18*d* shows the resulting table.

Example 7.3. Modify the last example using the starburst display segment interconnection scheme shown in Fig. 7.12 to create entries for the appropriate table, LCD_ALPH. This table should have entries to convert all ASCII codes between 20H and 5FH, that is, to convert the 6-bit ASCII subset which includes numbers, uppercase letters, and punctuation marks. The development of the table is shown in Fig. 7.19.

The actual multiplexing voltages produced by the Motorola LCD chips divide each refresh cycle, or frame, into four parts, as shown at the top of Fig.

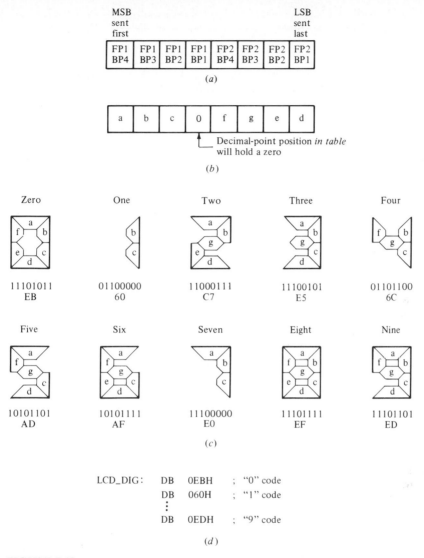

FIGURE 7.18
Development of LCD_DIG table. (*a*) Relationship between byte sent by microcontroller and LCD segment positions; (*b*) relationship between byte sent and seven-segment display segments; (*c*) segments to turn on for each digit and the resulting encoding; (*d*) table.

7.20. The display system clock has the slow (45 Hz or so) rate selected to meet the LCD specification. Each frame is divided into four quarter frames. The first quarter frame is used to select the segments which are connected together by the BP1 conductor. The second quarter frame is used to select the segments connected by the BP2 conductor, etc. The waveforms used to drive these four backplane conductors are shown in the figure. Note that each waveform can be thought of as consisting of two parts:

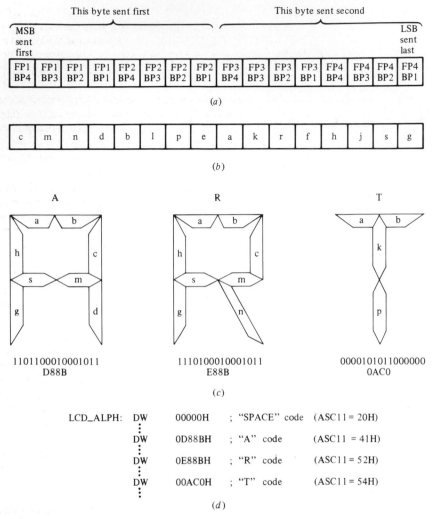

FIGURE 7.19
Development of LCD_ALPH table. (*a*) Relationship between bytes sent by microcontroller and LCD segment positions; (*b*) relationship between bytes sent and starburst display segments; (*c*) segments to turn on for each of several characters and the resulting encoding; (*d*) table.

During the quarter frame when it is to select segments in its row, the waveform raises to V_{CC} volts for the first half of the quarter frame and drops to 0 V for the second half of the quarter frame.

During the other three quarter frames, the waveform is a square wave oscillating between one-third V_{CC} and two-thirds V_{CC}.

A frontplane waveform, FP_j, also repeats every frame. Its first quarter frame is dedicated to turning on or off the segment in the row driven by BP1. Its second quarter frame is dedicated to turning on or off the segment in the row driven by

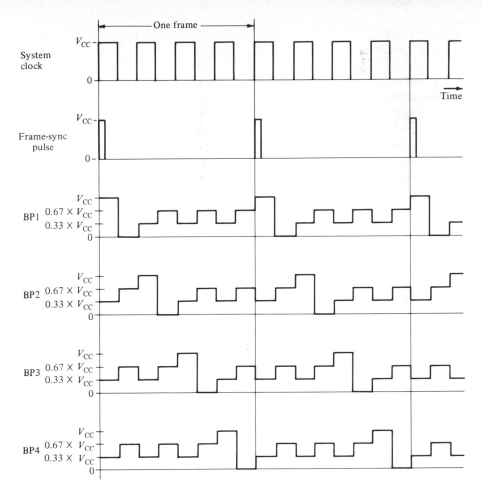

FIGURE 7.20
Backplane waveforms.

BP2, etc. If a segment is to be turned on, then the frontplane quarter-frame waveform is that shown in Fig. 7.21a. In this case, the voltage impressed across the LCD fluid, $BP_i - FP_j$, rises to V_{CC} during the first half of the quarter frame and drops to $-V_{CC}$ during the second half of the quarter frame.

For the quarter frame selected by a backplane waveform, the corresponding frontplane waveform to turn the segment *off* is shown in Fig. 7.21b. Note that in spite of having the same backplane waveform as in Fig. 7.21a, the voltage across the segment, $BP_i - FP_j$, is drastically different. It rises to only $V_{CC}/3$ during the first half of the quarter frame and drops to $-V_{CC}/3$ during the second half of the quarter frame.

The two quarter-frame waveform possibilities for a quarter frame *not* selected by a backplane waveform are shown in Fig. 7.21c. In either case, the voltage across the segment is one period of a square wave of amplitude $V_{CC}/3$.

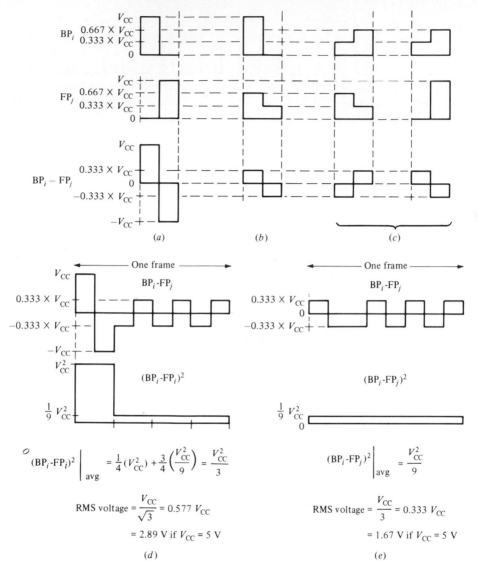

FIGURE 7.21
Determination of RMS voltages across on and off segments. (*a*) Selected quarter-frame on; (*b*) selected quarter-frame off; (*c*) deselected quarter-frame possible waveforms; (*d*) on segment RMS voltage determination; (*e*) off segment RMS voltage determination.

From these considerations, we can now determine the RMS value of the waveform across a segment that is to be turned on as well the RMS value of a waveform across a segment that is to be turned off. Figure 7.21*d* shows a typical waveform across an *on* segment. To get its RMS value, we must square the waveform, find the average value of the squared waveform, and then take the square root of this average value. The squared waveform and the remaining calculations are shown, leading to an RMS value of 2.89 V if $V_{CC} = 5$ V. Figure 7.21e shows a typical waveform across an *off* segment. It has an RMS value of 1.67 V if $V_{CC} = 5$ V.

In illustration of a frontplane waveform, Fig. 7.22 shows six cases. Figure 7.22a shows the waveform generated when all segments are to be turned off in the column selected by the FP_j driver. Figure 7.22f shows the waveform if three of the four segments are turned on.

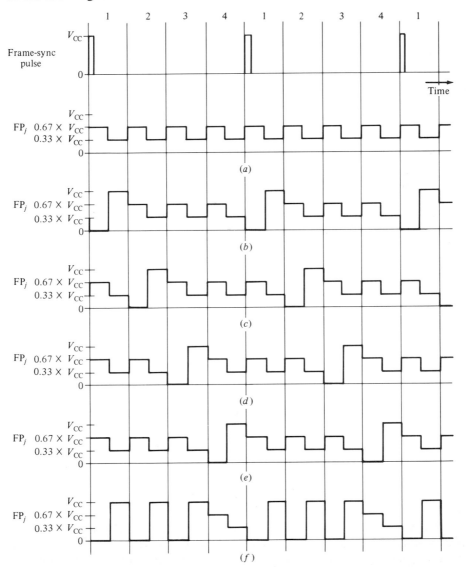

FIGURE 7.22
Frontplane waveforms. (a) All four segments off; (b) segment 1 (bottom segment) on; 2, 3, and 4 off; (c) segment 2 on; 1, 3, and 4 off; (d) segment 3 on; 1, 2, and 4 off; (e) segment 4 on; 1, 2, and 3 off; (f) segments 1, 2, and 3 on; 4 off.

To illustrate how these waveforms interact with a liquid crystal fluid, we will consider the display used in the energy management unit of Fig. 7.9. This display is manufactured by N. V. Philips of the Netherlands, which is served in the United States by Amperex Electronic Corp. (a North American Philips company) of Smithfield, Rhode Island. When used with Motorola's MC145000/ 5001 LCD drivers and ¼ multiplexing, the recommended power supply voltage is shown in Fig. 7.23. Thus, for operation over the temperature range from −20° to 50°C, a power supply voltage of 4.55 V will provide excellent contrast for on segments and negligible contrast for off segments. A power supply voltage *above* the recommended range will cause off segments to become slightly visible. A power supply voltage *below* the recommended range will reduce the contrast for on segments.

While a constant supply voltage serves well over a limited temperature range, the temperature compensation circuit of Fig. 7.24 provides for operation from −20°C to 70°C. It does this by generating an output voltage which has a −14 mV per degree Celcius temperature gradient. This is achieved by magnifying the −2 mV per degree Celcius temperature gradient of a silicon diode. The resulting fit to the recommended power supply voltage characteristic is shown in Fig. 7.25.

As one measure of the hardiness of a liquid crystal display, N. V. Philips specifies a life of over 100,000 hours for their displays. Of course, the display must be operated within specification to warrant such a long life. In particular, Philips specifies a maximum dc current component in the drive voltage across

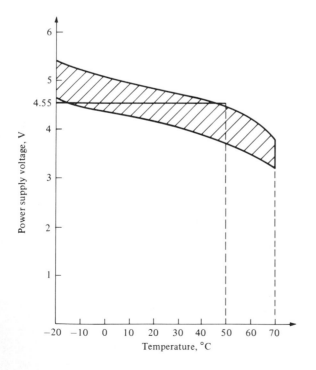

FIGURE 7.23
Recommended power supply voltage for a typical LCD display using 1/4 multiplexing. (*N. V. Philips.*)

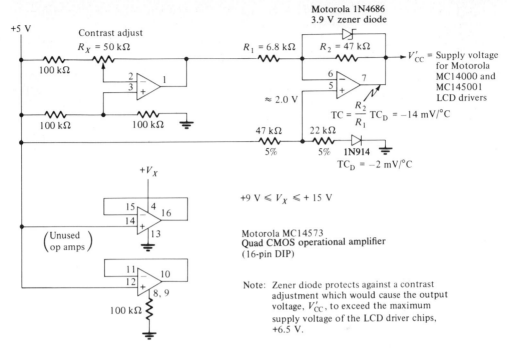

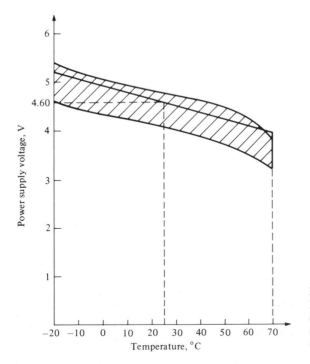

FIGURE 7.24
Temperature compensation circuit for Motorola MC145000 and MC145001 LCD driver chips.

Motorola 1N4686
3.9 V zener diode

$R_1 = 6.8$ kΩ $\qquad R_2 = 47$ kΩ

V'_{CC} = Supply voltage for Motorola MC14000 and MC145001 LCD drivers

≈ 2.0 V

$TC = \dfrac{R_2}{R_1} TC_D = -14$ mV/°C

47 kΩ $\quad$ 22 kΩ
5% $\qquad$ 5% $\quad$ 1N914
$\qquad TC_D = -2$ mV/°C

$+V_X$

$+9$ V ≤ V_X ≤ $+15$ V

Motorola MC14573
Quad CMOS operational amplifier
(16-pin DIP)

Note: Zener diode protects against a contrast adjustment which would cause the output voltage, V'_{CC}, to exceed the maximum supply voltage of the LCD driver chips, $+6.5$ V.

Power supply voltage, V

Temperature, °C

FIGURE 7.25
Recommended linear temperature compensation of -14 mV/°C for operation between -20 and $+70$°C.
(*N. V. Philips.*)

any segments of 100 mV. This is met by the specification of the Motorola MC145000/5001 parts of 50 mV, maximum.

The hurdles involved in obtaining a custom liquid crystal display are not too different from those involved in having a printed circuit board made. An LCD manufacturer must first be selected. Data sheets for LCD drivers generally list manufacturers of compatible displays. A manufacturer will request the following information:

1. A sketch of the information to be displayed. Note any critical dimensions. Maximum glass size and minimum viewing area (i.e., bezel opening). Size of image area elements.
2. Expected operating and storage environment. Is the application indoors or outdoors?
3. Viewing mode (reflective, transmissive, or transflective).
4. Connector type (dual-in-line or zebra strip).
5. Multiplex format, preferred drivers, voltage, and optimum viewing angle.
6. Projected timing and quantities of prototypes and production displays.

Typical lead-times on custom LCD development programs are:

1. Cost and engineering analysis: less than 2 weeks.
2. Specification drawing: 2 weeks.
3. Prototypes: 6 to 8 weeks after specification drawing approval.
4. Production quantities: 6 to 10 weeks after prototype approval.

The costs involved might run something like the following:

1. To develop the specification drawing, the artwork, and 10 prototypes for a reflective display of 3 × 5 in, $1000 to $3000.
2. For a production run consisting of 100 pieces, under $15 each. For larger quantities, under $10 each.

A liquid crystal display can be mounted by plugging the display's dual-in-line pins into PC board SIP (single-in-line package) sockets. A popular alternative makes use of a "zebra" strip,* as shown in Fig. 7.26. A pattern of conducting fingers on the display is mated to a corresponding pattern of conducting fingers on a printed circuit board by clamping the display to the PC board with the zebra strip in between. The zebra strip is a rubbery piece of material made up of thin conductors alternating with thin insulators. As you look down on a printed circuit board with the zebra strips in place, the stripes which give the zebra strip its name serve as conductors between the conducting fingers on the PC

---

* Manufactured by Tecknit, 129 Dermody Street, Cranford, New Jersey.

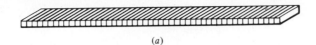

(a)

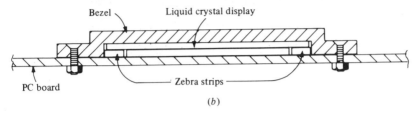

(b)

FIGURE 7.26
Mounting a liquid crystal display. (a) Zebra strip; (b) cross section showing zebra strips
clamped between display and PC board.

board and the display which will be laid upon it. The thin insulators in the zebra
strip prevent adjacent conductors on the PC board (and on the display) from
being shorted together.

7.4 CRT DISPLAYS

For the display of much alphanumeric information as well as the display of
information in graphic form, the CRT (cathode ray tube) display has long been a
favorite device for use by instrument designers. Various flat-panel technologies, such as dot-matrix electroluminescent panels, are making inroads into
applications for which the CRT used to be the sole contender. Nevertheless, on
a cost per character basis, it is hard to beat a CRT. For an instrument application, a little 5-in CRT such as that shown in Fig. 7.27 provides a versatile front-
panel display while still leaving plenty of room on the front panel for keyswitches.

The strength of the CRT for instrument applications has been enhanced
with the introduction of supporting chips which simplify the job of refreshing
the display as well as the job of interacting with a microcontroller. In this
section we will consider National Semiconductor's NS455-A12N Terminal
Management Processor (TMP) chip and the support it can give to designers
wanting to use a CRT display.

The TMP chip represents one wave into the future. It includes an 8048
core. This is the circuitry developed by Intel for their 8048 microcontroller. As
a consequence, the TMP chip is a *mask-programmable* part. The unit which we
will consider here (NS455-A12N) includes a program which National Semiconductor developed for use with a demonstration board to illuminate the TMP
chip's capabilities. It will let us use the chip as a low-cost low-parts-count CRT
controller.

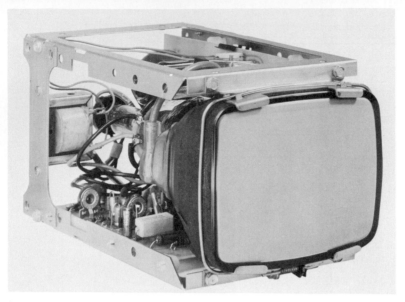

FIGURE 7.27
Five-inch CRT monitor. (*Motorola Inc.*)

The TMP chip also includes all the circuitry needed to refresh the CRT display automatically. It reads ASCII character codes from an external RAM chip and uses an on-chip character generator to convert each ASCII code to the dot pattern for that row of the character being refreshed. This refresh circuitry keeps the RAM very busy with accesses the entire time that the CRT is being scanned in its visible area. In fact, one of the jobs of the TMP program is to take new characters received via the on-chip UART and queue them up until the time when the CRT does a *vertical retrace* (i.e., when the electron beam is returned to the top of the screen). During the vertical retrace, the characters are taken out of the queue and written to the appropriate locations in the refresh RAM. The on-chip timing circuitry for the CRT generates the signals needed by the 8048 program to control when it writes to the refresh RAM.

The circuitry for a simple instrument CRT display is shown in Fig. 7.28. In addition to a handful of discrete parts, it includes only *four* integrated circuits! It will generate a display of 25 lines, each having 80 characters. For an instrument application using a 5-in CRT, 80-character lines are hard to read. The TMP chip can be sent an escape sequence (i.e., a control sequence of ASCII characters beginning with the ASCII escape character, coded 1BH in hexadecimal) to change to a double-width display mode. This changes the display into a 40-character-per-line, 25-line display which is clean and easy to read on a 5-in CRT. In an instrument environment, the 1000 character positions on the display provide a flexible means to show both measurement results and status information (i.e., annunciators).

Note that the CRT interface of Fig. 7.28 communicates with the micro-

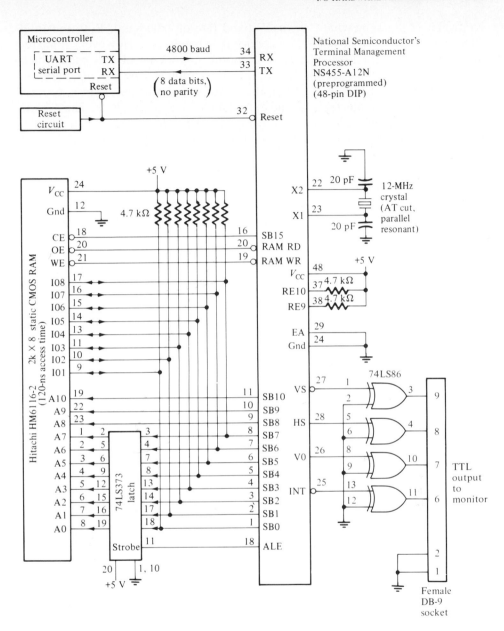

FIGURE 7.28
CRT interface.

controller at the somewhat "strange" baud rate of 4800 baud. Both the Intel 8096 and the Motorola 68HC11 support this baud rate. However, if faster updating of the CRT is desirable, or if another microcontroller is used which supports 9600 baud but not 4800 baud, then the circuitry shown in Fig. 7.29 can be added to that of Fig. 7.28. When the TMP program begins, it tries to read some DIP switches interfaced to the external data bus of Fig. 7.28. The 4.7-kΩ pullup resistors shown in that figure provide *default* values in lieu of DIP switches. The circuitry of Fig. 7.29a changes one bit of the value read, so that the TMP reads 10111111B instead of 11111111B. The program interprets this and sets up the internal UART to operate at 9600 baud.

The TMP chip and its preprogrammed ROM permit a variety of options to be selected by sending it appropriate escape sequences. Unfortunately, by selecting the double-width character display format (for obtaining 40 characters per line), we rule out most of these options. At reset time we must send the escape sequences shown in Fig. 7.30 to the TMP chip to put it into the mode that we want. The escape sequences are listed in the form of an ASCII string, terminated with the ASCII EOT (end of transmission) character, 04H. An algorithm such as that described in conjunction with Fig. 5.11 can be used to send the string to the TMP chip. We must follow the rules listed in Fig. 7.31 when using the TMP chip in the double-width character mode. Finally, we can

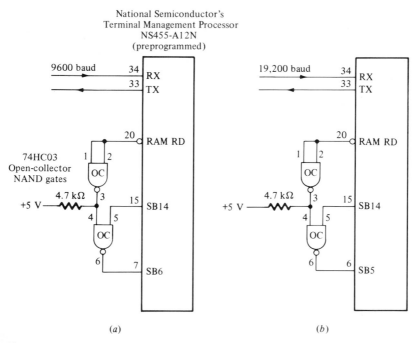

(a)   (b)

FIGURE 7.29
Addition to circuit of Fig. 7.28 required for changing baud rate. (a) 9600 baud; (b) 19,200 baud.

```
INIT_TMP:

DB    01BH      ;ASCII ESC -- beginning of an escape sequence
DB    047H      ;ASCII G -- the next byte will select display attributes
DB    0EFH      ;Select double-width characters
DB    01BH      ;ASCII ESC -- beginning of a second escape sequence
DB    04AH      ;ASCII J -- turn off status line on 25th row and clear screen
DB    004H      ;ASCII EOT -- this is used by interrupt service routine, not TMP
```

FIGURE 7.30
ASCII string used to initialize TMP chip.

use any of the control codes listed in Fig. 7.32*a* and any of the escape sequences listed in Fig. 7.32*b* to modify the display or the cursor position.

Example 7.4. Show the ASCII string, called PER_ANN, to write an eight-character annunciator, "PERIOD", in the lower-left corner of the CRT.

```
PER_ANN:  DB   1BH,4DH,20H,38H    ;Get to bottom left corner
          DB   0D0H,0D0H          ;P
          DB   0C5H,0C5H          ;E
          DB   0D2H,0D2H          ;R
          DB   0C9H,0C9H          ;I
          DB   0CFH,0CFH          ;O
          DB   0C4H,0C4H          ;D
          DB   0A0H,0A0H          ;blank
          DB   0A0H,0A0H          ;blank
          DB   04H                ;ASCII EOT to terminate
```

This example illustrates that the ASCII strings which we generate for use with this display would be much less cryptic and more ROM efficient if we let the UART interrupt service routine carry out a *translation* from characters which it reads from an ASCII string to the sequence of characters which it sends out to the display. It could look for control codes and escape sequences and send them on to the TMP chip unmolested. For displayable characters, it

---

1. The TMP chip deals with 80 RAM memory locations as it refreshes each row of the display. When using the double-width-character mode, we must deal with *two* of these 80 RAM memory locations for each character displayed. Furthermore, all *displayable* characters must be sent to the TMP with a one in the most-significant bit position of each byte sent.
2. Nondisplayable ASCII codes (e.g., carriage return, line feed, escape sequences, control codes) must be sent with a 0 in the most-significant bit position of each byte sent.
3. When all the characters for one line have been sent, send a carriage return (0DH) followed by a line feed (0AH) to return to the beginning of the next line.

FIGURE 7.31
Rules for using TMP chip in the double-width character mode.

would set the most-significant bit and then send the resulting byte to the TMP chip *twice*.

    A more far-reaching alternative will be available when National Semiconductor supports the TMP chip with a mask-programmed ROM version designed for use by instrument designers. Features that should be incorporated are:

1. Permit user access (via escape sequences) to the TMP chip's *two* character attribute registers. The attributes associated with the display of a character are picked from one or the other of these two attribute registers. The most-significant bit of the displayable character (stored with the ASCII code in the refresh RAM) is automatically used by the chip to pick between these two registers. The eight attributes available are: reverse video, half intensity, blink, double height, double width, underline, blank, and graphics.

    Since these are *global* attributes, we need to be able to select the double width attribute in both of them before we can use them both on a screen full of double-width characters. Given two sets available to us, we could have one or more blinking annunciators on an otherwise nonblinking display. At other times we could highlight parts of the display by using reverse video characters.

2. The ROM program should handle double-width characters automatically. When a displayable character is received, the program should look at the appropriate attribute latch. It should write the character into one, two, or four locations (for double-width double-height characters) so as to generate the correct display. This would permit the microcontroller to send just one character for each character displayed, regardless of the character's size.

3. An escape sequence should be added to turn the cursor off. A cursor is not a normal part of an instrument display.

4. An escape sequence should be added to set an overwrite mode for incoming characters. After setting the (invisible) cursor to the beginning of a field on the screen, this would permit the updating of the field with incoming characters without moving the rest of the characters on the display. Another escape sequence to switch to an insert mode would permit a choice between these two modes of updating the screen with new characters.

As users of the TMP chip, we actually do not need to wait for National Semiconductor to design and produce the ROMed version of the chip which we want. National supports the chip with information, an assembler, and other design tools, as well as the availability of the source file, assembled listing file, and object code file for the code used in their demo board* chip which we have been considering. Furthermore, while Fig. 7.28 represents a minimum configu-

---

* Using a computer with a modem, dial (408) 739-1162 for access to National Semiconductor's help line which provides free access to this software.

ASCII control code	Meaning
0DH	Carriage return
0AH	Line feed
1AH	Clear screen
12H	Cursor home (upper left corner)
10H	Cursor left
11H	Cursor right
0EH	Cursor up
0FH	Cursor down

(a)

Escape sequence	Meaning
1BH 54H	ASCII ESC ASCII T -- Erase from current cursor position to end of line
1BH 59H	ASCII ESC ASCII Y -- Erase from current cursor position to end of screen
1BH 4DH	ASCII ESC ASCII M -- Mcve cursor to position X,Y (X and Y are given below) Column position; 20H is leftmost double-character column position 6EH is rightmost double-character column position Row position; 30H is the top row; 38H is the bottom row
1BH 50H	ASCII ESC ASCII P -- Toggle cursor format; alternatives are: blinking block cursor, solid underline cursor, blinking underline cursor, solid block cursor
1BH 51H	ASCII ESC ASCII Q -- Run self-test diagnostic and then reset chip.
1BH 52H	ASCII ESC ASCII R -- Dump the current row via the UART output line
1BH 53H	ASCII ESC ASCII S -- Dump the entire screen via the UART output line

(b)

FIGURE 7.32
TMP control codes and escape sequences. (a) Control codes; (b) escape sequences.

347

ration for a CRT controller, the TMP chip can be used with *external* EPROM, connected as shown in Fig. 7.33, which permits us to write our own program. Even if we end up with a ROMed version of the chip, we have a path for undertaking the development and debugging of the code involved. Since the chip uses an 8048 core, we can even use a macro assembler for the Intel 8048 to assemble the code. If we use this approach, we need to define macros for all the entities which appear in the source code file which are not defined for the 8048.

A simpler alternative for using the TMP chip with our own code consists of *patching* the 2K of code which fits inside the TMP chip into 4K of code which resides in an external EPROM. Thus we avoid the issue of assembling

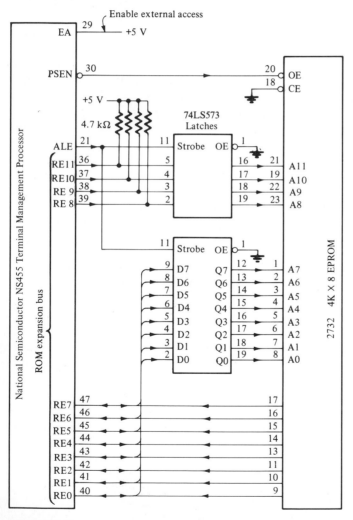

FIGURE 7.33
Replacing internal TMP program with a program in an external EPROM.

the bulk of the code. Instead, we look at the listing file and mark the changes which we want. Then, just where the code begins to do something we do not want, we insert a JUMP instruction down to the second half of the EPROM's memory space and insert the code which we want. When we are done with each segment, we insert a JUMP instruction back to the appropriate point in the original code. This alternative requires us to assemble only the *changes* to the code, plus the JUMP instructions. We can even *hand assemble* our changes. This is especially feasible if our patches do little more than change the default baud rate, handle double-width characters cleanly, and eliminate the cursor. The modifications are edited into the object code file, the EPROM is programmed, and the circuit modification of Fig. 7.33 is ready for testing.

7.5 D/A CONVERSION

A microcontroller can exert control over circuitry and devices in a variety of ways. It may need to switch an output on or off, in which case an interface driver may be needed to meet the current or voltage requirements of the load. Other devices may require a *proportional*, or analog, output. For example, a dc motor will produce an output torque which is proportional to the average value of the dc current flowing through it.

As pointed out in conjunction with Fig. 2.22, often a pulse-width modulated output can be used to produce a proportional output. This represents what is probably the simplest form of D/A (i.e., digital-to-analog) conversion for a microcontroller to implement. The Intel 8096 includes a pulse-width modulator so that it can drive one output in exactly this way. Once the waveform has been initiated, this on-chip facility avoids tying up any further CPU time with the process of generating the waveform.

For a microcontroller not having an on-chip pulse-width modulator, or for use with the Intel 8096 if more than one pulse-width modulated output is desired, we can implement this function with the help of the on-chip programmable timer facility.

> **Example 7.5.** Consider the use of the Motorola 68HC11's programmable timer to generate two independent pulse-width modulated outputs. One of these outputs might be used to drive a heater coil (interfaced with a suitable driver to handle the current and voltage involved). The other might be used to construct a high-resolution programmable power supply (interfaced through a low-pass filter to extract the dc component of the modulated waveform and a power operational amplifier to drive the load).
>
> The Motorola 68HC11 has *five* output compare registers, each of which can drive a separate output pin on port A. Furthermore, to support pulse-width modulated outputs, the designers of the 68HC11 set up output compare 1 so that it can affect any number of these same five output pins. Accordingly, we will use output compare 1 and 2 to drive the heater coil and output compare 1 and 3 to drive the programmable power supply output. We will also use the period of the timer's free-running counter as the period of the pulse-width modulated waveform. For

the Motorola 68HC11 this period can be set, by selecting any one of four possible prescaler values, to be about 33, 131, 262, or 524 ms. To generate pulse-width modulated outputs, the prescaler value which produces a 33-ms period is the best choice in that it will simplify the filtering required for the programmable power supply output. The two outputs can be operated as follows.

Heater coil

Output on bit 6 of port A.

Turned on by timer output compare 2.

Turned off by timer output compare 1.

Pulse width is determined by the number written into output compare register 2.

Programmable power supply

Output on bit 5 of port A.

Turned on by timer output compare 3.

Turned off by timer output compare 1.

Pulse width is determined by the number written into output compare register 3.

Initialization required

Output compare register 1 is initialized to FFFF.

Output compare registers 2 and 3 are also initialized to FFFF, turning off their outputs. This works because the 68HC11's timer has been designed so that the output compare 1 function will *override* any of the other output compare functions should they occur at the same time.

The timer control registers are set up so that when output compare 2 takes place, its output line will be set; when output compare 3 takes place, its output line will be set; when output compare 1 takes place, the lines associated with 2 and 3 will be cleared.

Interrupts from these three timers are left disabled.

The operation of the timer to generate these two pulse-width modulated outputs is illustrated in Fig. 7.34. Note that the average value of the outputs will be proportional to the *complement* of the numbers written into output compare registers 2 and 3. For example, the output will be high for $\frac{1}{65536}$th of each period if the complement of 1 (i.e., FFFEH) is written to output compare register 2 or 3. The maximum output occurs if the complement of FFFFH (i.e., 0000H) is written to output compare register 2 or 3. The output will be high in this case for $\frac{65535}{65536}$ of each period.

We have seen that the Motorola 68HC11 requires *no* CPU time to maintain modulated outputs. Furthermore, the conversion yields a high-resolution (16-bit) proportional output. For a different microcontroller having an on-chip timer which can affect an output pin, we can still generate an accurate proportional output using pulse-width modulation. However, we will require some CPU time to maintain the output. Furthermore, since we will require a new

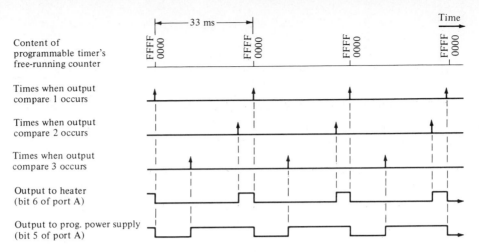

FIGURE 7.34
Use of the Motorola 68HC11's programmable timer to generate two pulse-width modulated outputs.

interrupt for each output change, we will be unable to approach the minimum and maximum limits of $1/65536$ and $65535/65536$ modulation. For many applications, this is not a requirement. Such applications can still take advantage of the high-resolution proportional output afforded by pulse-width modulation.

Example 7.6. As an example of this mode of operation, consider the Motorola 68HC11 and the use of a *single* output compare register to generate a pulse-width modulated output. Let us run the heater coil of the last example using *only* output compare 2.

For this operation, depicted in Fig. 7.35, we can again use the period of the timer's free-running counter to generate the period of the pulse-width modulated output. We will employ a 2-byte RAM variable, called MOD, to specify the modulation. That is, the mainline program will load MOD with a value to obtain an average output proportional to

$$MOD/65536$$

It will be used to load the output compare register 2 after every other interrupt. The output compare register will be cleared after the intervening interrupts. If the output on bit 6 of port A is set in one case and cleared in the other, then it will exhibit the desired pulse-width modulated output. The interrupt service routine to generate the desired output is shown in Fig. 7.36. Initialization requires that bit 6 be set up as an output from the timer and initialized to 0. When we are ready to turn on the pulse-width modulated output, we must enable interrupts from output compare 2 and load the variable MOD appropriately. In this case, a small MOD value produces a small dc output component.

Example 7.7. Consider now the addition of a second pulse-width modulated output for the Intel 8096 to supplement the dedicated modulator. Let us use the HSO.2 output for this purpose.

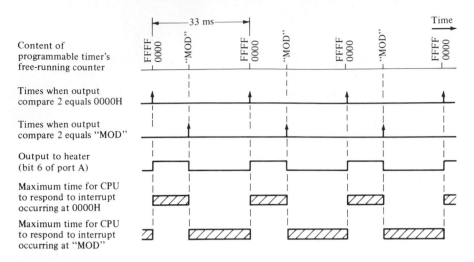

FIGURE 7.35
Use of a single timer output compare register to generate a pulse-width modulated output.

We can do this in a manner analogous to the last example. However, as discussed in detail in Sec. B.6 of Appendix B, the 8096 *buffers* its timed output events. This buffer can hold a command to set an output bit when the free-running counter reaches the value N. *At the same time*, it can hold another command to clear the same output bit when the free-running counter reaches the value $N+1$ (or any other time). This buffering permits superb control over outputs. In the case of this example, it means that we can generate a pulse-width modulated output ranging from $1/65536$ to $65535/65536$. The interrupt service routine is called HSO (since it services the high-speed output facility of the 8096's programmable timer) and is shown in Fig. 7.37. The bulk of HSO services all the *other* things which can call for action based upon the eight status bits in the IOS1 status register. However, when bit 4 is set, the interrupt service routine has been invoked because the 8096's timer 1 has overflowed. We will take advantage of this event to generate an output pulse once per period of timer 1, with a width proportional to MOD. The TIMER1 part of HSO uses the value of MOD to do any one of three things. If MOD is 0, it does nothing. This gives the mainline program an easy way to turn the modulated output entirely off; it need only load MOD with a value of 0. If MOD is nonzero but smaller than half its maximum value (i.e., less than 8000H), then TIMER1 raises the output on HSO.2 halfway through the period of timer 1 and lowers it again MOD ticks of the timer later. Since the period of timer 1 is 133 ms, the times when the HSO.2 is being changed will be restricted to well after the interrupt service routine has been completed (even if it is postponed by higher priority interrupts). It also ensures that the two changes of HSO.2 will be completed before the *next* timer overflow interrupt tells the output to do something else. In this way, we avoid having the commands produced by one invocation of the interrupt service routine interfering with the commands produced by the previous invocation of the interrupt service routine.

When MOD is larger than half its maximum value, then the value (10000H − MOD) is formed. The output on HSO.2 is *dropped to 0* for this many ticks of timer

```
;;;;;;;;;   INTERRUPT SERVICE ROUTINE FOR OUTPUT COMPARE 2   ;;;;;;;;;;;;;;;;;;;;;;;;;;
;   This Motorola 68HC11 interrupt service routine operates the pulse-width modulated output
;   on bit 6 of port A.
;   It generates an output with a dc component proportional to the 2-byte variable, MOD.
;   Recall that the Motorola 68HC11 automatically stacks all the CPU registers when an interrupt
;   occurs, so it is not necessary to do this within the interrupt service routine.
;;;;;;;;;;;;;;;;;;;;;;;;;;;;;;;;;;;;;;;;;;;;;;;;;;;;;;;;;;;;;;;;;;;;;;;;;;;;;;;;;;;;;;;;;;;;;

OC2:      LDAA  #01000000B  ;Clear timer interrupt flag for OC2
          STAA  TFLG1       ; by writing a 1 to the selected bit position in the timer flag register
          LDD   TOC2        ;Get output compare register; set Z flag if it equals 0
          BEQ   OC2_1       ;Branch if TOC2 = 0000H
          LDD   #0000H      ;Otherwise set TOC2
          STD   TOC2        ; to 0000H
          LDAA  TCTL1       ;Get timer control register
          ORAA  #11000000B  ; and set up so OC2 output line will be set to 1 when compare occurs
          STAA  TCTL1       ; while leaving other timer control bits unchanged
          BRA   OC2_END     ;Done
OC2_1:    LDD   MOD         ;Set TOC2
          STD   TOC2        ; equal to number in MOD
          LDAA  TCTL1       ;Get timer control register
          ANDA  #10111111B  ; and set up so OC2 output line will be cleared to 0 when compare occurs
          STAA  TCTL1       ; while leaving other timer control bits unchanged
OC2_END:  RTI               ;Return from interrupt (restoring all CPU registers)
```

FIGURE 7.36
Interrupt service routine for Example 7.6.

353

```
;;;;;;;;;    INTERRUPT SERVICE ROUTINE FOR HSO   ;;;;;;;;;;;;;;;;;;;;;;;;;;;;;;;;
; This Intel 8096 interrupt service routine services the high-speed output facility
; of the programmable timer.
; IOS1_COPY is a variable used to hold the content of a status register which is cleared when read.
;;;;;;;;;;;;;;;;;;;;;;;;;;;;;;;;;;;;;;;;;;;;;;;;;;;;;;;;;;;;;;;;;;;;;;;;;;;;;;;;;;
HSO:    PUSHF                           ;Save status
        LDB   INT_MASK,#xxxxxxxxB       ;Enable higher priority interrupts
        EI
        LDB   IOS1_COPY,IOS1            ;Get status register and execute service for each bit set
        JBS   IOS1_COPY,0,ST0           ;Software timer 0 has timed out; take action
HSO_1:  JBS   IOS1_COPY,1,ST1           ;Software timer 1 has timed out; take action
HSO_2:  JBS   IOS1_COPY,2,ST2           ;Software timer 2 has timed out; take action
HSO_3:  JBS   IOS1_COPY,3,ST3           ;Software timer 3 has timed out; take action
HSO_4:  JBS   IOS1_COPY,4,TIMER1        ;Timer 1 has overflowed; take action
HSO_5:  JBS   IOS1_COPY,5,TIMER2        ;Timer 2 has overflowed; take action
HSO_6:  JBS   IOS1_COPY,6,FIFO          ;High-speed input FIFO is full; take action
HSO_7:  JBS   IOS1_COPY,7,HOLD          ;High-speed input holding register is loaded; take action
HSO_8:  POPF                            ;Restore status
        RET                             ; and return

ST0:    .                               ;Take action for software timer 0
        .
        SJMP HSO_1                       ;Check next status bit
ST1:    .                               ;Take action for software timer 1
        .
```

```
        SJMP HSO_2

TIMER1:
        CMP  MOD,#0000H                   ;THIS IS THE PART FOR PULSE-WIDTH MODULATED OUTPUT
        JE   TIMER1_DONE                  ;If output is turned off (MOD = 0),
        CMP  MOD,#8000H                   ; then do nothing
        JH   BIG_MOD                      ;Otherwise treat small and large values of MOD differently
        LDB  HSO_COMMAND,#00100010B       ;Set HSO.2 output to 1 when timer 1
        LD   HSO_TIME,#8000H              ; reaches 8000H
        LDB  HSO_COMMAND,#00000010B       ;Clear HSO.2 output to 0 when timer 1
        ADD  HSO_TIME,MOD,#8000H          ; reaches MOD + 8000H
        SJMP TIMER1_DONE                  ;Done with TIMER1
BIG_MOD: LDB HSO_COMMAND,#00000010B       ;Clear HSO.2 output to 0 when timer 1
        LD   HSO_TIME,#8000H              ; reaches 8000H
        LDB  HSO_COMMAND,#00100010B       ;Set HSO.2 output to 1 when timer 1
        NEG  MOD
        ADD  HSO_TIME,MOD,#8000H          ; reaches 8C00H + (10000H - MOD)
        NEG  MOD                          ;Restore the original value of MOD
TIMER1_DONE: SJMP HSO_5                   ;Done

TIMER2:                                   ;Etc.
```

FIGURE 7.37
Interrupt service routine for Example 7.7.

1, starting at the midpoint of timer 1's period and ending before the end of the period. That is, HSO.2 is *raised* for MOD ticks of timer 1 since the timer has a period of 10000H ticks.

As has been pointed out, pulse-width modulation provides D/A conversion with extremely high resolution. It is suitable where the output does not change very fast. A heating coil is an excellent application since the thermal time constant between a temperature change in the coil winding and a temperature change in the device being heated is almost certainly much longer than the period of the timer. Thus it serves as an inherent low-pass filter. A power MOSFET drive circuit, discussed in Sec. 7.8, can translate the output from the microcontroller to the higher voltage and/or current requirement of the heater coil.

A high-resolution, programmable dc power supply fashioned from the pulse-width-modulated output of a microcontroller needs a different drive circuit from that for a heater coil. Figure 7.38*a* illustrates the use of a power operational amplifier for this purpose. The pulse-width-modulated output from the microcontroller is used to drive a CMOS gate (the 74HC04) so that it will pull up to +5 V and down to 0 V and so that it will do this through virtually the same source impedance in both cases. Most of the ac component of this waveform is then suppressed by the low-pass filter, leaving a dc voltage which can be controlled from 0 to +5 V. This filter is shown as an *RC* circuit. Because it drives the extremely high input impedance of a 741 operational amplifier, the circuit can employ a relatively small capacitance value together with a relatively high resistance value. If the values chosen do not suppress the ripple output sufficiently, then a more sophisticated low-pass filter can be substituted.

The op amp circuit translates the dc output of the filter to ±25 V, producing a *bipolar* power supply. If the more normal *unipolar* supply is desired, the resistor network surrounding the op amp can be modified appropriately. For the bipolar supply shown in Fig. 7.38*a*, the dc input is translated to a dc output as shown below the equivalent circuit in Fig. 7.38*b*. The 1 percent resistors shown in Fig. 7.38*a* are standard 1 percent values. If it is important for a MOD* value of 8000H to produce zero output voltage, then a potentiometer can be used as a trimmer.

Since power supplies are often called upon to drive resistive loads which include bypass capacitors, the circuit of Fig. 7.38*a* shows an inductor inserted between the op amp output and the load. This can be removed if it is not needed to keep the high-gain op amp circuit stable with a given load.

The Intersil power op amp shown in Fig. 7.38 includes a provision for adding current limiting resistors (shown as 0.68-Ω resistors) to protect its output against accidental short circuiting. Its dc gain of greater than 100 dB and its use of a standard 741 op amp to implement its input circuitry give it true op amp

---

* The modulation variable discussed previously.

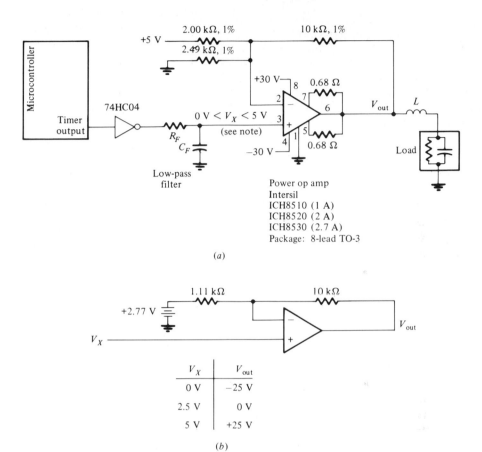

Note: Power op amp uses a
standard 741 op amp
for its input circuit,
so its input impedance
is extremely high.

(a)

(b)

V_X	V_{out}
0 V	−25 V
2.5 V	0 V
5 V	+25 V

FIGURE 7.38
Power op amp driver circuit for a bipolar programmable power supply constructed from the pulse-width-modulated output of a microcontroller. (a) Circuit; (b) equivalent circuit.

performance. It is packaged in the same TO-3 "can" used so often for power transistors and voltage regulators, where power dissipation and heat sinking are issues.

Note that if the desired output voltage, V_{out}, were encoded as a 2-byte, 2s-complement signed number, where

7FFFH represents the maximum positive value

0000H represents the zero output value

8000H represents the maximum negative value

then we need only add 8000H to this number to obtain the value of MOD needed to drive the pulse-width modulator circuit so as to obtain this voltage output. That is, the relationship between the output voltage and the value of MOD needed to produce it is a simple one.

An alternative way to generate proportional outputs is shown in Fig. 7.39a. Each Motorola MC144111 D/A converter chip produces four analog outputs which are buffered with MC14573 CMOS op amps. These op amps

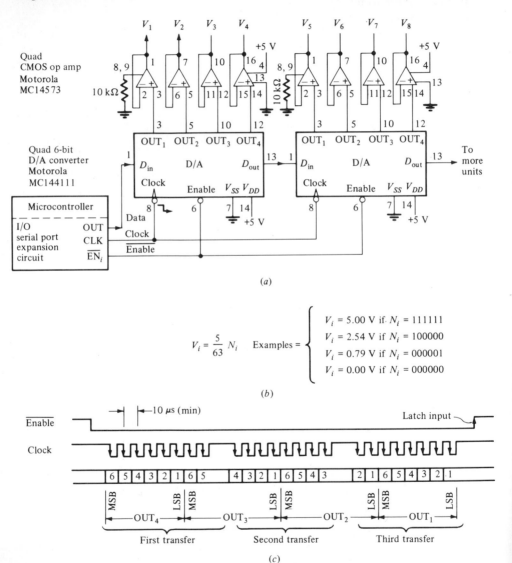

(a)

$$V_i = \frac{5}{63} N_i \quad \text{Examples} = \begin{cases} V_i = 5.00 \text{ V if } N_i = 111111 \\ V_i = 2.54 \text{ V if } N_i = 100000 \\ V_i = 0.79 \text{ V if } N_i = 000001 \\ V_i = 0.00 \text{ V if } N_i = 000000 \end{cases}$$

(b)

(c)

FIGURE 7.39
D/A conversion from I/O serial port. (a) Circuit; (b) equation; (c) timing for use with a single quad 6-bit D/A converter.

provide the high impedance load required by the D/A converters and the low output impedance required to drive anything. They also have the attractive capability of driving the output to within 50 mV of both the high and low supply voltages. Consequently, they provide satisfactory performance for many applications with no extra power supplies, giving a 0- to +5-V output range.

This circuit is particularly suitable for use with a microcontroller because it can make use of the microcontroller's I/O serial port. Each of the Motorola MC144111 chips includes four 6-bit D/A converters. The input to each converter is passed serially. The timing diagram of Fig. 7.39c illustrates how the active-low enable input is first driven low. Then for each MC144111 chip, the I/O serial port is used to execute three 8-bit transfers. If two MC144111 chips are being used, as shown in Fig. 7.39a, then six 8-bit transfers will move the required 48 bits out to the eight D/A converters. Then the enable input is raised, transferring the 48 bits from the shift registers to 48 bits of latches. In this way, the proportional outputs are updated upon command from the microcontroller, when the enable line is raised.

This circuit has a couple of wrinkles. First, the MC144111 is a CMOS part with a serial input which is slow enough to require care in using it. It will not operate reliably with an I/O serial port clock rate above 100 kHz. That is, the clock period of the I/O serial port must be at least 10 μs long. Furthermore, the clock signal must be low for at least 5 μs and high for at least 5 μs during each clock cycle.

Second, the clock input to the chip is negative-edge triggered. This leads to the same considerations which arose in conjunction with Figs. 7.4, 7.5, and 7.6.

Third, the circuit needs to have the four 6-bit digital outputs *packed* together into 3 bytes. Then they must be sent to the D/A converter chip *most-*significant bit first. For a microcontroller, such as the Intel 8096, which sends bytes out of its serial port *least*-significant bit first, the four digital output numbers can first be packed into 3 bytes. Then the bits of each byte can be flipped with the same table-lookup procedure* discussed earlier.

Example 7.8. Write an Intel 8096 subroutine, called PACK, which will pack the lower 6 bits of the four 1-byte variables DAC4, DAC3, DAC2, and DAC1 into 3 bytes having the format shown in Fig. 7.39c.

Assume that the four variables have 0s in their two most-significant bit positions. If this were not true, then we would need to AND 0s into these two bit positions of each variable before using them.

We need a temporary double-word variable for forming the output. Since we will be using the 8096's shift instruction for double words, the first of the 4 bytes making up the double word must have an address which is evenly divisible by 4. As shown in Fig. 7.40a, the label AL meets this requirement. AL, . . . , BH are RAM locations recommended by Intel for general use as scratch-pad regis-

---

* Using a 256-byte table in which, for example, the 00001010th entry is 01010000.

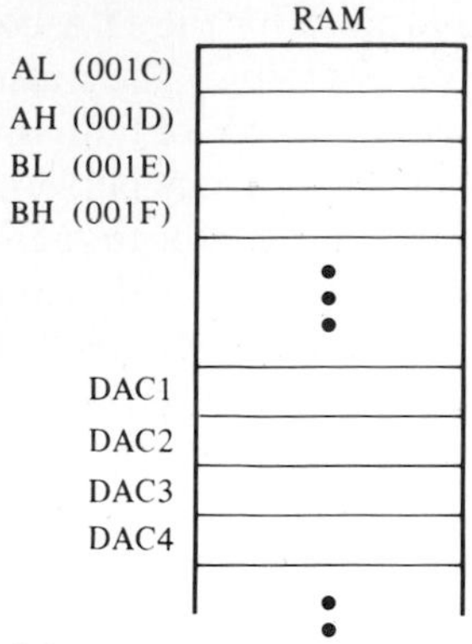

RAM

```
AL  (001C)  ┌──────────┐
AH  (001D)  │          │
BL  (001E)  │          │
BH  (001F)  │          │
            │          │
            │    ⋮     │
            │          │
    DAC1    │          │
    DAC2    │          │
    DAC3    │          │
    DAC4    │          │
            └──────────┘
                 ⋮
```

(a)

```
PACK:  LDB    AL, DAC4    ; Get DAC4 for OUT4 (Fig. 7-39)
       SHLL   AL, 6       ; 4-byte shift left 6 times
       ORB    AL, DAC3    ; Assume MS two bits equal zero
       SHLL   AL, 6
       ORB    AL, DAC2
       SHLL   AL, 6
       ORB    AL, DAC1
       RET
```

(b)

The first packed byte to be reversed and sent is BL.

The second packed byte to be reversed and sent is AH.

The third packed byte to be reversed and sent is AL.

(c)

FIGURE 7.40
Intel 8096 implementation of PACK subroutine for Example 7.8. (a) Memory map; (b) subroutine; (c) use of resulting bytes.

ters, as discussed in conjunction with Fig. 4.27. The subroutine assumes that the content of these can be changed. The first instruction of the PACK subroutine shown in Fig. 7.40b gets the 6 bits of DAC4 (which must be shifted out first) and puts them in the *least*-significant 6 bits of the 4-byte shift register made up of AL, . . . , BH. The second instruction shifts this 4-byte shift register to the left six places, filling in the six vacated bits with 0s. The third instruction gets the 6 bits which must be shifted out second and ORs them into the least-significant byte of the 4-byte shift register. This process continues until the most-significant bit of DAC4 has been shifted left 18 places, putting it in the most-significant bit position of the variable labeled BL in Fig. 7.40a. The PACK subroutine ends with the least-significant bit of DAC1 in the least-significant bit position of the variable AL. The bytes in AL, AH, and BL hold the result of the PACK subroutine. BH, which has been used by the SHLL shift instruction, is changed as a side effect of the PACK subroutine.

Upon returning from the PACK subroutine, we can flip the bits of the three bytes in BL, AH, and AL and then send each one to the D/A converter chip via the I/O serial port.

7.6 POSITION CONTROL WITH STEPPING ACTUATORS

Some applications of microcontrollers require mechanical positioning. In this section we consider two stepping actuators for implementing this function:

Stepper motors, or steppers, which cause a shaft to rotate

Linear stepping actuators, which cause a shaft to extend

If a stepper motor or a linear stepping actuator can do the job, then there is no easier way to implement mechanical positioning. A stepping actuator has the following characteristics:

High resolution without gearing. We will consider a stepper which has an inherent resolution of 48 steps per revolution (i.e., 7.5° per step). We can double the resolution to 96 steps per revolution simply by changing the sequencing of drive signals to the stepper. We will consider a linear stepping actuator with an inherent resolution of 0.002 in. Again, by changing the stepping regime we can double this resolution to 0.001 in. These are two examples of *low-cost* stepping actuators. More expensive models can be obtained with inherent resolutions of 1000 steps per revolution. With gearing, an arbitrary resolution can be achieved. However, for bidirectional position control, the high resolution obtainable without gearing means that position control can be achieved without introducing an error term because of gear backlash.

Fast positioning. The stepper motor we will consider can step at rates above 250 steps per second (i.e., above 5 r/s), starting and stopping on a dime. The linear actuator will also step at rates above 250 steps per second, which translates into 0.5 in/s. Thus, within a second it can move to any position within 0.5 in of its present position, stopping with a positional accuracy of 0.002 in. Again, these are low-cost actuators. Faster versions can step well over 1000 steps per second.

Position error does not accumulate. With no external torque (i.e., a torque applied to the shaft by some source such as a weight or spring loading) on the shaft of the stepper we will be considering, it will be within ±1° of its nominal angular position. Furthermore, if we cause it to step 243 steps clockwise and then step 243 steps counterclockwise, it will be back to its exact starting position, within ±1°.

High and low torque models are available. The models we will consider produce very little torque. They are excellent for positioning a lens or a mirror in an instrument or the air intake valve in the carburetor of an automobile engine. Other models are available from many manufacturers covering a wide range of torque outputs. There are even hydraulic stepper motors available that produce extremely high torque/inertia ratios and are suitable for high-speed position control applications.

Drive circuitry is simple and efficient. Because a stepping actuator is driven by turning the current in four motor windings on or off, the drive circuitry requires only transistor switches. Because each switch is either turned entirely on or entirely off, the switch dissipates power only to the extent that it is imperfect. It dissipates power when turned on because of the nonzero voltage drop across it. It dissipates power when it is switching from on to off and vice versa because it cannot do so instantaneously. Relative to a linear driver, such as the power operational amplifier discussed in the last section, a transistor switch is the model of efficiency!

Figure 7.41*a* shows the small (1.39 in diameter) stepper motor which has been alluded to above. It has the torque-speed characteristic shown in Fig. 7.41*b* and the other features listed in Fig. 7.41*c*. The "start-without-error" torque-speed characteristic shows how much of a load can be impressed upon the stepper motor and still have it start and stop on a dime. For example, with a load of 0.25 oz-in it can go from standstill to a stepping rate of about 320 steps per second without missing a step. It can be stepping along at 320 steps per second and stop dead without skidding over any extra step positions. The running* torque-speed characteristic shows how much of a load can be impressed upon it at a certain speed without it missing any steps. However, it must be carefully accelerated to this speed and carefully decelerated to a stop again if steps are not to be missed. This running torque-speed characteristic gives the actual torque available from the motor when it is running at a given speed. The vertical difference between the two curves represents the effect of accelerating the stepper motor's own rotor inertia. It can be used to estimate the effect of an *added* inertial load upon the start-without-error characteristic.

> **Example 7.9.** Determine the maximum start-without-error speed of this stepper motor if it is used to rotate a small mirror which doubles the moment of inertia of the motor's rotor by itself.
> We can construct a new start-without-error curve for this case by doubling the distance between the two original curves. This is shown in Fig. 7.42. However, we are really only interested in the speed at the point where the new curve drops to zero torque. This is about 290 steps per second. So we should operate the stepper motor at a speed somewhat slower than this.

> **Example 7.10.** Determine the maximum start-without-error speed of this stepper motor if we double the moment of inertia of the rotor by itself and also add a maximum friction load of 0.25 oz-in.
> The new start-without-error torque-speed curve for this case is shown in

---

* Also called *pullout*.

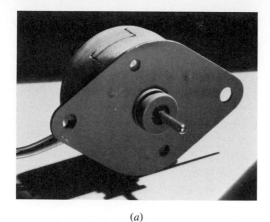

(a)

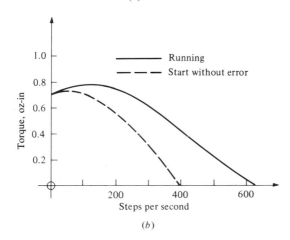

(b)

Airpax model K82201-P1 (bifilar winding)

Steps per revolution	= 48
Step angle	= 7.5°
Step angle tolerance	= ±1.0°
DC operating voltage	= 5 V
Resistance per winding	= 26 Ω
Inductance per winding	= 10 mH
DC current per winding	= 192 mA
Holding torque	= 1.4 oz-in
Rotor moment of inertia	= 2×10^{-4} g·m²
Diameter	= 1.39 in
Length, not including shaft	= 0.71 in
Shaft diameter	= 0.125 in

(c)

FIGURE 7.41
A small stepper motor. (*Airpax Corp.*) (*a*) Motor; (*b*) torque-speed characteristic; (*c*) other characteristics.

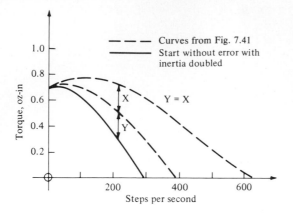

FIGURE 7.42
Start-without-error torque-speed curve for the stepper motor of Fig. 7.41 if the inertial load of the motor is doubled (relative to that of the motor's inertia).

Fig. 7.43. It drops to zero torque at about 220 steps per second. This is the speed below which we need to operate the motor, if we want to start and stop it on a dime without missing any steps.

Figure 7.44a shows a linear stepping actuator. It is available in a model with a maximum linear travel of 0.5 in and another model with a maximum linear travel of 1.875 in. It has the force-speed characteristic shown in Fig. 7.44b and the other characteristics listed in Fig. 7.44c.

Driving either a stepper motor or a linear stepping actuator requires the same circuitry (given the same voltage and current loads) and the same sequencing for turning the current on and off to each winding. The drive circuitry is illustrated in Fig. 7.45a and the sequencing of the two signals labeled X and Y to step in either direction is shown in Fig. 7.45b. For example, if XY presently equals 00, then to make the stepper motor move one step clockwise we must

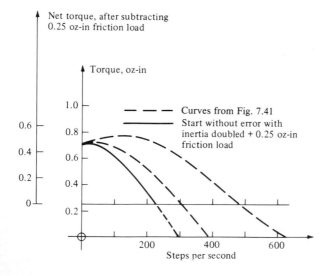

FIGURE 7.43
Start-without-error torque-speed curve for the stepper motor of Fig. 7.41 if the inertial load of the motor is doubled and if the motor is subjected to an 0.25 oz-in friction load.

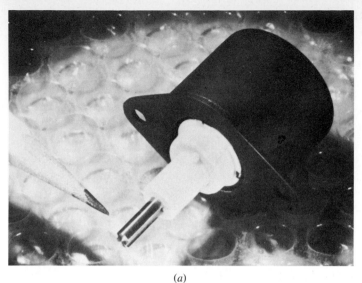

(a)

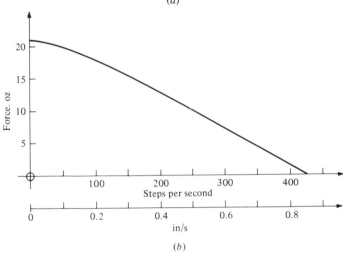

(b)

Airpax model K92121-P1 (bifilar winding)

Steps per inch	= 500
Linear travel per step	= 0.002 in
Maximum travel	= 0.5 in
Resistance per winding	= 15 Ω
Inductance per winding	= 5 mH
DC operating voltage	= 5 V
DC current per winding	= 333 mA
Diameter	= 1.06 in
Length, not including shaft	= 1.08 in

(c)

FIGURE 7.44
A linear stepping actuator. (*Airpax Corp.*) (*a*) Actuator; (*b*) force-speed characteristic; (*c*) other characteristics.

365

change XY to 01. To move a second step clockwise (after waiting $\frac{1}{200}$th s or so), we must change XY to 11. And so forth.

The driver must be able to handle the current and voltage involved. Furthermore, because the load is inductive, the drivers must be protected against the inductive voltage surge which occurs when a transistor switch tries to open up and discontinue the current flow through a winding of the stepper. The drivers shown in Fig. 7.45a include clamping diodes to alleviate this problem. Now when one of the drivers turns off, its diode permits the current which was flowing through the motor winding to continue, dying away at a rate determined by the L/R ratio of the motor winding. The voltage on the output of the driver is driven by the motor winding only slightly above +5 V, by an amount equal to the voltage drop across the diode. This protects the driver. Unfortunately, it

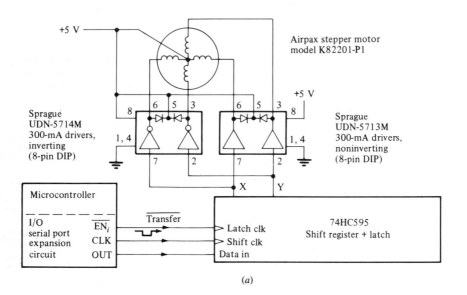

(a)

(b)

FIGURE 7.45
Drive circuitry for a stepping actuator. (a) Circuit using drivers which have built-in arc-suppression diodes; (b) stepping sequence.

also decreases the maximum stepping rate. Two popular alternatives are shown in Fig. 7.46. These circuits permit the driver voltages to rise a controlled amount above +5 V.

Example 7.11. Determine the voltage, current, and power specifications for a zener diode used in the circuit of Fig. 7.46a with the linear stepping actuator of Fig. 7.44.
Assume that the driver has a breakdown voltage of 50 V. If we use a 35-V zener diode, its voltage plus the +5 V of the supply will subject the driver to a maximum of 40 V.

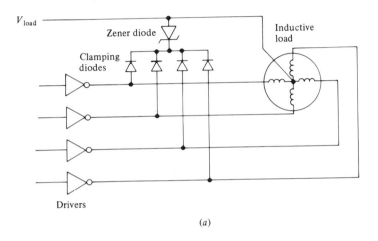

(a)

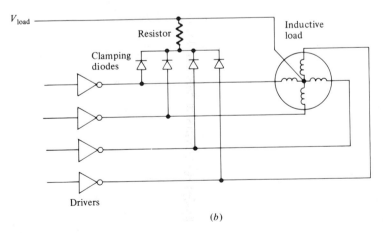

(b)

FIGURE 7.46
Drive circuits for inductive loads. (a) Clamping diodes plus zener diode; (b) clamping diodes plus resistor.

When one of the windings of this actuator has +5 V impressed across it, the dc current which flows in the winding will be 333 mA. At the moment the driver is turned off, the 333 mA which is flowing through the actuator's winding is diverted through the zener diode circuit. At this moment of switching, we never turn off more than one winding. Consequently, the maximum zener diode current is 333 mA. Its instantaneous power dissipation at that moment is $0.333 \times 35 = 12$ W. Its average power dissipation depends upon how fast the current dies away. However, we are adding the zener diode to increase the maximum stepping rate; therefore, we should calculate the average power dissipation at that rate. Presumably at that rate, the current will have died down to almost 0 mA by the time the motor takes another step. The voltage across the zener diode remains at +5 V as long as it is still doing its job (i.e., conducting). Therefore the average power dissipation when the stepper is stepping at its maximum rate can be expressed as

$$K \times 0.333 \times 35$$

where K represents the ratio of the average value over the peak value of the exponentially decaying current through the zener diode. If we assume that the exponential decays for two time constants before the next step takes place, then it will be down to 14 percent of the initial value and will have an average value of about 43 percent of the peak value of 333 mA. This gives an average power dissipation of 5.0 W.

If we look at Motorola's series of 5-W zener diodes, we will find a 5-W 33-V zener diode, 1N5364. This is a $2 part in small quantities. Assuming that the actuator is not going to be stepping continuously, the 5-W rating of the zener diode will be satisfactory without further derating.

Often a resistor is used in place of the zener diode, as in Fig. 7.46*b*. The goal is the same, to speed up the maximum stepping rate. This circuit uses a cheaper component, a resistor, in place of a more expensive one, the zener diode. Its resistance is chosen so that the surge voltage across each driver stays below the breakdown voltage of the driver.

Example 7.12. Repeat the last example, this time using a resistor instead of a zener diode.

If the driver can handle up to 50 V, then we might pick the resistor to impress no more than 40 V across the open-circuited driver. This means that the voltage across the resistor must be a maximum of 35 V so that it plus the supply voltage of +5 V will surge to a maximum of 40 V. The initial, maximum current through the resistor will be 333 mA. If we use an actual resistance value of 100 Ω, we will obtain an initial voltage of 33 V.

The instantaneous power dissipation at that moment will be $0.333 \times 33 = 11$ W. The instantaneous power will drop off more quickly than if a zener diode were used. As the current through the resistor decreases, the voltage across it will also decrease by a proportional amount. In fact if we again assume that the current decays out to two time constants before the motor steps again, then a calculation of the average power yields about 25 percent of the peak power instead of the 43 percent we found with the zener diode. Consequently, the average power in this case is about 3 W. We might use a standard 100-Ω 4-W resistor.

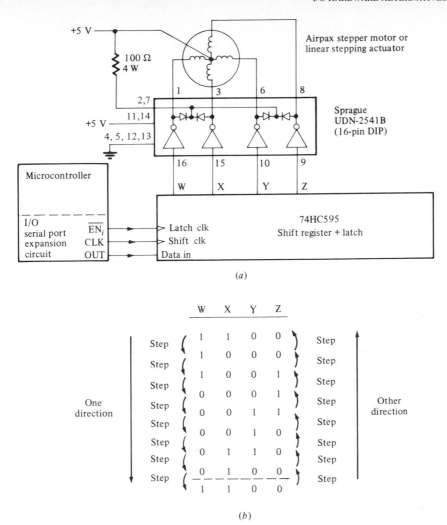

(a)

(b)

FIGURE 7.47
Doubling the resolution of stepping. (*a*) Circuit; (*b*) stepping sequence.

The stepper motor and linear stepping actuator which we have been considering employ four-coil *bifilar** windings. This is one of the two ways in which stator windings of stepping actuators are wound. Bifilar windings permit the current flow to be reversed along one of the two electrical axes by changing whichever of two wires is pulled down to ground by a transistor switch. A stepping actuator with bifilar windings also permits the step size to be halved, doubling the resolution of the motor. As shown in Fig. 7.47*a*, each winding must be driven not only by a separate driver but by a separate *signal* as well.

---

* Also called *unipolar* windings, as compared with two-coil bipolar windings.

Then instead of always turning on a *pair* of windings, one on each electrical axis, we can drive a single winding. If we think of the rotor as a dc magnet which aligns itself with the magnetic field set up by the currents in the windings, then driving Y and Z high grounds the top and right winding in Fig. 7.47*a*. We can think of this as setting up a magnetic field directed up and toward the right at 45°. The normal four-step switching sequence of Fig. 7.45*b* would move this stator magnetic field from 45° as follows:

$$45° \text{ to } 135° \text{ to } 225° \text{ to } 315° \text{ to } 45°$$

This sequence will produce four steps in one direction. For a motor with 12 pole pairs, like the Airpax stepper motor, each step of 90 electrical degrees translates into $90/12 = 7.5$ mechanical degrees.

    The drive circuitry of Fig. 7.47*a* plus the eight-step switching sequence of Fig. 7.47*b* changes the sequence of magnetic fields to the following:

$$45° \text{ to } 90° \text{ to } 135° \text{ to } 180° \text{ to } 225°, \text{ etc.}$$

Now each step of 45 electrical degrees translates into 3.75 mechanical degrees. The steps are no longer of equal torque. On the other hand, the rotor moves only half as far per step as before. Also, the magnetic field set up along each electrical axis is reversed every *four* steps instead of every two steps, as before. So the time constants involved in changing the drive currents would imply that doubling the stepping rate would lead to the same excitation conditions within the motor. All this is to say that there is a trade-off which affects the maximum stepping rate. Because of the reduced torque which occurs on every other step, the maximum stepping rate is not doubled. However, it is significantly higher than when stepping in the mode of Fig. 7.45.

    Driving a stepper motor implies that we must generate either the four-step switching sequence of Fig. 7.45*b* or the eight-step switching sequence of Fig. 7.47*b*. This is a good application for a table.

> **Example 7.13.** Develop the code for the Intel 8096 to update the output for a stepper motor being driven with the eight-step switching sequence of Fig. 7.47.
>
>   Assume that the stepper motor is driven from one of several 74LS595 shift registers daisy-chained to the I/O serial port. A copy of the content of each 74LS595 output is kept in a separate byte of RAM. For example, if we have three 74LS595s connected together to form a 24-bit shift register, then in RAM we will have three consecutive bytes to hold the copy of the port information. Each time that we want to update any of these 24 bits, we change the copy. Then we can call a subroutine which sends the 3 bytes out the I/O serial port. This scheme will update the outputs immediately.
>
>   A more straightforward scheme for handling a stepper motor and all the other outputs driven by the chain of 74LS595s is to generate the stepper motor output and to update the 74LS595s from within a tick clock interrupt service routine. If ticks are set up to occur every 5 ms,* then when steps are to be taken,

---

* In spite of our discussion of maximum stepping rate, many applications of stepper motors are served well with a reasonably fast rate, such as 100 or 200 steps per second.

they will occur every 5 ms (i.e., at a maximum stepping rate of 200 steps per second). As we discussed in Sec. 5.4 on program organization, we can give this tick clock interrupt service routine the lowest of priorities, reenabling all other interrupts while tick clock tasks are being carried out. In this way, we need not be troubled even if this interrupt service routine takes several milliseconds to complete.

We can maintain a 2-byte signed* number variable called NUMSTEPS to keep track of how many steps remain to be taken. If NUMSTEPS equals 0 when the tick occurs, then no step will be taken. If NUMSTEPS is positive, then a single step will be taken, perhaps in a clockwise direction, and NUMSTEPS will be decremented toward 0. A negative value of NUMSTEPS will cause a single counterclockwise step to be taken, and NUMSTEPS will be incremented toward 0.

This scheme simplifies the use of the stepper motor by the mainline routine. If it calls for 35 steps clockwise, it just *adds* 35 to the present content of NUMSTEPS. Even if a previous command to the stepper motor has not been completed, this will produce the correct final position after all stepping has been completed. In fact, the mainline routine can determine when stepping has been completed by monitoring the value of NUMSTEPS.

Given the tick clock interrupt service routine, we need to include a subroutine in it, called STEP, which will look at NUMSTEPS and update the RAM copy of the I/O serial port content. Let us assume that the stepper motor output occupies the upper 4 bits of a RAM location called IO3. This might signify the third byte sent out on the I/O serial port during an update of the 74LS595s. It will load the chip *closest* to the microcontroller. Since the 8096 sends out bytes least-significant bit first, this will load the four output bits closest to the Data in pin of this leftmost 74LS595.

Figure 7.48*a* shows a STEP_TBL which holds the 8 bytes that we will use to update the upper nibble of IO3. We need one additional 2-byte variable called STEP_PTR, which is a pointer into STEP_TBL. It keeps track of where we are in the eight-step switching sequence.

The actual code for the STEP subroutine is shown in Fig. 7.48*b*. It first checks the value of NUMSTEPS and then does a three-way branch. If NUMSTEPS equals 0, it returns from the subroutine. Otherwise, it does the appropriate changing of NUMSTEPS and STEP_PTR, depending upon the sign of NUMSTEPS. Finally it changes the upper 4 bits of IO3, leaving the lower 4 bits unchanged.

Upon the return from this subroutine, the tick clock interrupt service routine does whatever updating is needed for any other bits of the 74LS595s and then calls the subroutine which updates these 74LS595s. Since the 8096's I/O serial port is shared with the UART, we may want to disable interrupts while transmitting the 3 bytes out to the daisy-chained 74LS595s. We cannot afford to have a higher priority interrupt use the UART while we are trying to send these 3 bytes. Alternatively, we can set a flag when we begin the transfer of the 3 bytes and clear it at the end. This flag can be used by other interrupt service routines to determine whether the tick clock interrupt service routine is in the middle of the three transfers over the I/O serial port.

---

* Using 2s-complement code.

```
STEP_TBL:  DB    11000000B        ;Table for eight-step switching sequence of a stepper
           DB    10000000B
           DB    10010000B
           DB    00010000B
           DB    00110000B
           DB    00100000B
           DB    01100000B
           DB    01000000B
```

(a)

```
;;;;;;;;;;  STEP SUBROUTINE  ;;;;;;;;;;;;;;;;;;;;;;;;;;;;;;;;;;;;;;;;;;;;;;;;;;;;;
; This subroutine updates the upper nibble of IO3 with the output needed by the stepper motor.
;;;;;;;;;;;;;;;;;;;;;;;;;;;;;;;;;;;;;;;;;;;;;;;;;;;;;;;;;;;;;;;;;;;;;;;;;;;;;;;;;;;;;

STEP:      CMP   NUMSTEPS,#0000H          ;Are there steps to be taken?
           JE    STEP-2                   ; No, then return
           JLT   STEP_NEG                 ;Is NUMSTEPS negative?

STEP_POS:  DEC   NUMSTEPS                 ;No, then decrement NUMSTEPS
           INC   STEP_PTR                 ; and increment pointer
           CMP   STEP_PTR,#STEP_TBL+8     ; checking to see if we went past the end of table
           JC    STEP_1                   ;   no; go on
           LD    STEP_PTR,#STEP_TBL       ;   yes; reset pointer and go on
           SJMP  STEP_1

STEP_NEG:  INC   NUMSTEPS                 ;For NUMSTEPS < 0, increment NUMSTEPS
           DEC   STEP_PTR                 ; and decrement pointer
           CMP   STEP_PTR,#STEP_TBL-1     ; checking to see if we went past beginning of table
           JH    STEP_1                   ;   no; go on
           LD    STEP_PTR,#STEP_TBL+7     ;   yes; reset pointer to end of table and go on

STEP_1:    ANDB  IO3,#00001111B           ;Clear upper nibble of IO3
           ORB   IO3,[STEP_PTR]           ; and OR in the table value

STEP_2:    RET                            ;Return
```

(b)

FIGURE 7.48

An Intel 8096 STEP subroutine to effect a single step of a stepping actuator. (a) Table containing the eight-step switching sequence; (b) the STEP subroutine.

    While bifilar windings simplify stepper motor drive circuitry, they do not maximize stepping rate. The alternative of having a single winding for each electrical axis means that each end of a winding must be driven by a driver which can pull the line high or pull it low. The high-performance SynchroStep motor shown in Fig. 7.49 has two internally unconnected windings per phase (i.e., a total of four windings) so that it can be treated as *either* a bifilar-wound (i.e., uni*polar*-wound) motor or as a bi*polar*-wound motor. For highest performance, the two windings for each phase are connected in parallel and driven as a bipolar-wound motor.

    This motor employs an unusual state-of-the-art design. Its rotor is a thin disk rather than a cylinder, giving it an exceptionally high torque-to-inertia ratio. The rotor is a rare earth–cobalt permanent magnet structure with a pattern of 25 pole pairs. The magnetic circuit exhibits much lower losses than would be true of a comparable iron rotor, a significant factor at stepping rates of 10,000 steps per second or so.

    The recommended drive circuit for this motor is shown in Fig. 7.50a. It employs H-bridge drivers for each of the two phases. As shown in Fig. 7.50b, the bridge functions to force current through a winding in either direction—or to turn off the current in the winding. By using a supply voltage (i.e., 24 V) which is much higher than the steady-state voltage which can be impressed across a winding (3.6 A $*$ 0.35 Ω = 1.3 V), the circuit *forces* the current in the windings to change quickly. It avoids burning out the motor by using a chopper to cut off the H-bridge drivers as the current exceeds 3.6 A and to turn them on again as the current decays below 3.6 A. The diodes shown in Fig. 7.50b permit the current in the windings to flow continuously even though the current through the H-bridge transistors is being chopped on and off.

    The translator circuit shown in Fig. 7.50b accepts a step input which must be pulsed once for each step to be taken. It also accepts a CW/CCW control input which specifies the direction of stepping for the applied step pulse inputs. Finally, it has a HALF/FULL control input which is tied low if we want the motor to take full steps and tied high if we want it to take half steps.

    Given this motor, this translator, and a 24-V power supply, the torque-speed curve for the motor is given in Fig. 7.50c. What is shown is the running, or pullout, torque. To get this torque, the motor must be gently accelerated up to speed. In fact, to get the high performance of which the motor is capable, it must be carefully accelerated and decelerated. Note that if this is done under interrupt control by a microcontroller, then when the motor is being stepped at 10,000 steps per second, the microcontroller is handling a new interrupt every 100 μs. Because of this, there is a market for dedicated microcontrollers to drive such stepper motors. The characteristics of one of them are highlighted in Fig. 7.51. This unit is actually an Intel 8048 microcontroller, programmed to serve as an "intelligent ramping stepper motor controller." Not only does it respond to each command sent to it, but it can also accept a *program* of commands, including wait instructions and jump instructions. In this way, for example, we might program a sequence of motions which repeats over and over·

(a)

Model:	USS-52-007 SynchroStep
Size:	2 in diameter
	1.25 in long (excluding shaft)
Weight:	7 oz
Rotor inertia:	$0.17*10^{-3}$ oz-in-s^2
Winding resistance:	0.7 Ω
Phase resistance:	0.35 Ω (windings in parallel)
Phase current:	3.6 A, maximum
	(into the two windings in parallel)
Phase inductance:	0.6 mH
Full step angle:	3.6°
Steps per revolution:	100 full steps
	200 half steps
Recommended drive:	
Manufacturer:	PMI Motion Technologies
Model:	BSD-40, 24 V
Power required:	24 V dc ± 10 percent at 4 A
Output current:	3.6 A per phase

(b)

FIGURE 7.49
SynchroStep disk-rotor stepper motor. (*PMI Motion Technologies*.) (*a*) Motor and rotor; (*b*) characteristics.

again. We can also synchronize the operation of several CY525 controllers, even when each is running a program of commands.

7.7 POSITION ENCODING

The stepping actuators in the last section permit fast, fine-grained position control. What a stepping actuator does *not* do is provide an absolute position reference. The addition of a *slotted optical switch*, such as the one shown in Fig. 7.52*a*, solves this problem. It can be used to detect when a moving part passes a specified point, as in Fig. 7.52*b*. The optical switch consists of a light-emitting diode which is constantly energized. It shines a focused beam upon a phototransistor through a carefully shaped aperture. If the beam is not inter-rupted, then it turns on the phototransistor, making the output of the phototran-sistor in Fig. 7.52*c* go low. If the beam is interrupted, the phototransistor gets no light and turns off. The pullup resistor raises the output to +5 V.

Companies which make slotted optical switches make them for a variety of conditions. For fine position resolution, the width of the slot can be made narrow, the LED beam focused, and the aperture (i.e., the opening to the

(*a*)

FIGURE 7.50
SynchroStep bipolar chopper drive. (*PMI Motion Technologies.*) (*a*) Drive; (*b*) functional dia-gram; (*c*) torque-speed curve for USS-52-006 SynchroStep motor driven by BSD-40, 24-V bipo-lar chopper drive (3.6 A per phase, parallel windings).

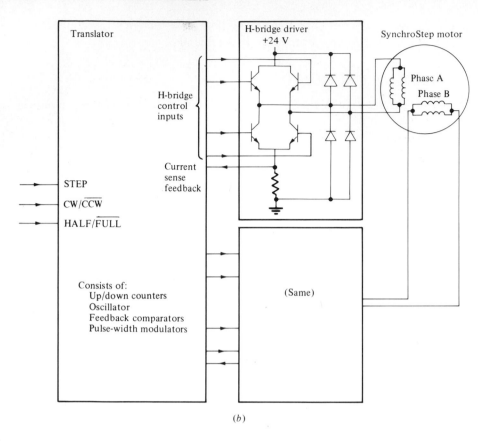

Translator

H-bridge driver
+24 V

SynchroStep motor

Phase A

Phase B

H-bridge
control
inputs

H-bridge control inputs

Current
sense
feedback

STEP

CW/CCW

HALF/FULL

Consists of:
Up/down counters
Oscillator
Feedback comparators
Pulse-width modulators

(Same)

(b)

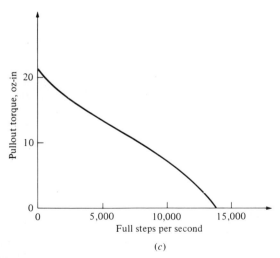

(c)

FIGURE 7.50   (continued)

Interfaced via an 8-bit port with handshake.

Twenty-seven high-level commands, which include:

Programmable maximum stepping rate (slew rate) (up to a 10,000 steps per second)

Programmable linear acceleration and deceleration rate

Programmable direction

Programmable selection of full or half stepping

Programmable absolute position

Programmable relative position

Programmable to step continuously

FIGURE 7.51
CY525 Intelligent Ramping Stepper Motor Controller. (*Cybernetic Micro Systems.*)

phototransistor) shaped as a narrow slit. The unit specified in Fig. 7.52 is optimized in this way. Its high-impedance output needs to drive a high-impedance input circuit such as the CMOS Schmitt-trigger inverter shown. The Schmitt trigger sharpens up the output. Even if the optoswitch output should slowly change, the Schmitt trigger output will *snap* from high to low or from low to high. Furthermore, its hysteresis means that the optoswitch can be turned on partially, with its output in the active region of the phototransistor, and yet the output of the Schmitt trigger will not dither between 1 and 0. If it switches as the phototransistor output rises past 2.4 V, then it will not switch back again until the phototransistor output drops below 1.8 V.

For other applications a more robust output may be desired. In this case a slotted optical switch can be selected which has a larger aperture and a Darlington output configuration (where the phototransistor drives another transistor so as to obtain a gain equal to the product of the gains of the two transistors). If the optoswitch is located some distance from the circuitry which it will drive, then it can be selected to include the Schmitt trigger as a built-in component. For special mounting requirements, slotted optical switches come in a variety of configurations. The unit shown in Fig. 7.52 includes two mounting holes. Other units have one. Others have a hole for mounting to a surface perpendicular to that for the unit shown in the figure. Some small units have no mounting holes. They can be mounted with epoxy glue, or their leads can be soldered to a PC board. All in all, slotted optical switches represent a versatile solution to the problem of obtaining a zero-reference position for a stepper motor or a linear stepping actuator.

Sometimes position control is required and the actuator cannot be a stepper motor. Or the position to be controlled is removed from the actuator through gearing which exhibits backlash. In such cases, an incremental shaft-angle encoder represents the lowest cost solution to obtaining position control. The unit shown in Fig. 7.53*a* represents an interesting solution to many encod-

(a)

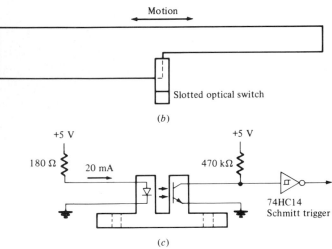

FIGURE 7.52
Slotted optical switch (*TRW, Optoelectronics Division.*) (*a*) Unit; (*b*) application as a zero-reference position sensor; (*c*) circuit; (*d*) characteristics.

(a)

Hewlett-Packard three-channel incremental optical encoder kit

Model No. HEDS-5010 Option F03

Power:	5 V at 60 mA (max)
TTL-compatible outputs:	
1 output	2.4 V (min) at 40 μA
0 output	0.4 V (max) at 3.2 mA
Diameter:	1.13 in
Depth:	0.70 in
Shaft diameter:	⅛ in
Resolution:	256 cycles per revolution (see Fig. 7.54)
	1024 counts per revolution
	(256 cycles per revolution × 4 counts/cycle)
Maximum count frequency:	130,000 Hz
Moment of inertia:	0.4 Gcm^2
Termination:	10-conductor ribbon cable
Recommended interface:	74LS14, hex Schmitt-trigger inverter
Other resolutions available:	100, 192, 200, 360, 400, 500, 512
	(cycles per revolution)
Other shaft diameters available:	2 mm, 3 mm, ⅚₃₂ in, ³⁄₁₆ in, ¼ in, 4 mm, 5 mm

(b)

FIGURE 7.53
A three-channel incremental encoder. (*Hewlett-Packard Co.*) (*a*) Exploded view of unit; (*b*) specifications.

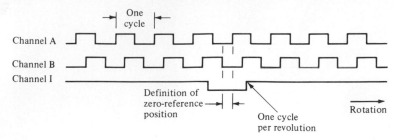

FIGURE 7.54
Incremental encoder output waveforms.

ing problems. Hewlett-Packard* has fashioned a unit available in the form of a *kit*. In this way they optimize the mounting alternatives while minimizing the moment of inertia introduced by the use of the unit. The *code wheel* (i.e., the moving part) is a small disk with a hole in the center. As noted in the specifications of Fig. 7.52*b*, the hole size comes as an option, so as to provide a snug fit on a shaft having any one of eight diameters. The encoder is useful for applications where it can be mounted directly on a motor which has a two-sided shaft extension or where it can be mounted on a remote bearing support at the end of a shaft. To use this unit, the encoder body is attached to the mounting surface with screws and RTV (a silicone rubber which provides a lubricating medium while aligning the unit and acts as an adhesive thereafter). Then the code wheel is epoxied to the shaft and the end plate snapped into place. Finally, the shaft is rotated and the outputs monitored with a scope to align the encoder body. Hewlett-Packard says that this entire process can take less than 30 s in a high-volume application. They support the procedure with appropriate tools (i.e., a gap setter, a centering cone, and a tool kit) as well as explicit instructions in their *Optoelectronics Designer's Catalog*.

The unit shown in Fig. 7.53 produces TTL-compatible outputs (which should be interfaced through Schmitt triggers, in the same way as the optoswitch just discussed). The two-channel output shown in Fig. 7.54 permits direction sensing. The third channel, which is an option in this and other incremental encoders, produces a zero-reference position.

To take advantage of the maximum resolution of an incremental encoder, we must sense when either the channel A or the channel B output changes. The circuit of Fig. 7.55*a* uses the delay of an *RC* circuit together with an exclusive-OR gate to generate a pulse out of a transition, as illustrated in Fig. 7.55*b*. The circuit also uses 74HC14 Schmitt-trigger inverters on each of the encoder outputs. The pullup resistors convert the encoder's TTL compatible outputs to CMOS compatible inputs.

For a microcontroller which has an input which can generate an interrupt when the input changes (i.e., when the input exhibits either a rising edge or a

---

* As well as several other encoder manufacturers.

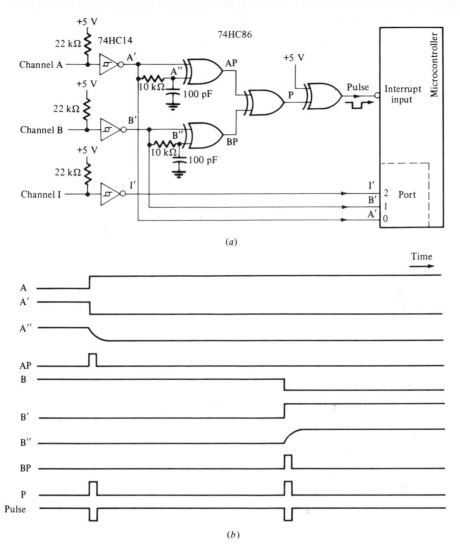

FIGURE 7.55
Incremental encoder interface. (a) Circuit; (b) timing diagram.

falling edge), the simpler circuit of Fig. 7.56 can be used. Note that the pro-grammable timer inputs of the Motorola 68HC11 and of the Intel 8096 can be set up to serve in this capacity. The time captured by the timer when the edge occurs is ignored in this application. All we want out of the timer is its ability to generate an interrupt from either edge of the input. Each time the interrupt occurs, an encoder transition is signaled. The interrupt service routine reads the port and converts the values read into a change in absolute position.

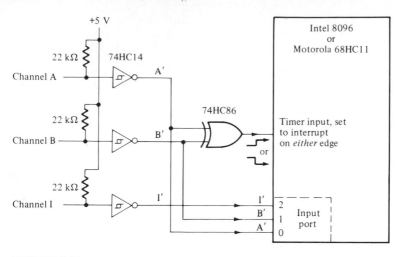

FIGURE 7.56
Alternative incremental encoder interface to a microcontroller which can be set up to sense both rising and falling edges.

Example 7.14. Develop the Motorola 68HC11 code to handle the mechanical positioning configuration shown in Fig. 7.57. This consists of a dc motor to which a three-channel incremental encoder is attached. The dc motor drives a lead screw which in turn drives the piece to be positioned, translating the rotational motion of the motor and encoder to a linear motion. The slotted optical switch is used with the lead screw in exactly the same way as the zero-reference output is used by the incremental encoder. We will assume that its output, called channel J, goes high during just one of the turns of the lead screw. The zero-reference position is thus defined by low outputs from the three encoders and a high output from channel J. These signals are inverted in the interface circuitry to form A', B', I', and J'. Then A' and B' are exclusive-ORed to obtain an interrupt input. One of the 68HC11's programmable timer inputs, IC1, is shown being used to generate the interrupt from either edge of this signal, as discussed above.

While the specifications of Fig. 7.53*b* indicate a maximum output frequency of 130 kHz from the encoder, this represents an extremely fast rate for positioning applications, which translates into 130 *r/s*! If we really care about such a rate, we need the help of a microcontroller's pulse accumulator facility just to count pulses. This might be done in an application where an incremental encoder with a single output channel is used as a tachometer to measure speed and where motion is only in one direction.

For more normal speeds we must determine a maximum rate at which interrupts will ever occur. Then we can decide whether to use a port on the chip or a port built out of a shift register and connected to the I/O serial port for accessing the four inputs. We can also determine whether the priority normally assigned to this interrupt source is satisfactory, changing the priority if necessary. For example, a maximum turning rate of 1 r/s will generate interrupts no faster than every millisecond. At that rate, we can probably live with an arbitrary

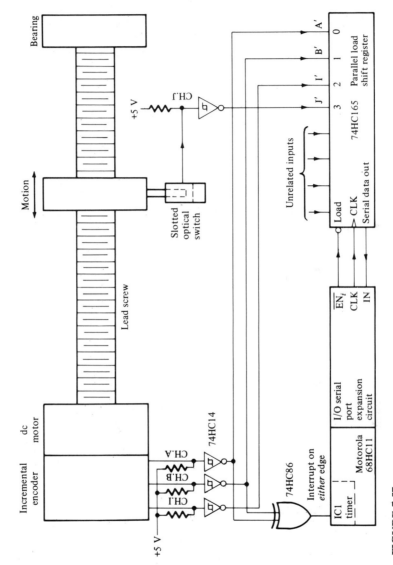

FIGURE 7.57
Positioning system.

383

priority and read the three inputs via the I/O serial port. If we find that we are using too large a percentage of CPU time dealing with the encoder via the I/O serial port, we can revise accordingly. However, many positioning applications will execute a position change only sporadically. The microcontroller can probably afford to be dedicated to position control at such times.

For simplicity, let us assume that the three buffered and inverted encoder outputs are connected to a parallel-load shift register and read in via the I/O serial port, as shown in Fig. 7.57. Assume that we have a subroutine called READX to read this shift register into a RAM location labeled PORTX. The inverted encoder outputs become PORTX bits as follows:

```
A'                    PORTX, bit 0
B'                    PORTX, bit 1
I'                    PORTX, bit 2
```

The buffered and inverted output of the slotted optical switch is connected as:

```
J'                    PORTX, bit 3
```

Each time that the encoder output changes, the timer's IC1 interrupt service routine reads the inputs and puts what it reads into a queue for later consumption. This is shown in Fig. 7.58a. The tick clock interrupt service routine discussed in Sec. 5.4 can be used to dump the queue, using the values obtained to update a 2-byte signed variable, POSITION. The maximum output rate from the encoder together with the tick time determine the size of the queue needed.

By having part of the job done by the IC1 interrupt service routine and the rest of the job done by the tick clock interrupt service routine, we reduce the latency seen by *other* interrupt sources. The IC1 interrupt service routine is executed with interrupts turned off, but it runs fast. In contrast, the tick clock interrupt service routine is executed with interrupts *reenabled*. Consequently, even though its execution may take some time, this does not affect the latency of other interrupt sources.

The ENCODER subroutine shown in Fig. 7.58c carries out the operation of converting the queue output into changes in absolute position. If we think of the two bits of information coming from channels A and B as *phase* information, then this subroutine just compares the phase of the last encoder output with the phase of the present encoder output. Moving a single count away from the last reading will produce a new reading which differs in a single bit. If *both* bits have changed, then we have missed a change of at least one count and have produced an error. This is noted by putting a nonzero value into the variable ENC_ERR. If *neither* bit has changed, then we are probably dithering at the boundary between two counts. A change has occurred, triggering an interrupt. However, when we actually serviced the interrupt, we found that it had changed back again (and lost an interrupt in the process). The hysteresis of the Schmitt triggers should preclude this from actually happening. However, if it does, then no harm is done. The table of Fig. 7.58b is used by the ENCODER subroutine to help sort out these possibilities.

Using this incremental encoder handler requires some special initialization at start-up time. We must drive the dc motor so as to make the device being positioned cross the zero-reference position. Once this has been done, the error flag ENC_ERR must be cleared. Since the ENCODER subroutine only sets this error flag and never clears it, we can check it periodically to see if an error has

```
;;;;;;;;;;; IC1 INTERRUPT SERVICE ROUTINE ;;;;;;;;;;;;;;;;;;;;;;;;;;;;;;;;;;;;;
;
; This interrupt service routine reads the input port to which an incremental encoder's three
; outputs and a slotted optical switch are attached.
; It queues this up for later, leisurely processing.
;
; The IC1 programmable timer input must be set up to interrupt on either edge.
;
; No CPU registers need be set aside at the beginning and restored at the end. The 68HC11 does
; this automatically.
;;;;;;;;;;;;;;;;;;;;;;;;;;;;;;;;;;;;;;;;;;;;;;;;;;;;;;;;;;;;;;;;;;;;;;;;;;;;;;;;;

IC1:    LDAA    #00000100B      ;Clear interrupt flag for IC1
        STAA    TFLG1           ; (see Fig. A.22 in Appendix A)
        BSR     READX           ;Call subroutine to read serially expanded port X in Fig. 7.57
        PUTB    PORTX, QENCODER ;Use macro to put its value into a queue
        RTI

                                        (a)


ENC_TBL:    DB    0        ;00 -> 00    Interrupt occurred but no change noted
            DB    1        ;00 -> 01    CW
            DB    2        ;00 -> 10    CCW
            DB    3        ;00 -> 11    Error

            DB    2        ;01 -> 00    CCW
            DB    0        ;01 -> 01    Interrupt occurred but no change noted
            DB    3        ;01 -> 10    Error
            DB    1        ;01 -> 11    CW
```

                                                                    (continued)

FIGURE 7.58
Incremental encoder handler. (a) Interrupt service routine; (b) table used for updating POSITION and ENC_ERR; (c) tick clock subroutine.

```
DB      1       ;10 -> 00   CW
DB      3       ;10 -> 01   Error
DB      0       ;10 -> 10   Interrupt occurred but no change noted
DB      2       ;10 -> 11   CCW

DB      3       ;11 -> 00   Error
DB      2       ;11 -> 01   CCW
DB      1       ;11 -> 10   CW
DB      0       ;11 -> 11   Interrupt occurred but no change noted
```

(b)

```
;;;;;;;;;   ENCODER SUBROUTINE FOR "TICK CLOCK" INTERRUPT SERVICE ROUTINE   ;;;;;;;;;;;;;;;;;;;
;
; This subroutine dumps the encoder queue, using bytes put into the queue by the IC1 interrupt
; service routine to update a 2-byte signed POSITION variable which represents the absolute
; encoder position as a 2s-complement number.  POSITION = 0000H when A', B', and I' all equal 1
; while the inverted output of the slotted optical switch, J', equals 0.
; It assumes that the inputs are located as follows: bit 0 = A', bit 1 = B', bit 2 = I', and
; bit 3 = J'.
; Errors are reported with a nonzero value in ENC_ERR.
; Assume caller has saved CPU registers.
;;;;;;;;;;;;;;;;;;;;;;;;;;;;;;;;;;;;;;;;;;;;;;;;;;;;;;;;;;;;;;;;;;;;;;;;;;;;;;;;;;;;;;;;;;;;;;;;
ENCODER: GETB   A, QENCODER   ;Check whether queue is empty; if not, get byte
         BNE    ENC_1         ;Carry on unless queue is empty (i.e., if Z flag is set)
         RTS                  ;Return if queue is empty
ENC_1:   LDY    POSITION      ;Use Y to work with value of POSITION
         ANDA   #00001111B    ; and get rid of extraneous bits
         EORA   #00000111B    ;Invert lower three bits; A = 00H if at the zero-reference position
         PSHA                 ;Save this value for test at end
         LDAB   ENC_OLD       ;Get old phase information
```

FIGURE 7.58 (continued)

```
ANDA   #00000011B   ;Drop zero-reference information, leaving only new phase information
STAA   ENC_OLD      ;Replace old phase information with new phase information
LSLB                ;Shift old phase information to left
LSLB                ; two places
ORAB   ENC_OLD      ; and then OR in new phase information to form offset into table
LDX    #ENC_TBL     ;Get base address of table
ABX                 ; and add offset
LDAB   0,X          ; and then get table entry
BEQ    ENC_END      ;Entry = 0; new phase = old phase; interrupted but no change occurred
CMPB   #1           ;Entry = 1; CW change
BNE    ENC_1        ; Skip if not 1
INY                 ; Decrement POSITION for CW
BRA    ENC_END
ENC_1:  CMPB   #2           ;Entry = 2; CCW change
BNE    ENC_2        ; Skip if not 2
DEY                 ; Decrement POSITION for CCW
BRA    ENC_END
ENC_2:  LDAB   #0FFH        ;Entry = 3; Report error
STAB   ENC_ERR
ENC_END: PULA                ;Restore present encoder bits and check for zero-reference position
TSTA                ; setting zero flag if at zero-reference position
BNE    ENC_3        ;Skip if not at zero-reference position
CPY    #0000H       ;Have any counts been lost since last time at zero-reference point?
BEQ    ENC_3        ;If not, then update POSITION and return
LDAB   #0FFH        ;Otherwise, report error
STAB   ENC_ERR
LDY    #0000H       ; and restore POSITION = 0000H
ENC_3:  STY    POSITION     ;Save Y in POSITION
ENC_4:  BRA    ENCODER      ; and look for another byte in queue
```

(c)

FIGURE 7.58 *(continued)*

387

occurred. If so, then the software should question any positioning undertaken since the last time the error flag was checked. It can drive the dc motor so as to reinitialize at the zero-reference position again and then continue normal operation.

One interesting configuration for an incremental shaft-angle encoder is shown in Fig. 7.59a. It is known by several names:

Digital potentiometer

RPG (short for rotary pulse generator)

It is designed to be used like a multiple-turn potentiometer on the front panel of an instrument. For an example, look back at Fig. 1.16, showing Hewlett-Packard's digitizing oscilloscope. The signal generator of Fig. 7.7 shows another application. As can be seen by comparing the specifications of Figs. 7.59b and 7.53b, the unit is a specially packaged two-channel version of Hewlett-Packard's incremental encoder.

In a potentiometer application, rate of rotation and direction are the all-important performance parameters. Consequently, the interface circuit and software can be simplified. Note from Fig. 7.54 that the state of channel B when a rising edge occurs on channel A gives sufficient information to determine the direction in which the encoder is turning. Furthermore, when turning continuously in one direction, the encoder will produce one rising edge per cycle. Figure 7.60 illustrates the interface simplification. The 68HC11's IRQ interrupt input is shown in use since it can be set up to detect falling edges only. Channel A is used to generate the interrupt. The interrupt service routine reads channel B and then increments or decrements an absolute position variable, POSITION, depending upon whether the bit read is a 0 or a 1.

This simplified scheme reduces the possible resolution available from the encoder by a factor of four. Thus, even though this encoder and that of Fig. 7.53 both have a resolution of 256 cycles per revolution, this interface and software produce 256 counts per revolution whereas the previous interface and software produced 1024 counts per revolution. In its role as a digital potentiometer, 256 counts per revolution gives excellent fine-grain adjustment capability.

In a digital potentiometer application, rate of turn and direction of turn can be obtained by having the tick clock interrupt service routine read the POSITION variable and form the difference from one tick to the next.

The relatively low-cost way to make an *absolute* shaft encoder having a small size and a resolution of 256 counts per revolution or so is to use mechanical brushes. These brushes ride upon tracks of an encoding disk which are broken up into alternating conducting and insulating segments. The conducting segments are connected to ground. With each brush tied to +5 V through a pullup resistor, the encoder generates outputs which can be interpreted as 1s and 0s. Litton's pin-contact shaft encoders, illustrated in Fig. 7.61, employ a pair of redundant pins on each track to improve reliability.

(a)

Hewlett-Packard panel mount digital potentiometer

Model No. HEDS-7501

Power:	5V at 40 mA (max)
TTL-compatible outputs:	
1 output	2.4 V (min) at 40 μA
0 output	0.4 V (max) at 3.2 mA
Diameter:	1.13 in
Depth:	0.70 in
Mounting:	⅜ in threaded bushing (same as a standard potentiometer)
Shaft diameter:	¼ in (same as a standard potentiometer)
Resolution:	256 cycles per revolution (see Fig. 7.54) 1024 quanta per revolution (256 cycles per revolution × 4 quanta per cycle)
Rotation speed:	300 r/m (max)
Termination:	10-conductor ribbon cable
Recommended interface:	74LS14, hex Schmitt-trigger inverter

(b)

FIGURE 7.59

Panel mount digital potentiometer. (*Hewlett-Packard Co.*) (*a*) Unit; (*b*) specifications.

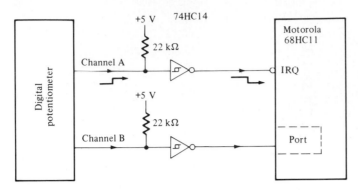

FIGURE 7.60
Digital potentiometer interface.

While such encoders are available with a binary output, a *Gray code* encoder has one major advantage over a binary code encoder. There is no synchronization problem. To see what this means, consider reading the 8-bit output of a binary encoder into a latch at *just the moment* that the output changes from 01111111 to 10000000. With all bits changing, the value latched can turn out to be *any* value, from 00000000 to 11111111. Even though such a synchronization error may occur only rarely, our software should nevertheless take special precautions against the possibility of such an error occurring. Otherwise each wildly erroneous reading may cause a drastic, momentary perturbation in performance.

No such synchronization problem occurs with a Gray code encoder. As the encoder changes from one encoded position to the next, exactly one bit of the output code changes. This is illustrated in Fig. 7.62*a* which shows a 4-bit Gray code side by side with its binary code equivalent. Since only one bit changes at any time, we can read a Gray code encoder while it is moving with complete reliability. If it happens to be on a boundary between two encoded positions, the Gray code encoder will produce an encoding of one of those two positions, exactly as it should.

If we strobe the encoder output into a shift register used as an extended input port, as in Fig. 6.30, then we can read it into the microcontroller and convert it to binary code. The conversion algorithm is developed, in the 4-bit case, in Fig. 7.62*b*. It uses exclusive-OR functions to produce each bit of the output.

> **Example 7.15.** Assume that we have expanded the input port capability of a Motorola 68HC11 by connecting three 74HC165 shift registers to the I/O serial port, as in Fig. 6.30. Also assume that we have written a subroutine called INPUT

FIGURE 7.61
Pin-contact shaft encoders. (*Encoder Division, Litton Systems, Inc.*) (*a*) Units; (*b*) specifications.

(a)

Some encoding options:
Model GCC11-08P1 — Single revolution, 256 counts/r, Gray code
Model VNB11-13P1 — 512 revolutions, 256 counts/r, V-scan binary

Mechanical specifications:	
Size: Single revolution	1.1 in diameter $\times$ 1.4 in long
512 revolutions	1.1 in diameter $\times$ 3.0 in long
Shaft	0.125 in diameter $\times$ 0.5 in long
Synchro mount	Size 11
Starting torque	0.5 oz-in maximum
Running torque	0.4 oz-in maximum
Moment of inertia	0.02 oz-in^2 maximum
Slewing speed (to avoid damage)	3600 r/m maximum
Operating speed (for reading encoder)	200 r/m maximum
Useful life	10×10^6 revolutions minimum

Electrical specifications:	
Output circuitry	Pin contacts
Output logic levels	
Logic 1	Closed contact
Logic 0	Open contact
Recommended operating voltage	5 to 15 V dc
Recommended operating current	Up to 3 mA/bit

Environmental:	
Operating temperature	$-55°C$ to $+85°C$
Storage temperature	$-65°C$ to $+125°C$

(b)

391

Gray code

g3	g2	g1	g0
0	0	0	0
0	0	0	1
0	0	1	1
0	0	1	0
0	1	1	0
0	1	1	1
0	1	0	1
0	1	0	0
1	1	0	0
1	1	0	1
1	1	1	1
1	1	1	0
1	0	1	0
1	0	1	1
1	0	0	1
1	0	0	0

Binary code

b3	b2	b1	b0
0	0	0	0
0	0	0	1
0	0	1	0
0	0	1	1
0	1	0	0
0	1	0	1
0	1	1	0
0	1	1	1
1	0	0	0
1	0	0	1
1	0	1	0
1	0	1	1
1	1	0	0
1	1	0	1
1	1	1	0
1	1	1	1

(a)

$$b3 = g3 = g3$$
$$b2 = g2 \oplus b3 = g3 \oplus g2$$
$$b1 = g1 \oplus b2 = g3 \oplus g2 \oplus g1$$
$$b0 = g0 \oplus b1 = g3 \oplus g2 \oplus g1 \oplus g0$$

(b)

FIGURE 7.62
Four-bit Gray code. (a) Comparison with binary equivalent values; (b) Gray-to-binary conversion.

to read the three registers into three RAM locations, IN1, IN2, and IN3. Finally, assume that we have connected an 8-bit Gray code shaft encoder to the third shift register. Calling INPUT loads IN3 with the Gray code encoder output, with the most-significant Gray code bit in the most-significant bit position of IN3. Actually, for the 8-bit Gray code encoder of Fig. 7.61, INPUT loads IN3 with the complement of IN3 since the encoder output is available in active-low form.*

The instruction sequence to load the 68HC11's accumulator A with the binary representation of the shaft angle position is shown in Fig. 7.63. If we are dealing with several 8-bit shaft angle encoders, we might write this as a macro. Note that this algorithm must work with a *copy* of the encoded output. If the encoder were connected directly to one of the microcontroller input ports, then reading the port successively during the conversion process (rather than just once before the conversion is begun) would lead to the same synchronization problem that arises when reading the output of a binary encoder.

---

* Note that Fig. 7.61b states that logic 1 is represented by a *closed* contact, producing an output of 0 V.

```
GRAY_BIN:  CLRA
GRAY_1:    EORA    INP3
           LSR     INP3
           BNE     GRAY_1
```

FIGURE 7.63
Motorola 68HC11 instruction sequence to convert the output of an 8-bit Gray code encoder to its binary equivalent.

With a microcontroller on our side, the conversion is so simple that it makes a Gray code encoder very attractive for absolute position encoding. It solves the synchronization problem which can occur when reading a moving encoder.

Absolute position encoders having a resolution much above 8 bits do not use the Gray code encoding technique. A Gray code encoder requires that all pins (or all optical sensors, for an optical encoder) be aligned to within $\pm\frac{1}{2}$ count. As the number of bits increases, the required alignment becomes impossible to produce and maintain. *V-scan encoding* represents an ingenious solution to this problem. As illustrated in Fig. 7.64*a*, this encoding scheme uses a binary pattern on the encoding disk and *two* sets of pins (or optical sensors). The least-significant bit, b0, determines a *read line*. When the encoder is read, the value obtained will be the value which would be produced *if* all of the pins were positioned along this read line (with perfect alignment).

To understand how a V-scan encoder gets around the alignment problem, consider Fig. 7.64*b*. Each segment on track *n* has a width which is labeled as β. Each segment on track $n+1$ has a width of 2β. It is a characteristic of binary code that when one of the tracks changes from a 0 segment to a 1 segment in the direction of increasing binary numbers, *none* of the higher-order tracks changes. Figure 7.64*b* shows the read line centered in the middle of a 0 segment on track *n*. The leading pin on track $n+1$ is offset to the right of this read line by $\beta/2$. When the read line is right in the middle of the 0 segment on track *n*, the leading pin on track $n+1$ is *right in the middle* of a segment. Note that as the read line ranges over the entire 0 segment on track *n*, the leading pin on track $n+1$ never gets closer to the edges of its segment than $\beta/2$. This ranging of the read line on track *n* and of the leading pin on track $n+1$ is depicted in Fig. 7.64*b* by double-ended arrows. *The alignment of the leading pin on track $n+1$ can be off by any value less than $\beta/2$ and still read the correct read-line value.*

The positioning of the lagging pin on track $n+1$ follows the same argument. Referring to Fig. 7.64*c*, we see that when the read line ranges over a 1 segment on track *n*, the lagging pin on track $n+1$ never gets closer to the edges of its segment than $\beta/2$. This results in the placement of the leading and lagging pins shown in Fig. 7.64*a*.

To read a V-scan encoder, we read b0. Then if b0 = 0, we read b1lead to get the value of b1 along the read line. If b0 = 1, we read b1lag to get the value of b1 along the read line. On the basis of the value read for b1 along the read line, we carry out the same selection for the pins on the b2 track. That is,

If bn-1 = 0, then read b$n$lead to get the value of bn.
If bn-1 = 1, then read b$n$lag to get the value of bn.

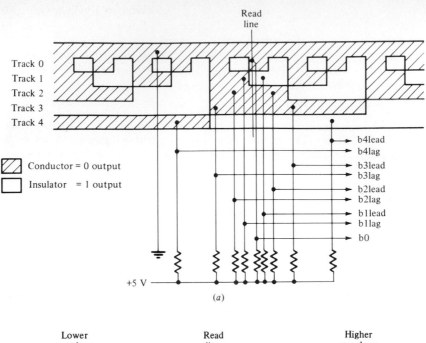

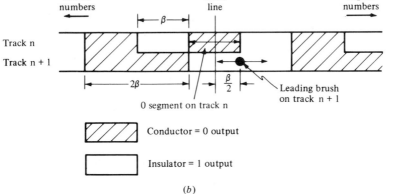

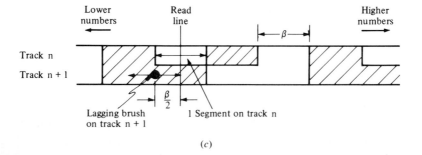

FIGURE 7.64
V-scan encoding; (*a*) Pattern of brushes, pins, or optical sensors; (*b*) positioning of leading brushes; (*c*) positioning of lagging brushes.

For tracks on a circular encoding disk, this leads to a wonderful result. The track for the least-significant bit is placed at the outer edge of the disk. The tracks for bits of increasing significance are placed toward the center of the disk. As the read line reaches from the position of the b0 pin toward the center of the disk, the alignment of pins on successively higher order tracks is relaxed exponentially.

Example 7.16. Consider a binary encoder which uses the V-scan technique to encode the angular position over one revolution as a 10-bit binary number. Determine the angular tolerance required for the positioning of each pin.

The pin for the least-significant bit, b0, determines the position of the read line. It serves as a reference for the alignment of the remaining pins. The track of the encoding disk used by this b0 pin divides one revolution into 1024 segments. Each segment extends over $360/1024 = 0.35°$. The remaining pins must be aligned as follows:

Pins	Angle from read line	Tolerance
b1lead and b1lag	0.17°	±0.17°
b2lead and b2lag	0.35°	±0.35°
b3lead and b3lag	0.70°	±0.70°
b4lead and b4lag	1.41°	±1.41°
b5lead and b5lag	2.81°	±2.81°
b6lead and b6lag	5.63°	±5.63°
b7lead and b7lag	11.25°	±11.25°
b8lead and b8lag	22.50°	±22.50°
b9lead and b9lag	45°	±45°

This says that the pins for the innermost track must be aligned to their ideal positions to within ±1/8th of a revolution. Aligning pins for V-scan encoders is a minimum-wage job!

By the same token, the tolerances which must be held during the manufacture of the encoding *disk* become successively relaxed the closer a track is to the center of the encoding disk.

Another ramification of the relaxation of the tolerances for higher order bits arises in the manufacture of multiple-turn absolute shaft encoders. A 512-revolution, 256-count-per-revolution unit is shown in Fig. 7.61*a*. It is the unit on the right and is really two V-scan encoders geared together. One of the encoders encodes a single revolution as an 8-bit number. The output from the selected pin for its most-significant bit (b7lead or b7lag) is used to select the leading or lagging pin for the least-significant bit of the second encoder which encodes revolutions. The backlash in the gearing has no effect as long as it still permits the least-significant bit of the second encoder to meet its tolerance requirements.

Self-decoding V-scan encoders include internal logic to carry out the required decoding of the leading and lagging pins to obtain the binary output. However, the usual combinational logic approach produces the same synchronization problem discussed earlier for an encoder which is moving while it is being read. On the other hand, if the microcontroller can send the encoder a pulse to *latch* the data from the pins, then the internal logic can decode the latched data and give an accurate binary output. It is important to latch the leading and lagging pin data *before* it is decoded, not afterward.

The usual interface to a V-scan encoder latches the outputs from all of the pins and reads them into the microcontroller for decoding in software. If we use shift registers attached to the I/O serial port of the microcontroller, as in Fig. 6.30, then the initial pulse which loads the shift registers takes care of the synchronization problem.

Example 7.17. Consider the use of a 256-revolution, 256-count-per-revolution V-scan binary encoder with the Intel 8096. This might be considered as an alternative way to encode the position of the device in Fig. 7.57. When the decoding to a binary number has been completed, the position will be encoded as a 16-bit number.

A 16-bit V-scan encoder has 31 signals to be read. They might be shifted into the microcontroller and then manipulated there. Figure 7.65a shows an alternative approach, using one of the 8096's bidirectional I/O ports* to access the encoder output, one bit at a time, via a 32-input data selector circuit. This circuit, and the algorithm of Fig. 7.65b, assume that the encoder output is available in positive-true form (i.e., that a 1 is represented by a high output and a 0 by a low output). Note that bits 4 and 5 of the I/O port's lines are unused by this application. If they are used elsewhere as *inputs*, then our dealings with the port are simplified in that we can simply write 1s to these lines whenever we write to the port.

In the circuit of Fig. 7.65a, the 74LS139 decoder is used to enable the three-state output of just one of the data selectors. In this way, we create a 32-input data selector out of four 8-input data selectors.

Because the selected bit from the encoder at each step must be put into the most-significant bit of a 16-bit result register (BX), it is helpful to read the data selector output into bit 7 of the port. Furthermore, because this same bit is used in forming the selection address for the next bit, the algorithm is simplified somewhat if bit 6 of the port is driven with this value.

The algorithm of Fig. 7.65b emits successive 4-bit binary numbers on bits 0 to 3 of port 1, thereby selecting

$$b_i\text{lag or } b_i\text{lead, for } i = 0, 1, \ldots , 15$$

Which value is selected at each stage is determined by the output on bit 6 of port 1. The algorithm loops 16 times, picking up 1 bit of the resulting binary number each time through the loop. These bits are entered into bit 15 of the BX

---

* Refer to Fig. B.9 in Appendix B for information on how to use this quasi-bidirectional I/O port.

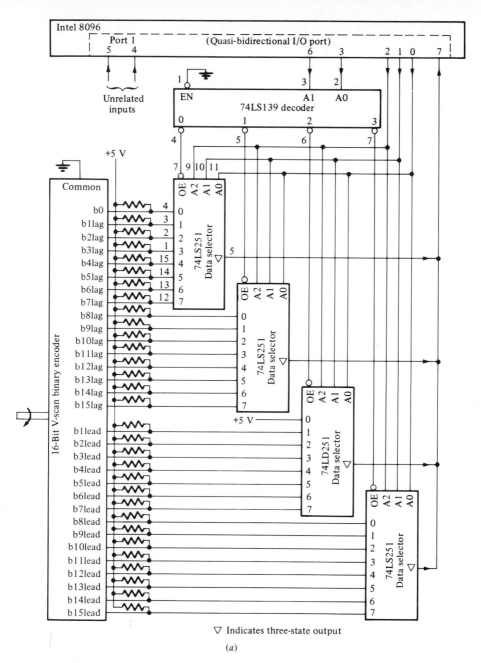

▽ Indicates three-state output

(a)

FIGURE 7.65
V-scan code conversion using the Intel 8096. (a) Brush selection circuitry; (b) V_SCAN subroutine to implement the conversion.

```
;;;;;;;;;  V_SCAN SUBROUTINE  ;;;;;;;;;;;;;;;;;;;;;;;;;;;;;;;;;;;;;;;;;;;;;;
; This subroutine assumes that the 31 outputs from a 16-bit V-scan binary encoder drive the
; interface circuit of Fig. 7.65a.  Of the "8096 scratch-pad registers" of Fig. 4.27, it uses
; AL for brush selection, AH as a scratch-pad register, and BX to form the 16-bit binary result.
;
;;;;;;;;;;;;;;;;;;;;;;;;;;;;;;;;;;;;;;;;;;;;;;;;;;;;;;;;;;;;;;;;;;;;;;;;;;;;;;;

V_SCAN:     CLR     BX                  ;Clear result register
            CLR     AX                  ;Initialize brush selection register and scratchpad register

V_AGAIN:    ORB     AL,AH               ;AL now contains selected brush value in bit 6
            ORB     AL,#10110000B       ;Write 1 to bits of port 1 used as inputs
            STB     AL,P1               ;Update port 1

            INCB    AL                  ;Increment brush selection value
            ANDB    AL,#00001111B       ; and force upper bits of AL to 0
            JE      V_DONE              ;Quit when lower four bits of brush address roll over to 0000

            SHR     BX,#1               ;Shift result register to make room in bit 15
            LDB     AH,P1               ;Read port 1
            ANDB    AH,#10000000B       ; and force all but bit 7 (i.e., selected brush value) to 0
            ORB     BH,AH               ; so we can move value of selected brush into bit 15 of result

            SHRB    AH,#1               ;Move selected brush value to bit 6 of AH
            ANDB    AH,#01000000B       ;Save only bit 6, forcing other bits to 0
            SJMP    V_AGAIN             ;Repeat

V_DONE:     RET
```

(b)

FIGURE 7.65 *(continued)*

register and shifted to the right. After 15 shifts, b0 (the first bit read) will have been moved to bit 0 of the BX register.

7.8 IC DRIVERS

Thus far in this chapter we have seen a variety of integrated circuit drivers used to support the driving of some relatively low-power devices. In this section we consider some drivers which translate the output signals from a microcontroller into the higher voltage and current requirements posed by a variety of higher power devices.

The *power MOSFET*, whose development has advanced in great strides since 1979, is a key power-switching device. International Rectifier, a major manufacturer of power MOSFETS, has been joined by many other manufacturers, with devices having voltage ratings reaching toward 1000 V and continuous current ratings reaching toward 100 A. With such ratings, the power dissipation of the driver itself becomes a major issue. Probably the most popular package is the TO-220 shown in Fig. 7.66a. It is easy to mount on a printed circuit board, both electrically and for heat sinking. The higher power TO-3 package is shown in Fig. 7.66b. Its use with a heat sink to switch maximum power is illustrated in Fig. 7.66c.

In contrast to a bipolar transistor switch, which is current-driven, the power MOSFET is *voltage-driven*. Like a CMOS input, the only current load imposed upon the device which drives a power MOSFET is a leakage current, on the order of microamperes. However, it does need an input voltage of at least 10 V or so to switch the output. The 5-V output from a microcontroller is not enough.

The Motorola MC14504B hex CMOS-to-CMOS level shifter shown in Fig. 7.67 translates the 0- to +5-V swings on one of its inputs to a 0-V to V_{DD} volt swing on the corresponding output, where V_{DD} is the supply voltage for the output drivers. It can be any value up to +18 V. Consequently, this level shifter can use a standard +12- or +15-V supply to serve this role. If such a supply is not already being used in the instrument or device being designed, then this voltage can be derived from the higher voltage supply for the load that we want to control. With a quiescent supply current of less than 15 μA, the MC14504B can use the simplest of regulator circuits. One possibility, shown in Fig. 7.68, uses a low-power zener diode to regulate V_{DD} down to 10 V.

When a power MOSFET is turned on by driving its gate voltage up to +10 V or so, the power-handling part of the device between its drain and source pins can be characterized by a low resistance. Since the load current dissipates power in this resistance, a power MOSFET designed to handle a large current will have an extremely low on-state resistance.

Example 7.18. Consider International Rectifier's power MOSFETs* having a 50-V rating (their lowest voltage rating) and a TO-220 package. The specifications for

---

* Which they call HEXFETs.

(a)

(b)

(c)

FIGURE 7.66
Power MOSFET packages. (*International Rectifier.*) (*a*) TO-220 package; (*b*) TO-3 package; (*c*) TO-3 package with heat sink.

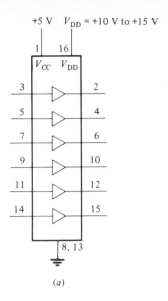

+5 V V_{DD} = +10 V to +15 V

1   16

V_{CC} V_{DD}

3   2

5   4

7   6

9   10

11   12

14   15

8, 13

(a)

V_{DD} = +18 V, max

Input current   = 1 μA, max

Output current (high) = 3.0 mA at V_{out} = 13.5 V min
                                     with V_{DD} = 15 V

Output current (low) = 3.0 mA at V_{out} = 1.5 V max
                                     with V_{DD} = 15 V

(b)

FIGURE 7.67
Motorola MC14504B CMOS-to-CMOS hex level
shifter. (a) Pinout; (b) specifications.

their six units are listed in Fig. 7.69. Determine the on-state power dissipation
when each one is handling its rated continuous drain current. Also determine the
drain-source voltage drop under this condition.

How much drain current can be handled by the part depends upon how
efficiently we cool the part. If we bolt the TO-220 package to a surface which
efficiently carries away heat (perhaps with the help of a fan), then the case can be
held at a lower temperature than if the part stands in still air above a PC board.
For the purposes of this example, let us consider the power dissipation with the
case temperature permitted to rise to 100°C.

For the heftiest part, the IRFZ40, the power dissipation when it is handling
32 A is

$$P = I^2R = 32 \times 32 \times 0.028 = 29 \text{ W}$$

This is well below the specification of 50 W, when the case temperature is permitted to rise to 100°C. The drain-source voltage drop under this condition is

$$V = IR = 32 \times 0.028 = 0.90 \text{ V}$$

The values for all six of the power MOSFETs of Fig. 7.69 are presented in Fig.
7.70.

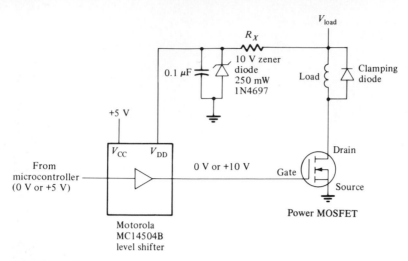

FIGURE 7.68
Simple zener diode regulator circuit to derive V_{DD} from supply for load, if $V_{LOAD} > 10$ V.

It can be worthwhile to pick an oversized power MOSFET in order to simplify the cooling requirements for a design. This tactic may eliminate the need for a fan.

Example 7.19. Consider the application of each of the power MOSFETs of Fig. 7.69 in a design which could be met by the least of these, the IRFZ22. Determine the power dissipation and voltage drop in each case.
    The IRFZ22 can handle 9 A continuously if the case temperature is permitted to rise to 100°C. As seen in Fig. 7.70, it will dissipate 10 W and exhibit a voltage drop of 1.08 V. We want to determine the power dissipation and voltage

Part number	On-state resistance, Ω	Continuous drain current		Pulsed drain current, A	Maximum power dissipation	
		100°C case, A	25°C case, A		100°C case, W	25°C case, W
IRFZ40	0.028	32	51	160	50	125
IRFZ42	0.035	29	46	145	50	125
IRFZ30	0.05	19	30	80	30	75
IRFZ32	0.07	16	25	60	30	75
IRFZ20	0.10	10	15	60	16	40
IRFZ22	0.12	9	14	56	16	40

FIGURE 7.69
International Rectifier 50-V, power MOSFET specifications. TO-220 package.

Part number	Power dissipation, W	Voltage drop, V
IRFZ40	29	0.90
IRFZ42	29	1.02
IRFZ30	18	0.95
IRFZ32	18	1.12
IRFZ20	10	1.00
IRFZ22	10	1.08

FIGURE 7.70
Maximum power dissipation and voltage drop for the power MOSFETs of Fig. 7.69 when the case temperature is allowed to rise to 100°C.

drop for each of the parts of Fig. 7.69 as it handles 9 A continuously. The results are shown in Fig. 7.71. Note that because of its much lower resistance, the IRFZ40 part dissipates only one-quarter of the power of the IRFZ22 in this application.

For relatively low power applications, four power MOSFETs are available in a dual-in-line package. This provides for ease of handling.

Example 7.20. Consider the stepper motor shown in Fig. 7.72. If we drive it with the circuit of Fig. 7.73, then determine the power dissipation and voltage drop across each power MOSFET when the MOSFET is conducting.

The four power MOSFETs in this circuit are packaged in a 14-pin DIP. With an on current of 0.30 A per winding of the stepper, each driver will exhibit a voltage drop of

$$V = 0.30 \times 1.0 = 0.30 \text{ V}$$

and a power dissipation of

$$P = V \times I = 0.30 \times 0.30 = 0.09 \text{ W}$$

This looks like an easy solution to the driver problem. With such low power dissipation, the power MOSFET DIP can be treated like other ICs in a circuit, with no special heat sinking.

Part number	Power dissipation, W	Voltage drop, V
IRFZ40	2.3	0.25
IRFZ42	2.8	0.32
IRFZ30	4.1	0.45
IRFZ32	5.7	0.63
IRFZ20	8.1	0.90
IRFZ22	9.7	1.08

FIGURE 7.71
Power dissipation and voltage drop for the power MOSFETs of Fig. 7.69 when each is passing a current of 9A.

(a)

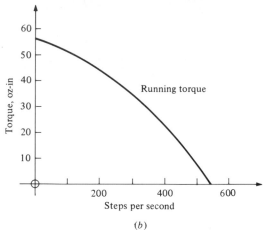

(b)

Airpax model 4SH24A56S (bifilar winding)

Steps per revolution   = 200

Step angle   = 1.8°

Step angle tolerance   = ±5%

DC operating voltage   = 24 V

Resistance per winding   = 79 Ω

Inductance per winding   = 150 mH

DC current per winding   = 0.30 A

Holding torque   = 85 oz-in

Rotor moment of inertia   = 1.67×10^{-2} g·m²

Diameter   = 2.2 in

Length, not including shaft   = 2.2 in

Shaft diameter   = 0.250 in

Weight   = 21.5 oz

(c)

FIGURE 7.72
A medium-size stepper motor. (*Airpax Corp.*) (*a*) Motor; (*b*) torque-speed characteristic; (*c*) other characteristics.

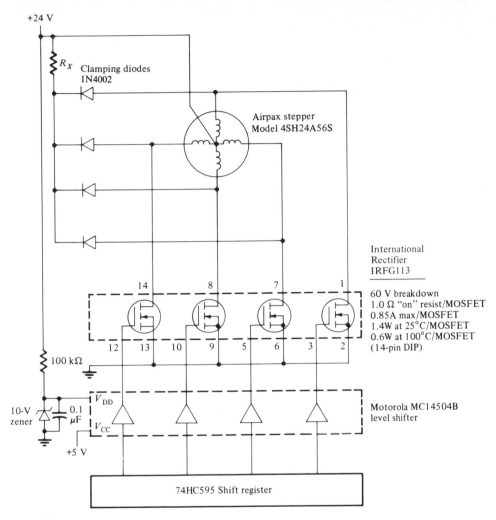

FIGURE 7.73
Stepper motor drive circuit.

Power MOSFETs provide a general solution to the driver problem, handling awkward voltage and current requirements wherever they arise. For a more limited range of output requirements, Sprague makes a parallel-output shift register, their UCN-4821A, which operates much like the 74HC595 which we discussed earlier. Its characteristics are listed in Fig. 7.74. It is ideal for many microcontroller applications because it combines a CMOS input circuit with a bipolar output circuit which includes eight Darlington open-collector drivers. The stepper motor of Fig. 7.72 can be driven with the circuit of Fig. 7.75, reducing the parts count relative to the drive circuit of Fig. 7.73.

DC motor control poses another interesting drive problem. If we embed a dc motor in an H-bridge circuit (like that employed in the stepper motor drive

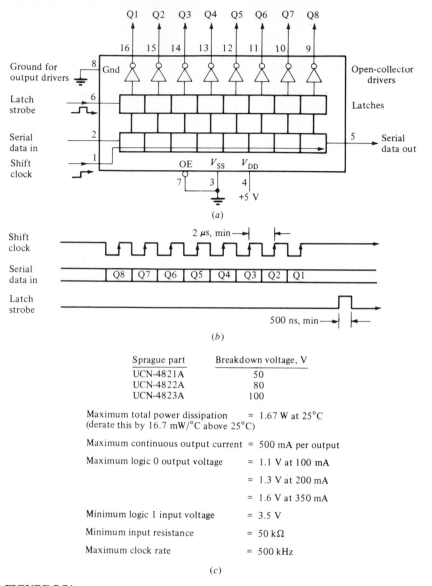

(a)

(b)

Sprague part	Breakdown voltage, V
UCN-4821A	50
UCN-4822A	80
UCN-4823A	100

Maximum total power dissipation   = 1.67 W at 25°C
(derate this by 16.7 mW/°C above 25°C)

Maximum continuous output current = 500 mA per output

Maximum logic 0 output voltage   = 1.1 V at 100 mA

                                             = 1.3 V at 200 mA

                                             = 1.6 V at 350 mA

Minimum logic 1 input voltage   = 3.5 V

Minimum input resistance   = 50 kΩ

Maximum clock rate   = 500 kHz

(c)

FIGURE 7.74
Sprague UCN-482X serial input, parallel output drivers. (*a*) Pinout; (*b*) timing diagram; (*c*)
specifications.

circuit of Fig. 7.50*b*), then we can control both speed and direction. The circuit
of Fig. 7.76*a* shows a low-parts-count design. The two Texas Instruments half-
H drivers are designed to work together, drawing their power from the supply
for the dc motor (of up to 40 V). As shown in the data of Fig. 7.76*b*, they have
inputs which are both TTL and CMOS compatible. The circuit can handle dc
motors with current ratings of up to 2 A.

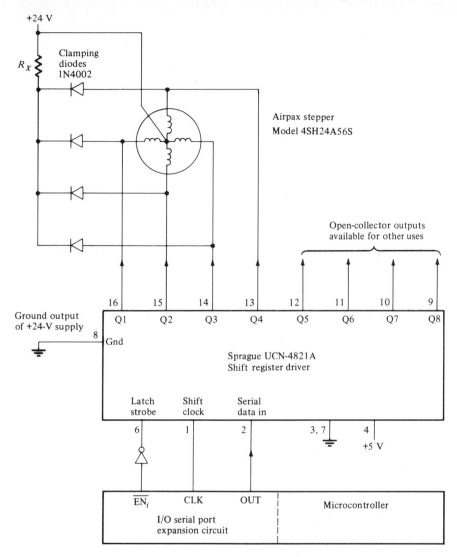

FIGURE 7.75
Alternative stepper motor drive circuit.

If the dc motor is driven with the pulse-width modulation output from a microcontroller* as shown in Fig. 7.76a, then a given modulation will turn on the full supply voltage to the motor (less about a 2-V drop in the drivers) for the on part of each modulation cycle and then put the drivers in the high-impedance state for the off part of each cycle. When the drivers are turned off, the current in the dc motor coasts through the diodes built into the driver parts. Conse-

---

* Refer to Sec. 7.5.

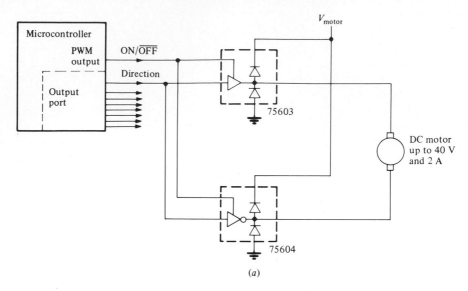

(a)

Manufacturer:	Texas Instruments
Part numbers:	SN75603
	SN75604
Part designation:	High-current half-H drivers
Package:	5-lead TO-220
Maximum supply voltage, V_{CC}:	40 V
Maximum continuous output current:	±2 A
Maximum power dissipation:	6.25 W at (or below) 125°C case temperature
Minimum high-level input voltage:	2.0 V
Maximum high-level input current:	10 μA
Maximum low-level input voltage:	0.8 V
Maximum low-level input current:	20 μA
Typical high-level output voltage:	V_{CC}-0.9 V at 1 A
	V_{CC}-1.2 V at 2 A
Typical low-level output voltage:	0.9 V at 1 A
	1.2 V at 2 A
High-impedance-state output current:	100 μA

(b)

FIGURE 7.76
DC motor drive circuit using a PWM output from a microcontroller. (a) Circuit; (b) characteristics of driver chips.

quently, the dc motor responds to the percent of modulation in the same manner that it would to a constant dc voltage equal to that same percent of the dc voltage supply (less about 2 V). Its speed increases as the percent of modulation increases. The *direction input* determines whether the current which is turned on to the motor flows in one direction or the other, thereby controlling whether the motor turns clockwise or counterclockwise.

If we seek position control rather than speed control, then we can combine the drive circuit of Fig. 7.76a with the feedback circuit of Fig. 7.57. To reach a desired position, the pulse-width modulator output can be turned on to a maximum (by driving the on/off signal high) until we approach the desired position. Then we can apply maximum braking by reversing "direction." Finally, we can use a linear control algorithm to home in on the final position. Control of the modulation and direction are now based upon the difference between the desired position and the actual position (and how fast that difference is changing, etc.).

AC power control is attractive for applications requiring appreciable power-handling capability. AC power can be applied directly to a load, without the need for a power supply. A *triac** can be interposed between the ac power line and the load. It can be controlled by a microcontroller to turn the power to the load on and off.

When switching high voltages with a microcontroller, one of the threats which arises is the risk of having the high voltage work its way back to the sensitive circuitry of the microcontroller. To minimize this risk, we can do several things:

> Use optoisolation between the 5-V microcontroller circuitry and the 115-V ac power circuitry. The *only* connection between the two circuits will then be the light from an LED (driven by the microcontroller) falling upon a phototransistor (in the circuit which drives the triac).
>
> Put epoxy, or tape, over all high voltage traces on a PC board and shrink tubing over all PC board edge-connector contacts having high voltages. This will reduce the chances of damage when probing a PC board or an edge connector with a scope probe or a multimeter probe during debugging. Without such protection, a probe inadvertently shorting the wrong two traces together during debugging can fry every single 5-V chip on a board!

Another potential problem which can arise when switching ac power is the generation of radio-frequency interference (RFI). By turning on the ac power to a resistive load only as it passes through 0 V, RFI can be eliminated. Using a zero-crossing detector with an inductive load helps to reduce the generation of RFI even if it does not eliminate it.

---

* An electronic ac power switch.

The circuit of Fig. 7.77*a* shows the CMOS output of a 74HC595 (used as an output port) serving as a 15-mA driver for the LED in an optically isolated triac driver. While this is an unusually high load current for a 74HCxxx part, the part can supply it. In doing so, the output voltage may rise to 1.0 V or so. With a maximum voltage drop of 1.5 V across the LED in the triac driver, the 120-Ω resistor in the LED circuit will ensure a current of 15 mA, thereby guaranteeing that the MOC3031 will turn on.

The MOC3031 triac driver is shown driving a compatible triac, Motorola's MAC3030-4. This unit includes an *RC* snubber circuit across its output to limit the rate of change of the voltage across the triac. A triac does not handle transient voltage changes very well; it can be triggered by such a change even when its input circuitry is telling it to remain turned off.

As shown in Fig. 7.77*b*, we can handle significant amounts of power with a simple triac circuit. The bigger units use the same TO-220 package shown in

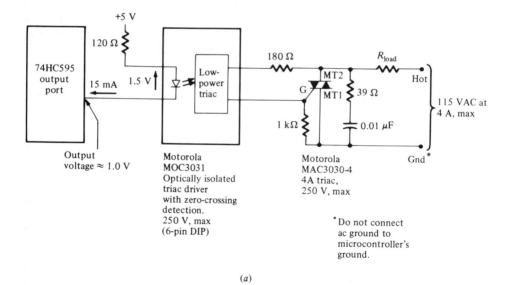

(*a*)

Triacs compatible with MOC3031 (250 V, max)

Motorola no.	RMS amperes, max	Case
MAC3030-4	4	TO-126
MAC3030-8	8	TO-220
MAC3030-15	15	TO-220
MAC3030-25	25	TO-220

(*b*)

FIGURE 7.77
Controlling power to an ac load. (*a*) Circuit; (*b*) compatible triacs.

(a)

General Instrument 74OL6000

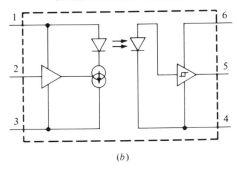

(b)

Pin 1 = Input V_{CC} Pin 6 = Output V_{CC}

Pin 2 = Data in   Pin 5 = Data out

Pin 3 = Input Gnd   Pin 4 = Output Gnd

(c)

4.5 V ≤ Input V_{CC}, Output V_{CC} ≤ 5.5 V

Input characteristics:

   Low-level input current   <   0.4 mA at 0.8 V max

   High-level input current   <   40 μA at 2.0 V min

Output characteristics:

   Low-level output current >   16 mA at 0.6 V max

   High-level output current >   0.4 mA at 2.4 V min

Propagation delay < 100 ns

(d)

FIGURE 7.78

Optocoupled TTL buffer. (*General Instrument*.) (*a*) Symbol; (*b*) equivalent circuit; (*c*) pin configuration; (*d*) electrical characteristics.

Fig. 7.66*a*. The smaller TO-126 package mounts in much the same way as the TO-220 package to take advantage of easy mounting and effective heat sinking.

This circuit can be used to drive heaters, 115-V light bulbs, ac motors, power relays, etc. For switching 230-V power, Motorola offers the MOC3041 with the same characteristics as the MOC3031 of Fig. 7.77*a*, but with a peak voltage of 400 V. Likewise, they offer a MAC3040-xx family of parts analogous to the parts listed in Fig. 7.77*b* but able to handle a peak voltage of 400 V.

Another application of optocoupling is the electrical isolation of two parts of a circuit that must pass digital signals between them. The multimeter of Fig. 1.19 represents an example of this, where its front end needs to be electrically isolated in order to permit high-quality "guarded" measurements. It uses a microcontroller in this front-end circuitry both for control and to serialize the data, to minimize optocouplers.

The little six-pin DIP characterized in Fig. 7.78 represents a state-of-the-art optocoupler for this purpose. Its input and output circuitry are designed to make it look like a TTL buffer. Consequently, it fits into a microcontroller circuit with no extra drivers needed. If the output is to drive a CMOS device, then a 10-kΩ pullup resistor to +5 V should be connected to the output.

7.9 SUMMARY

In this chapter we have considered many ways to handle the inputs and outputs to a microcontroller. This has augmented the treatment of inputs from analog transducers via an A/D converter in Chap. 2, the treatment of timed input and output events in Chap. 3, the treatment of keyswitches in Chap. 5, and standard serial interfaces in Chap. 6.

We began the chapter with a discussion of the display of information. For a very small amount of information (e.g., a few digits of data plus a few annunciator lights) we saw that LED displays could handle the task with ease. However, as the amount of information to be displayed increases, the LED approach becomes prohibitively expensive. In contrast, a custom LCD display becomes a very viable alternative. Because *custom* anything may sound intimidating, much of the emphasis of the section on LCD displays was intended to show that a product can include an LCD display with about the same amount of complication as using a printed circuit board. We found that the interface to both LED and LCD displays could use standard chips having built-in I/O serial ports.

For the display of even more information, a CRT presents an economical alternative. The display becomes bright and crisp for applications which can handle the bulk of a CRT. A little 5-in display can present a very readable array of up to 40 by 24 characters. We saw the impact which is being made by the development of application-specific integrated circuit (ASIC) chips. National Semiconductor's Terminal Management Processor puts the complexity of driving a CRT into just a handful of chips. Its built-in UART serial port makes interfacing with a microcontroller easy. Its organization around an internal microcontroller core means that it can be customized with a new feature set by reprogramming its internal ROM. In addition, the chip has been designed so that the internal ROM can be disabled and replaced by an external ROM or EPROM.

The generation of analog voltage outputs is important to many applications. For some of them, a dc voltage output can be replaced by a pulse-width modulated output having the same dc component. The microcontroller can generate the PWM output directly.

The addition of a D/A converter provides a ready solution when a dc voltage is actually needed. If the D/A converter is updated from within the service routine for a real-time clock, then a sampled-data output waveform can be easily produced. In another application, a dc voltage output which is produced at initialization time can be just what is needed to translate the calibration data stored in an EEPROM into a calibrating voltage for use by an instrument. In keeping with the general goal of keeping the parts count low, we looked at a D/A converter which includes an I/O serial port. It provides a simple solution to many analog voltage output requirements. Alternatively, D/A converters which will hang on a microcontroller's expansion bus are plentiful, providing many choices for resolution and updating rate.

Many applications of microcontrollers include a requirement for position control, which is met easily and inexpensively, under many circumstances, by a stepping actuator. It has the advantage over dc motors and most other positioning actuators in having an inherent D/A capability. That is, a given number of steps translates into a well-defined, precise change in position.

Stepping actuators also have the advantage over other positioning actuators of being driven by *switches*. For power considerations (i.e., power supply requirements, minimization of heating in the drivers, elimination of the need for a fan, etc.), a switch approaches the ideal as a driver. When it is turned off, it dissipates virtually no power. When it is turned on, its power dissipation is proportional to whatever small *on* voltage it exhibits. We considered some drivers in both Secs. 7.6 and 7.8.

A recurring position-sensing requirement of many microcontroller applications is the need to sense a stop position or a zero-reference point. For example, a plotter which uses stepper motors for positioning needs a single reference position on each axis. Given those positions, it can precisely define, and get to, *any* position. In some applications, motors have the potential for destruction if permitted to run on past an end point of travel. Sensing a stop or end position gives the microcontroller the information it needs to prevent such an event. Slotted optical switches provide a low-cost, effective means for sensing position in all of these applications. They are available from myriad manufacturers with myriad configurations to handle myriad mounting alternatives and sensitivity requirements.

Incremental encoders give a higher degree of position sensing with what is actually a rather simple mechanism. They require the microcontroller to respond to *every* change in position, from one quantum position to the next. Consequently, they are almost universally responded to under interrupt control. We saw that the interface from an incremental encoder must generate an interrupt signal and have two outputs read. If the incremental encoder incorporated a zero-reference output, then it had to be read also. For the specialized application of presenting the user of an instrument with a potentiometerlike knob for adjusting input parameters, a rotary pulse generator (RPG) permits an even simpler interface. While it is no different from any other incremental encoder (requiring the sensing of position change and direction), this application can throw away the last two bits of positioning resolution and thereby simplify both the interface circuitry (which is already quite simple) and the interrupt service routine's algorithm.

Position sensing with an incremental encoder requires that the shaft whose position is being monitored by the encoder be rotated at power-on time. It also potentially subjects the application to erroneous position readings if counts are ever lost. Absolute shaft encoders solve these problems by encoding the position as an N-bit number. While they do not require interrupt control in the same sense that an incremental encoder does, the least-significant bit of the output can be used as an interrupt signal if such control is desired.

The code employed by an absolute shaft encoder has a larger impact upon encoder use than is apparent to the unsuspecting. That is, a binary encoder is *not* the simplest encoder to use. That this is true arises as a result of trying to read a number reliably when several bits are changing at the same time and at times which are unsynchronized to the reading operation. It also arises because of the need to align many pickups, whether conductive or optical, to form the code pattern for the absolute position.

For a 256-count or 512-count, single-turn absolute shaft encoder, the simplest one to use is a Gray code encoder. It is simple because the number of output lines which must be read into the microcontroller is just eight or nine. The algorithm to convert from Gray code to binary code is short and fast. In contrast, an N-bit binary encoder will have $2N - 1$ output lines. Not only does the number of lines which must be read in to the microcontroller double, the algorithm to handle them significantly increases in complexity.

For absolute shaft encoders having a higher resolution and for multiple-turn shaft encoders, Gray code is not an available option. The positioning tolerance required for each optical or conductive pickup is eased by using V-scan encoding. However, this leads to the more complicated interface and algorithm alluded to above.

This chapter concluded with examples of IC drivers for handling current, voltage, and power requirements of a load which cannot be handled by the microcontroller directly. The marriage of electromechanical actuators to built-in ICs, which handle their peculiar drive requirements effectively, will simplify interfacing in the future. Even now we are seeing devices of this type. For example, the SynchroStep stepper motor drive circuit shown in Fig. 7.50 puts the *intelligence* of the circuit into a single IC. That chip includes

An up/down counter to change the output phase requirements corresponding to each step

An analog comparator to determine whether the stepper motor current is above or below the desired value

An oscillator to set a chopping (i.e., modulation) frequency

A pulse-width modulator to turn the voltage on and off repeatedly across a winding so as to build it up and hold it at the desired value.

A power MOSFET provides a close-to-ideal power switch. Its input draws virtually no current. Its output can switch many amperes of load current.

Its output can also withstand a severe load voltage when the current is turned off. However, a power MOSFET needs a larger input voltage range between the off state and the on state than 5 V. Consequently, a microcontroller output must be buffered to drive a power MOSFET. We saw how this could be done and the benefits which could be gained by using an oversized power MOSFET in an application in order to avoid a heatsink and/or a fan.

Another example of the integration of logic with drive circuitry was illustrated in Fig. 7.75. A shift register with high-drive-capability outputs was shown which could be connected to the microcontroller via its I/O serial port.

For dc motor control, we saw two driver ICs which are designed to work together to form an H-bridge driver. This circuit permits us to control the current in either direction through a load, as well as to turn it off. This circuit could be used with the pulse-width-modulation output of a microcontroller to control the speed of a dc motor. With a suitable incremental encoder to monitor the motor's position and a suitable algorithm, we could achieve closed-loop position control.

AC power control by a microcontroller raises the specter of total destruction, as the high voltage of the ac power line inadvertently reaches back into a logic circuit and fries all of the 5-V chips. There are many ways in which this "reaching back" can occur. However, a common way to avoid problems is to isolate the ground of the internal logic circuitry from the ac ground. The transformer in a power supply serves exactly this function (in addition to its role of producing a low ac voltage output from a high ac voltage input). An optoisolator can achieve the same isolation function in a small DIP package for a low-level logic signal. Its combination with a triac permits this small device to switch ac power on and off. By including a zero-crossing detector in its circuitry, it can also minimize the RFI generated during the switching of the ac power.

PROBLEMS

7.1. *LED annunciators.* In this problem we will consider the effect of variations in LED voltage drop upon the current from LED to LED in Fig. 7.1. Assume that the current-limiting resistors are each exactly 680 Ω, that the 74HC595 outputs are 0.20 V, and that the supply voltage is exactly 5.0 V.

(*a*) Determine the LED current if the LED voltage drop is 2.2 V.

(*b*) Determine the LED current if the LED voltage drop is 1.5 V.

(*c*) Determine the ratio of the current in part *a* to the current in part *b*. Because of the nonlinearity of the LED's luminosity-versus-current characteristic, the ratio of the brightnesses of LEDs with these currents will be approximately 2.5 times the ratio of the currents.

7.2. *Multiplexed seven-segment LED display.* If the circuit of Fig. 7.4 is driven by the Intel 8096, which transmits bytes out of its I/O serial port least-significant bit first, then show the meaning of each of the bits of each of the bytes to be sent to the MC14499 chip. Do this with a diagram analogous to Fig. 7.5*b*.

7.3. *Byte-inverting table.* The previous problem illustrated the value of a 256-byte table which can be used to swap the order of the bits of a word. For example, a peripheral device may require us to send bytes to it most-significant bit first. If we are using the Intel 8096, or some other microcontroller with an I/O serial port which transmits data least-significant bit first, then the table can be used to swap the bits. The nth entry in the table must contain the number n, with its least-significant bit written in the most-significant bit position (e.g., the 10100011th entry of the table must be 11000101).

(*a*) Given such a table, called INVERT, show the Intel 8096 code to take the content of a byte in RAM (e.g., the content of AL) and swap the order of its bits.

(*b*) Determine how long this takes if each cycle takes 0.25 μs.

(*c*) Do you have a suggestion for creating this table which is better than the brute force way of writing out 256 ''define byte'' statements? What would you do? Can you write a BASIC or Pascal or C program to create the *source* file for the table?

7.4. *Byte-inverting algorithm.* Minimizing execution time, write a subroutine for the Intel 8096 called INVERT which will invert the content of the byte in AL, putting the result into AH. Note that successive right shifts of AL will move the bits of AL into the carry bit, least-significant bit first. Also note that an add with carry of AH to itself will shift AH to the left and load the carry bit into the least-significant bit position of AH. Determine how long this subroutine takes to execute if each cycle takes 0.25 μs.

7.5. *Multiplexed seven-segment LED display.* Write a subroutine called DISPLAY which will take the four-digit number stored in 2 bytes of RAM labeled SEVEN-_SEG and store it out to the display of Fig. 7.4. Assume the digits are packed together as in the second and third byte of Fig. 7.5*b*. Also assume that the decimal points of the displays are left disconnected.

As shown in Fig. 7.5*c*, the minimum clock period of the I/O serial port's clock must be 4 μs. In each case below, set up the serial port for a suitable clock rate and then make the transfer of 2 bytes. Do not transfer the second byte until the I/O serial port indicates that it has completed the first transfer.

(*a*) Write DISPLAY for the Motorola 68HC11 assuming that the display circuitry is connected to I/O serial port 0 in the expansion circuit of Fig. 6.33. Assume that at the time that DISPLAY is called, one of the UART serial ports of Fig. 6.33 may be in use, so do not change bits 3 and 4 of port C. Minimize (and determine) the execution time of DISPLAY.

(*b*) Rewrite DISPLAY for the Motorola 68HC11 with only one change to the conditions of part *a*. Assume that DISPLAY will never be called when a UART is in use, so that bits 3 and 4 of port C may be changed. Minimize (and determine) the execution time of DISPLAY.

(*c*) Rewrite DISPLAY for the Intel 8096. Assume that the INVERT table of Problem 7.3 has already been used to swap the bits of the 2 bytes to be displayed. Assume that the display circuit is connected to I/O serial port 0 of Fig. 6.31 and that none of the other serial ports is in use.

7.6. *LCD display.* For either the Intel 8096 or the Motorola 68HC11, write the interrupt service routine described in Example 7.1. Assume that when the interrupt occurs, the I/O serial port must be first checked to ensure that any previous use has been completed. If not, then wait until the flag is set saying it has been. Then set up to transfer data at the maximum rate. Use the expanded I/O serial port 0 for this interface.

7.7. *LCD_DIG table.* In this problem we will consider some variations on the LCD-_DIG table discussed in Example 7.2.

    (*a*) Complete the table of Fig. 7.18*d*. However, instead of expressing each byte in hexadecimal form, express it in binary form.

    (*b*) For this part, rewrite the LCD_DIG table (again in binary form) for use by the Intel 8096 which transmits bytes least-significant bit first.

    (*c*) For this part, rewrite the LCD_DIG table (in binary form) for use by the Motorola 68HC11 assuming that the display will be flipped over from that for Example 7.2. That is, assume that B1 will be the top backplane driver and B4 will be the bottom backplane driver. Also assume that FP1 will be the leftmost frontplane driver.

7.8. *Starburst alphanumeric encoding.* For the labeling of the segments of Fig. 7.12 and the order of bits shown in Fig. 7.19, show the encoding of the following characters both as a 16-bit binary number and as a four hexadecimal digit number.

    (*a*) The letter F.

    (*b*) The number 7. Use segment d as the bottom segment.

    (*c*) The number 7. Use segment r as the bottom segment.

7.9. *Multiplexed LCD display waveforms.* Figure 7.20 shows the backplane waveforms for the Motorola MC145000 chip. Figure 7.22 shows frontplane waveforms for the same chip which will turn on selected segments in a column driven by one of the frontplane drivers.

    (*a*) Draw a frontplane waveform, FP$_j$, which will turn segment 1 on, segment 2 on, segment 3 off, and segment 4 off (where segment 1 is driven by BP1, etc.).

    (*b*) Below this and aligned with it, show the BP1 backplane waveform.

    (*c*) Below this (and aligned with it), show the waveform across the segment driven by FP$_j$ and BP1.

    (*d*) Repeat parts *a, b,* and *c* for BP3 and the segment driven by FP$_j$ and BP3.

7.10. *LCD display temperature compensation.* Show the modification you would make to Fig. 7.24 to add another potentiometer for varying the temperature coefficient from about -25 to -35 mV/°C. What value potentiometer is needed in your circuit modification to cover this range, more or less?

    To analyze an op amp circuit, always note that if the circuit is operating properly, then the output of the op amp is doing whatever is necessary to force its differential input to 0 V. Furthermore, essentially zero current will flow into these op amp inputs. Therefore, an op amp circuit can be analyzed by drawing an equivalent circuit with the op amp removed and with 0 V between the nodes which drove its differential input. Finally, the zener diode acts like an open circuit during normal operation (i.e., as if it is not even in the circuit).

7.11. *CRT interface.* The circuit of Fig. 7.28 communicates with a microcontroller at the somewhat unusual baud rate of 4800 baud. The circuit of Fig. 7.29*a* modifies the behavior of the 8-bit data bus, SB0, . . . , SB7 at initialization time. The NS455 tries to read various set-up conditions, presumably from DIP switches which are enabled onto the data bus, at that time. The 4.7-kΩ pullup resistors shown in Fig. 7.28 provide default values of 1 in place of any DIP switch values.

    In this problem we will explore the possibility of replacing the circuit of Fig. 7.29*a* with a *pulldown* resistor tied between pin 7 (SB6) and ground. The NS455 chip specifies a maximum leakage current of 100 μA on pin 7 (when it is not acting like an output). It also specifies a maximum input low voltage of 0.8 V. The Hitachi 6116 CMOS RAM chip specifies a maximum leakage current of 10 μA. A Motorola MC74HC373 latch with a maximum input leakage current of 1 μA might be used. This means that the voltage across the pulldown resistor must not rise above 0.8 V when $100+10+1 = 111$ μA is flowing through it.

The pulldown resistor must have a low enough value to satisfy the total leakage current. On the other hand, it must not have a value so low that the chips which have to drive the line *high* cannot do so. The Hitachi 6116 RAM chip requires that an input high voltage be greater than 2.2 V. The National Semiconductor NS455 TMP chip requires that an input high voltage be greater than 2.0 V. The 6116 can supply 1.0 mA (minimum) and still hold the output voltage above 2.4 V. The NS455 chip can supply 125 μA (minimum) and still hold the output voltage above 2.4 V.

Sort through all of this data and determine whether the circuit of Fig. 7.29*a* can be replaced by a resistor. If so, what range of resistance values will work? If not, what combination of parameter values and circuit conditions makes up the weak link?

7.12. *CRT controller.* In Example 7.4, we wanted to use the CRT to display 40 double-width characters instead of 80 normal-width characters. This is what is needed to make a readable display on a little 5-in CRT for the front panel of an instrument. To do this, we first had to initialize the NS455 chip by sending it the string of characters shown in Fig. 7.30. Then each ASCII character to be displayed had to be sent to the NS455 chip twice.

In this problem we want to write an interrupt service routine for the Motorola 68HC11 or the Intel 8096 which will permit us to pass strings to the interrupt service routine *without* doubling the displayable characters and without setting their most-significant bit to 1. We will let the interrupt service routine do these things for us. With this interrupt service routine, the PER_ANN string of Example 7.4 can be written as

```
PER_ANN:   DB   1BH,4DH,20H,38H   ;Get to bottom left corner
           DB   "PERIOD "         ;Displayable message
           DB   04H               ;ASCII EOT to terminate
```

Write an interrupt service routine which will accept messages in this format, sort through the possibilities of Fig. 7.32, and send the next appropriate character.

7.13. *Pulse-width modulation.* Example 7.6 illustrates how a single output compare register of a programmable timer could be used to generate a pulse-width-modulated output. It required an interrupt for each change of the output. Accordingly, each change could not be followed immediately by another change. This made modulation values near 0 and 100 percent modulation difficult to achieve.

The bipolar power supply circuit of Fig. 7.38 achieves an output voltage of 0 V with a 50 percent pulse-width-modulated output from the microcontroller. If we want to limit the modulation to the range 25 to 75 percent, then

(*a*) Show the modified circuit, corresponding to Fig. 7.38, which will generate an output voltage between -25 V and $+25$ V.

(*b*) We can use a 2-byte signed variable, VOUT, to represent the output voltage as a signed number. For either the Intel 8096 or the Motorola 68HC11, show the code to convert VOUT to the value of MOD which will be used by the interrupt service routine. If VOUT = 0000H, MOD must give 50 percent modulation. What should its value be? What value of VOUT gives 75 percent modulation? What is the corresponding value of MOD? For values of VOUT greater than this, MOD should remain at the value just obtained. List a partial table of MOD versus VOUT values, hitting all the critical points so we can see what the conversion algorithm must do.

(c) For this bipolar power supply application to pulse-width modulation, what are the latency requirements for the interrupt which handles this output? That is, what leeway is made possible by limiting the modulation to the 25 to 75 percent range?

7.14. *D/A conversion*.

(a) Write the PACK subroutine of Example 7.8 for the Motorola 68HC11. Use RAM locations labeled with the same names as in Fig. 7.40. Minimize execution time.

(b) Compare the execution time for this subroutine with that listed in Fig. 7.40*b* for the Intel 8096.

7.15. *Stepper motors*. For the stepper motor of Fig. 7.41, determine its maximum start-without-error speed if it drives a frictional load of 0.4 oz-in. Show your work clearly.

7.16. *Protection circuits for inductive loads*. In Examples 7.11 and 7.12, we needed to compute the average power dissipated in a circuit element which is subjected to a periodically recurring, exponentially decaying waveform. In this problem you will compute the definite integrals involved.

(a) For a zener diode which passes a current waveform which exponentially decays from an initial value of I_{peak} out to two time constants and then repeats this sequence forever, determine the average power dissipated in a zener diode if its zener voltage is V_z. Show all work clearly.

(b) For a resistor which passes a current waveform which exponentially decays from an initial value of I_{peak} out to two time constants and then repeats this sequence forever, determine the average power dissipated in the resistor. Assume the voltage across the resistor is initially equal to V_{peak}. Show all work clearly.

7.17. *STEP subroutine*. For the stepper motor drive circuit of Fig. 7.45*a*, write a STEP subroutine for full stepping analogous to that of Fig. 7.48*b* (using a 2-byte signed number variable, NUMSTEPS, just as in Example 7.13). Do this for the Motorola 68HC11.

7.18. *Stepping with a zero reference*. Modify the STEP subroutine of the last problem to work with the slotted optical switch of Fig. 7.52. In addition to the variable NUMSTEPS, keep a 16-bit signed-number variable called POSITION. Before taking a step, determine if the sign of POSITION and the value of the output of the slotted optical switch are the same. (You can assume that STEP will be called in a tick clock interrupt service routine which has already read the output of the optoswitch and the other bits of an expanded input port into bit 7 of a variable called IO4.) If the sign of POSITION is *not* the same as the optoswitch output, then make POSITION = 0000 if the optoswitch output is 0 and make POSITION = FFFF (i.e., a value of -1) if the optoswitch output is 1. This simple scheme will result in POSITION being aligned with the optoswitch output as soon as the optoswitch output changes from 1 to 0 or from 0 to 1.

7.19. *Absolute position from an incremental encoder*. Using the Intel 8096, write the interrupt service routine to do the job of Example 7.14. For this purpose, one of the programmable timer inputs (e.g., HSI.0 of Fig. B.20 in Appendix B) can be used to obtain an interrupt on either edge of an input.

7.20. *Digital potentiometer.* Referring to Fig. 7.60, we want to write the interrupt service routine which is called when the digital potentiometer's channel A edge occurs. For either the Intel 8096 (using its EXTINT pin; refer to Fig. B.16 in Appendix B) or the Motorola 68HC11 (using its IRQ pin; refer to Fig. A.15 in Appendix A), write the interrupt service routine to read a bit of an input port (specify the bit you want to use) and either increment or decrement the 16-bit 2s-complement number POSITION based upon the state of this bit. Minimize and determine the execution time for this interrupt service routine. For the Intel 8096, in determining this time, assume that none of the other resources which share this same vector are being used.

7.21. *Algorithm to change the direction of an encoder output.* Refer to Fig. 7.62.
(a) Consider that we have an encoder which emits binary numbers. The binary output increases as the encoder shaft turns clockwise. How would you modify the binary output to derive an alternative binary number which increases as the shaft turns counterclockwise?
(b) Repeat part *a* for a Gray code encoder.

7.22. *Gray-to-binary conversion.* Rewrite the algorithm of Fig. 7.63 for the *9-bit* Gray code number which has already been read into the 2-byte RAM location labeled INP34 (with the seven most-significant bits set to 0). Minimize execution time (without using a 512-byte table).

7.23. *Gray-to-binary conversion.* Example 7.15 points out that the algorithm to convert from Gray code to binary code by shifting and exclusive-ORing *must not* use successive reads of the Gray code encoder. Rather it must read the encoder *once* and then work with a copy (so that no bit changes during the conversion). Using the two successive Gray code numbers 01000000 and 11000000 which convert to the binary numbers 01111111 and 10000000, show how a changing Gray code output which is used in the conversion process can produce a drastically erroneous result.
(a) Is it possible to produce a binary output of 00000000 (when the output should really be either 01111111 or 10000000)? If so, then show what would have to happen, step by step through the conversion process.
(b) Is it possible to produce a binary output of 11111111? If so, again show how.
(c) Is it possible to produce *any* binary number output between 00000000 and 11111111 (when the binary output should really be either 01111111 or 10000000) by changing the Gray code number at just the wrong time during the conversion process? If so, then explain.

7.24. *Gray-to-binary conversion.* Given a *9*-bit Gray code encoder, show how *you* would like the bits aligned after reading a copy into memory. Then write the conversion routine for going to binary code using the Intel 8096.

7.25. *V-scan encoder conversion.* Rewrite the V_SCAN subroutine of Fig. 7.65 for the Motorola 68HC11. Be sure to state where the subroutine expects the input and output to be.

7.26. *U-scan encoding.* A variation on the V-scan approach to reading the binary output of an encoder is called U-scan. It again uses the least-significant bit to determine a read line. It also uses leading brushes and lagging brushes. However, the leading brushes are all offset from the read line by an equal amount. Likewise, the lagging brushes are all offset on the other side of the read line by this same amount. Then depending upon the state of the least-significant bit, all of the leading brushes or all of the lagging brushes are read.

(a) How big should the offset from the read line be for the leading and the lagging brushes?

(b) What is the tolerance on the brush positions?

(c) What is the advantage of U-scan over V-scan?

(d) What is the disadvantage of U-scan over V-scan?

7.27. *Equivalent circuits.* To analyze what is happening in a circuit which connects a driver to a load, it is helpful to replace the driver by two Thevenin equivalent circuits, one when the output is driven high and one when it is driven low. Consider the level shifter circuit of Fig. 7.67.

(a) Knowing that the output drops (a maximum of) 1.5 V, from 15 down to 13.5 V, when the output is supplying 3 mA of load current, show the high output Thevenin equivalent circuit. It consists of a voltage source and a resistor in series. Your job is to determine the values of the voltage source and the resistor.

(b) Show the low output Thevenin equivalent circuit.

(c) The MC14504B level shifter, with $V_{DD} = 10$ V, lists typical unloaded output rise and fall times of 50 ns. If it is used to drive an IRFZ40 power MOSFET, then the load is mainly capacitive and is specified as 3000 pF, maximum. Determine the time constant of the circuit, with the power MOSFET driven by the level shifter. Determine the rise and fall times of the power MOSFET input. That is, determine the time for the output to change from 10 to 90 percent of the final output value. Do this by first assuming that the MC14504B has zero rise and fall times, calculate what the RC circuit will do in this case, and then add in the 50 ns due to the MC14504B. This is not exact, but it is a simple way to get a good idea of what is going on.

Rise and fall times on power MOSFET inputs can be important because it is during this transient period, as power is turning on and off, that the power MOSFET is operated in its active region. With both voltage and current having large values during this time, the momentary power dissipation goes way up. If the switching rate is high, this momentary power dissipation can contribute significantly to average power dissipation. The rise and fall times affect power dissipation because the power dissipated during the switching of the power MOSFET is the integral of the instantaneous power. If we could halve the switching time, we would halve the power dissipation.

7.28. *Zener diode regulator circuit.* Consider the circuit of Fig. 7.68. Since the load on the level shifter has such a high impedance, the average supply current is not much higher than its quiescent value of 15 μA.

(a) If V_{load} in Fig. 7.68 is 24 V, then determine the value of R_x which will provide regulation to the level shifter chip for a supply current of up to 100 μA.

(b) Determine the power dissipation in R_x.

(c) Determine the power dissipation in the zener diode if the level shifter actually draws essentially 0 μA. Does this worst-case power dissipation meet the zener diode's specification of 250 mW, maximum?

(d) Determine the power dissipation in the zener diode if the level shifter actually draws 100 μA.

7.29. *Power MOSFETs.* Rework Example 7.18 assuming that each MOSFET is force-cooled with a heat sink and a fan to hold its case temperature below 25°C.

7.30. *Power MOSFETs.* What duty cycle is necessary for the IRFZ40 power MOSFET of Fig. 7.69 to dissipate 125 W with a periodic current of either 0 or 160 A? Draw the current waveform and mark the percentage of each cycle during which the output can be turned on to 160 A.

To use the power MOSFET in this way, we must keep the case cooled to 25°C. We must also keep the frequency sufficiently high so that it need deal with only the average value. If we were to explore the power-versus-frequency issue further, we would need to use the derating curves for the power MOSFET available in its data sheet.

7.31. *Serial clock rate.* Consider the Sprague UCN-4821A of Fig. 7.74. This part can be used in place of the 74HC595 if we want to expand microcontroller output ports and also want more drive capability than the 74HC595 can provide. However, its timing may require some care.

(*a*) If the UCN-4821A is driven from the I/O serial port expansion circuit for the Motorola 68HC11, then can the serial port be run at its maximum clock rate? If not, then which option of the 68HC11 can be used and how long will this 1-byte transfer actually take?

(*b*) Repeat part *a* for the Intel 8096.

7.32. *Power dissipation.* The Sprague UCN-4821A of Fig. 7.74 requires some calculations to translate its power dissipation specifications into load handling capability. This is partly because the chip contains eight drivers, and one factor to be considered is the heating of the whole chip.

(*a*) Using the derating factor of 16.7 mW/°C above 25°C, determine the maximum total power dissipation for the chip if the temperature is permitted to rise to 100°C?

(*b*) Using the data of Fig. 7.74*c*, draw a curve of maximum logic 0 output voltage versus load current.

(*c*) If all eight drivers drive equal current loads, then how much current can each driver handle if the case is permitted to rise to 100°C?

(*d*) Repeat part *c* if the case temperature is maintained below 25°C?

(*e*) What Fahrenheit temperature corresponds to 25°C?

7.33. *DC motor drive.* The circuit of Fig. 7.76 illustrates the simplicity of an interface made possible when an IC manufacturer identifies a broad application and decides to support it with a chip. Assume that we have a signed number (2s complement code) called VMOTOR. We want to derive the two outputs shown in Fig. 7.76 so as to drive the dc motor with a voltage having a dc component proportional to VMOTOR.

(*a*) Do this for the Intel 8096 using its PWM output. Assume that VMOTOR is a 1-byte number.

(*b*) Do this for the Motorola 68HC11 using OC1 and OC5 to obtain a PWM output. Assume that VMOTOR is a 2-byte number.

7.34. *CMOS drive capability.* The circuit of Fig. 7.77 uses the 74HC595 as a driver for an LED which requires a minimum of 15 mA to ensure turning on its associated triac. The specifications for a CMOS chip such as the 74HC595 provide data on its drive capability to achieve logic levels on its output. We can use this data to determine its drive capability for a circuit like Fig. 7.77. The specifications for the 74HC595 list a maximum dc output current per pin of 35 mA.

(*a*) With a supply voltage of 4.5 V, a low output can drive 6.0 mA while rising no higher than to 0.26 V (at 25°C). With a supply voltage of 4.5 V, a high output can drive 6.0 mA while dropping no lower than 3.98 V (at 25°C). Use this data to determine the output resistance of the 74HC595 when it is driving the output low and when it is driving the output high.

(b) With a supply voltage of 6.0 V, a low output can drive 7.8 mA while rising no higher than 0.26 V (at 25°C). With a supply voltage of 6.0 V, a high output can drive 7.8 mA while dropping no lower than 5.48 V (at 25°C). Use this data to determine the output resistance of the 74HC595 when it is driving the output low and when it is driving the output high.

(c) Interpolate between the results obtained in parts a and b to show the Thevenin equivalent circuits for the output of a 74HC595 with a 5-V supply voltage. Show a circuit when the output is driven low and another circuit when the output is driven high.

(d) The circuit of Fig. 7.77 shows the 74HC595 driving an LED in series with a pullup resistor to +5 V. Could the 74HC595 have driven the other side of the LED in series with a pulldown resistor to ground? With a worst-case drop across the LED of 1.5 V (at 15 mA), determine the highest value of a resistor which will still supply 15 mA to the LED in each case. Assume that the power supply must be maintained within ±10 percent and determine these resistance values under worst-case conditions. Show your circuit in each case. Discuss any merits of one circuit over the other.

REFERENCE

Manufacturers' data sheets and application notes provide a wealth of up-to-date information on available devices and ways to use them. Just as *IC Master* serves as an excellent guide to integrated circuits, so *EEM*, the *Electronic Engineers Master* catalog, does for all kinds of parts including motors, encoders, switches, and displays. It is published by Hearst Business Communications, Inc., UTP Division, 645 Stewart Ave., Garden City, NY 11530. Phone: (516)222-2500. It is free to subscribers of *Electronic Products* magazine which, in turn, is free to qualified electronic designers. A designer becomes qualified by being a person who is responsible for specifying the purchase of the kinds of parts which are advertised in the magazine and by filling out the application postcard included in every issue of the magazine. Design engineers with a company affiliation, whether large or small, are regularly qualified. So are independent consultants.

CHAPTER
8

DEVELOPMENT TOOLS

8.1 OVERVIEW

Developing application software for a microcontroller requires development tools. While the chip itself is small and relatively inexpensive, the tools needed to make it useful are not. This situation is not unlike that faced by the integrated circuit manufacturer, where expensive automated IC-manufacturing facilities are needed for the production of inexpensive "jelly bean" parts. In this chapter we explore some of the alternatives available to the developer of microcontroller application software.

We begin by looking at the process of developing the hardware needed for the application. By looking at the considerations which arise, we can see the value of developing special, simple test programs to exercise the hardware, independent of the software which will eventually be needed for interacting with this same hardware.

Next we consider a "top-down design" approach to overall software development together with a "bottom-up design" approach to the development of device drivers. We can hide the complexities of interacting with specific devices in driver software so that the top-down view of a device will see a clean software interface.

424

As we develop application code, we need to verify its performance. We first consider the role which can be taken by a simulator. This tool permits us to exercise our software in an environment where both the microcontroller's operation and its I/O interactions are simulated on another computer. In spite of the artificiality of this approach, it offers the benefit of excellent *control* over the I/O environment. We can often ask questions more easily of a simulator than of the actual target system because we usually have better control over I/O interactions when we can contrive them in a simulator. On the other hand, because a simulator depends upon our representation of I/O interactions, it can only give answers to questions which we think to ask.

Next we look at a low-cost approach to emulating the behavior of a microcontroller. Our desire is to get into the microcontroller chip while it is plugged into the target system and have it execute application software while we monitor its performance. Getting into the microcontroller can be approached by building up a separate emulator circuit which then plugs into the microcontroller socket in the target system. The emulator circuit operates the microcontroller in the expanded mode so as to gain access to the internal address bus and data bus. The circuit employs extra RAM to hold the application software during development and a monitor program to permit us to monitor and control the microcontroller environment. The circuit must also include rebuilt ports to replace those lost when the microcontroller chip was put into the expanded mode. This approach allows us to run application software in the target system environment. However, it gives us tools of only limited value for monitoring the performance in real time.

Having seen all of the pieces needed for the development of a microcontroller-based product, we next consider the power of an integrated approach to the entire development cycle, using a full-featured development system. Such a system can expedite the process of making a software change and then verifying the performance of the changed software on the target system. Since this is done again and again in the development of a product, the tight coupling among the parts of a development system can ease the designer's job dramatically. Furthermore, the inclusion of a logic analyzer to monitor the performance of the system's emulator gives us an outstanding tool for debugging interrupt-driven real-time software.

In contrast to the power of the emulator and logic analyzer in a full-featured development system, we look next at a significantly lower-cost unit which is designed to work in conjunction with a personal computer. Many of the features available in the full-featured development system environment can be obtained in other ways.

We close this chapter by considering the programming of application software into a microcontroller part. During development it is valuable to have a quick, easy, and relatively low cost way to do this. Manufacturers of microcontrollers are essentially building this programming capability right *into* the microcontroller. With hardly any extra hardware, we can program these chips easily.

8.2   HARDWARE DEVELOPMENT STEPS

The development of a product which uses a microcontroller chip to organize and control its activities involves several phases. The hardware needed for input and output must be selected. A microcontroller chip must be chosen which can handle the I/O and processing requirements. Perhaps the microcontroller must be augmented with extra RAM chips or peripheral controller chips or I/O expander chips in order to meet the requirements of the job. In general, an instrument design requires circuitry to undertake jobs which the microcontroller cannot handle directly, either because the required circuitry runs too fast (e.g., the data gathering and the triggering circuitry of a logic state analyzer) or because it carries out extensive analog data handling (e.g., the switching and characterizing of low-level telephone signals in a phone company tester for its channel-multiplexing equipment). Other simple devices may consist of little more than the microcontroller chip plus output driver chips plus the I/O devices themselves.

A flowchart of some of the considerations which arise in the development of this *target-system** hardware is shown in Fig. 8.1. During the early stages of a development project, a hardware prototype must be built. This might take the form of a wirewrapped board (or boards) which supports the required hardware interactions with I/O devices. It becomes the test bed for the application software which must be developed.

The next stage of product development requires the writing of *little* hardware test programs to verify that the hardware interactions work. If we are using a multiplexed seven-segment LED display, then the test program might write a number out to the display repeatedly so it will be apparent whether it works. If it does not, then the repeated writes give us the conditions needed to debug the hardware using an oscilloscope to look for faulty signal levels or a logic state analyzer to look for faulty sequencing of signals, or missing signals.

Much development time will be spent generating application software. However, the above step of testing the hardware with *special*, simple test programs is not to be bypassed. The alternative of trying to debug unproven hardware with application software represents a long, tedious process using the wrong tools.

8.3   APPLICATION SOFTWARE DEVELOPMENT

Application software development can begin as soon as the system specifications have been completed, concurrent with the development of a hardware prototype. As development progresses, we may go back and relax some of the specifications which we find are not cost-effective to implement. We may add

---

* That is, the system to be designed, as compared with the development system used to support the design process.

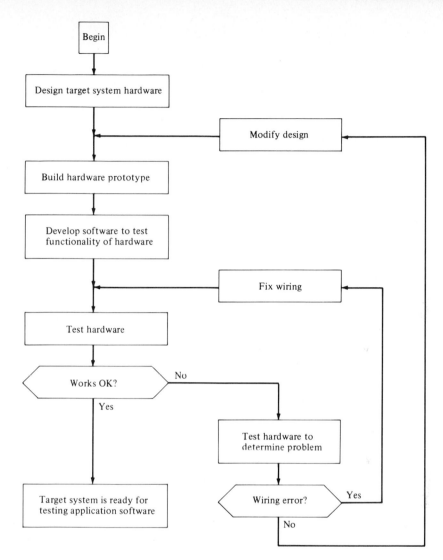

FIGURE 8.1
Flowchart for target system hardware development.

features which had not occurred to us at the start of the project or which turn out to be cost-effective enhancements (e.g., self-test capability).

As software routines are written, they can be tested to some extent even before the target-system hardware is ready. Algorithms which must run on the target microcontroller, but which do not involve hardware interactions, can certainly be tested early. An example of such an algorithm would be the key-switch-parsing algorithm discussed in Sec. 5.6. A flowchart for developing such algorithms using *top-down design* is shown in Fig. 8.2.

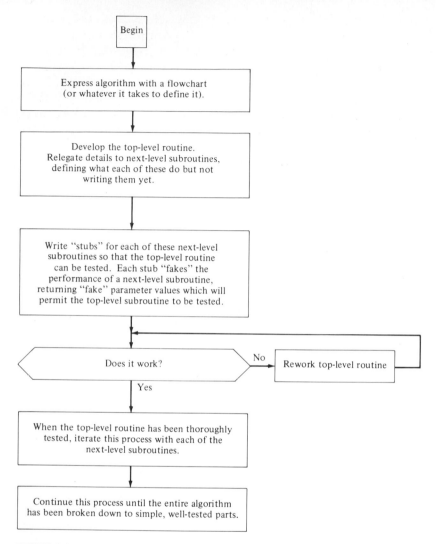

FIGURE 8.2
Flowchart for top-down software development for algorithms which do not require hardware interactions.

As portions of the target-system hardware become ready, the software which interacts with each portion can be developed and tested. A *bottom-up* approach to the development of device drivers will permit the complexities of each kind of hardware interaction to be buried in a clean software interface. For example, in Sec. 5.4 we considered how the interactions with a stepper motor could be implemented with a STEP subroutine. This STEP subroutine is executed every time a TICK_CLOCK interrupt service routine is invoked, perhaps every 10 ms. It looks at a signed-number variable called NUMSTEPS and takes a step if NUMSTEPS does not equal 0. Then it decrements (or increments) NUMSTEPS toward 0. Given this stepper-motor device-driver mechanism, our subsequent interactions with the stepper motor will consist of adding any changes in position into NUMSTEPS.

Another bottom-up design of a device driver was discussed in Sec. 5.3 in conjunction with driving a CRT display through a UART. We had the UART's interrupt service routine pick up the next character to be sent to the CRT from an ASCII string pointed to by a pointer which it maintained. When it got to the end of a string, it would look to see if another pointer had been passed to it in a queue from the mainline program. If so, then it would send the first character of that string. If not, then it would disable transmit register empty interrupts. Given this CRT device-driver mechanism, a write to the CRT consists of putting into a queue a pointer to an ASCII string. Each time this is done, we must also make sure that transmit register empty interrupts are enabled. Our CRT device-driver interface might consist of a SEND_TO_CRT subroutine which takes a pointer passed to it and does these two operations. A flowchart for the kinds of considerations which arise in this process of developing device drivers is illustrated in Fig. 8.3.

At this point, we need to check worst-case timing considerations for I/O devices, to make sure that each device will get serviced, even when other devices are requesting service at the same time. Now that each interrupt service routine has been written, we need to determine:

How long it takes to execute.

How long a latency period it can handle, i.e., the period of time from when it requests service until it receives service.

The maximum rate at which it will request service.

How long it will leave interrupts disabled for higher priority devices.

Given these times, we are armed with the data needed to resolve interactions *between* interrupts and to determine whether a worst-case sequence of interrupts will result in an interrupting device not getting serviced within its acceptable time. The techniques of Sec. 3.6 can be used to resolve these issues. If a worst-case problem exists, we are armed with the information we need to go back and either augment the hardware or speed up some aspect of the software.

Once I/O devices have been insulated from the rest of the application software with carefully written device drivers, the design can proceed in the top-down fashion of Fig. 8.2 to meet the overall system specifications. For an instrument driven by keyswitch input sequences, the keyswitch parser determines what action to take at any moment. Consequently, the parsing routine serves as the key element in the mainline program for tying together the variety of tasks which can be undertaken by the instrument. As each task is completed, its performance in conjunction with the prototype target-system hardware can be checked. In this way, the application software can be built up and tested, task by task, until every task has been completed.

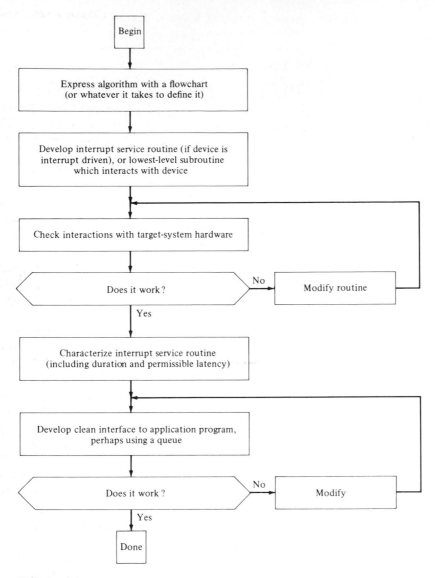

FIGURE 8.3
Flowchart for bottom-up software development of device-driver algorithms.

On occasion it seems that in spite of all our careful efforts, the systems we design still go awry. A condition is probably arising which we have failed to take into account. Or some part of our hardware or software may not be behaving in exactly the way we had understood it to behave. At such times we need debugging tools which can help us to *isolate* the fault and then determine its cause. It is at such times that powerful logic-analysis tools can probably help us whereas simple debugging tools may fail us.

8.4 SIMULATORS

The development of application software is a phase of the design process which is supported by a wide variety of tools. For any but the simplest of applications, it is the phase which is the most uncertain in terms of predicting a time to completion. In fact, this time to completion can be drastically affected by the tools at hand.

In this section we consider a debugging tool which has been made feasible by the proliferation of IBM PC computers in offices and laboratories everywhere. Given an IBM PC, or a PC look-alike, the opportunity exists to put together a relatively low-cost development system with software and hardware designed to work with the PC.

We need some way to debug our application software. We must be able to view its execution and see its response to external and internal events. In support of this, *simulators* are programs which run on another machine (e.g., the IBM PC) and simulate the operation of the microcontroller. An *emulator*, on the other hand, permits us to connect our development environment to the target system hardware and to run it. It emulates the behavior of the microcontroller in the target system while permitting us to load programs and to monitor and control the innards of the (emulated) microcontroller chip. An emulator permits us to carry out debugging in the actual target-system environment whereas a simulator must deal with a reconstructed version of reality.

Cybernetic Micro Systems supports the development of application software for a variety of microcontrollers, including the Intel 8096, with assemblers and simulators which run on the IBM PC computer. Using a program editor, we write the application software in assembly language, which in turn is assembled to the object code which is eventually executed by the microcontroller in the target system. However, before we get to that stage, the simulator can be used to verify the performance of the code.

The Cybernetic Micro Systems simulator, Sim8096, turns the IBM PC display into a multiple-windowed view into the operation of the 8096 as it executes application software. As shown in Fig. 8.4, the display consists of seven windows:

> *Code window*. Source code, with labels and symbolic arguments, is displayed as it is executed.
>
> *Register window*. The major registers and flags have their current states displayed. In addition to the program counter and status bits, this includes the ports, the counters in the programmable timer, and pointers used to establish what is to be displayed in the memory window.
>
> *Memory window*. This is used to monitor, and possibly modify, a group of registers, RAM variables, or port bits during the execution of code. For example, if we are debugging code which responds to

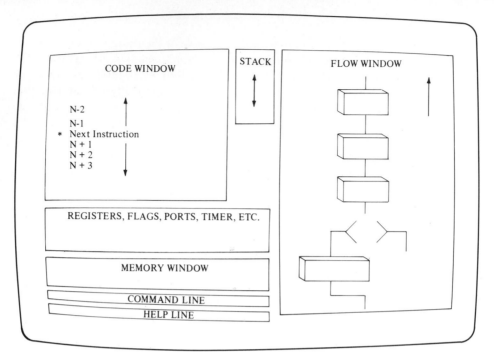

FIGURE 8.4
Sim8096 user interface. (*Cybernetic Micro Systems, Inc.*)

an external interrupt (e.g., a rising edge on bit 2 of port 2), then we can set the memory window to display this one bit as follows:

```
Port 2.2 = 1 Ext. Interrupt
```

While our program is single stepped or is executed at a rate which we can follow on the screen of the IBM PC, we can hit the 0 and 1 keys on the keyboard and change this bit, causing an interrupt. In this way, we can monitor the interactions of our program with *any* event which involves a single bit.

In like manner we can use the memory window to monitor and change a byte at a time (e.g., the 8 bits of a port). This mechanism is Sim8096's simplest method for checking the response of a program to I/O interactions.

Stack window. This shows the content of the stack during program execution.

Flow window. The control flow is shown as a program executes. Only labels and branches are shown, but they are automatically formatted into a flowchart, using words which have been embedded into the source file (in the comment field associated with a label or a branch instruction). The result is a superb, intuitive view of the "big picture" as a program executes.

Command window. As a simulator command is entered from the keyboard, it appears in this one-line window.

Help window. This one-line window gives brief syntax help for the present command. It also can be used to scroll through all of the simulator commands, serving as a miniature manual.

For executing an internal algorithm (e.g., a Gray-code-to-binary-code conversion algorithm), single stepping is an appropriate debugging technique. For longer internal algorithms, we can set breakpoints and have the simulator execute instructions quickly until a breakpoint is reached. At that point we can check if the desired states of various variables have been reached. If so, then we can go on. If not, we can rerun the algorithm with a breakpoint which occurs earlier. In this way, we can use a divide and conquer strategy to home in on any problem areas which exist in the execution of an algorithm.

Virtually all simulation and emulation tools permit the use of breakpoints to troubleshoot internal algorithms. A more telling test of a simulator or an emulator is the ease with which we can use it to debug code which involves complex I/O interactions. Sim8096 is quite powerful in this regard. Debugging code which involves complicated I/O interactions will only find errors which arise in response to whatever I/O interactions we choose to simulate or emulate. The Sim8096 simulator includes several mechanisms for simulating I/O interactions. We have already discussed how a selected bit or byte can be changed as the code is running. Going a big step further, we can define up to 10 macros of Sim8096 commands. These are executed by pressing a keyboard function key. For example, the macro for function key 1 might be defined with the following keyswitch sequence:

```
!1  p1=7f,p0=f7,t5,; P0 and P1 setup ‹CR› ‹CR›
```

Now, any time function key 1 is pressed, the input to port 1 will be set to 7f and the input to port 0 will be set to f7. Then five instructions will be executed and traced. The comment "P0 and P1 setup" will appear in the command window.

Sim8096 even has the commands needed to write *interactive* macros. For example, consider the macro defined by:

```
!2  w2.7,p3=45,/p1.7,/w2.7,p1.7 ‹CR› ‹CR›
```

When function key 2 is subsequently pressed, the macro will wait until the simulator executes a program instruction which sets the output on bit 7 of port 2. Then the macro command will load input port 3 with 45 (hex) and clear input bit 7 of port 1. Then the macro command will again wait until a program instruction clears the output on bit 7 of port 2. When that occurs, the macro command will set input bit 7 of port 1.

A normal reaction to the use of a simulator versus an emulator is the feeling that the former operates in a make-believe world whereas the latter deals with the reality of the actual target system. However, there is another side to the issue. When using an emulator, it is sometimes difficult to get the target system to produce the set of conditions needed as a stimulus for the microcontroller in order to debug a section of the application program. For example, if we want to test the effect of three unrelated interrupt sources occurring at the same time, we may have difficulty in getting the target-system hardware to produce this condition. For a simulator, this is a trivial problem. Any simulator will provide us with the ability to inject inputs at any point in the execution of an application program.

The Sim8096's register window includes a CC register. This does not correspond to an actual 8096 register. Rather, it displays a cycle count as instructions are executed. Each count corresponds to 0.25 μs, or one internal clock period, for an 8096 run with a 12-MHz crystal. This CC register is exactly what is needed to check the times which are important for determining whether interrupt servicing will be reliable (i.e., will meet the timing requirements of Sec. 3.6).

The Sim8096 simulator permits a further use of its macro command capability. Often for debugging we want to have input sequences which are synchronized to the program. For example, we may want to have a voltage input to an A/D converter which causes the A/D input to increase steadily until it reaches a specified value. Then it should decrease at a different rate until it reaches another value. To do this, the Sim8096 simulator permits its macro commands to be defined and invoked in the *source code*. For example, early in the program we can *define* macro commands 3 and 4 with the following *comments* (i.e., comments as far as the assembler is concerned, but commands when they get passed to the simulator):

```
;!3 A3+47,; Analog voltage rising
;!4 A3-25,; Analog voltage falling
```

Then at the point where we want the voltage on the channel 3 input to increase, we insert the "comment"

```
;#3
```

which adds 47 to whatever the value of the input on channel 3 has been (each time that the instruction which has this as its comment is executed). It will also display

```
Analog voltage rising
```

in the command window until overwritten by the execution of another embedded macro command or until we write a command from the keyboard.

To use this, we can introduce a test into the main loop of the application program which clears a flag called AnStat (analog state) when the analog input on channel 3 exceeds a specified upper threshold and sets AnStat when the analog input on channel 3 drops below a specified lower threshold. We then also test AnStat and if it is set, we execute

```
NOP                ;#3
```

to increase the analog voltage by 47. If AnStat is set, then we execute

```
NOP                ;#4
```

to decrease the analog voltage by 25. Figure 8.5 illustrates the display for a typical example.

One final capability of the Sim8096 simulator should be pointed out. If we embed the following macro command definition into our source code

```
;!5  pl=?,; Port 1 loaded
```

then when the subsequent instruction having a "comment" of

```
;#5
```

is executed, the embedded macro command number 5 will be invoked. It will stop execution and prompt us with

```
P1=?
```

After we enter a value and continue execution, the command window will display

```
PORT 1 LOADED
```

These examples serve to illustrate how a powerful simulator can be used to exercise and debug application software. We have seen that by writing embedded macro commands into the source code, we can make virtually any sequence of I/O interactions occur. Given that, we can then check and debug the performance of our application software in this environment.

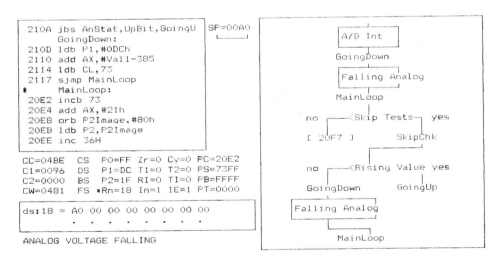

FIGURE 8.5
Sim8096 display showing the effect of an embedded macro command.

8.5 LOW-COST EMULATION

As we saw in Sec. 6.3, both the Intel 8096 and the Motorola 68HC11 can be operated in an expanded mode, giving us access to the address-bus–data-bus structure. This is the opportunity needed to build a low-cost emulator which will permit us to do the following things:

Provide RAM memory for the microcontroller, for use as program memory and as data memory, and for holding variables needed by a monitor program (which implements the low-cost emulator's commands).

Reconstruct ports (i.e., those ports which were given up by the microcontroller when it was configured to bring out its internal address-bus–data-bus structure).

Provide communications between the microcontroller's monitor program and a personal computer serving as a low-cost development system. For a stand-alone low-cost emulator, this might be implemented with a serial port (but not the microcontroller's serial port since we want to leave that port available for target system use). For a board plugged into a slot in an IBM PC computer, the communications can take place through the board's edge connector.

Provide the ability to download object code from the personal computer to the low-cost emulator's RAM for execution by the microcontroller.

Permit the display and changing of RAM variables and CPU registers from the personal computer.

Permit single stepping through application software, displaying the effect of each instruction on CPU registers.

Permit the execution of application software at full speed* from an arbitrary point in a program until a breakpoint is reached. The monitor program supports the setting, displaying, and clearing of breakpoint addresses.

The stand-alone low-cost emulator shown in Fig. 8.6 is Motorola's M68HC11EVM (evaluation module). It requires connections to +5-, +12-, and −12-V power as well as an RS-232 serial interface to either a terminal or a personal computer. Code can be downloaded over this same port or, alternatively, via a separate RS-232 serial port. The second serial port might be used, for example, if the low-cost emulator were controlled by a terminal. The second port can then be connected to a computer. For downloading code from the computer, the monitor program can be put in a mode of operation whereby the keys pressed on the terminal are echoed to the computer and where the characters sent back from the computer are echoed to the terminal display. After

---

* For the Motorola 68HC11. For the Intel 8096, external accesses require several more cycles than internal accesses.

FIGURE 8.6
Motorola M68HC11EVM low-cost emulator. (*Motorola Inc.*)

initiating the transfer of object code from the computer, the microcontroller's monitor program can intercept the downloaded code and store it into RAM for subsequent execution.

At the lower left-hand corner of the board in Fig. 8.6 are two 34-pin Berg connectors (i.e., two 2 × 17 arrays of pins spaced on 100-mil centers). They connect to the target system with a ribbon cable that plugs into either a 48-pin DIP socket or a 52-contact plastic-leaded chip-carrier socket for a Motorola 68HC11. For initial development work with a wirewrapped target-system board, the connection can be made with ribbon cables that plug into Berg connectors on the target-system board.

The Motorola board of Fig. 8.6 includes a second pair of 34-pin Berg connectors to be used if the target system uses the 68HC11 in the expanded mode. It also has sockets for both the 48-pin and the 52-pin versions of the 68HC11, for programming the internal EEPROM.

The big 64-pin DIP shown in Fig. 8.6 is a specially packaged 68HC11. The extra pins permit a few extra lines to be brought out from the chip to help with the implementation of the low-cost emulator functions.

In contrast with the complexity of the board in Fig. 8.6 (which uses 53 integrated circuits!), a low-cost emulator *can* be a rather simple circuit. For

FIGURE 8.7
Motorola M68HC11EVB low-cost emulator. (*Motorola Inc.*)

example, consider Motorola's M68HC11EVB evaluation board shown in Fig. 8.7. It includes a monitor program called BUFFALO (Bit Users Fast Friendly Aid to Logical Operation). The BUFFALO monitor can actually be used with several alternative circuits. The printed circuit board shown in Fig. 8.7 employs a 68B50 UART chip. Alternatively, the circuit shown in Fig. 8.8 employs a 68681 DUART (dual UART) and consists of just seven integrated circuits. Either one provides a simple and low-cost way to develop code for the 68HC11. However, the simplicity of the circuit requires a user to make some accommodations when writing code to use it. This contrasts with the board of Fig. 8.6, where extra circuitry has been added so that the *exact* code written for the ultimate, single-chip 68HC11 implementation is what is also used during debugging. In terms of the circuit of Fig. 8.8, the following accommodations must be made:

1. Since BUFFALO resides in addresses E000 to FFFF (where our application program will ultimately go), the program must be assembled into addresses occupied by the 8K × 8 RAM, extending over addresses 8000 to 9FFF. This is done by inserting an

        ORG   8000H

   just before the beginning of the program code. Eventually, when the application program has been developed and debugged, it can be

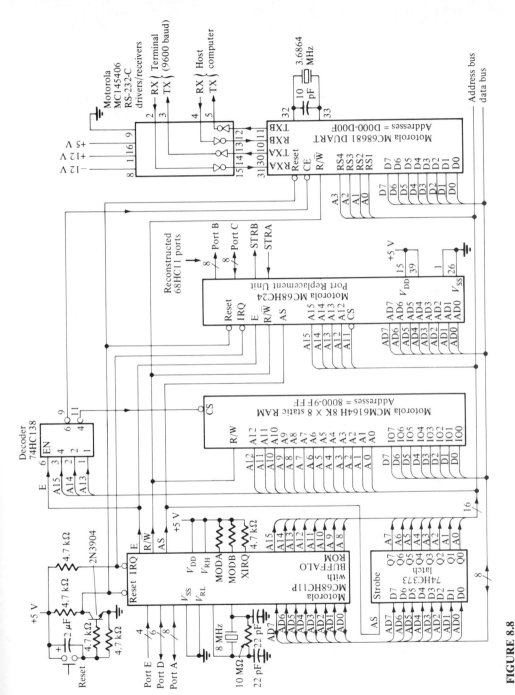

FIGURE 8.8
Motorola 68HC11 low-cost BUFFALO emulator circuit.

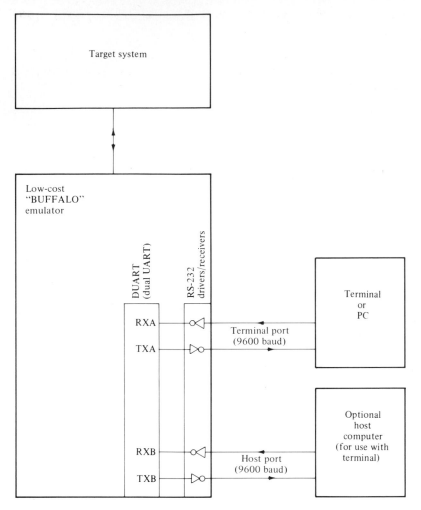

FIGURE 8.9
Connections between target system, low-cost emulator, and terminal and/or computer.

reassembled into addresses E000 to FFFF (for mask-programming into the ROM of a 68HC11) by changing the above ORG pseudoinstruction to

```
ORG   0E000H
```

2. Whereas the 68HC11 has RAM extending from addresses 0000 to 00FF, BUFFALO uses the RAM in addresses 0040 to 00FF. Consequently, the application program can only use 0000 to 003F. To get the remaining bytes of RAM, addresses 9F40 to 9FFF of the 8K × 8 RAM might be used during development. After the application program has been debugged, it can be reassembled with

RAM extending over addresses 0000 to 00FF for the mask-programmed 68HC11.

3. The interrupt vectors which actually extend over addresses FFD6 to FFFF must be replaced with a jump table extending over addresses 00C4 to 00FF. The jump table takes the form:

```
ORG   00C4H
JMP   UART_INTERRUPT_SERVICE_ROUTINE
JMP   IO_SERIAL_PORT_INT_SERV_ROUTINE
JMP   PULSE_ACCUM_INPUT_EDGE_INT_SERV_ROUTINE
JMP   PULSE_ACCUM_OVERFLOW_INT_SERV_ROUTINE
JMP   TIMER_OVERFLOW_INT_SERV_ROUTINE
JMP   TIMER_OUTPUT_COMPARE_5_INT_SERV_ROUTINE
        .
        .
        .
JMP   ILLEGAL_OPCODE_INT_SERV_ROUTINE
JMP   COP_INT_SERV_ROUTINE
JMP   CLOCK_MONITOR_INT_SERV_ROUTINE
```

The BUFFALO monitor vectors each interrupt through the corresponding entry in this table. After the application program has been debugged, this table can be removed and the actual vectors assembled into addresses FFD6 to FFFF with

```
ORG   0FFD6H
FDB   UART_INTERRUPT_SERVICE_ROUTINE
        .
        .
        .
FDB   CLOCK_MONITOR_INT_SERV_ROUTINE
FDB   MAINLINE ;Reset vector
```

If we are willing to put up with these few complications, we obtain an extremely low-cost emulator for the 68HC11.

The BUFFALO monitor program supports the test circuit of Fig. 8.9. The low-cost emulator board can be interposed between a terminal and a host computer. This is done by disconnecting the RS-232 cable which goes to a host computer from the back of the terminal and connecting it to the "host" port on the BUFFALO board. A second RS-232 cable is then connected between the "terminal" port on the BUFFALO board and the back of the terminal. The BUFFALO monitor program now permits two modes of operation:

The normal BUFFALO *monitor mode*, in which the user at the terminal controls the 68HC11 interactions with the target system using BUFFALO commands.

The *transparent mode*, in which the user at the terminal appears to be connected directly to the host computer. This is useful for editing and assembling application software and then downloading it to the BUFFALO board.

To enter transparent mode (where terminal appears to be connected directly to host computer), type the BUFFALO command:

```
HOST
```

(*a*)

To return to the BUFFALO monitor, press the reset pushbutton, or type

```
<control> A
```

(*b*)

To download code from host, type the BUFFALO command

```
LOAD <host download command>
```

where <host download command> is the command you must send to the host computer to instruct it to send a file to the terminal. For a UNIX machine, this might be

```
LOAD cat myfile
```

where "myfile" must be an object code file, expressed in standard Motorola S record format. Assemblers for Motorola microcontrollers can generate object code in this format.

(*c*)

FIGURE 8.10
BUFFALO commands. (*a*) Entering the transparent mode; (*b*) returning to the BUFFALO command mode; (*c*) downloading code from a host computer.

To illustrate what is involved, Fig. 8.10 shows how each mode is entered and how object code is downloaded to the BUFFALO board.

The monitor programs for both the EVM board of Fig. 8.6 and the EVB board of Fig. 8.7 actually support downloading from *either* port. If we are communicating with the board from a personal computer rather than from a terminal, then it is useful to be able to have the downloaded file also come in over the same serial line as is used for interacting with the monitor program. Communications software, such as Crosstalk,* supports such operations. It permits the personal computer to be in the terminal mode while talking to the monitor program (to send the LOAD command). Then we can escape so that our keyboard input tells Crosstalk to download a file. When the download is complete, we continue in the terminal mode with commands to the monitor.

We can get by with a simple power supply since the circuit of Fig. 8.8 includes very few chips and since the CMOS parts dissipate very little power.† The wall plug-in unit shown in Fig. 8.11 provides +5 V at 750 mA as well as ±12 V at 150 mA.

---

* A software product for the IBM PC produced by Microstuf.
† For example, the 68HC11 draws less than 20 mA of 5-V power.

FIGURE 8.11
Wall plug-in power supply, Model MPPS 512. (*D & B
Power, Inc.*)

Both of the low-cost emulator boards of Figs. 8.6 and 8.7 leave the connections to the target system up to the user. They both include Berg strips for
making these connections. However, the pins on neither board are organized
so that a simple ribbon cable can be made to go from a Berg strip (to the
emulator board) on one end to a 48-pin DIP plug* (to the target system socket)
on the other. One possible solution is to make up a little wirewrapped board
which accepts a ribbon cable from the emulator board via its own Berg strip.
The signals from this Berg strip are then routed to a wirewrap socket which
matches the microcontroller socket in the target system. Finally a plug-to-plug
cable† connects the pins of this socket to those of the target system socket.

For the 8096, Intel supplies a low-cost emulator, the iSBE-96, shown in
Fig. 8.12. This unit is built on a standard Multibus-size board so that it can be
housed in, and draw power from, an Intel development system. Alternatively,

---

* Or a 52-lead "quad pack" plug.
† For example, a short 48-pin ribbon cable with DIP plugs on each end, for the DIP version of either
the 68HC11 or the 8096.

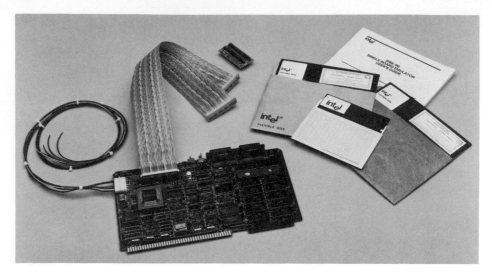

FIGURE 8.12
Intel iSBE-96 low-cost emulator. (*Intel Corp.*)

it is supplied with rubber feet for use as a stand-alone unit. One of its serial ports can be used to connect either to a personal computer or to an Intel development system for interacting with a target system through monitor commands and for downloading code. Alternatively, a second serial port is available so that the board can be interposed between a terminal and a host computer. These are the same two options which are available with the Motorola units of Figs. 8.6 and 8.7. The board offers the same kinds of interactions: examining and changing registers and memory, single stepping, and running to a breakpoint. It uses two 50-pin Berg connectors on the board to connect two ribbon cables to either of two little adapter boards. One of these boards provides an interface for the 48-pin DIP version of the 8096 while the other provides for the 68-pin plastic leaded chip carrier.

The 8096 offers a similar possibility for building up a minimum-chip version of the low-cost emulator, analogous to that of Fig. 8.8. It will involve somewhat more circuitry if it employs the same 16-bit data bus as is used internally in the single-chip 8096. Alternatively,* an external 8-bit data bus can be used to simplify the hardware at the expense of execution speed. Also, the reconstruction of ports 3 and 4 (which are given up in order to bring out the multiplexed address-bus–data-bus structure) is most easily carried out if each port is used solely for input or for output. An input port can be implemented with an octal three-state buffer and an output port can be implemented with an octal flip-flop, as discussed in Sec. 6.3.

---

* As pointed out in Sec. B.2 of Appendix B.

In this section we have looked at a way to run and debug microcontroller software using *very* simple circuitry (e.g., the seven-chip circuit of Fig. 8.8). Before concluding, we need to explore the *weakness* of the debugging tools which this approach offers. As a specific example, we look at how the BUFFALO monitor implements single stepping and breakpoints, the two main tools available for exploring the execution of application software.

The BUFFALO monitor expects a breakpoint to be set with a command such as

```
BREAK 801C
```

which sets a breakpoint at hex address 801C, or

```
BREAK 801C 801E 8166 8169
```

which sets four breakpoints (the maximum number which BUFFALO can handle at one time). Breakpoints can be individually cleared with

```
BREAK -8166
```

or all cleared with

```
BREAK -
```

As part of the GO instruction to begin the execution of the application program, the monitor program sets aside the content of the breakpoint address and then stores an SWI software interrupt instruction at this address in the program.* The application program is then run at full speed until an SWI instruction is reached and executed. At that point, the BUFFALO monitor regains control. It displays the CPU registers as they were at the moment that the SWI instruction occurred. It also restores the application program bytes to the program in place of the SWI instructions (so that a memory dump of the program area will show the correct code, not code with embedded SWI instructions).

Looking at Figs. 3.5 or A.15, either of which shows all of the 68HC11 interrupts, we see that the software interrupt instruction is nonmaskable and that it has the very lowest priority of all interrupts. This is just what we need for implementing breakpoints. It means that even if interrupts are turned off (i.e., if the I bit is set in the condition code register), we will not go past a breakpoint without stopping. It also means that all interrupts will get serviced even though we are using breakpoints for debugging.

Single stepping is implemented in the BUFFALO monitor with the help of the 68HC11's timer output compare 5 interrupt. Consequently, if the application program uses this interrupt, then we must forgo use of the single-stepping command. Actually, single stepping is really part of the larger TRACE command which can take up to 255 steps, displaying the CPU registers after each step. For example,

---

* This is why breakpoints do not work with code stored in ROM or EEPROM.

TRACE 15

will execute 15 instructions, while

TRACE

will execute a single instruction (i.e., will single step). In response to the trace command, the BUFFALO monitor sets up the timer output compare 5 interrupt to occur one cycle after it has restored the user state of the CPU registers. The intention is to execute exactly one user instruction and then have BUFFALO regain control because of the interrupt. In an interrupt-driven application program (i.e., just the kind of program which a microcontroller ought to be using), this mechanism for implementing single stepping has several weaknesses:

1. It is designed for debugging code in which interrupts play no part (e.g., for debugging an arithmetic algorithm). Interrupts are enabled by the TRACE instruction *even if the application program has disabled interrupts*. Consequently, critical regions of code, in which interrupts are intentionally turned off, are violated by a TRACE instruction.

2. While the intent of the BUFFALO monitor is to permit the execution of exactly one user instruction, it probably will not do this in an interrupt-driven environment. Looking again at Fig. 3.5, we see that many interrupt sources have a higher priority than the timer output compare 5 interrupt source. Consequently, while BUFFALO sets up to execute a single user instruction, what will *actually* be executed is all of the higher priority interrupt service routines which are pending. For example, if we are using the real-time clock interrupt, then every time we type TRACE, a real-time clock interrupt will be pending. Its interrupt service routine will be executed as well as the interrupt service routines for all other pending interrupt sources having a higher priority than timer output compare 5. When BUFFALO finally regains control and displays the CPU registers, they will be *exactly* what they were before the execution of the TRACE instruction. In another words, it looks like something is wrong in that BUFFALO will no longer single step, but instead locks up on the present instruction. Note that the same thing happens whether we are trying to single step through a mainline routine or an interrupt service routine. In either case BUFFALO's enabling of interrupts opens the door to pending interrupt sources, to the *exclusion* of the code we intended to single step through.

There is another way in which a low-cost emulator can implement single stepping. It is used by the monitor program of the Intel low-cost emulator board of Fig. 8.12 as well as the Motorola EVM board of Fig. 8.6. When a step command is executed, the monitor can modify the user program and insert a software-interrupt instruction to follow the next instruction to be executed. After the next instruction has been executed, the software interrupt will be executed, reverting control to the monitor. The monitor then restores the user program and adjusts the user program counter content to the address which held the software-interrupt instruction. If the next user instruction to be executed is a conditional branch, then *two* software interrupts are inserted, one in place of each of the two instructions which can follow execution of the conditional-branch instruction.

Some monitor programs which implement single stepping in this way offer two versions of single stepping. Normal stepping also replaces the first instruction of each interrupt service routine with a software interrupt so that single stepping will show everything, interrupt service routines and all. On the other hand, "super stepping" does *not* make this replacement. If interrupts are enabled at the point in the user program where a super step command is given, then the service routines for *all* pending interrupts will be executed, followed by the single next instruction.

Super stepping can also be designed to insert the software breakpoint in the instruction which follows *in line* after a subroutine call, rather than as the first instruction of the subroutine. In this way, a normal step will drop down into the subroutine whereas a super step will execute the subroutine at full speed, stopping upon the return from the subroutine.

The breakpoint approach to single stepping discussed here does permit us to use single stepping in an interrupt-driven microcontroller environment. Nevertheless, the quality of the debugging we can carry out in this way is limited, as real-time interrupts occur during the non–real-time execution of program code.

A low-cost emulator can run a user program and still return control to the monitor in two ways:

By the setting of a breakpoint

By single stepping

However, only one way (i.e., breakpoints) provides completely satisfactory operation in an interrupt-driven environment. That is, only breakpoints provide for the real-time execution of the user program (up to the breakpoint).

In this section, we have seen that the cost savings realized when using a *low-cost* emulator is reflected in *low-quality* debugging tools in an interrupt-driven environment. And since most microcontroller applications make extensive use of interrupts to sequence a variety of activities, careful thought must be given to *how* debugging of software will be carried out. In the next two sections, we consider the benefits which can accrue to us if we have quality tools on our side at debugging time. Nevertheless, we should remember that many fine products employing microcontrollers have been developed with little more than the low-cost emulators discussed in this section.

8.6 FULL-FEATURED DEVELOPMENT SYSTEMS

In this section we consider Hewlett-Packard's highly regarded 64000 microprocessor development system, shown in Fig. 8.13. While most development sys-

FIGURE 8.13
Full-featured development system. (*Hewlett-Packard Co.*)

tems support a single user, the HP development system shares resources so that the efforts of a design *team* are supported well. That is, each person on the design team has access to the latest update of software for a project since the hard disk is shared among the several workstations in the system. The multiple-station approach also has the benefit of sharing expensive resources like a fast printer and a highly reliable hard disk drive among all the users of the system.

The developers of the system shown in Fig. 8.13 have given much thought to how design engineers do what they do. In particular, they have taken the design iteration cycle

Make a software change

Assemble (or compile) and link

Verify performance on target system

and made it into a closely coupled, tight loop. When this loop is easy and quick to traverse and when the tools available at each stage provide powerful support, then the development system actually stimulates the design process.

To see how the elements of this loop support the design process, let us look first at how the user perceives the operating system. When the HP 64000 system is turned on, it presents the user with the options available by listing them along the bottom edge of the CRT. For example, the user might see

```
edit   compile   assemble   link   emulate   prom_prog   run   ----ETC--
```

Just below the bottom edge of the CRT are eight *softkeys*, one for each option. When the "---ETC--" softkey is pressed, a new set of options is written along the bottom edge of the CRT. In this way the eight softkeys serve to call up any one of 20 or so options.

In like manner, when the edit softkey is pressed, the user is led into a full-featured screen editor. The editor commands are presented to the user as softkeys. Each command may have a variety of options. For example, the following list command allows the user to obtain a hardcopy listing of the file being edited, extending from the present cursor position to the end of the file:

```
list printer thru end
```

The operating system presents the options available to the user with *directed syntax softkeys*. For example, after the list softkey is pressed, the word "list" appears on the command line and the softkeys are changed to the two options

```
<FILE> printer
```

The user can type in the name of a file which will be created to hold the listing. Alternatively, the printer softkey can be pressed, which will add the word "printer" to the command line. In either case, the next set of options for the command line will be provided by the softkeys.

Hewlett-Packard, as a company, is known throughout the industry for the excellence of its manuals. However, one of the interesting consequences of having directed syntax softkeys is that users of the HP 64000 system rarely refer to *any* of the manuals provided with the system. Since the options available to a user are displayed at each step, the manuals do not have much to add.

For both the newcomer to the HP 64000 system and the experienced user, the screen editor and its use of directed syntax softkeys provide a smooth path for creating microcontroller source code as well as for editing the source file to remove bugs found through emulation. The development system supports assembly language development for a variety of microprocessors. It also supports the compilation of code written in either Pascal or C, again for a broad group of microprocessors. Since this development system is not dedicated to a specific microcontroller chip, it can be upgraded to support code development and debugging for new microcontrollers as they appear. The upgrade requires a change of an emulator board and its associated "pod." The pod sits next to the target system and has a short cable and probe which connect into the microcontroller socket in the target system. A longer ribbon cable connects the pod back to the emulator board in the workstation.

The HP 64000 family of assemblers provides support for many multiple-chip microprocessors and single-chip microcontrollers. They include a powerful macro capability which can find many uses (e.g., in support of queues, as discussed earlier, in Sec. 5.2). In addition each assembler on the 64000 has been written using HP's user definable assembler. We can use this user definable assembler to create an assembler for a new microcontroller even before HP provides support for it. To the extent that the new microcontroller is similar to an already supported microcontroller, the upgrade can be rather simple. The macro capability gives us an alternative way to achieve the same goal.

The HP 64000 assemblers produce *relocatable* code. The *linker* can take any number of relocatable files and link them together to obtain the object code which runs on the target microcontroller. When working on the extensive software for a project, this formation of the object code in two steps permits us to make a change in one of the source files, reassemble it, and then link it with the remaining (already assembled) relocatable files for the rest of the software. This is much faster than reassembling a single massive file for the entire project. For smaller projects, we can put all of the code into a single source file and then use a procedure file to carry out the assembly and linking process. In this way, the procedure file makes the assembler and linker look like an assembler which is designed to produce an object code file from a single source file.

To get an idea of what is involved in using relocatable files, consider Fig. 8.14. This shows the format of two source files which will be assembled and then linked together. Instead of having an ORG assembler directive preceding the assignment of RAM to variables, a source file will use a DATA assembler directive for this purpose. The linker is told where RAM begins. When it links files together, it takes the first file and assigns its RAM variables starting at the beginning of the available RAM. The second file's RAM variables follow the first's. In a similar fashion, relocatable files use a PROG assembler directive to tell the linker that what follows is intended to be assembled to addresses where the microcontroller has program memory (i.e., ROM, EPROM, or EEPROM). In this way, RAM and ROM are effectively used in spite of the expansion or contraction of any of the files as software development and debugging progresses.

Earlier we discussed the three steps making up a design iteration cycle. We have seen the support which the HP 64000 gives to the first two steps:

Making software changes
Assembling and linking

The third step,

Verifying performance on the target system

is supported in a variety of ways. Our code can be loaded into the emulator and run. Furthermore, a *logic analyzer* board in the workstation monitors the microcontroller's address bus and data bus. (They are made accessible in much the same way as for the low-cost emulators of the last section. The microcon-

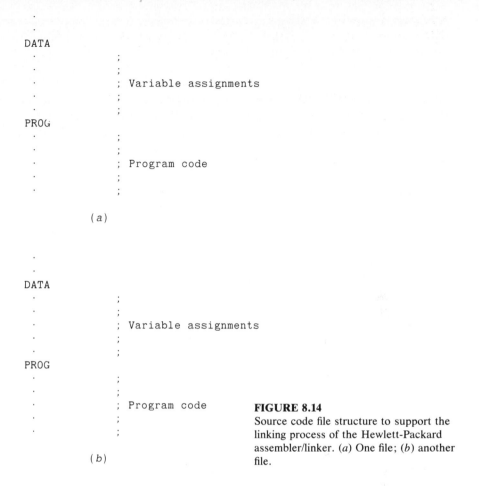

```
        .
        .
   DATA
        .               ;
        .               ;
        .               ;  Variable assignments
        .               ;
        .               ;
   PROG
        .               ;
        .               ;
        .               ;  Program code
        .               ;
        .               ;

            ( a )
```

```
        .
        .
   DATA
        .               ;
        .               ;
        .               ;  Variable assignments
        .               ;
        .               ;
   PROG
        .               ;
        .               ;
        .               ;  Program code
        .               ;
        .               ;

            ( b )
```

FIGURE 8.14
Source code file structure to support the linking process of the Hewlett-Packard assembler/linker. (*a*) One file; (*b*) another file.

troller is operated in an expanded mode so that the emulator can gain access to the internal buses. The ports are then reconstructed to make the emulator look like the single-chip microcontroller.)

The logic analyzer gives us the power to trace the performance of the microcontroller at any point in its execution of code. After a trigger condition has been set up, it will collect the 256 memory transactions which take place on the microcontroller's bus when the trigger condition occurs. In fact, the logic analyzer can be set up to collect the 256 memory transactions which *lead up* to the trigger condition, or the 256 memory transactions which occur after the trigger condition, or the 256 memory transactions centered about the trigger condition. After the data has been collected, it can be disassembled so that we can debug source code. The CRT display shown in Fig. 8.15 illustrates what a trace looks like.

For debugging interrupt-driven microcontroller software, a huge gap exists between having only the breakpoint debugging capability of the low-cost

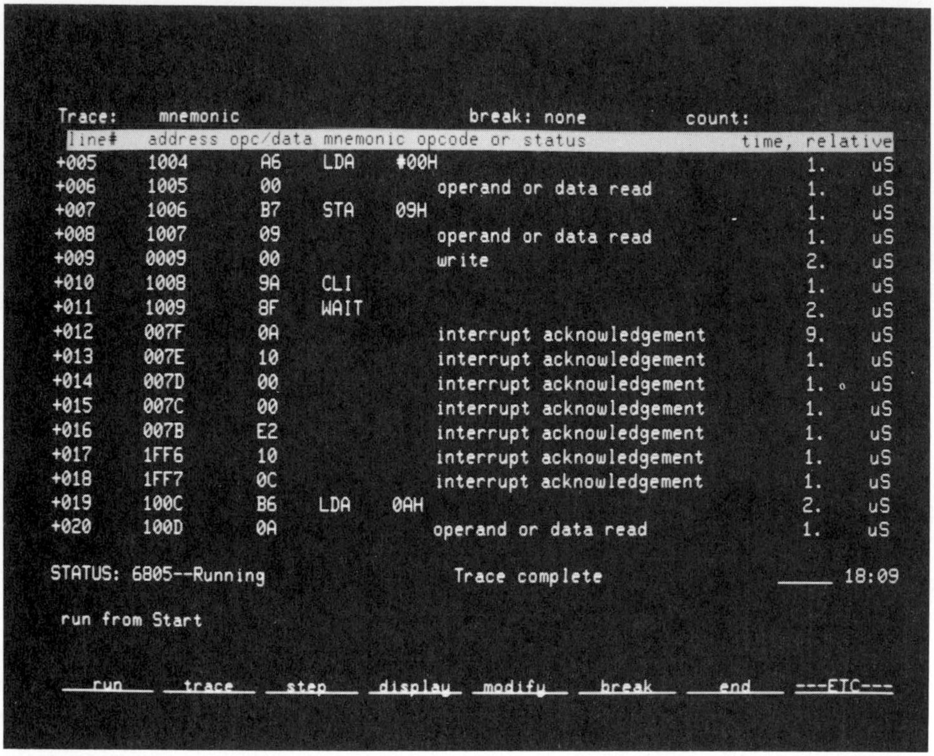

```
Trace:     mnemonic                           break: none          count:
line#    address opc/data mnemonic opcode or status              time, relative
+005     1004      A6    LDA    #00H                                  1.      uS
+006     1005      00           operand or data read                 1.      uS
+007     1006      B7    STA    09H                                   1.      uS
+008     1007      09           operand or data read                 1.      uS
+009     0009      00           write                                2.      uS
+010     1008      9A    CLI                                         1.      uS
+011     1009      8F    WAIT                                        2.      uS
+012     007F      0A           interrupt acknowledgement            9.      uS
+013     007E      10           interrupt acknowledgement            1.      uS
+014     007D      00           interrupt acknowledgement            1.      uS
+015     007C      00           interrupt acknowledgement            1.      uS
+016     007B      E2           interrupt acknowledgement            1.      uS
+017     1FF6      10           interrupt acknowledgement            1.      uS
+018     1FF7      0C           interrupt acknowledgement            1.      uS
+019     100C      B6    LDA    0AH                                   2.      uS
+020     100D      0A           operand or data read                 1.      uS

STATUS: 6805--Running                    Trace complete                     18:09

run from Start


____run____trace____step____display__modify____break____end____---ETC---
```

FIGURE 8.15
A trace display for the development system of Fig. 8.13. (*Hewlett-Packard Co.*)

emulators of the last section and having a logic analyzer to take snapshots of CPU activity. The logic analyzer built into a full-featured development system like that of Fig. 8.13 is particularly helpful. Since the development system is an integrated unit, the coordination needed for collected data to be disassembled and displayed appropriately is built in. Also, the probing of the address-bus–data-bus structure has already been handled by the HP designers of the development system. Consequently, loading considerations and timing considerations have already been handled, as well as just making sure that each of the logic analyzer's probes are connected to the appropriate point. The alternative of using a stand-alone logic analyzer and connecting each of its inputs to a microcontroller pin with a grabber can lead to a rat's nest of wiring like that shown in Fig. 8.16.

The HP 64000 development system also includes a software performance analyzer capability. This is useful for looking for bottlenecks in code which seems to work correctly most of the time, but not all of the time. For example, the software analyzer can be set up to tell us the percentage of CPU time spent

FIGURE 8.16
Probing a microcontroller chip with a logic analyzer's grabbers. (*Rand Renfroe*)

in each interrupt service routine and in carrying out specific algorithms in the mainline program. An example of this is shown in Fig. 8.17. A designer who has been living with a microcontroller-based design for an extended period, and who would seem to have a pretty good idea of how the CPU spends its time, is almost invariably surprised when the design is subjected to the software performance analyzer.

Given the insight gained from a software performance analyzer, the designer can go back and use the logic analyzer to monitor the activity of key variables. Instead of capturing bus activity when a trigger condition occurs, the logic analyzer can be set up to capture the successive execution of specific events. For example, it can be set up to capture the successive writes to a specific port or the successive writes to a specific RAM variable. In this way the software performance analyzer (which gives global performance clues) lends support to the logic analyzer (which obtains a microscopic view of specific events). Furthermore, at each step of the way, hardcopy listings of any of the displayed outputs can be obtained. For anyone who has had the experience of losing track of what has been found out and what clues this gives for further tests, this hardcopy echoing of a debugging session is most welcome.

In addition to giving assembler support, compiler support, and emulation support for a specific microcontroller, the HP system permits the programming of parts. The PROM programmer board installed in the workstation together with the personality module for a specific microcontroller chip (located to the right of the keyboard) permits the application software to be installed in an actual microcontroller chip's PROM, EPROM, or EEPROM.

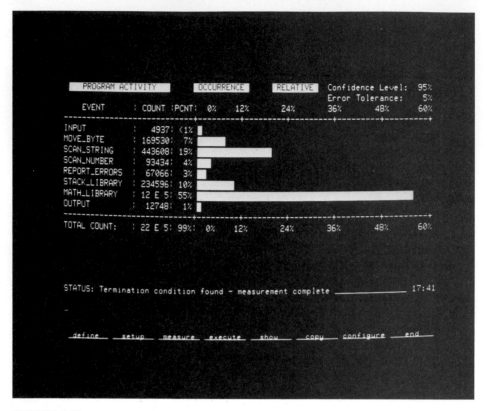

FIGURE 8.17
Memory activity measurement made with a software performance analyzer on the development
system of Fig. 8.13. (*Hewlett-Packard Co.*)

8.7 EMULATOR/LOGIC ANALYZER DEVICES

The broad acceptance of the IBM PC as a standard around which products can
be built has made relatively low-cost emulation and logic analysis possible. As
we discussed in the last section, the logic analyzer in a full-featured development system gives the designer an outstanding view of what the microcontroller is doing. This ability to capture data and display it can also be achieved in a
unit which uses an IBM PC for display, for command entry, and for formatting
captured data.

A particularly intriguing unit is Orion Instruments' Universal Development Laboratory (UDL), shown in Fig. 8.18. The designers of this unit have
developed a universal hardware interface for a large number of microprocessors. It includes a plug for a ROM socket, which picks up the data bus and most
of the address bus lines. Then it includes a DIP clip to probe the remaining
address and read/write control lines on the microprocessor itself. Furthermore,

FIGURE 8.18
An emulator/logic analyzer unit. (*Orion Instruments Inc.*)

by having access to the microprocessor through the ROM address space, the UDL presents a rather noninvasive addition to the target system. That is, the target system runs with the microprocessor in its socket, being clocked by its own clock. The UDL monitors *all* bus transactions. It exerts control by the program instructions which it presents to the microprocessor in the ROM address space. In a sense, it exerts control like the BUFFALO monitor discussed in conjunction with the circuit of Fig. 8.8. However, with more sophisticated hardware control, the monitor instructions are switched into and out of the ROM address space. Consequently, the monitor does not consume any of the target system ROM or RAM space. Code is written for the target system using the target system's ROM and RAM addresses, not some intermediate addresses (as had to be done for the circuit of Fig. 8.8).

For software development for a microcontroller chip operating in the single-chip mode, the UDL must gain access to the internal bus. One way this can be done for the 48-pin DIP version of the Motorola 68HC11 is with a little five-chip adapter which is built up as a simplified version of the low-cost emulator of Fig. 8.8. The adapter board can include a 48-pin DIP socket which is *partially* connected to various parts of the circuit of Fig. 8.8. Then a 48-pin ribbon cable with 48-pin DIP plugs on each end* can be used to connect this into the target system. The ports from the 68HC11 and the 68HC24 need to be connected to this 48-pin socket as well as V_{SS}, V_{RL}, V_{DD}, V_{RH}, XIRQ, IRQ, and RESET. On the other hand, the MODA and MODB pins must *not* be brought from the adapter circuit's 68HC11 chip to this 48-pin DIP socket since we do not want the target system forcing the adapter circuit into the single-chip mode. The target system's 8-MHz crystal circuitry must be duplicated on the adapter board since it needs to be located within inches of the 68HC11 chip. The XTAL and EXTAL pins are not connected to the 48-pin DIP socket going to the target system.

On the adapter board, the DUART circuitry is eliminated and the static RAM is replaced by a ROM socket which is addressed over the range E000 to FFFF. As discussed in Sec. A.2 of Appendix A, the 68HC11's EEPROM includes a bit to disable the internal ROM, thereby permitting the ROM space (including the reset vector and the interrupt vectors) to be mapped to this external socket. The reset circuitry as well as the pullup resistors on IRQ and XIRQ should also be eliminated on the adapter board. These three pins of the adapter's 68HC11 must be tied to the 48-pin DIP socket going to the target system. Furthermore, the target system *must* be built with passive pullup circuitry on these pins so that the target system's driving of these pins can be overridden by the UDL unit.

For the Intel 8096, gaining access to the internal bus means operating the chip in the expanded mode, adding two EPROM sockets, and then reconstructing ports 3 and 4. A circuit for doing this is shown in Fig. 8.19. The port reconstruction scheme used in this figure is not an exact reconstruction, just a simple one. Each port is dedicated to being either an input port or an output port by its associated DIP switch. Furthermore, the actual 8096 ports have open-collector outputs, meaning that they can pull the output lines low but need the addition of pullup resistors in order to be used as outputs. A port reconstruction which is functionally equivalent to the 8096's port 4 requires more chips and is shown in Fig. 8.20. A port 3 equivalent would replace the active-low WRPH signal with the WRPL signal of Fig. 8.19. It would also connect to the lower half of the multiplexed bus, AD7 . . . AD0.

Building up a little adapter board for the Intel 8096 entails the same kinds of considerations as were discussed for the 68HC11 concerning the connections into the target system. Since the adapter board must be powered from the target system, obviously power and ground connections must be carried over. *Separate* lines should be run for the A/D converter's reference voltages, V_{ref} and Angnd (so that the current which powers the adapter circuit will not cause a voltage drop in the reference voltages).

For building up a wirewrapped version of either a low-cost emulator or circuitry similar to that of Fig. 8.19, a problem arises in mounting the microcontroller chip. If the 48-pin DIP version of the microcontroller is used, then two

---

* Or two 24-pin ribbon cables with DIP plugs.

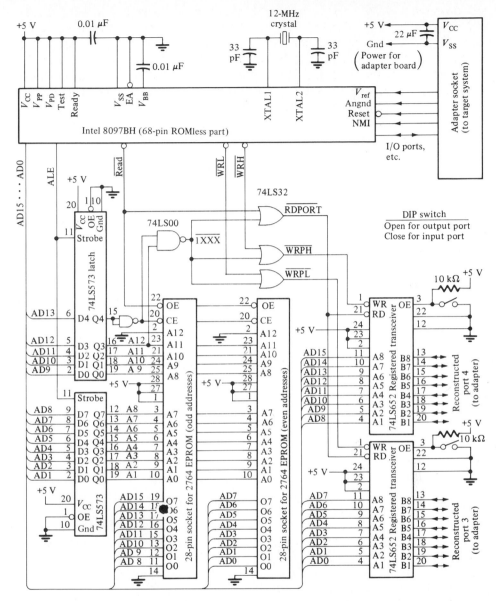

FIGURE 8.19
Nine-chip emulator of a single-chip Intel 8096 for use with Orion Instruments' UDL probe.

standard (600-mil-wide) 24-pin DIP wirewrap sockets can be mounted together (on a development board having holes drilled for mounting wirewrap sockets). On the other hand, if the 68-pin plastic-leaded chip-carrier version of one of the Intel 8096 family of parts is used, then an adapter such as that shown in Fig. 8.21 can be used.

Because of the universality of the hardware of the Orion UDL, it is

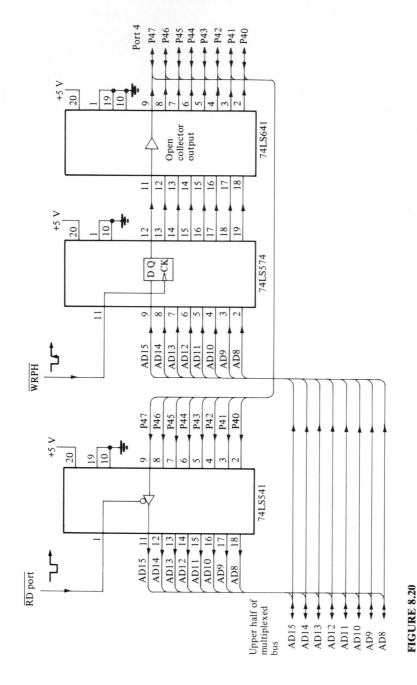

FIGURE 8.20

A functionally equivalent reconstruction of the Intel 8096's bidirectional port 4 (which has an open-collector output).

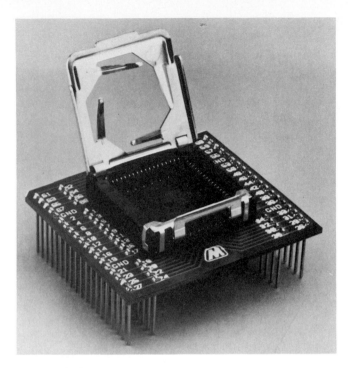

FIGURE 8.21
Adapter for mounting a 68-pin PLCC package on a wirewrap board. (*Methode Electronics, Inc.*)

significantly less expensive than other emulator/logic analyzer units. Its customization comes through the software which is downloaded from the IBM PC to the UDL for doing monitor-type jobs. The software which runs on the IBM PC must also be customized for disassembling the instructions which the CPU has executed and for customizing the display of CPU registers.

The logic analyzer built into the UDL unit is 48-bits wide and can collect 170 memory transactions. Since 48 input lines are more than are needed to monitor just the address-bus–data-bus structure, the extra lines can be used to probe other parts of the circuit (e.g., the ports of the microcontroller chip). An example display of a trace is shown in Fig. 8.22.

One key to the potential power of a logic analyzer is its ability to set up and use complex triggering conditions. As the microcontroller executes hundreds of thousands of instructions per second, we want to be able to trigger on just that condition which warrants our attention. The UDL unit permits us to trigger after four successive conditions have been met on the 48 lines. It permits triggering on any access within a specified address range, or on any access outside a specified address range. It permits triggering on the ANDing of simultaneously occurring conditions and the ORing of alternative conditions. An arbitrary delay can be interposed after triggering occurs and before data is collected. It can collect only those events which meet the trigger condition.

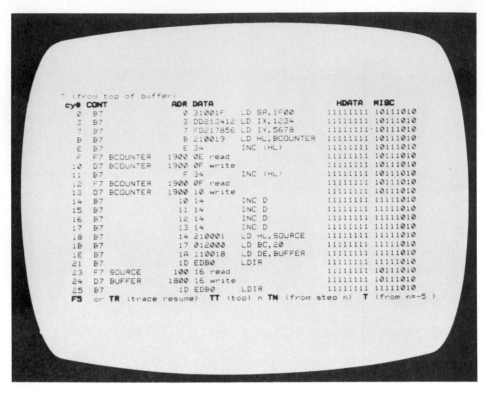

FIGURE 8.22
Trace generated by the unit of Fig. 8.18. (*Orion Instruments Inc.*)

Alternatively, it can collect this trigger condition plus the one, two, three, or four cycles which follow the trigger condition.

As an example of the power of a good logic state analyzer, consider the case in which an algorithm works fine when run by itself. However, when this algorithm is run in conjunction with the complete instrument or device software needed by an application, one of its variables becomes corrupted. The Orion UDL permits us to trigger only on writes to the variable and to collect not only the data written to that address but also the memory transaction which takes place during the following CPU cycle. In this way, we can determine the address from which the following instruction is fetched and determine the offending part of the program.

We cannot evaluate the cost effectiveness of a tool without knowing its cost. At 1987 prices, the Orion UDL costs $3000, including everything needed for one of the 20 or so families of microprocessors which they support. The upgrade needed to support a second microprocessor was only about $500 (which includes an analyzer cable and the disassembler/debugger software for the new microprocessor).

In this section we have considered a tool which is designed to work with

FIGURE 8.23
Datavue 25 desktop computer. (*Quadram Corp.*)

an IBM personal computer. It is worth being aware that there are many IBM
look-alike personal computers which will run the same software and yet may
have some advantageous features. For example, the unit shown in Fig. 8.23 has
the advantage in a lab environment of not dominating a workbench. Its serial
port can communicate with development hardware while its parallel port can be
used to dump listings and traces to a printer. Even though it has only a single
floppy disk drive, its built-in RAM drive software together with 640K of RAM
give it much of the flexibility and responsiveness of a more expensive, and
expansive, desktop computer having a hard disk drive.

8.8 EPROM AND EEPROM PROGRAMMING

As shown in Sec. B.13, Intel supports the 8096 family of parts with EPROM
versions. As shown in Sec. A.14, Motorola supports the 68HC11 with several
parts having various sized EEPROMs (up to 8K). This will permit a company to

FIGURE 8.24
Microcontroller with a piggybacked EPROM. (*Zilog, Inc.*)

build prototypes and early production versions of a design without committing to mask-programmed ROMs.

Another option permitted by some microcontroller manufacturers is to piggyback a standard EPROM into a socket riding on top of the microcontroller chip, as shown in Fig. 8.24 for Zilog's Z8 microcontroller. This requires the manufacturer to have a special "bond-out" part, in which the address bus lines, data bus lines, read line, and chip select line needed by the EPROM are brought out and made available in this way. While the microcontroller package thus becomes a relatively expensive package, it allows us to take advantage of low-cost EPROMs and the EPROM programmers available to program them. For example, the Orion UDL discussed in the last section includes as a standard feature an EPROM programmer for standard EPROMs extending from 2K × 8 up to 64K × 8.

Another approach to this problem is used by Hitachi in their ZTAT (zero turn-around time) versions of their 6301 and 6305 series of microcontrollers. While Hitachi also has EPROM (and ROM) versions of these microcontrollers, they have reduced the packaging cost significantly by also packaging the EPROM version in an inexpensive plastic package. By eliminating the window (needed to expose the chip to ultraviolet light for erasing the EPROM), they have eliminated a major packaging expense (as well as eliminating the ability to erase the EPROM). In effect, the ZTAT part includes a PROM, not an EPROM.

The Intel 879xBH* family of parts are the EPROM versions of the 8096 family. They can be programmed in any of several ways. An EPROM programmer for an 879xBH chip can apply addresses, data, and programming pulses to the part, just as is done with an EPROM. This is the mode which Hewlett-Packard supports with their 64000 development system. However, for someone who has a programmer for standard EPROMs, Intel also provides an autoprogramming mode. To use it, a designer first programs the application software into a standard 2764A (8K × 8) EPROM. Then this 2764A is inserted into a socket in the circuit of Fig. 8.25. The 8796BH will then execute an internal program (which is hidden from view under normal operation) which copies the 2764A's content into the 8796 EPROM. This autoprogramming mode of operation is invoked when the 8796BH part comes out of reset with 1100 applied to bits 7..4 of port 0, and with +12.75 V applied to the V_{PP} programming supply voltage pin and to the active-low EA (external access) input.

As pointed out in Sec. B.2 of Appendix B, the Intel 8096BH family of parts permits the expanded mode to be operated with either a 16-bit external data bus or an 8-bit external data bus, depending upon the state of the BUSWIDTH pin. By using an 8-bit external data bus during EPROM programming, we not only eliminate a latch chip but we can also use a single 2764A EPROM to hold the content targeted for both even and odd EPROM addresses in the 8796BH part.

The Motorola 68HC11 permits its own version of a minimum parts circuit for programming the internal EEPROM. The circuit is shown in Fig. 8.26. This circuit uses a bootstrap operating mode of the 68HC11 to load a program into the internal 256-byte RAM and then to execute it. The program is loaded into RAM via the 68HC11's UART serial port (operating at 1200 baud with an 8.0-MHz crystal). The first byte sent should be the hexadecimal number FF. The following byte received by the 68HC11 will be put into address 0000 and the one after that into address 0001. The 68HC11 continues to wait for bytes to be received via its UART until it has received the 256 bytes needed to fill its RAM on page 0. After the character for address 00FF has been received, the CPU begins executing the program beginning at address 0000.

As discussed in detail in Sec. A.10 of Appendix A, the 68HC11 must program its own EEPROM. This is controlled by setting and clearing selected bits in the 68HC11's PPROG register. Each byte is written to the appropriate EEPROM address. A latch holds both data and address while the EEPROM programming voltage is applied for 10 ms. This process is repeated for every location to be programmed.

The program which was bootstrapped into RAM reads successive bytes

---

* See Fig. B.41 of Appendix B.

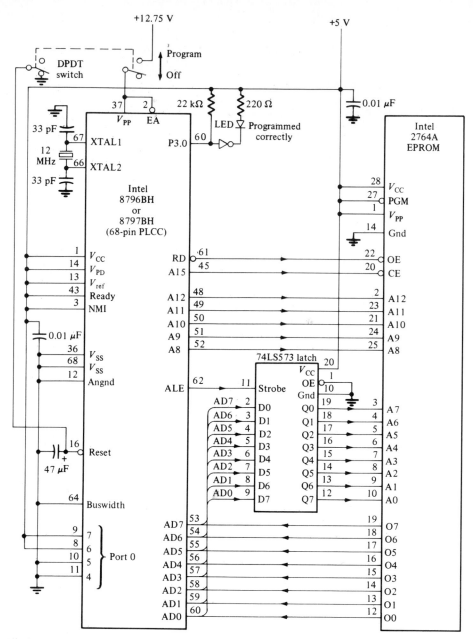

FIGURE 8.25
Intel 8796 autoprogramming circuit.

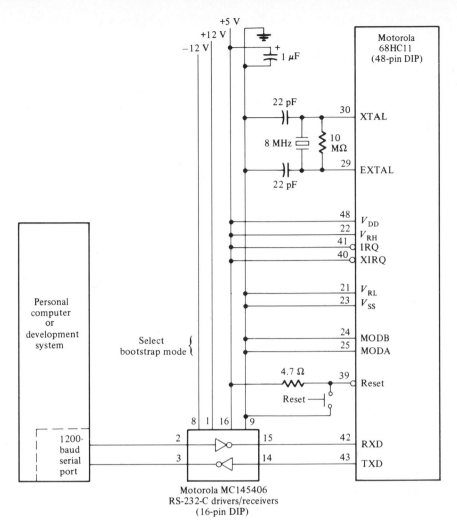

FIGURE 8.26
Motorola 68HC11 EEPROM programming circuit.

from the personal computer to be programmed into the EEPROM. If an assembler on the personal computer has created what Motorola calls an "S record" file, then this can be sent to the bootstrapped program in RAM. That program must sort through the S record file to pull out addresses and data appropriately. Since the programming of each byte of the EEPROM takes 10 ms while the time to receive each successive byte is 8.33 ms, a flow control mechanism is required to tell the computer to pause until the programming catches up. The bootstrapped program might put received bytes into a queue and send either of 2 bytes (for example, XOFF and XON characters) back to the computer to tell it when to pause and when to resume the sending of bytes. When programming has been completed, a message can be sent back to the personal computer to signal this event.

PROBLEMS

The support given to microcontrollers by a low-cost emulator board like the one shown in Fig. 8.7 offers extensive opportunity for project work. Such a board costs in the neighborhood of $100 and is thus attractive to many students and designers for their own experimenting and learning. Furthermore, the board is probably supported with a free assembler which will run on a personal computer. Finally, it may even be supported by the availability of the source code for the monitor program, perhaps over a dial-up phone line. Thus, Motorola supports the use of their M68HC11EVB low-cost emulator board, shown in Fig. 8.7, with either the IBM PC or the Apple Macintosh. By dialing up the 1200-baud line at 1-512-440-FREE, anyone can download the version of the 68HC11 assembler to run on one of the above computers, as well as the BUFFALO monitor program source code, and as well as the source code for various algorithms (e.g., floating-point arithmetic routines). The problems which follow give added support to development with a low-cost emulator like the EVB board.

8.1. *Timing macros.* One of the features which tends not to be implemented in a low-cost emulator board is the ability to time events. In this problem, we consider some macros which can be defined in an application program and then easily used to measure the time taken to execute the code beginning at one instruction and ending just prior to the execution of any other instruction.

(a) Define two macros, called STARTTIME and STOPTIME. STARTTIME stores what the programmable timer's free-running counter will hold when the CPU executes the first cycle of the first instruction which follows the STARTTIME macro. STOPTIME figures out what the free-running counter held during the first cycle of the first instruction in the STOPTIME macro. It then forms the difference between these two times and stores this difference in a 2-byte CPU register. To use these two macros, we invoke them at the two points in the source code file between which we want to determine the execution time. Then when we test the program, we can set a breakpoint just after STOPTIME, run the program up to the breakpoint, and examine the time interval stored in the 2-byte CPU register. This scheme works for measuring time intervals which are shorter than a complete cycle of the timer's free-running counter.

(b) Rewrite the STARTTIME and STOPTIME macros as subroutines. In this case, invoking them will entail the insertion of subroutine calls into the source file for the program we are testing.

(c) Set up the programmable timer's overflow interrupt service routine to increment a 1-byte variable, TCNT_EXT, each time the free-running counter rolls over from FFFF to 0000. Modify STARTTIME and STOPTIME to form a *3-byte* time interval. How long must a time interval be in order to be too long for this modified scheme?

8.2. *Latency time macros*. Randy Abler, at Georgia Tech, suggests the following scheme for measuring the time between when an interrupt source requests service and when it gets service. It uses one of the microcontroller's input capture registers or output compare registers to mark the time when the interrupting event occurred. Consequently, the scheme addresses programmable timer latency questions directly. It can also be used to measure the latency associated with *any* edge-triggered interrupt input to the microcontroller by tying that input to an input capture input while the latency time test is carried out.

(*a*) Create a macro called TRAM which defines two 2-byte RAM variables, TMAX and TMIN. Randy's scheme measures latency repeatedly and captures the minimum value of latency into TMIN and the maximum value into TMAX. This macro is invoked in the RAM area of an application program to set aside 4 bytes of RAM for use during latency tests.

(*b*) Create a macro called TCLR which loads TMIN with FFFF and TMAX with 0000. This initialization is such that the loaded values will necessarily be superseded by actual values when the program is run. For example, if the first measurement of latency comes up with a value of 0014, then after that first measurement, TMIN and TMAX will both contain 0014. If the second measurement gives a latency of 000C, TMIN will then contain 000C and TMAX will contain 0014.

(*c*) Create a macro called RDTIME which has passed to it the name of the output compare register or the input capture register to be used in the measurement. For example, referring to Fig. A.22 for the Motorola 68HC11, the measurement of latency for the input capture 3 programmable timer input would be made by imbedding the macro

```
RDTIME  TIC3
```

as the first instruction in the input capture 3 interrupt service routine.

This macro reads the programmable timer's free-running counter and derives the time when the CPU first began to fetch the instruction which read the free-running counter. For the Motorola 68HC11 using extended addressing to read TCNT into accumulator D with the LDD instruction, this means subtracting 3 (because the LDD instruction reads TCNT three cycles after its execution is begun, rather than during the first cycle of its execution). Now subtract the content of the register used as the macro's parameter (TIC3 in the example above). The result is the actual latency for the interrupt source. Finally, update TMAX and TMIN appropriately, given this value.

For an interrupt source which interrupts repeatedly, this test gives terrific insight into what is going on. Using it, Randy has even found some undocumented good features about the 68HC11. For example, in responding to several interrupt requests, the CPU does not actually sort out which device requested service until the last two cycles of the stacking procedure. A request from a high priority interrupt service routine occurring even as late as this will be immediately honored.

Latency is measured by running the program using these macros imbedded in the source file. After it has run long enough so that many interrupts have occurred, the reset button is pushed and TMAX and TMIN are examined.

(*d*) Some assemblers permit the use of "include" files. If you are using such an assembler, then put all three of these macro definitions into a file called TIME. Then their definition can be added to the source file for your application software with a line like

```
INCLUDE TIME
```

Note that no code is added to your application software by the macro definitions made with the TIME file. Code is added when they are invoked for testing purposes with the inclusion of the three lines

```
TRAM
```

in the RAM variable definition area of your program,

```
TCLR
```

in the initialization part of your program, and

```
RDTIME    <timer register name>
```

as the first instruction of an interrupt service routine.

8.3. *Board modification for enlarging the monitor program.* The problems which follow make additions to the monitor program. If the monitor is stored in a 2764 (8K × 8) EPROM (as it is for Motorola's EVB board), then determine what modification is needed to the board to insert a 27128 (16K × 8) EPROM into the socket intended for the 2764. That is, what trace(s) on the board must be cut? What new connection(s) must be made? This is an interesting opportunity because the two EPROMs are almost identical in their pinouts. They are also virtually identical in price (approximately $4.00 in single quantities), in spite of one having twice the storage of the other. It's a wonderful era to be a digital designer!

8.4. *Monitor modification to relocate monitor variables.* If the monitor program does as Motorola's BUFFALO monitor does, using the microcontroller's RAM for its own variables, then any application program will not be able to use the entire internal RAM. In this, it will differ from the final application program for a single-chip microcontroller application which needs the entire RAM. Since the low-cost emulator necessarily has *additional* RAM on the board (to hold the application program which is to be tested), change the ORG before the RAM allocation part of the monitor program so that the monitor variables are relocated to high addresses in this external RAM. Reassemble the monitor program, program it into an EPROM, insert the EPROM onto the low-cost emulator board, and test the result.

8.5. *New monitor commands.* Look at the source file for the monitor program for a low-cost emulator to determine how it parses the command line and then branches to the code to be executed in response to a command. More than likely, there is a table of commands in which each entry consists of a string (which is compared with the requested command) and an address of the routine to be executed when that command is requested. In this problem, we provide the hooks needed to add, and easily test, new features to the monitor program.

(*a*) Add three new entries, called TEST1, TEST2, and TEST3, into this table of commands. Dedicate three 16-bit RAM locations, VECTOR1, VECTOR2, and VECTOR3, for use by the monitor program as it deals with these commands.

(*b*) Add whatever code is needed in the monitor program so that when one of these three commands is invoked, the monitor will execute, as a subroutine, the code whose address is stored in the corresponding vector dedicated in part *a*.

(*c*) Reassemble the monitor program. Test the result by making TEST1 set a bit on an output port, TEST2 clear the same bit, and TEST3 toggle the bit. That is, write a little program consisting of the three subroutines, ORGed into RAM available on the low-cost emulator board. The program should also load their addresses in VECTOR1, VECTOR2, and VECTOR3. The assembled code is downloaded to the emulator board. It is executed by responding to the monitor's prompt with

```
TEST1  <return>
```

or

```
TEST2  <return>
```

or

```
TEST3  <return>
```

and watching what happens to the bit of the output port in each case.

8.6. *New monitor commands.* Use your solution to the last problem to explore how the monitor program picks up a parameter. For example, the BUFFALO monitor for Motorola's EVB low-cost emulator board executes the command

```
MD 0020 003F
```

by displaying memory locations 0020 to 003F. In executing this command, the monitor first finds a match for MD. Then the called subroutine has a pointer passed to it which points to the beginning of the string which makes up the operand part of the command. If the parameter is optional, the subroutine will look for a carriage return (ASCII-encoded as the hexadecimal number, 0D) to determine if the line ends with no parameter. Otherwise, the monitor probably has subroutines for converting the parameter appropriately. For example, if an address is expected, then a subroutine is probably already available in the monitor program which reads the string making up the operand and converts it to an address, loading a CPU register with its value.

(*a*) Create a new command, implemented with TEST1 of the last problem, which is invoked with

```
TEST1    <address>
```

and which clears the content of the specified address to 00.

(*b*) Create another command, implemented with TEST2, which is invoked with

```
TEST2    <address>    <binary number>
```

and which complements those bits in the content of the selected address which correspond to the position of 1s in the binary number parameter. For example,

```
TEST2    0014  00000001
```

should complement the least-significant bit of the content of hex address 0014.

8.7. *New monitor treatment of software interrupts.* Look at how the monitor implements breakpoints. As discussed toward the end of Sec. 8.5, Motorola's BUF-

FALO monitor inserts a software interrupt instruction, SWI, in place of the instruction where the breakpoint is to be inserted. Then when the CPU gets to the SWI instruction, it vectors back to the monitor program which interprets what to do. A simple monitor program may expect the SWI instruction (or its equivalent, for another microcontroller) to be used for nothing but breakpoints.

(*a*) For this problem, you are to modify the monitor program to use the following data structure:

```
SWI_RAM    RMB     16      ;Storage for 8 program counter values
           RMB     16      ;Storage for 8 vector values
           RMB     8       ;Storage for 8 opcodes
```

When a software interrupt occurs, the program checks the first eight 2-byte RAM locations to see which one matches the content of the program counter stacked by the execution of the SWI instruction. Associated with each 2-byte RAM location is another 2-byte RAM location which contains the address of a subroutine to be called to take proper action. Finally, each of the eight entries also has associated with it the opcode in the application program which was replaced by the SWI instruction.

(*b*) Now modify the breakpoint commands so that they make use of the first four of the eight entries of this data structure to implement up to four breakpoints.

(*c*) Implement a single-stepping command, S, using the fifth of the eight entries. That is, when a single-step command is executed, the address and the opcode of the user instruction *which follows the next one* are inserted into the data structure, along with the address of a new single-stepping subroutine. Then the opcode is replaced by a software interrupt instruction in the user program. To do this, a 256-byte table needs to be added to the monitor. This table is entered with an opcode as an offset into the table. The table entry is the number of bytes in the instruction having that opcode. For those instructions having a 2-byte opcode, the table entry can steer the program to use an appropriate second table to get the number of bytes in the instruction. Thus, the Motorola 68HC11 would require 4 × 256 = 1024 bytes in the monitor EPROM to serve this function. This is of little consequence, given the extra 8K of EPROM introduced by the modification of Problem 8.3.

When the software interrupt instruction is executed, the monitor program checks through the eight program counter values in the data structure, sees the match, and vectors off to the single-step routine. This routine reinitializes the fifth entry in the data structure. Then it displays the CPU registers, followed by the prompt for the next command. The monitor then waits for this next command to be entered.

(*d*) Implement a super-stepping command, SS, using the sixth of the eight data structure entries. Super stepping is discussed near the end of Sec. 8.5. This implementation is a relatively small modification of the single-stepping command which provides a useful variation in performance.

(*e*) Implement a TIME command which is invoked with:

```
TIME    ‹start address›    ‹stop address›
```

This command uses the seventh and eighth data structure entries. It inserts a software interrupt at the two addresses (each of which must hold the first byte of an instruction in the user program). When a software interrupt occurs which leads to the seventh entry, the equivalent of the STARTTIME macro of Problem 8.1 is executed. When a software interrupt leads to the eighth entry, the equivalent of STOPTIME is executed except that execution of the user program remains stopped. The display can show

```
TIME = ‹decimal equivalent of time interval›
```

followed by the prompt for the next command.

8.8. *New monitor commands to oversee interrupt activity.* For this problem we will add three new monitor commands:

```
EIM     Enable interrupt monitoring
DSPM    Display result of previous monitoring
DIM     Disable interrupt monitoring
```

The EIM command makes a copy of the interrupt vectors. Then it modifies the user program so that a call of each interrupt service routine executes an instruction to increment a variable set aside as a counter for interrupts from this source. Then the interrupt service routine vectors off to the original interrupt service routine. What this does is to modify the normal execution slightly so that each servicing of an interrupt results in the incrementing of a counter dedicated to that interrupt source.

When enough execution has taken place to be interesting, the low-cost emulator board is reset, and then the DSPM command is executed. This command reads all of the counters, converts them to BCD, and then displays them, suitably formatted. For example, for the Motorola 68HC11, this display might look like:

```
IRQ          000
RTI          000
TIC1         005
TIC2         027
TIC3         000

    .
    .
    .

SPI          003
SCI          000
```

The DIM command simply restores the original interrupt vectors.

8.9. *New monitor command to profile CPU activity.* For this problem you are to create a new command with the following syntax:

```
ACTIVITY   ‹lower address›   ‹upper address›
```

This command is to determine the percentage of time which the CPU spends executing code which ranges between the two addresses specified. For example, if a 68HC11 mainline program and its subroutines extend from C000 to C123 on the EVB board of Fig. 8.7 (which has RAM for the user program from C000 to DFFF), then we can determine how much time is spent in the mainline program by executing

```
ACTIVITY   C000   C123
```

followed by

```
G C000
```

to begin the execution of the program. To carry out such a test, we need to engage a nonmaskable interrupt so that we can sample CPU activity *independent* of interrupt activity. Consequently, our development board can benefit from the addition of a removable jumper between a programmable timer output and a nonmaskable interrupt input. Given this, then the execution of this command is to start the programmable timer output so that it will generate a square wave output to the nonmaskable interrupt input with a period of perhaps a millisecond. When the nonmaskable interrupt occurs, its interrupt service routine checks the stacked program counter and adds 1, using BCD addition, to a four-digit number IN-RANGE, if the program counter content lies within the specified range. The interrupt service routine also decrements another 2-byte variable, COUNT, which was initialized to 1000 (decimal). When COUNT gets down to 0, we know that 1000 samples of CPU activity have been checked. The nonmaskable interrupt service routine can display the message

```
XX.X% OF CPU TIME SPENT BETWEEN YYYY AND ZZZZ
```

where XX.X is the number in INRANGE, converted to ASCII, while YYYY and ZZZZ are the values used in the address checks. Finally the nonmaskable interrupt service routine can turn off the output square wave from the programmable timer, clean up the stack, and jump back to the monitor program at the point where it generates the prompt on the display.

8.10. *Remapping of 68HC11 registers on the EVB board.* The 68HC11 defaults to having its RAM located at addresses 0000 to 00FF and its registers located at addresses 1000 to 103F. As discussed in the text, the 68HC11 permits RAM and registers to be remapped, providing that the remapping is done within the first 64 clock cycles after the chip comes out of reset. We found some advantage in remapping both RAM and registers to page 0 of the memory space. Change the BUFFALO monitor source code to do this. Reassemble and program an EPROM to hold the new version of the monitor. Install the new monitor program on the EVB board and test it.

REFERENCE

For an excellent book of experiments using the 8096 and Intel's iSBE-96 low-cost emulator board, refer to *The 16-bit 8096: Programming, Interfacing, Applications* by Ron Katz and Howard Boyet [Microprocessor Training Inc., 14 East 8th Street, New York, NY 10003. Phone: (212) 473-4947].

MOTOROLA 68HC11

A.1 OVERVIEW

In this appendix we first consider alternative ways to configure the 68HC11 (e.g., single-chip mode versus expanded mode). Next, we look at each of the on-chip resources, the options available when using each one, how to initialize the resource, and how to interact with it during normal operation. Finally, we detail the instruction set, listing the internal clock cycles and bytes taken by each instruction for each addressing mode.

A.2 CONFIGURING THE 68HC11

To make the 68HC11 operate in its normal single-chip mode, the two pins, MODA and MODB, must be held low and high, respectively, when the chip comes out of its reset state (i.e., when the voltage on the $\overline{\text{Reset}}$ pin goes high). This is illustrated in Fig. A.1a. In this mode, the internal address bus and data bus are inaccessible, with as many pins on the chip as possible dedicated to I/O functions. For the 48-pin DIP package, this means that 38 pins* are available for input-output functions, while for the 52-lead quad pack, 42† pins are available for I/O.

   The 68HC11 can be operated in the expanded multiplexed mode discussed in Chap. 6 by holding *both* the MODA and MODB pins high when the

---

* Thirty-four general-purpose I/O lines plus two interrupt inputs and two handshake control lines.

† Thirty-eight general-purpose I/O lines plus two interrupt inputs and two handshake control lines.

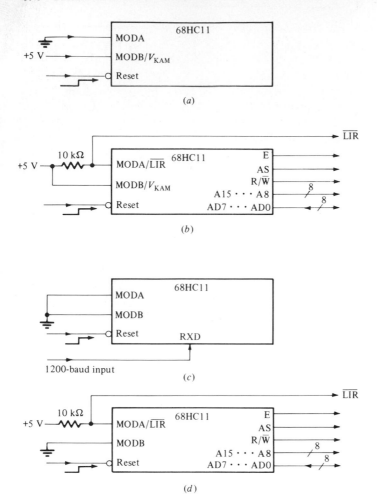

FIGURE A.1
Modes of 68HC11 operation. (*a*) Single-chip mode; (*b*) expanded multiplexed mode; (*c*) boot-strap mode; (*d*) manufacturing test mode.

chip comes out of its reset state, as shown in Fig. A.1*b*. In this case two of the 8-bit ports (ports B and C) available in the single-chip mode become no longer available and are converted to bringing out the internal address bus and data bus. In addition, the two handshaking lines associated with port C in the single-chip mode (STRA and STRB) are no longer available as such and are converted to an address strobe output and a read/write output (R/$\overline{\text{W}}$), used to support data transfers to external chips. Finally, the MODA line serves double duty. After its job at reset time is completed, it serves as an output, going low during the first cycle of each instruction. This is of great help to users of logic analyzers during debugging since it helps sort out each instruction from the next one.

For both of these normal operating modes, the MODB line serves two functions. At reset time, it works with MODA to define the mode of operation as discussed above. In addition, it can serve as a V_{KAM} (keep alive memory) voltage supply during power down. A battery connected to this pin will supply power just to the internal RAM when the voltage on the normal +5 V supply pin drops more than a diode drop (approximately 0.75 V) below the voltage on the V_{KAM} pin.

The bootstrap mode is entered at reset time by grounding both the MODA and MODB lines. This can be a terrific asset for debugging a board controlled by the single-chip 68HC11 since it means that any number of short test programs can be loaded into the 68HC11's RAM (one program per bootstrap loading of the RAM). To carry out such a test, it is only necessary to ground MODB, connect the RXD serial input to an RS-232 port from a personal computer or other test instrument,* and then raise the $\overline{\text{Reset}}$ line to initiate the bootstrap loading operation.

If the 68HC11 is being clocked from an 8-MHz crystal, then the short test program is loaded at 1200 baud, where each character is formatted with a start bit, eight data bits, and a stop bit. The personal computer must transfer exactly 257 bytes, beginning with an opening character of FF (hex), followed by the 256 bytes which will load the internal RAM, beginning at hexadecimal address 0000. When the 257th byte has been transferred, the 68HC11 will load its program counter with 0000 and will begin executing the bootstrapped test program. Consequently, even if the test program consists of less than 256 bytes, it must be filled out with the extra bytes needed to bring the total number of bytes transferred up to 257. It is the reception of the 257th byte which triggers the 68HC11 to begin execution of the test program.

The final mode, shown in Fig. A.1*d*, is a manufacturing test mode. It is used by Motorola during the manufacture of 68HC11 chips to test for proper operation. It can also be used during incoming inspection of 68HC11 chips. In this way, chip performance can be verified before the chip is installed in the application hardware.

Another configuration choice for the 68HC11 consists of where to position the internal RAM and the registers in the memory space. The default choice puts the starting address for the internal RAM at 0000 and the starting address for the 64-byte internal register space at 1000 (hex). If all 256 bytes of RAM are known to be needed, then this choice allows access to all RAM variables using *direct* addressing (i.e., 2-byte instructions).

In one major respect a better choice is to map the 64-byte internal register space down to page 0 (beginning at 0000) and move the RAM to some other location, such as 1000 to 10FF (hex). This permits the internal registers to be accessed using direct addressing. Even more important, it permits the *bit-manipulation* instructions (i.e., BSET and BCLR) as well as the *bit-testing*

---

* Via an interface chip, like one of those shown in Fig. 6.23.

instructions (i.e., BRSET and BRCLR) to use direct addressing to access the assorted bits in the internal registers associated with I/O functions. This is especially valuable because *extended* addressing is not an available addressing mode for these bit-manipulation and bit-testing instructions; we either use direct addressing or indexed addressing, or else we forgo the use of these instructions. Because bit manipulation is so important in a microcontroller environment, this mapping option is especially valuable. While it means that all RAM variables must now be accessed with either extended or indexed addressing, that is not too bad an alternative since virtually all of the 68HC11's instructions (other than the bit-manipulation and the bit-testing instructions) are still available with either of these addressing modes. The "cost" of this memory mapping option is an extra cycle in the execution time of every instruction which accesses RAM as well as an extra byte in the program for each such instruction.

If our application requires no more than 192 bytes of RAM, then the best choice for configuring RAM and registers is to map them *both* to page 0. When this is done, addresses 0000 to 003F access the internal registers while addresses 0040 to 00FF access the internal RAM. Now *all* internal resources (except for the EEPROM and ROM) can be accessed using direct addressing.

Configuring the memory map of the 68HC11 for RAM and registers must be done during the first 64 internal clock cycles after the chip comes out of the reset state. During this time we must write to the INIT register, located at address 103D (hex). The upper "nibble" of the INIT register will then become the upper hex digit of the (four-hex-digit) address of RAM. The lower nibble of the INIT register will become the upper hex digit of the address of the registers. For example, to map the registers to addresses 0000 to 003F and RAM to addresses 1000 to 10FF, we need to execute the following instruction sequence at the beginning of our application program (so that it will be executed immediately after coming out of reset):

```
LDAA #10H        ;RAM = 1xxx and Registers = 0xxx
STAA INIT        ;where INIT  EQU  103DH (initially)
```

As another example, we can map both RAM and registers to page 0 with the single instruction:

```
CLR INIT         ;RAM = 0xxx and Registers = 0xxx
```

As pointed out earlier, each subsequent access to a register address (e.g., 0003) will enable the register and not the RAM. Consequently, even though it appears that a contention problem will arise between RAM and registers during reads from any address between 0000 to 003F, such is not the case. The designers of the chip have disabled RAM for any addresses where RAM and registers overlap. *Throughout this appendix, we assume that the registers have been remapped so that they have page 0 addresses* (whether or not the RAM remains on page 0 or has been remapped to 10xx addresses).

The OPTION register, shown in Fig. A.2, is another register which con-

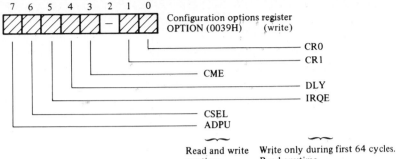

CR1, CR0 set the timeout period for the COP (watchdog) timer.

DLY   determines how fast the 68HC11 resumes operation
         after exiting from the STOP state.

IRQE   determines whether the IRQ pin is falling-edge-sensitive
         or low-level-sensitive.

CME   enables a clock monitor circuit which will cause a reset
         if the 68HC11's clock rate ever drops below 10 kHz
         (or if the clock stops).

CSEL   selects the clock source for the A/D converter and
         the EEPROM charge pump.

ADPU   enables (or disables, and powers down) the on-chip
         analog-to-digital converter.

FIGURE A.2
OPTION register.

tains configuration bits that can be changed within the first 64 cycles of operation after the 68HC11 comes out of reset. Thereafter these selected bits become
read-only. The other bits in the OPTION register can be changed at any time.

The notation we use in describing the operation of each 68HC11 register
by means of a figure requires some explanation. Motorola designed the 68HC11
so that all of its *control* registers can be read as well as written to. However, as
illustrated in Fig. A.2, a control register will be marked with

```
Configuration options register
OPTION (0039H)          (write)
```

when in fact it is actually

```
Configuration options register
OPTION (0039H)     (read,write)
```

This is done to quickly identify a control register whose bits are all set under
program control. In comparison,

```
Parallel I/O control register
PIOC (0002H)       (read,write)
```

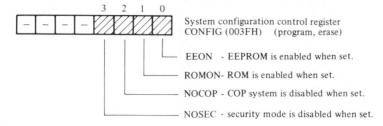

FIGURE A.3
CONFIG (configuration) register, which is implemented as an EEPROM register.

is intended to identify a register* which consists of both control bits (set and cleared under program control) and one or more status bits which cannot be set by writing to the register, but only by the operation of an on-chip facility.

A register which serves solely as a *status* register is identified as illustrated† by the example:

```
Timer interrupt flag register 1
TFLG1 (0023H)                  (read)
```

Each of its bits can be set only by an on-chip facility, not by the execution of a write to the register.

Several configuration options within the 68HC11 are selected or deselected with bits stored in the internal EEPROM. These options have the advantage of requiring neither the use of input pins to the chip nor initial program instructions. However, they imply that a chip may come up in an unexpected mode. These EEPROM bits are all contained in the CONFIG register shown in Fig. A.3. While this register has an address of 003FH (and its content can be read just like any other register), it is written to or erased with the programming or erasing procedure needed by other EEPROM cells. In its erased state, the lower 4 bits of this register contain 1s, enabling both the internal ROM and EEPROM as well as disabling any unexpected resets from the COP (watchdog timer) system and disabling the security system (which, when enabled, permits only single-chip modes to be selected). In the single-chip mode, the designers of the 68HC11 made sure that regardless of the state of the CONFIG register, the ROM at the top of memory will be enabled. Since the reset vector is located at FFFE and FFFF (hex), the CPU will begin execution with the program counter loaded as intended.

The complete register map for the 68HC11 is shown in Fig. A.4. The registers can be divided into the following groups (using addresses obtained

---

* Refer to Fig. A.7.
† See Fig. A.19.

```
PORTA    EQU   0000H   ;I/O Port A
;                      ;Address 0001H is reserved
PIOC     EQU   0002H   ;Parallel I/O control register
PORTC    EQU   0003H   ;I/O port C
PORTB    EQU   0004H   ;Output port B
PORTCL   EQU   0005H   ;Alternate latched port C
;                      ;Address 0006H is reserved
DDRC     EQU   0007H   ;Data direction register for port C
PORTD    EQU   0008H   ;I/O Port D
DDRD     EQU   0009H   ;Data direction register for port D
PORTE    EQU   000AH   ;Input port E
CFORC    EQU   000BH   ;Timer compare force register
OC1M     EQU   000CH   ;Output compare 1 mask register
OC1D     EQU   000DH   ;Output compare 1 data register
TCNT     EQU   000EH   ;Timer counter register
;                      ;   (2 bytes)
TIC1     EQU   0010H   ;Timer input capture 1 register
;                      ;   (2 bytes)
TIC2     EQU   0012H   ;Timer input capture 2 register
;                      ;   (2 bytes)
TIC3     EQU   0014H   ;Timer input capture 3 register
;                      ;   (2 bytes)
TOC1     EQU   0016H   ;Timer output compare 1 register
;                      ;   (2 bytes)
TOC2     EQU   0018H   ;Timer output compare 2 register
;                      ;   (2 bytes)
TOC3     EQU   001AH   ;Timer output compare 3 register
;                      ;   (2 bytes)
TOC4     EQU   001CH   ;Timer output compare 4 register
;                      ;   (2 bytes)
TOC5     EQU   001EH   ;Timer output compare 5 register
;                      ;   (2 bytes)
TCTL1    EQU   0020H   ;Timer control register 1
TCTL2    EQU   0021H   ;Timer control register 2
TMSK1    EQU   0022H   ;Timer interrupt mask register 1
TFLG1    EQU   0023H   ;Timer interrupt flag register 1
TMSK2    EQU   0024H   ;Timer interrupt mask register 2
TFLG2    EQU   0025H   ;Timer interrupt flag register 2
PACTL    EQU   0026H   ;Pulse accumulator control register
PACNT    EQU   0027H   ;Pulse accumulator count register
SPCR     EQU   0028H   ;SPI control register
SPSR     EQU   0029H   ;SPI status register
SPDR     EQU   002AH   ;SPI data I/O register
BAUD     EQU   002BH   ;SCI baud rate generator
SCCR1    EQU   002CH   ;SCI control register 1
SCCR2    EQU   002DH   ;SCI control register 2
SCSR     EQU   002EH   ;SCI Status register
SCDR     EQU   002FH   ;SCI data register, consisting of
RDR      EQU   002FH   ;SCI receive data register (for reads)
TDR      EQU   002FH   ;SCI transmit data register (for writes)
```

FIGURE A.4
68HC11 registers (with addresses shown *after* mapping them to page 0).

```
ADCTL   EQU   0030H   ;A/D control/status register
ADR1    EQU   0031H   ;A/D result register 1
ADR2    EQU   0032H   ;A/D result register 2
ADR3    EQU   0033H   ;A/D result register 3
ADR4    EQU   0034H   ;A/D result register 4
;                     ;Address 0035H is reserved
;                     ;Address 0036H is reserved
;                     ;Address 0037H is reserved
;                     ;Address 0038H is reserved
OPTION  EQU   0039H   ;Configuration options register
COPRST  EQU   003AH   ;Arm/reset COP timer circuitry
PPROG   EQU   003BH   ;EEPROM programming control register
HPRIO   EQU   003CH   ;Highest priority I-bit interrupt register
INIT    EQU   003DH   ;RAM and I/O mapping register
TEST1   EQU   003EH   ;Factory test control register
CONFIG  EQU   003FH   ;System configuration control register
```

FIGURE A.4 (*continued*)

after remapping the registers to page 0, as described above):

Addresses	Function
0000-000A	General purpose I/O ports
000B-0027	Programmable timer facility
0028-002A	Serial peripheral interface (serial I/O port)
002B-002F	Serial communication interface (UART)
0030-0034	A/D converter
0039-003F	Configuration registers

A.3   GENERAL-PURPOSE I/O PORTS

The general purpose I/O ports take lines which *could* be used in conjunction with one of the special purpose I/O facilities on the chip (e.g., the timer or the UART) and make those lines available for general use if the special purpose I/O facility is not needed in a specific application. For example, Fig. A.5 shows port A being shared between the programmable timer, the pulse accumulator, and general purpose I/O lines. The three inputs on bits 0, 1, and 2 can actually serve double duty. Each can serve as a timer input, permitting the *time* of a transition on that input to be recorded and to generate an interrupt, if desired. In addition, by reading port A, we can sense these inputs directly.

The port A outputs on bits 3 to 6 can be controlled by writes to the port itself (at address 0000). Alternatively, we can set up the programmable timer to control any number of these outputs. A write to port A does not affect any output lines which have been previously set up to be controlled by the timer.

Bit 7 of port A is a special case. It defaults to being an input bit, read by reading port A (at address 0000). At the same time, it can serve as the input to a

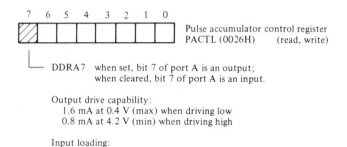

Pulse accumulator control register
PACTL (0026H)   (read, write)

DDRA7   when set, bit 7 of port A is an output;
        when cleared, bit 7 of port A is an input.

Output drive capability:
1.6 mA at 0.4 V (max) when driving low
0.8 mA at 4.2 V (min) when driving high

Input loading:
10 μA (max) on port A, bit 7
1 μA (max) on port A, bits 0, 1, 2

FIGURE A.5
Optional uses for port A.

pulse accumulator facility, for counting pulses over a given interval of time or for counting time (i.e., clock pulses) over an interval determined by the input pulse. As shown in Fig. A.5, bit 7 of PACTL, the pulse accumulator control register (at address 0026H) serves as a data direction bit. This bit is cleared at reset time, making bit 7 of port A into an input. If we set bit 7 of PACTL, bit 7 of port A becomes an output, to be controlled by writes to port A unless the programmable timer has been set up to control this output. We discuss in detail the workings of the programmable timer and the pulse accumulator in Sec. A.6.

In contrast to port A, port B is simply an output port, as shown in Fig. A.6. Writes to port B (at address 0004H) change the output lines accordingly. Reads of port B will read what has been previously written to port B. As was pointed out earlier, when the 68HC11 is operated in the expanded mode, port B is lost. In that case its lines serve as the upper half of the address bus.

Port C is a bidirectional port. By writing 1s into selected bits of the data direction register, DDRC, selected I/O lines of port C are set up as outputs. The remaining I/O lines (which have corresponding 0s in DDRC) default to being inputs.

The parallel I/O control register, PIOC, can be used to set up a wide

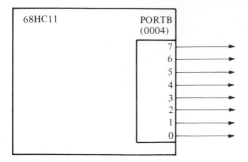

Output drive capability:

1.6 mA at 0.4 V (max) when driving low
0.8 mA at 4.2 V (min) when driving high

FIGURE A.6
Port B.

variety of options for the use of port C and for the use of the handshaking lines (STRA and STRB). For example, port C is shown set up as a *handshaking output port* in Fig. A.7. In this mode, to change the output port, PIOC is first read and then the write is made to PORTCL (at address 0005) rather than PORTC (at address 0003). While a write to either address will actually change the output port, only the write to PORTCL will clear the strobe A interrupt status flag, STAF (bit 7 of PIOC) and drive the STRB output handshake line low. When the device connected to port C sees that STRB has gone low, it knows that new output data is available. After it reads this data and is ready for another byte, it drives the STRA input handshake line low, indicating that it acknowledges receiving the output data. This will set the STAF bit, cause the STRB output handshake line to go high, and optionally generate an interrupt (assuming we want to handle the data transfers under interrupt control). The STRB line will remain high, to indicate that the output data is no longer valid, until the port is written to again. If we ever want to *read* the data which has been written to port C, we should read PORTC, not PORTCL. PORTCL holds a *latched* version of what is out on port C. The latching occurs on the falling edge of the STRA input handshake line, *even when* the port is being used as an output port. Consequently, PORTCL *may* hold the correct value of what has been written to the port, but only if the external handshaking device has already read the data written to the port and has carried out its part of the handshake.

While we have looked at the setup for a handshaking output port which uses active-low handshaking lines, the designers of the chip actually permit either or both handshaking lines to be active high. The selection is made by the initialization of the PIOC register.

Port C can also be set up as a *handshaking input port*, as shown in Fig. A.8. In addition to initializing the DDRC and PIOC registers as shown, a read of PIOC followed by a read of PORTCL will clear the STAF flag as well as drive the STRB output handshake line to its low, asserted state, indicating "Ready for new input data." Then, after interrupts are enabled, a device will handshake in a byte of data by putting it on port C and then driving the STRA

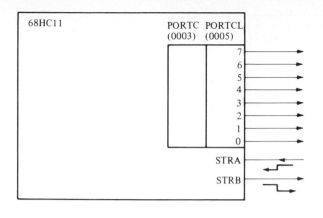

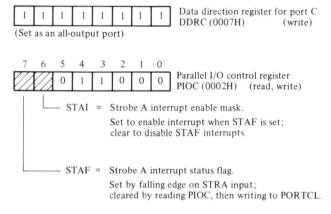

FIGURE A.7 in bold below, with image 2 containing the register diagrams:

| 1 | 1 | 1 | 1 | 1 | 1 | 1 | 1 |

(Set as an all-output port)

Data direction register for port C
DDRC (0007H)   (write)

7 6 5 4 3 2 1 0

| ///// | 0 | 1 | 1 | 0 | 0 | 0 |

Parallel I/O control register
PIOC (0002H)   (read, write)

STAI = Strobe A interrupt enable mask.
Set to enable interrupt when STAF is set;
clear to disable STAF interrupts.

STAF = Strobe A interrupt status flag.
Set by falling edge on STRA input;
cleared by reading PIOC, then writing to PORTCL.

Output drive capability: same as port B.

FIGURE A.7
Port C used as a handshaking output port.

line low. This will set the STAF bit and generate an interrupt. It will also automatically drive the STRB output handshake line back to its high, inactive state. When the 68HC11 responds to the interrupt, it reads PIOC and then reads PORTCL to get the byte of data, to clear the STAF flag, and to drive the STRB output handshake line low, again to indicate "Ready to accept more data." When the external device sees STRB low, it will present a new byte of data to port C and will pulse STRA again.

While PORTC *could* be read, this would not trigger the handshaking mechanism. On the other hand, if the input data is not stable for some nanoseconds after the input handshake occurs, then PORTCL will latch the wrong data. In this case, *both* PORTC and PORTCL should be read in the interrupt service routine. PORTC is read to obtain the data on port C as it is at the time of reading. PORTCL is read to carry out the handshake.

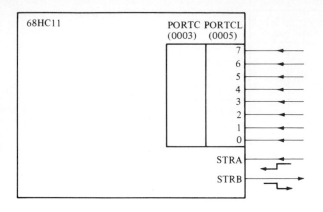

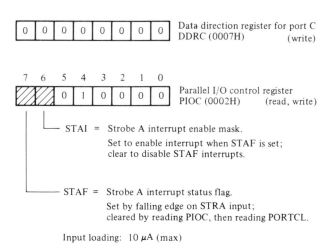

0	0	0	0	0	0	0	0

Data direction register for port C
DDRC (0007H)   (write)

7 6 5 4 3 2 1 0

		0	1	0	0	0	0

Parallel I/O control register
PIOC (0002H)   (read, write)

STAI = Strobe A interrupt enable mask.

Set to enable interrupt when STAF is set;
clear to disable STAF interrupts.

STAF = Strobe A interrupt status flag.

Set by falling edge on STRA input;
cleared by reading PIOC, then reading PORTCL.

Input loading: 10 μA (max)

FIGURE A.8
Port C used as a handshaking input port.

Another use of the STRA and STRB lines is shown in Fig. A.9. Instead of full handshaking, this application associates a strobe line with each of two separate ports. Port B, which is necessarily an output port, uses STRB as an output strobe, set up to produce a 1-μs-long strobe pulse* each time data is written to port B. Port C is set up as an input port, and STRA is used to generate an interrupt whenever a falling edge occurs on that line. In this case, port C and the STAF bit are dealt with in exactly the same way as they were for Fig. A.8.

As is the case for port B, when the 68HC11 is operated in the expanded mode, port C is lost as well as the STRA and STRB handshake lines. The port C lines become the lower half of the address bus multiplexed with the data bus.

---

* When the 68HC11 is clocked with an 8-MHz crystal.

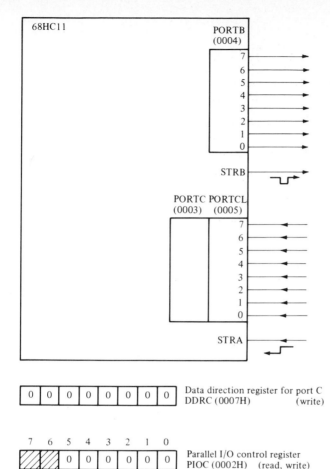

FIGURE A.9
Use of strobe lines with separate ports.

The STRA and STRB lines become AS (address strobe) and R/$\overline{\text{W}}$ (read/write) lines.

Port D has six lines, some of which may be available as general purpose I/O lines if the serial I/O capabilities of the 68HC11 are underutilized. As shown in Fig. A.10a, two of these lines are associated with the serial communication interface (UART). The other four are associated with the serial peripheral

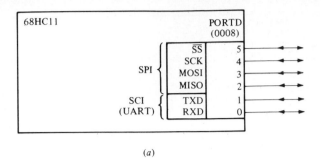

(a)

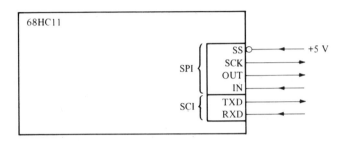

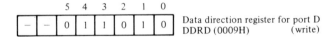

Output drive capability: same as port B
Input loading: 10 μA (max)

(b)

FIGURE A.10
Port D. (a) General configuration; (b) setup of DDRD required for using SPI and SCI.

interface (I/O serial port). Figure A.10b shows the setup required of DDRD (the data direction register for port D, located at address 0009H) when the SPI and SCI capabilities are both used. Note that it is not enough to deal *just* with the registers associated with the SPI and SCI; if the lines which are intended to be outputs do not have 1s in DDRD, the signals coming from the SPI and SCI will never make it all the way to the output pins of the chip.* The serial peripheral interface is shown in Fig. A.10b set up as a "master" device. This is the setup used to expand the 68HC11 for interacting with a variety of I/O devices, as discussed in Chaps. 6 and 7. The alternative of using the SPI as a "slave"

---

* The input data to the SPI will be received *regardless* of the DDRD bits. On the other hand, the DDRD register is overridden if the SCI function is enabled.

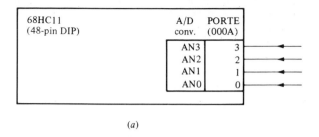

Input loading: 1 μA (max)

(b)

FIGURE A.11
Port E. (*a*) 48-pin DIP package;
(*b*) 52-pin PLCC package.

device requires that the DDRD be loaded so as to make SS, SCK, and MOSI inputs and to make MISO an output.

The SCI* and the SPI both come up disabled at reset. Consequently, if one or both of these utilities are not used, then the corresponding lines of port D can be used as either inputs or outputs. Simply load DDRD appropriately and then read PORTD (at address 0008H) to access the inputs and write to PORTD to change the outputs.

The remaining port of the 68HC11 to be discussed is port E, shown in Fig. A.11. While any number of these lines may be used as analog inputs, the remaining lines are available as general purpose (digital) inputs with no special setup required. That is, a read of PORTE (at address 000AH) will treat *all* of the lines as digital inputs, even those actually being handled with the A/D converter. However, a read of PORTE while an A/D conversion is taking place will impair the accuracy of the conversion.

---

* Except in the bootstrap mode.

A.4   RESETS AND SELF-PROTECTION OPTIONS

When the $\overline{\text{Reset}}$ pin on the 68HC11 is driven low by an external source, the chip is forced into its reset sequence of operations, *regardless* of what it is doing. Pulling the $\overline{\text{Reset}}$ line low will invariably get the attention of the 68HC11!

The $\overline{\text{Reset}}$ pin is a bidirectional line. Not only can we pull it low to force a reset sequence, it can pull *itself* low as it carries out one of its self-protection mechanisms. Because of this, *our* reset circuitry must be designed for pulling the line low at reset time and for providing a high impedance pullup at all other times.

In Chap. 2 we discussed power-down considerations and looked at how we could provide both battery backup and reliable startup and shutdown as ac power is turned on and shut off. However, suppose our application does not require battery backup. Even in this case, the only way to be absolutely certain of ensuring the integrity of EEPROM data is to keep the CPU from thrashing as it powers up and down. After all, if we have code which is designed to erase part or all of the EEPROM as well as write to it, then we cannot afford to have a powering-down CPU do a faulty load of the program counter with an address into this code while it is still trying to execute instructions.

Most power-up circuits avoid a thrashing CPU by using an *RC* circuit in the reset circuit which has a longer time constant than the relatively short time constant of the power supply. While this helps power up, it does nothing for power down. The reliable alternative is to control the $\overline{\text{Reset}}$ pin by sensing the supply voltage and using a *voltage threshold* to drive the $\overline{\text{Reset}}$ pin high or low. Such a circuit is shown in Fig. A.12.

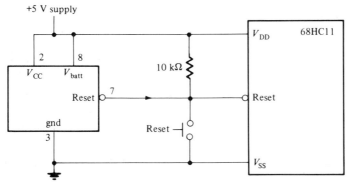

Maxim Integrated Products
  MAX690
CMOS reset generator
   8-pin DIP
Threshold voltage = 4.65 V

FIGURE A.12
Voltage-threshold-sensing reset circuit.

As the 68HC11 comes out of reset, the CPU does the following things automatically:

1. It sets the X, I, and S bits in the condition code register so that interrupts are initially disabled and so that the stop mode is initially disabled.
2. The INIT register is initialized to 01H. This puts RAM at 0000 to 00FFH and the registers at 1000 to 103FH.
3. All bidirectional I/O lines are configured as inputs.
4. All interrupt sources are initially disabled. All facilities which can drive I/O pins are initially disabled.
5. The CPU fetches the restart vector from FFFE and FFFFH, loads this into the program counter, and begins executing instructions.

As pointed out earlier, the CPU may or may not find the internal EEPROM and ROM because of the states of some bits in the CONFIG register which is implemented in EEPROM. For the same reason, it may or may not come up with the COP (computer operating properly) facility enabled, again because the enabling or disabling is done with an EEPROM bit in the CONFIG register.

The watchdog timer (COP) facility provides a second reset mechanism. The registers associated with the watchdog timer are shown in Fig. A.13. Any of the four timeout periods may be selected, if done so right when power first comes up.* Then, to avoid having the watchdog timer reset the CPU, our system software must execute the following sequence periodically, with the time between executions of the sequence *never* exceeding the selected timeout period.

```
LDAA #55H
STAA COPRST
COMA
STAA COPRST
```

In this way, if the 68HC11 ever malfunctions and stops executing the intended software (for whatever reason), the watchdog timer reset sequence will not be executed, the watchdog timer will time out, the 68HC11 will be reset, and the $\overline{\text{Reset}}$ pin will be pulled low (so that any other chips connected to this $\overline{\text{Reset}}$ line will also be reset). The 68HC11 will be initialized to the same state as if the $\overline{\text{Reset}}$ pin had been pulled low by an external source. However, it will vector to the service routine whose address is stored in addresses FFFA and FFFBH. If the response desired is identical to that for a normal reset, then it is only necessary to have the content of FFFA and FFFBH be the same as that for the reset vector in FFFE and FFFFH.

The 68HC11 includes a clock monitor option which has the ability to reset

---

* Within the first 64 internal clock cycles after reset.

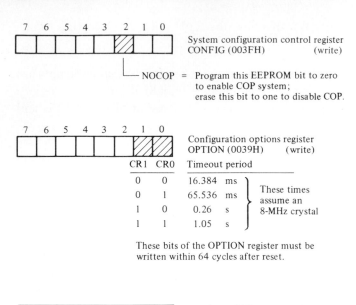

System configuration control register
CONFIG (003FH)   (write)

NOCOP = Program this EEPROM bit to zero
to enable COP system;
erase this bit to one to disable COP.

Configuration options register
OPTION (0039H)   (write)

CR1	CR0	Timeout period	
0	0	16.384 ms	These times assume an 8-MHz crystal
0	1	65.536 ms	
1	0	0.26 s	
1	1	1.05 s	

These bits of the OPTION register must be
written within 64 cycles after reset.

Arm/reset COP timer circuitry
COPRST (003AH)   (write)

Write 55H followed by AAH to COPRST before
timeout occurs, to reset watchdog timer
(so that watchdog timer will *not* reset 68HC11).

15	0	
Watchdog timer		FFFAH
reset vector		FFFBH

FIGURE A.13
68HC11 watchdog timer (COP).

the chip if the clock ever slows down excessively or stops. As shown in Fig.
A.14, it is enabled by writing a 1 into bit 3 of the OPTION register and disabled
by clearing this bit. This can be done at any time. In particular, if an application
makes use of this feature as well as the STOP function (which stops the clock),
then before executing the instructions to stop the 68HC11, we will want to
disable the clock monitor function. Then after beginning operation again, we
may want to reenable the clock monitor function.

A clock failure will be detected if the CPU's internal clock rate ever stops,
or if it drops below 10 kHz. It will not be detected for any clock rate above 200
kHz. It may or may not be detected if the clock rate drops to a rate somewhere
in between. If a clock failure is detected, then the reset sequence will occur,
culminating with the loading of the clock monitor reset vector into the program
counter from addresses FFFC and FFFDH. Again, this action will also result in

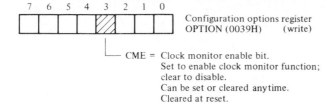

FIGURE A.14
68HC11 clock monitor.

the $\overline{\text{Reset}}$ pin being pulled low from within the 68HC11 so that any other chips tied to this line will also be reset.

A.5   INTERRUPTS

The 68HC11 uses an interrupt-handling mechanism which is clean cut in the sense that almost every possible source of an interrupt possesses a unique interrupt vector. That is, when an interrupt occurs, we automatically get to the interrupt service routine which will handle the specific needs of the device or facility needing service. The interrupts are listed in Fig. A.15 in the order of their priority. The top five interrupt sources respond to events which call for drastic action. We have already considered the first three of these, each of which resets the 68HC11 and then vectors to its own restart program.

An illegal opcode trap interrupt is invoked any time the CPU tries to fetch an instruction and does not recognize what it has fetched as a legitimate instruction. Since this cannot be turned off, either in the CPU or with a local mask, it makes sense not to ignore this feature, but at least to treat it in the same way as a reset. For example, we might handle all four of the top interrupt vectors in the following way (unless we write a specific routine to be executed for any one of them):

```
ORG    0FFF8H
FDB    START      ;Vector for illegal opcode trap
FDB    START      ;Vector for COP failure
FDB    START      ;Vector for clock failure
FDB    START      ;Vector for reset
```

where FDB is Motorola's form double byte assembler directive which tells the assembler to store the address (or value) represented by the operand as a 2-byte

Interrupt source	Vector address	CC register mask	Local mask	Local flag	Number of sources with this vector
Reset	FFFE,FFFF	None	None	None	1
COP clock monitor fail (reset)	FFFC,FFFD	None	CME	None	1
COP failure (reset)	FFFA,FFFB	None	NOCOP	None	1
Illegal opcode trap	FFF8,FFF9	None	None	None	1
XIRQ	FFF4,FFF5	X bit	None	None	1
Selected I-bit-maskable interrupt	(see below)	I bit	(................see below................)		
IRQ pin or parallel I/O	FFF2,FFF3	I bit	None	None	2
			STAI	STAF	
Real-time interrupt	FFF0,FFF1	I bit	RTII	RTIF	1
Timer input capture 1	FFEE,FFEF	I bit	IC1I	IC1F	1
Timer input capture 2	FFEC,FFED	I bit	IC2I	IC2F	1
Timer input capture 3	FFEA,FFEB	I bit	IC3I	IC3F	1
Timer output compare 1	FFE8,FFE9	I bit	OC1I	OC1F	1
Timer output compare 2	FFE6,FFE7	I bit	OC2I	OC2F	1
Timer output compare 3	FFE4,FFE5	I bit	OC3I	OC3F	1
Timer output compare 4	FFE2,FFE3	I bit	OC4I	OC4F	1
Timer output compare 5	FFE0,FFE1	I bit	OC5I	OC5F	1
Timer overflow	FFDE,FFDF	I bit	TOI	TOF	1
Pulse accumulator overflow	FFDC,FFDD	I bit	PAOVI	PAOVF	1
Pulse accumulator input edge	FFDA,FFDB	I bit	PAII	PAIF	1
SPI serial transfer complete	FFD8,FFD9	I bit	SPIE	SPIF	1
SCI serial system	FFD6,FFD7	I bit	TIE	TDRE	5
			TCIE	TC	
			RIE	OR,RDRF	
			ILIE	IDLE	
SWI	FFF6,FFF7	None	None		1

FIGURE A.15
68HC11 interrupt sources, listed by priority, with highest priority listed at the top.

constant. In this example, START is the label used to identify the instruction where our code is to begin execution at power-up time.

XIRQ is the last of the five interrupts calling for drastic action. If it is enabled, then it will be invoked when the $\overline{\text{XIRQ}}$ pin is pulled low. Since a *low level* and not a *falling edge* is sensed, the XIRQ interrupt service routine must interact with the hardware which pulled the line low and restore it to a high level before the RTI instruction is executed, returning from the interrupt service routine. Furthermore, the $\overline{\text{XIRQ}}$ input *requires* a pullup resistor (perhaps 100 kΩ to the supply voltage) for proper operation. At reset time, the X bit (bit 6 of the condition code register) is set, disabling this interrupt source. It will remain disabled until we use the TAP instruction to load the condition code register with the content of accumulator A. Thus, if we want to use this feature of the 68HC11, we would first do whatever setup is called for and then execute the following instructions:

```
TPA
ANDA #10111111B
TAP
```

Once the X bit has been cleared, software cannot set it again. In this sense, this interrupt source can be a nonmaskable interrupt. Unlike the nonmaskable interrupt input to many processors, this one has the advantage of giving the processor a chance to set up the conditions needed for the correct servicing of this source *before* such an interrupt can occur.

A nonmaskable interrupt is often used to respond appropriately, when power goes down unexpectedly, as discussed in Chap. 2. An external power-sensing circuit may be included to detect that the supply voltage has dropped below a certain point and to warn the 68HC11 by pulling the $\overline{\text{XIRQ}}$ pin low. In the last milliseconds before power drops to the point where CPU operation becomes unreliable, the CPU can protect itself as appropriate. It might put the chip in a low power STOP state and then use a battery to maintain this low power to the chip, thereby protecting the contents of its internal RAM as well as the state of all of its registers. Alternatively, if battery backup is applied only to the MODB/V_{KAM} pin,* then just the internal RAM contents will be preserved, with a further reduction in power consumption.

Referring back to Fig. A.15, after these first five drastic action interrupt sources, we get to the interrupt sources which are maskable by the I bit (bit 4 of the condition code register). At reset time the I bit is set, disabling the CPU from responding to any of these sources. Furthermore, each of the local mask bits is cleared at reset time, disabling each of these sources from generating an interrupt. Part of the initialization routine required for an application consists of setting up selected interrupt sources, enabling these sources, and then enabling the CPU to receive them by executing a CLI instruction (which clears the I bit).

When one of these interrupts occurs (with I=0, enabling the CPU to receive interrupts), the CPU does the following:

1. It completes the instruction it is presently executing.
2. It stacks all of the CPU registers (other than the stack pointer).
3. It sets the I bit, disabling further interrupts while in the interrupt service routine.
4. It vectors to the appropriate interrupt service routine.

Within the interrupt service routine:

1. The code is executed to carry out the appropriate action for this device or facility.

---

* See Fig. A.1.

2. The flag set when the interrupt occurred must be cleared if a subsequent interrupt from this source is to be handled. Alternatively, the local mask for this source must be cleared if no further interrupts from this source are to be handled.

3. An RTI instruction must be executed.

The RTI instruction restores the previous CPU register contents from the stack. In so doing, it reenables maskable interrupts by clearing the I bit (by restoring the condition code register from the stack).

It is at this moment of coming out of an interrupt service routine and re-enabling interrupts that the 68HC11 sorts out the priority of pending interrupts. If two or more sources of interrupts have asked for service while interrupts were disabled, then when interrupts are reenabled, the CPU will vector to the service routine for the highest priority pending interrupt. As long as all interrupt service routines are kept short, then the effect of this response to interrupts is not much different from the response obtained by a chip like the Intel 8096 which reenables *higher priority* interrupts from *within* an interrupt service routine. The Motorola approach has an advantage over that used by the Intel 8096 in that almost all of the interrupt service routines know what they have to do without polling. In contrast, the Intel 8096 typically carries out a polling routine at the beginning of an interrupt service routine in order to determine which of several possible events caused the interrupt so that it can respond accordingly.

While the priority of the Motorola 68HC11 interrupts is fixed, we can elevate any *one* of the I-bit-maskable interrupts to the highest priority. This is done by initializing the lower 4 bits of the HPRIO register (at address 003CH) with a value selected from the table in Fig. A.16. The default value selects the IRQ-parallel I/O interrupt source (as it should so as to leave the priority of Fig. A.15 intact, should we choose not to use this feature).

The $\overline{\text{IRQ}}$ pin on the 68HC11 can be used as either a low-level-sensitive or as a falling-edge-sensitive input, as shown in Fig. A.2 and again in more detail in Fig. A.17. This $\overline{\text{IRQ}}$ pin requires a pullup resistor for proper operation. If the prescribed event occurs and if interrupts are enabled in the CPU (i.e., if the I bit is cleared), then the interrupt will be handled since there is no local mask.

The vector for this IRQ interrupt is shared with the interrupt discussed earlier in conjunction with port C and the STRA input. The PIOC register holds the 3 bits associated with this interrupt source, shown in Fig. A.18. The STAF flag is set when the interrupting event occurs. It is cleared as discussed earlier. If just one of these two interrupt sources is used, then no polling is required to distinguish between the two. On the other hand, if both are used, then we can determine whether STRA caused the interrupt by testing the STAF flag. If not, then we service IRQ. If so, then we service STRA. In this latter case we can still wonder whether an IRQ interrupt occurred also, and we need some way to determine this. If the IRQ interrupt is set up to be level sensitive (i.e., if IRQE=0 in the OPTION register) and if the device driving $\overline{\text{IRQ}}$ low holds the

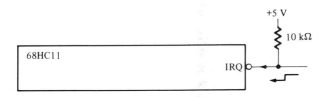

7	6	5	4	3	2	1	0	
–	–	–	–	▨	▨	▨	▨	Highest priority I-bit interrupt register HPRIO (003CH)　　　(write)

PSEL3	PSEL2	PSEL1	PSEL0	
0	0	0	0	Timer overflow
0	0	0	1	Pulse accumulator overflow
0	0	1	0	Pulse accumulator input edge
0	0	1	1	SPI serial transfer complete
0	1	0	0	SCI serial system
0	1	0	1	Reserved (default to IRQ)
0	1	1	0	IRQ pin or parallel I/O
0	1	1	1	Real-time interrupt
1	0	0	0	Timer input capture 1
1	0	0	1	Timer input capture 2
1	0	1	0	Timer input capture 3
1	0	1	1	Timer output compare 1
1	1	0	0	Timer output compare 2
1	1	0	1	Timer output compare 3
1	1	1	0	Timer output compare 4
1	1	1	1	Timer output compare 5

FIGURE A.16
Setting the highest priority interrupt source.

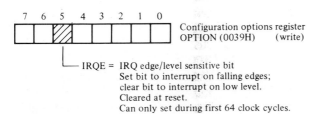

FIGURE A.17
IRQ interrupt input.

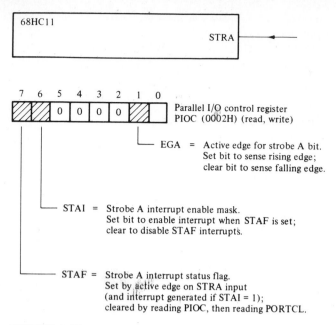

FIGURE A.18
Parallel I/O interrupt input.

line low until we take some action, then after servicing the STRA interrupt we simply return from the interrupt service routine and the IRQ interrupt will reassert itself. This time our servicing should find the STAF flag clear and so we service IRQ.

The SWI interrupt listed at the bottom of Fig. A.15 is the software interrupt. It is invoked whenever the CPU executes the SWI instruction (having opcode 3FH). Because of its low priority, it will be put off by *any* other pending interrupt. On the other hand, the CPU will always get to a SWI instruction and execute its interrupt service routine since there is no way to mask it. This instruction is used by the monitor program for Motorola's EVM development board, discussed in Chap. 8, to help implement both breakpoints and single stepping. To insert a breakpoint into a user program, the monitor program inserts a SWI instruction in place of the user instruction at the point where the breakpoint is to occur. This SWI instruction ensures that control will be returned to the monitor program, which can then sort out what to do next. In the case of single stepping, the monitor program knows what instruction of the user program is supposed to be executed next. It uses the opcode for that instruction to determine whether it is a 1-byte instruction, a 2-byte instruction, or whatever. Then it inserts an SWI instruction *after* this next instruction. In this way when the monitor program turns control of the CPU over to the user program, it knows that control will be returned to the monitor program after the user program has executed all pending interrupt service routines (if the user state has interrupts enabled) plus the one instruction pointed to by the user's program counter contents which was supposed to be executed next.

We will discuss all of the other interrupts of Fig. A.15 as we discuss the programmable timer, the serial peripheral interface, and the serial communication interface.

A.6 PROGRAMMABLE TIMER

The 68HC11's programmable timer is a versatile, well thought out facility offering many opportunities to measure and control timing. As was illustrated in Fig. A.5, the timer shares I/O pins with port A.

The output compare function represents one of the fundamental features of the timer. The designers of the 68HC11 realized they had a good thing and put *five* of them on the chip! The registers associated with one of these, timer output compare 5, are shown in Fig. A.19. With six registers involved, there are obviously a variety of options available. The first register shown, TMSK2, has 2 bits in it which must be configured within 64 cycles after reset. These 2 bits set up a prescaler for the free-running counter TCNT. Since both the prescaler and the free-running counter are used by all five of the output compare functions and all three of the input capture functions, the selection made for these 2 bits of TMSK2 will affect all measured and controlled times. The default values for these 2 bits are 00. Thus TCNT is clocked twice every microsecond,* giving the highest possible resolution both for measuring and for controlling time intervals. It means that the free-running counter will count through a complete cycle of 65,536 counts every 32.768 ms. While time intervals longer than this can be measured and controlled by counting cycles of TCNT, it is easier to pick a different prescaler value if this leads to all measured and controlled time intervals being less than one complete cycle of the free-running counter. A prescaler of 16 will mean that the 65,536 counts of TCNT take over 0.5 s. This same prescaler value of 16 will cut the resolution on the measured and controlled time intervals to 8 μs.

While the prescaler and free-running counter are shared by all of the timer's inputs and outputs, the output compare register 5 and the other bits shown in Fig. A.19 are employed solely by the OC5 function. Every time the free-running counter (which is clocked continuously) equals TOC5, the timer circuitry sets the OC5F bit in TFLG1. Even if this facility is not being used, the timer circuitry will come back and set the OC5F bit again and again. Of course, once the bit is set, it remains set until we take the appropriate action to clear it. The action taken by the timer when OC5F is set depends upon how we have initialized the TMSK1 and TCTL1 registers. If we want to generate an interrupt when the compare takes place, then we need to set the OC5I bit in TMSK1. In addition, whether or not an interrupt takes place, when this compare occurs,

---

* Assuming an 8-MHz crystal clock frequency.

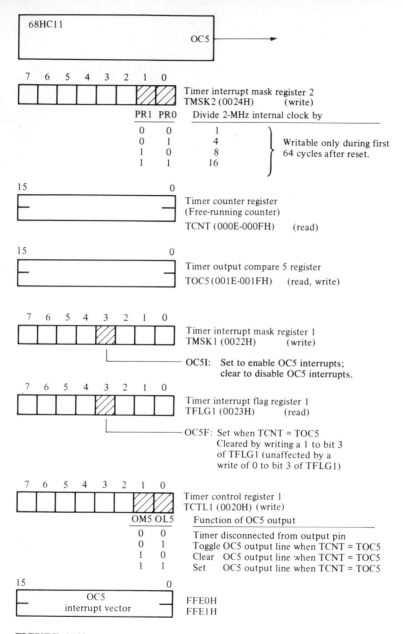

FIGURE A.19
Timer output compare 5.

we can cause the OC5 pin to be set, cleared, toggled, or left alone, according to the 2-bit value of the OM5 and OL5 bits in the TCTL1 register. In Chap. 2 we saw some of the ways this capability could be used. Here it is only necessary to realize that the bit-setting and bit-clearing instructions, BSET and BCLR, are particularly useful for changing TCTL1 and TMSK1 without changing any of the other timer capabilities incorporated in these two registers. Clearing the OC5F bit of the TFLG1 register is another story. Since the TFLG1 register contains other flag bits which can be set at any time, the Motorola designers have come up with a 100 percent reliable mechanism for clearing one flag bit and being absolutely certain of leaving the other flag bits alone. A write to the TFLG1 register is carried out with a 1 in the bit position for the flag to be cleared and with 0s in the other bit positions. Thus, to clear the OC5F flag, the following instructions are executed:

```
LDAA #00001000B
STAA TFLG1
```

When the hardware of the timer senses that TFLG1 is being written to, it clears any bits of TFLG1 for which there are 1s on the data bus and leaves the remaining bits of TFLG1 alone.

Note that one of the options for this timer is to generate an interrupt after a known time interval, using OM5=0, OL5=0. This use of the timer does not make use of the OC5 output pin on port A. Consequently, that pin (bit 3 of port A, as seen in Fig. A.5) can be used for any other purpose, with the BSET and BCLR instructions operating on the PORTA register address.

The registers and bits associated with output compare 2, 3, 4, and 5 are shown in Fig. A.20. They each function in the same way as has just been discussed.

Output compare 1 differs from the other four in that it can affect all five outputs, not just its own output. The registers and bits associated with it are shown in Fig. A.21. The new feature is held in registers OC1M and OC1D. Motorola has named these bits so as to align with the bits of port A. Thus OC1D3 and OC1M3 are aligned with bit 3 of port A, which is the output for OC5. When a compare takes place between TCNT and TOC1, the data in OC1D3 will be written to bit 3 of port A (i.e., to the OC5 output) if the OC1M3 bit equals 1.* If the OC1M3 bit equals 0, then bit 3 of port 3 will be unaffected by the compare. We can use this capability to generate up to four pulse-width-modulated outputs, on OC5, OC4, OC3, and OC2, as discussed in Sec. 7.5. These outputs can serve as extremely high-resolution D/A converters in many applications without even needing the support of interrupts. After the initial setup, it is only necessary to write a new value into TOC5 (for example) to change and maintain a new value of modulation on the OC5 output. The DC component of this modulated output will take on any one of 65,536 values.

---

* The action desired on the output lines when this compare occurs can be made to happen prematurely, under program control, by writing 1s to the selected port bit positions in the CFORC register (address = 000BH).

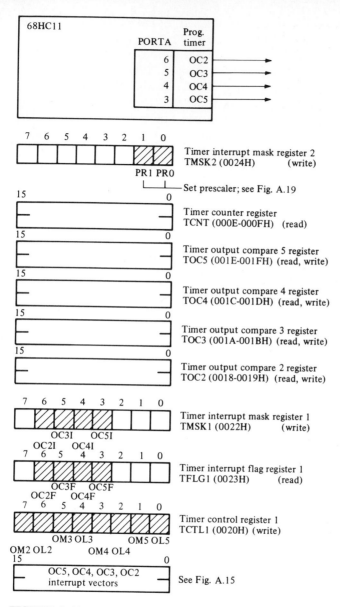

FIGURE A.20
Timer output compare 2, 3, 4, and 5.

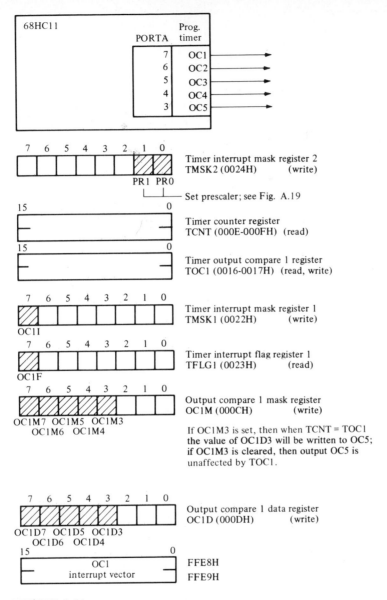

FIGURE A.21
Timer output compare 1.

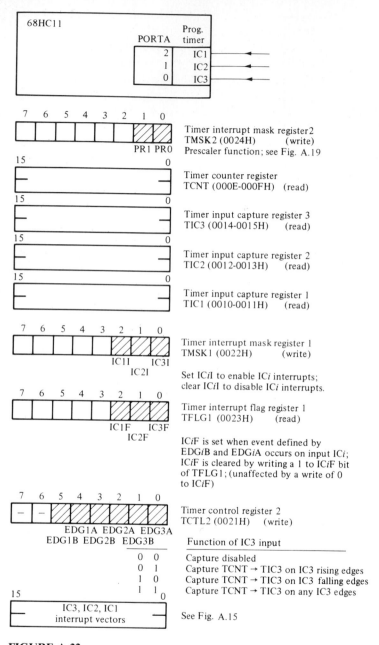

Timer interrupt mask register 2
TMSK2 (0024H)   (write)
Prescaler function; see Fig. A.19

Timer counter register
TCNT (000E-000FH)   (read)

Timer input capture register 3
TIC3 (0014-0015H)   (read)

Timer input capture register 2
TIC2 (0012-0013H)   (read)

Timer input capture register 1
TIC1 (0010-0011H)   (read)

Timer interrupt mask register 1
TMSK1 (0022H)   (write)

Set ICil to enable ICi interrupts;
clear ICil to disable ICi interrupts.

Timer interrupt flag register 1
TFLG1 (0023H)   (read)

ICiF is set when event defined by
EDGiB and EDGiA occurs on input ICi;
ICiF is cleared by writing a 1 to ICiF bit
of TFLG1; (unaffected by a write of 0
to ICiF)

Timer control register 2
TCTL2 (0021H)   (write)

Function of IC3 input

0 0	Capture disabled
0 1	Capture TCNT → TIC3 on IC3 rising edges
1 0	Capture TCNT → TIC3 on IC3 falling edges
1 1	Capture TCNT → TIC3 on any IC3 edges

See Fig. A.15

FIGURE A.22
Timer input capture 1, 2, and 3.

The input capture function of the timer is a little simpler to understand because there is no interaction between the three input capture facilities. As shown in Fig. A.5 and again in Fig. A.22 the three inputs usurp bits 0, 1, and 2 of port A. If any of the three input timers are not used, then the corresponding bit of port A can be used as a general purpose input line. In fact, the input lines can be read at PORTA while also being used to generate interrupts and to capture the time of input transitions via the IC1, IC2 and IC3 functions.

Two bits of the TCTL2 register are used to specify which of three possible events each timer will sense: a rising edge, a falling edge, or either edge. (If the timer is not to be used at all, then this, the default possibility, is coded with these 2 bits being cleared to 0.) When the selected edge occurs, the flag for that timer is set in the TFLG1 register. In addition, the content of the free-running counter TCNT at that instant is transferred to the appropriate timer input capture register, TICi. In this way, the time of an input edge can be measured down to the nearest 0.5 μs. If the corresponding TMSK1 bit has been initialized to 1 to enable interrupts from this source, then this input edge will trigger an interrupt of the CPU. Within the interrupt service routine the content of the input capture register is read and stored in RAM. The flag bit is cleared just as is done for each of the OCiF bits in the same TFLG1 register, as discussed in conjunction with Fig. A.19. That is, to clear IC1F we execute the instructions

```
LDAA  #00000100B
STAA  TFLG1
```

Finally, if we are trying to measure the width of a positive-going pulse, then we can use the BSET and BCLR instructions on TCTL2 to select a rising edge for the first event and then a falling edge for the second event. The difference between the contents of the timer input capture register for these two events yields the pulse width.

This technique yields the measurement of pulse widths that are less than a complete cycle of 65,536 counts of the free-running counter, TCNT. However, if the pulse width is extremely short, then there may not be enough time, after the leading edge of the pulse occurs, to reinitialize the timer in order to catch the trailing edge of the pulse. In such a situation, two of the three timers working together can do what a single timer cannot. For example, IC1 can be initialized to sense a rising edge while IC2 can be initialized to sense a falling edge. Then the input to be measured is connected to *both* IC1 and IC2. The counters are initialized with

```
BCLR  TCTL2,00100100B   ;Clear bits 2 and 5
BSET  TCTL2,00011000B   ;Set bits 3 and 4
LDAA  #00000110B        ;Clear IC1F and IC2F
STAA  TFLG1
BSET  TMSK1,00000010    ;Enable IC2 interrupt
```

When the interrupt occurs, if both flags have been set again and if TIC2 is larger than TIC1, then the pulse width has been captured. Otherwise, the timers can be reinitialized to give it another try.

For very long pulse widths, longer than 65,536 counts of the free-running counter (even including the $\frac{1}{16}$ prescaler), we need to keep track of how many times TCNT rolls over from FFFF to 0000. The mechanism for this is shown in Fig. A.23, as applied to the IC3 input. A 1-byte variable, which might be called TCNTEXT (TCNT Extension), is allocated in RAM and used to append bits 23..16 to TCNT. At startup, the timer overflow interrupt is initialized to enable interrupts from this source. The timer overflow interrupt service routine does the following minimal job:

```
TIMEROVERFLOW: INC   TCNTEXT          ;Increment variable
               LDAA  #10000000B       ;Clear flag
               STAA  TFLG2
               RTI                    ;and return
```

In response to the occurrence of the leading edge of the pulse on IC3, the IC3 interrupt service routine forms the 3-byte variable, TL, for the time of its occurrence. The lower 2 bytes of TL are taken from the input capture register TIC3, while the upper byte of TL is taken from TCNTEXT.

Next we call a subroutine, ICEXT (input capture extension), to ensure that the correct value of TCNTEXT has been loaded into the upper byte of TL. The potential problem which ICEXT sorts out occurs when the lower 2 bytes of TL hold a value close to the FFFFH to 0000H overflow point. Since ICEXT is executed in an interrupt service routine, interrupts will be disabled during its

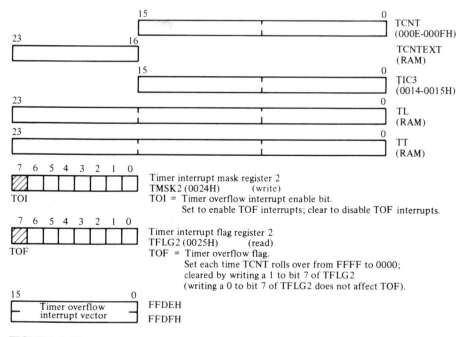

FIGURE A.23
Extending the effective size of TIC3 to measure long pulse widths on IC3.

execution. This will prevent the timer overflow interrupt service routine from being executed in the middle of ICEXT. There are two conditions which require the upper byte of TL to be changed:

1. If TIC3>=0 (considering it as a signed number) AND TOF=1, then increment the upper byte of TL.
2. If TIC3<0 AND TCNT>0 AND TOF=0, then decrement the upper byte of TL.

In the first case, the capture occurred after the counter rolled over. However, TCNTEXT has not yet been incremented so as to correspond correctly to TCNT. Therefore the value of the upper byte of TL must be incremented to fix this.

The second case handles the condition where the capture occurred before the counter rolled over. Since the capture occurred, the counter has rolled over, setting the TOF flag. However, since TOF is now clear, evidently the TIMEROVERFLOW interrupt service routine has already been executed, incrementing TCNTEXT to one count higher than the value which would correspond correctly to TIC3. Therefore the value of the upper byte of TL must be decremented.

The IC3 interrupt service routine concludes by clearing the IC3F flag in TFLG1 and then reinitializing the EDG3B and EDG3A bits in TCTL2 (see Fig. A.22) to capture the trailing edge of the pulse.

When the interrupt for the trailing edge of the pulse occurs, the ICEXT subroutine is again called to ensure that the correct value is captured in TT, a 3-byte variable used to hold the time captured at the trailing edge. Note that ICEXT deals with TT in this case and TL in the last case. The subroutine can do this if X is loaded with a pointer to the correct parameter (the upper byte of TT or TL) before it is called.

After forming TT, we are ready to conclude the pulse width computation. The pulse width equals TT − TL, where binary subtraction is used on the two 24-bit numbers.

While any one of the five output compare registers can be used as a software interrupt to generate periodic interrupts, the 68HC11 includes a real-time interrupt capability for just this purpose. As shown in Fig. A.24, these interrupts can be set up to occur approximately every 4, 8, 16, or 33 ms. The starting address of the interrupt service routine is stored at FFF0 and FFF1H, as shown in both Figs. A.15 and A.24. In addition to doing whatever tasks must be done at the real-time rate (e.g., stepping a stepper motor once), the interrupt service routine need only clear the RTIF flag with

```
LDAA #01000000B
STAA TFLG2
```

before returning.

The final feature of the programmable timer is its pulse accumulator capa-

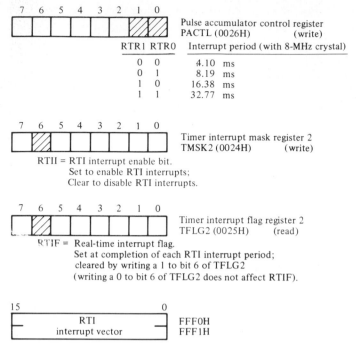

FIGURE A.24
Real-time interrupts.

bility. While it offers several options, its main contribution is that it permits an accurate measurement of the time taken by a number of pulses in an input pulse train *without* having to service each pulse with a separate interrupt. In conjunction with Fig. 3.18 (and in general terms), we considered the determination of the time that a 144-tooth gear takes to make a single revolution. A variable reluctance pickup produces a magnetic field which changes strength as the magnetic path is changed. The movement of each gear tooth beneath the pickup changes the magnetic path. A coil in the pickup generates an output voltage pulse as the coupled magnetic field changes. In this way, each gear tooth produces an output pulse. We can square up the waveform with a Schmitt trigger and then let it drive *both* IC3 and PAI, the pulse accumulator input, as shown in Fig. A.25. Both IC3 and PAI can be set up to respond to rising edges of the input signal. When the next rising edge occurs, the time of occurrence of this edge is saved in RAM from TIC3. Then the pulse accumulator counter, PACNT, is preset with $256 - 144 = 112$. An interrupt is enabled to occur when the pulse accumulator counter, PACNT, overflows. Interrupts are *not* enabled from IC3. The input capture register, TIC3, will record a new value on each rising edge of the input waveform, *regardless of the fact that its IC3F flag is not being cleared each time*. After exactly $144 + 1$ rising edges, the pulse accumulator counter will overflow and cause an interrupt. The interrupt service routine reads TIC3 and subtracts the previous value (stored in RAM) from this value to

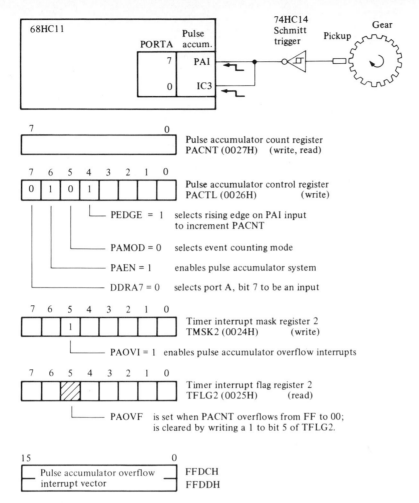

FIGURE A.25
Timing a number of pulses with only one interrupt.

determine the time between 145 edges (i.e., the time for 144 pulse periods), or one revolution of the gear.

If the pulse accumulation feature is not needed in an application and if the OC1 output compare function is also not needed (both of which use bit 7 of port A), then this line can serve as an edge-sensitive input, generating an interrupt with its own interrupt vector. This option is illustrated in Fig. A.26. Alternatively, as shown in Fig. A.5, this pin can be used as either a general-purpose input line or a general-purpose output line (as bit 7 of port A).

A.7   SERIAL PERIPHERAL INTERFACE

The I/O serial port in the 68HC11 is called the serial peripheral interface, or SPI. There are three registers which are involved in SPI data transfers. These

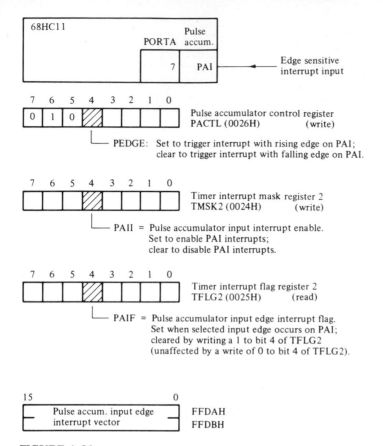

FIGURE A.26
Using the pulse accumulator input as an edge-sensitive interrupt input.

are shown in Fig. A.27. The SPI control register, SPCR, sets the transfer time. The default time of 8 μs is the transfer time of choice unless the SPI is transferring data to a slow CMOS chip which cannot be clocked as often as every microsecond. The other choices solve that problem. The SPCR register also specifies the phasing and polarity of the clock relative to the data, using one of the choices shown in Fig. A.28. If the 68HC11's use of the SPI facility is to expand its I/O capability as discussed in Chaps. 6 and 7, then the DWOM bit should remain 0 so that port D output pins pull up as well as pull down. In a multiple-microcontroller environment, the DWOM bit is set so that port D output pins can only pull down. Pullup resistors on these output pins then permit the bussing of corresponding output pins from several microcontrollers.

Setting the SPE bit is what dedicates I/O pins to the serial peripheral interface, rather than to port D, with which it shares pins, as shown in Fig. A.29. This figure depicts the SPI register use for the common application in which the serial peripheral interface is used for I/O expansion. In this applica-

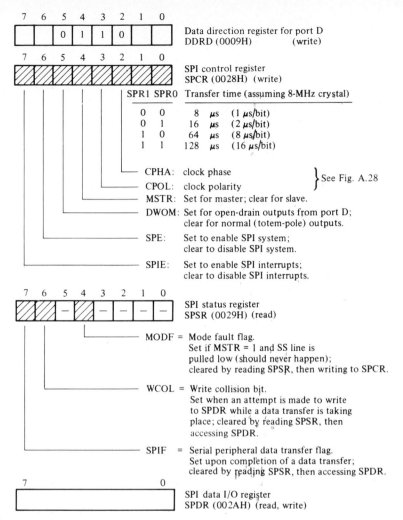

```
        7   6   5   4   3   2   1   0
       [       | 0 | 1 | 1 | 0 |       ]     Data direction register for port D
                                             DDRD (0009H)          (write)

        7   6   5   4   3   2   1   0
       [ ///////////////////////// ]         SPI control register
                                             SPCR (0028H)  (write)
```

SPR1 SPR0   Transfer time (assuming 8-MHz crystal)

0	0	8 μs	(1 μs/bit)
0	1	16 μs	(2 μs/bit)
1	0	64 μs	(8 μs/bit)
1	1	128 μs	(16 μs/bit)

CPHA: clock phase   } See Fig. A.28

CPOL: clock polarity

MSTR: Set for master; clear for slave.

DWOM: Set for open-drain outputs from port D; clear for normal (totem-pole) outputs.

SPE: Set to enable SPI system; clear to disable SPI system.

SPIE: Set to enable SPI interrupts; clear to disable SPI interrupts.

```
        7   6   5   4   3   2   1   0
       [ ////|///| — |///| — | — | — | — ]    SPI status register
                                              SPSR (0029H) (read)
```

MODF = Mode fault flag. Set if MSTR = 1 and SS line is pulled low (should never happen); cleared by reading SPSR, then writing to SPCR.

WCOL = Write collision bit. Set when an attempt is made to write to SPDR while a data transfer is taking place; cleared by reading SPSR, then accessing SPDR.

SPIF = Serial peripheral data transfer flag. Set upon completion of a data transfer; cleared by reading SPSR, then accessing SPDR.

```
        7                           0
       [                             ]        SPI data I/O register
                                              SPDR (002AH) (read, write)
```

Writing to SPDR (with SPIF cleared) in the *master* SPI initiates the simultaneous transmission and reception of a byte.

Writing to SPDR (with SPIF cleared) in a *slave* SPI loads SPDR in anticipation of a transfer initiated by the master.

Reads of SPDR actually read the buffer shown in Fig. A.29.

After the slave's SPIF flag is set, the slave's CPU must read SPDR before another 8-bit transfer has been completed. Otherwise an (unflagged) overrun error will occur.

FIGURE A.27
Serial peripheral interface (SPI) registers.

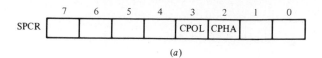

(a)

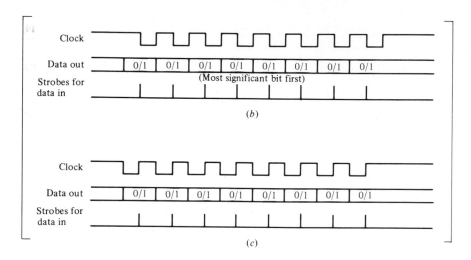

(b)

(c)

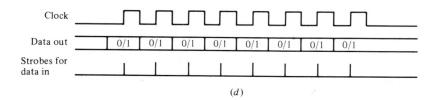

(d)

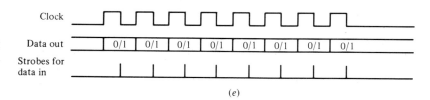

(e)

FIGURE A.28
Motorola 68HC11 serial peripheral interface timing modes. (a) SPCR, the control register in which mode is selected; (b) CPOL=1, CPHA=0 (preferred mode for serial *input*); (c) CPOL=1, CPHA=1 (preferred mode for serial *output*); (d) CPOL=0, CPHA=0; (e) CPOL=0, CPHA=1.

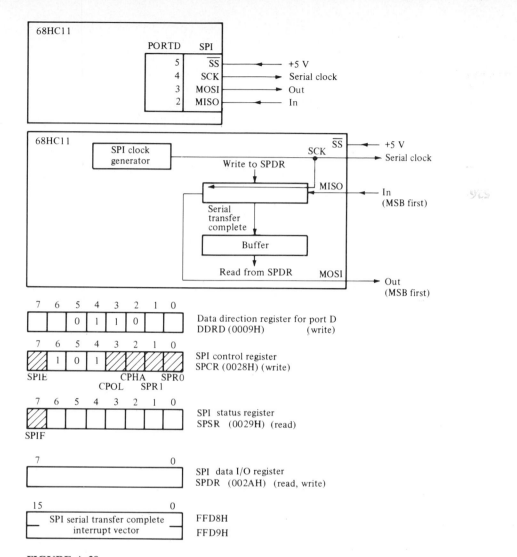

FIGURE A.29
I/O expansion use of serial peripheral interface.

tion, the SPI is set up as a master, which means that its SPI clock generator drives the SCK line. As we shall see shortly, when the SPI is set up as a slave, its SCK line serves as an input for the clock from the master.

Figure A.29 shows the $\overline{\text{Slave Select}}$ ($\overline{\text{SS}}$) pin tied to +5 V. While it would seem that this is all that would be needed to tell the hardware that this is a master device, such is not actually the case. The MSTR bit of the SPCR register works together with the $\overline{\text{SS}}$ pin to make this designation. The combination of the $\overline{\text{SS}}$ pin and the MSTR bit of the SPCR register comes into its own in the multiple-microcontroller environment, to be discussed shortly.

To use the SPI facility at all, it is necessary to set the SPE bit. In addition (and as was pointed out earlier in this appendix), to use the SPI system it is also necessary to set bits in the data direction register for port D (which shares the same lines) in order to establish the necessary lines as outputs. Since the DDRD bits are cleared at reset time, we can configure port D for use by the SPI as a *master* with

```
BSET    DDRD,00011000B
```

For I/O expansion use of the serial peripheral interface, we need to select the transfer time and the clock phasing and polarity. Also, if we desire an interrupt after each transfer has been completed, then the SPIE bit needs to be set.

In the SPSR register, the only bit normally dealt with is the SPIF flag bit which signifies that a transfer has been completed. The SPDR register actually represents two distinct devices. Writes to SPDR are writes to the shift register itself. On the other hand, reads of SPDR do not read from the shift register, but rather from a buffer register which is automatically loaded from the shift register upon the completion of a data transfer.

The master-slave interconnection of two or more 68HC11s via the serial peripheral interface represents an intriguing capability of the chip. The interconnections for three chips (one master and two slaves) are shown in Fig. A.30*a* and the register contents shown in Fig. A.30*b* and *c*. Each of the two I/O lines, MISO and MOSI, are selected to be either an input line or an output line, depending upon whether the chip is a master or a slave. Thus, the MOSI (master-out–slave-in) lines become inputs in the slaves while the MISO (master-in–slave-out) lines become outputs. Furthermore, the slaves lose the ability to drive the SCK line, and their SCK lines become inputs for receiving the clock signal from the master. In this way, the data transfers are synchronized by the SPI clock in the master 68HC11.

Even if there is only one slave, it still makes sense to have the master drive the slave's $\overline{\text{SS}}$ input. When the master drives this line low, the slave senses this and resynchronizes so that the next 8 bits transferred are treated as 1 byte. That is, the $\overline{\text{SS}}$ input *frames* the data transfer. For multiple-byte messages, this dropping and raising of the slave's $\overline{\text{SS}}$ input is not really necessary between each byte of the message. On the other hand, if the slave's $\overline{\text{SS}}$ input were simply tied to ground, then if the master and the slave ever got out of synchronization (so that bit 0 of the master's transferred byte were treated as bit 1 of the slave's received byte, for example), then the master and slave would have no way to resynchronize, short of resetting the two chips!

If this approach is used for sharing the overall microcontroller task in an instrument or device between two 68HC11s, then the protocol used between the two chips is user-defined. Assuming that the slave is thoroughly interrupt-driven, then any multiple-byte messages sent from the master to the slave must deal with the slave's latency before the slave reads the first byte transferred. To help with this, the priority of the SPI interrupt in the slave might be raised from near bottom to the top of the (maskable) interrupts by loading the lower 4 bits of the HPRIO register with 0011. Also, if ASCII code is used for all transfers, then messages might end with the ASCII code for a carriage return (0DH). After sending the first byte of the multiple-byte message, the master might wait for the worst-case latency time of the slave and then assume that the slave will wait

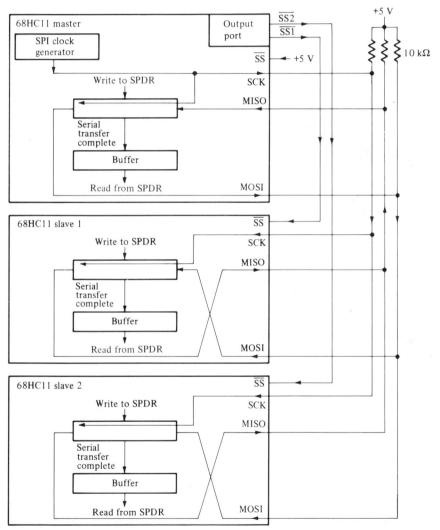

Maximum transfer rate = 1 byte every 8 μs

(*a*)

FIGURE A.30
Master-slave use of SPI facility to interconnect several 68HC11s. (*a*) Interconnections; (*b*) registers in master; (*c*) registers in slaves.

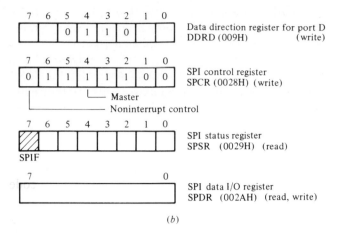

7	6	5	4	3	2	1	0
		0	1	1	0		

Data direction register for port D
DDRD (009H)   (write)

7	6	5	4	3	2	1	0
0	1	1	1	1	1	0	0

SPI control register
SPCR (0028H) (write)

— Master
— Noninterrupt control

7	6	5	4	3	2	1	0
//							

SPIF

SPI status register
SPSR (0029H) (read)

7							0

SPI data I/O register
SPDR (002AH) (read, write)

(b)

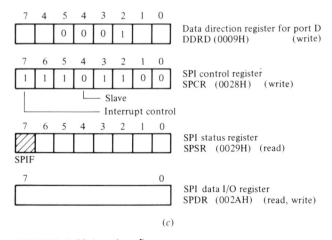

7	4	5	4	3	2	1	0
		0	0	0	1		

Data direction register for port D
DDRD (0009H)   (write)

7	6	5	4	3	2	1	0
1	1	1	0	1	1	0	0

SPI control register
SPCR (0028H)   (write)

— Slave
— Interrupt control

7	6	5	4	3	2	1	0
//							

SPIF

SPI status register
SPSR (0029H) (read)

7							0

SPI data I/O register
SPDR (002AH)   (read, write)

(c)

FIGURE A.30 (*continued*)

in its SPI interrupt service routine to read the following bytes at full speed into a buffer, waiting to interpret the received message until the CR (carriage return) code has been received. If the message calls for a response from the slave, then the master might continue to send repeated CR codes to the slave. In this way, if the slave cannot respond immediately without upsetting the desired operation for the device or instrument, then the master will know this by the sequence of CR codes which keep coming back to it (after being shifted through the slave SPI's shift register). When the slave finally responds, the master might assume that the slave has buffered its response so that it can keep up with the master, clearing its SPIF flag and reloading its SPDR each time the SPIF flag is set by the master.

If the slave has to put off the master in this way, then it has one tricky moment with which to contend. When it is ready to send its first byte, it must wait until after the master has completed one transfer and then load SPDR before the master begins the next transfer. Otherwise a "write collision" will occur (signaled to the slave by the setting of the WCOL bit in the SPSR register). Recognizing the potential problem leads to its solution. Before making its first transfer, the slave simply reads SPSR and then *reads* SPDR to clear the SPIF bit. Then it waits with

```
WAIT:     BRCLR    SPSR,10000000B,WAIT
```

until the SPIF bit is set again before writing the first byte of the response message. It too might terminate its multiple-byte response with the ASCII code for a carriage return.

This discussion has emphasized the master-slave use of the SPI facility because it is so dependent upon the exact behavior of the 68HC11. However, the big application of the SPI facility, for most users, is to expand I/O capability. This has already been explored in thorough detail in Chap. 6. A variety of applications have been developed in Chap. 7.

A.8 SERIAL COMMUNICATIONS INTERFACE

In Sec. 6.5 we discussed what a UART (universal asynchronous receiver transmitter) does and the registers involved with its use, in general terms. The UART in the 68HC11 is termed its serial communications interface (SCI). While it uses just two pins of the chip, usurping bits 0 and 1 of port D, the SCI facility offers many options. In this section we look at what must be done to set it up to obtain "normal" UART performance. Then we look at how it can be set up for the master-slave interconnection of several 68HC11s. In comparison with the use of the SPI for interconnecting multiple 68HC11s, the SCI uses fewer pins on the chip (i.e., only two) but can support a maximum transfer rate only one-eighth of that of the SPI.

The registers involved with the normal use of the SCI are shown in Fig. A.31. As is the case for the SPI, use of the SCI requires that the data direction register for port D be initialized with bit 1 set in order for the transmit output of the SCI to make it out of the chip. The SCI baud rate generator register permits a variety of "standard" baud rates to be obtained, even given any one of several crystal clock rates. With the standard 8-MHz crystal, Fig. A.31 shows what must be loaded into the BAUD register to obtain the maximum supported rate (125,000 Hz, for interconnecting multiple 68HC11s) as well as to obtain a variety of standard rates. For normal use, the SCCR2 register contains 6 bits which must be set as shown and also the RIE and TIE bits for enabling and disabling interrupts. When transmitting or receiving a string of successive bytes of data, interrupt control makes sense because even at the maximum normal baud rate of 9600 baud, a new character is sent or received only every millisec-

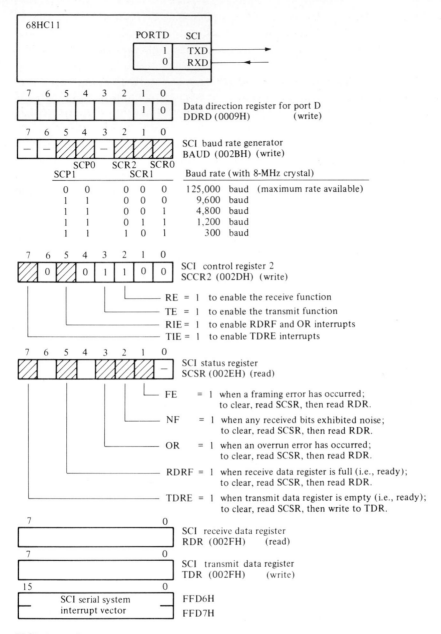

FIGURE A.31
Registers associated with normal use of the serial communications interface (SCI).

ond. Interrupt control permits us to use most of every millisecond productively doing other things, rather than simply waiting until the SCI says that the last transfer has been completed and that the CPU can take action with the SCI again.

The SCI status register, SCSR, contains the TDRE flag which is set every time the transmit side of the SCI is ready for another character to be sent. It also contains the RDRF flag which becomes set every time that the receive side of the SCI holds another character which it has received. In addition there are three flags (OR, NF, and FE) which can be checked with each received character if we want to detect and respond to any of three different types of transmission errors. Finally, note that the receive data register and the transmit data register share the same register address, namely 002FH. Since one of these registers is a read-only register while the other is a write-only register, their sharing of the same address does not cause any great complication.

The SCI is supported by a single interrupt vector even though there are several independent sources of interrupts. When an SCI interrupt occurs, the interrupt service routine must test the following flags:

```
SCI_INT_SERV:   BRCLR   SCSR,00101000B,SCI_SKIP1
                BSR     SERVE_RECEIVE
SCI_SKIP1:      BRCLR   SCSR,10000000B,SCI_SKIP2
                BSR     SERVE_TRANSMIT
SCI_SKIP2:      RTI
```

where the SERVE_RECEIVE subroutine handles the response to either the RDRF flag or the OR flag being set, and where the SERVE_TRANSMIT subroutine handles the response to the TDRE flag being set. Since these are the only enabled sources of SCI interrupts, this polling routine will respond appropriately whether the receive side of the SCI, the transmit side of the SCI, or *both* sides of the SCI request service.

For multiple-microcontroller communication using the SCI facility, the circuit of Fig. A.32 illustrates the inherent simplicity of the interconnections, given one master and two slaves (or any number of slaves, for that matter). Because the slaves share the driving of the line over which they transmit data (going back to the master's RXD input), the slave TXD pins must be set up to be open-drain outputs. Figure A.32 shows the master's TXD pin as having an open-drain output also, requiring the 10 kΩ pullup resistor. While the master could be operated with the normal totem-pole output, the circuit of Fig. A.32 is just as good, except for the need for an extra pullup resistor.

The 68HC11's SCI facility supports *two* quite distinct approaches for multiple-microcomputer communication. The problem which they each solve is this. When the master sends out a multiple-byte message, each slave needs to know whether the message is directed to it. Each slave also needs to know when one multiple-byte message ends and another begins. In this section we explore the details of one approach. The interested reader can seek out the other approach from the Motorola 68HC11 manuals.

The registers involved are shown in Fig. A.33, and their use is the same

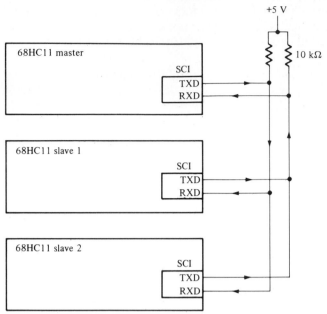

Maximum transfer rate = 1 byte every 88 μs

FIGURE A.32
Master-slave use of SCI facility to interconnect several 68HC11s.

for both the master and the slave. An extra bit is tacked onto each transmitted "word." If the data being sent is nothing but 7-bit ASCII encoded characters, then the eighth bit can serve as the extra bit. On the other hand, if 8-bit bytes of data are being transmitted, then a ninth data bit can be sent. This choice between eight and nine data bits is handled by the M bit in the SCCR1 register. If nine data bits are dealt with, then the ninth bit during a transmission becomes whatever has been written to bit 6 of the SCCR1 register (the T8 bit). In similar fashion, the ninth bit during a reception will be loaded into bit 7 of the SCCR1 register (the R8 bit).

When the master sends out a multiple-byte message, it sets the last data bit of the first byte. Each slave gets an interrupt when it receives this byte of data. If the master intended the message to go to slave 1, then this first byte might be the number 1. Slave 1 makes no change to its control register contents and will be interrupted by each successive byte of the message. On the other hand, slave 2 sets the RWU (receiver wakeup) bit in the SCCR2 register. By doing this, it will not be interrupted by the remaining bytes of the multiple-byte message. However, when the first byte of the *next* multiple-byte message is received, the one in the last data bit will clear the RWU bit in the SCCR2 register and will cause an interrupt in that slave. The slave which received the last multiple-byte message (and which was therefore interrupted as each byte of that message arrived) can know that this is the first byte of a *new* message by looking at the last data bit of each received byte to see if it is set to 1.

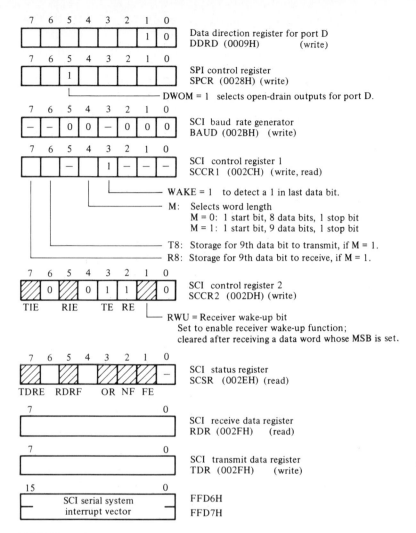

FIGURE A.33
Registers associated with master-slave use of the serial communications interface (SCI).

The master-slave protocol requires the slaves not to transmit except in response to a query from the master. In this way, the master is never surprised by what it receives. Also, it never witnesses a collision as two slaves try to drive their shared transmit line at the same time.

A.9 ANALOG-TO-DIGITAL CONVERTER

The A/D converter capability on the 68HC11 chip actually consists of an eight-channel analog multiplexer,* a sample and hold circuit, and an 8-bit succes-

---

* Four channels for the 48-pin DIP-packaged version of the 68HC11.

sive-approximation A/D converter which carries out a conversion in 16 μs. The converter makes use of two pins on the chip, V_{RH} and V_{RL}, to define high and low *reference voltages*. An input voltage equal to the V_{RL} input converts to 00 while an input voltage equal to the V_{RH} input converts to FF (hex). These reference voltage inputs can be set to any values in the range of the power supply voltage for the chip itself.* This permits us to carry out "ratiometric" conversions, where the input voltage is compared to a steady (or slowly changing) reference voltage, to form a ratio between the two. It also permits us to use a reference voltage which is more accurate, and perhaps less noisy, then the supply voltage for the chip, creating a more accurate conversion.

The pins and the registers associated with the A/D converter are shown in Fig. A.34. As was discussed in Chap. 2, the 68HC11's A/D converter always collects readings in groups of four samples. If the MULT bit in the ADCTL register is cleared to 0, then a single channel of the multiplexer is selected and four readings of this one channel are carried out, one every 16 μs.†

The channel selection is made as shown in Fig. A.35, which shows the 4 bits to be loaded into the lower 4 bits of the ADCTL register. The first conversion goes into the ADR1 register, the second into ADR2, the third into ADR3, and the fourth into ADR4.

Upon the completion of the fourth conversion, the CCF flag in the ADCTL register is set. If the SCAN bit was previously loaded with a 0, then the conversion process stops and the conversion results remain indefinitely in the four ADRi registers. On the other hand, if the SCAN bit was previously set to 1, then another cycle of four conversions is begun immediately after the CCF flag has been set. Since the content of ADR1 will be replaced by a new measurement 16 μs after the CCF flag is set, we need to read the content of ADR1 within 16 μs. We have a little more leeway for reading the other three registers, after the CCF flag is set and before they will each be updated.

This A/D converter facility is good for giving a quick reading of the analog input on any channel. Since the conversion process begins when we write to the ADCTL register, we need only write the selected channel number to ADCTL, wait 16 μs, and read ADR1. For example, to read channel 1 into accumulator A, we might execute the following code:

```
         LDAA #00000001B      ;Select channel 1
         STAA ADCTL           ;and begin conversion
         LDAB #6              ;1 μs
WAIT:    DECB                 ;6 × 1 μs
         BNE  WAIT            ;6 × 1.5 μs
         LDAA ADR1            ;Get result
```

If we want to sample one channel continuously and at the maximum sampling

---

* Actually, from 0.1 V *below* the "0-V" power supply input up to 0.1 V *above* the "+5-V" power supply input, with $V_{RH} > V_{RL}$
† With an 8-MHz crystal.

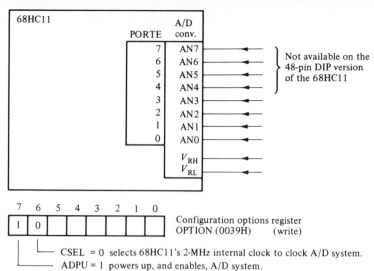

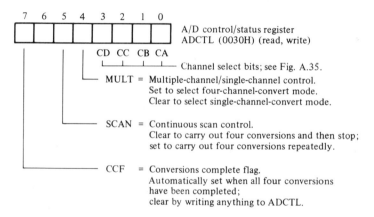

A conversion sequence is initiated by a write to ADCTL.

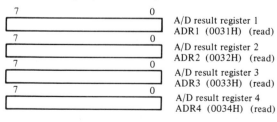

FIGURE A.34
A/D converter.

MULT = 0 (Single-channel selection)

CD	CC	CB	CA	Channel selected
0	0	0	0	AN0
0	0	0	1	AN1
0	0	1	0	AN2
0	0	1	1	AN3
0	1	0	0	AN4
0	1	0	1	AN5
0	1	1	0	AN6
0	1	1	1	AN7
1	1	0	0	V_{RH} pin ⎫
1	1	0	1	V_{RL} pin ⎬ Self-test support
1	1	1	0	$V_{RH}/2$ ⎭

MULT = 1 (Four-channel selection)

CD	CC	Result
0	0	AN0 goes to ADR1 AN1 goes to ADR2 AN2 goes to ADR3 AN3 goes to ADR4
0	1	AN4 goes to ADR1 AN5 goes to ADR2 AN6 goes to ADR3 AN7 goes to ADR4
1	1	V_{RH} goes to ADR1 V_{RL} goes to ADR2 $V_{RH}/2$ goes to ADR3

FIGURE A.35
A/D converter—channel select options.

rate of four new samples every 64 μs, then we almost necessarily have to turn off interrupts and pay close attention to the A/D converter.

If we want to sample one channel continuously but at a slower rate, we can again turn off interrupts and then use a timer under flag control to initiate data collection. Each time the timer times out, we execute the six-instruction sequence listed above to collect a sample. After handling the sample (e.g., saving it in a buffer or adding it into an average), we augment the timer's output compare register and clear its flag to set the next time interval.

Because this last approach repeatedly tests the timer's flag in a tight loop, waiting for the flag to become set, we should make that loop take as little time as possible. For example, if we use the timer's OC5, then the test can be made with

```
WAIT:            BRCLR    TFLG1,00001000B,WAIT
```

This instruction takes six cycles to execute. Consequently, we will tend to get a 3-µs *jitter* in our sampling intervals as the flag becomes set during any one of the six cycles of the BRCLR instruction. We can remove the sampling jitter by doing two things:

1. Make our procedure to collect a sample and handle it take a fixed amount of time. If the procedure has no branches in it and if interrupts are turned off, it will meet this requirement.
2. Instead of reading TOC5, the output compare 5 register, to update the time, read TCNT. This will get the *present* time. If we add a constant to this value and store it in TOC5, then the timeout will be synchronized to these instructions. Accordingly, the flag will become set during the same cycle of the BRCLR instruction every time.

Of course, while the sampling jitter has been reduced to zero, the value to be added to TCNT to update TOC5 to obtain a desired sampling interval is now more difficult to determine since it must take into account the execution time of all of these instructions.

The above considerations have dealt with fast sampling. Probably an equally common requirement is to sample several analog inputs periodically and at a rate which is fast by thermal or electromechanical standards, but not by microcontroller standards. For example, if the microcontroller samples the dc voltage output of a temperature transducer, then any rate faster than a few samples per second probably represents overkill. Under these circumstances, the programmable timer can be used under interrupt control to set the sampling interval. If other interrupts put off the immediate servicing of the timer's interrupt service routine, then the sampling jitter which results will be inconsequential. For example, if the timer's interrupt service routine is put off by even as much as 200 µs while the sampling rate is five samples per second, then the jitter is no more than one part in one thousand, or 0.1 percent.

A.10   EEPROM PROGRAMMING AND USE

The normal 68HC11 chips have a 512-byte EEPROM extending over hex addresses B600 to B7FF. Some higher-priced parts have the internal ROM replaced by a 2-kbyte EEPROM extending over addresses F800 to FFFF. Other even higher-priced parts have the internal ROM replaced by an 8-kbyte EEPROM extending over addresses E000 to FFFF. The EEPROM is designed to be written to and erased during normal operation, with nothing more than the normal +5-V supply applied to the chip. Alternatively, we may want to use the bootstrap mode to install and run an EEPROM-loader program from RAM. This, in turn, can then load a *test* program into the EEPROM since the test program can be bigger when handled in this way than if our test programs are constrained by the size of the on-chip RAM. In this case, the bootstrap mode (discussed at the beginning of this appendix) is first used to install a 256-byte EEPROM-loader program which will do two things:

1. It must read further data coming in over the serial communication interface using a flow control protocol.*
2. It must erase the EEPROM and then program it with this further data.

While this program must be 256 bytes long in order to work with the bootstrap loader, a few of these bytes need to be "filler" bytes, and not part of 1 and 2 above. The locations where these filler bytes are stored on page 0 will become a RAM buffer for the data to be programmed into the EEPROM. However, since the data comes in serially at 1200 baud (i.e., 1 byte every 8.33 ms) and since each byte requires 10 ms to be programmed into the EEPROM, not much of a buffer is needed.

Since what is to be programmed into the EEPROM probably comes from the output of a 68HC11 assembler, the 256-byte program must handle the object code format generated by the assembler. The assembler for a Motorola microcontroller most likely generates the object code in what is called "S-record" format, which includes miscellaneous bytes of address and formatting information in addition to the actual code to be programmed into the EEPROM.

Motorola is supporting the 68HC11 with a dial-up line in Austin, Texas, over which users can obtain free software (e.g., a cross-assembler to run on a PC and a BASIC interpreter to run on the 68HC11). The EEPROM loader discussed above will surely be one such available program.

The registers involved with using the EEPROM and with its erasure and programming are shown in Fig. A.36. As was discussed earlier, the CONFIG register (which is an EEPROM register) contains an EEON bit which enables or disables the use of the EEPROM. To use the EEPROM, the EEON bit must be set. We discuss how to program the CONFIG register shortly.

Since the EEPROM actually needs something like +19 V applied to it for either programming or erasing, it may seem surprising that there is no "programming voltage" pin on the chip. What Motorola has done has been to include a dc-to-dc converter to generate this programming voltage within the 68HC11 chip itself. The EEPGM bit of the PPROG register is used as a switch

---

* For example, the commonly used XON-XOFF protocol would have the 68HC11 send an XOFF ASCII control character (13H) to tell the personal computer to stop sending data. Then when the 68HC11 is ready for more data it would send an XON ASCII character (11H) to the personal computer to tell it to resume sending data.

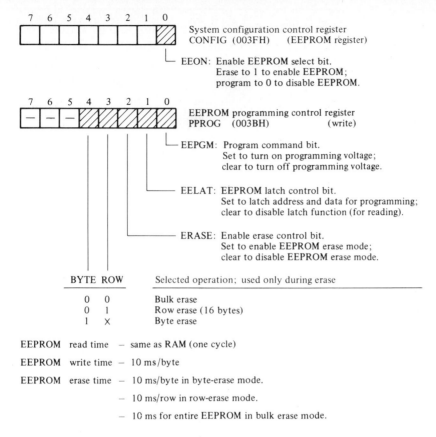

FIGURE A.36
EEPROM registers and characteristics.

to connect this programming voltage into the EEPROM array during programming and erasing.

Since the on-chip EEPROM programming circuitry needs to hold the address and data bus inputs to the EEPROM while the programming voltage is applied to the selected EEPROM cells, latches are interposed on these EEPROM inputs during programming and erasing. If the EELAT bit is set, then any write to an EEPROM address will be latched and held until another write takes place, or until the EELAT bit is cleared.

All of the programming and erasing options can be carried out with the same short subroutine. It is only necessary to pass alternative sets of parameters to the subroutine to select the option desired. The parameter values and the subroutine, called EEPROM, are shown in Fig. A.37. Note that for an EEPROM bit, a 0-to-1 change requires the bit to be erased whereas a 1-to-0 change requires the bit to be programmed.

Operation	Load A with	Load B with	Load X with
Erase all of EEPROM		00000110B	B600H (i.e., an EEPROM address)
Erase CONFIG register		00000110B	003FH (i.e., address of CONFIG)
Erase a row of EEPROM		00001110B	Any address within the row
Erase a byte of EEPROM		00011110B	Address of byte to be erased
Program a byte of EEPROM	Data	00000010B	Address of byte to be programmed
Program CONFIG register	Data	00000010B	003FH (i.e., address of CONFIG)

(a)

```
EEPROM:     STAB    PPROG       ;Set up latch mode and selected erase or program mode
            STAA    0,X         ;Latch selected EEPROM address (and data, if programming)
            INC     PPROG       ;Set EEPGM (this works because PPROG can be read)
            LDY     #2857
WAIT:       DEY                 ;Wait for 10 ms
            BNE     WAIT
            DEC     PPROG       ;Clear EEPGM
            CLR     PPROG       ;Return EEPROM to read state
            RTS
```

(b)

FIGURE A.37
EEPROM subroutine to carry out EEPROM program or erase operations. (a) Parameters required by each operation; (b) EEPROM subroutine.

A.11 WAIT AND STOP MODES

Being a CMOS chip, the 68HC11 dissipates relatively little power, by microcontroller standards. The supply current to the chip when it is running in the single-chip mode is specified to be less than 20 mA.* This compares with a maximum current of 300 mA in the previous generation 68701 NMOS microcontroller.

While the normal power dissipation of the 68HC11 is relatively low, the chip possesses two modes of standby operation which significantly decrease this power dissipation. The *WAIT* mode, to be discussed shortly, is specified for a maximum of 5 mA, one-fifth the maximum running current. The *STOP* mode has been specified in early data sheets for the 68HC11 as a maximum of 300 μA. However, indications are that a more realistic maximum for the chip in the STOP mode will be something like 10 μA!

The WAIT mode is entered by executing the WAI instruction. In this mode, the internal oscillator keeps running and the power dissipation depends

---

* With all programmable I/O lines configured as inputs and with no dc loads on any ports, which would, of course, contribute directly to the supply currrent.

upon what on-chip facilities (e.g., timer, SPI, SCI) are active when the WAIT mode is entered. If the I bit is set (to disable maskable interrupts) and the COP watchdog timer is disabled, then the timer system will be automatically turned off to reduce power consumption. The CPU stops executing instructions.

The WAIT mode is exited in either of two ways. An unmasked interrupt can be used to begin operation again with that interrupt's service routine. When the RTI instruction at the end of the interrupt service routine is executed, operation will return to the instruction following the WAI instruction.

Alternatively, the WAIT mode can be exited by pulsing (1-0-1) the $\overline{\text{Reset}}$ pin. In this case, the chip will go through its normal startup procedure.

The STOP mode is entered by executing a STOP instruction after clearing the S bit (bit 7) in the condition code register with the sequence

```
TPA
ANDA #01111111B
TAP
STOP
```

The purpose of the S bit is to reduce the chances of a runaway CPU accidentally executing a STOP instruction and halting all operation of the chip, including the watchdog timer (which otherwise could have provided for the recovery to normal operation). A STOP instruction might be inadvertently executed if the program counter content were corrupted so as to point into a table of constants. If one of the constants just happened to be the opcode of the STOP instruction, then this operation of the CPU could lead to the execution of a STOP instruction.

In the STOP mode the clock is stopped, halting all activity within the chip. Recovery from STOP may be accomplished from any one of three pins (while the other two are left inactive): the $\overline{\text{Reset}}$ pin, the $\overline{\text{XIRQ}}$ pin, or the $\overline{\text{IRQ}}$ pin. The latter will only cause an exit from the STOP mode if maskable interrupts are enabled (i.e., if I=0 in the condition code register). In each case, there is a restart delay of 4064 internal clock cycles, or about 2 ms before operation resumes. This time is built into the chip to permit the clock oscillator to stabilize before being used.*

Recovery from STOP via the $\overline{\text{Reset}}$ pin will carry out the normal startup sequence of operations. Recovery via the $\overline{\text{IRQ}}$ pin will vector through the IRQ interrupt service routine. This permits operation to resume with the instruction which follows the STOP instruction (after executing the IRQ interrupt service routine).

Recovery from STOP via the $\overline{\text{XIRQ}}$ pin offers two options. If the X bit (bit 6 of the condition code register) is cleared (enabling XIRQ interrupts), then recovery proceeds with the normal response to an XIRQ interrupt (stacking CPU registers and executing the XIRQ interrupt service routine). On the other

---

* If an external clock is used, then a DLY (delay) bit in the OPTION register should also be cleared. In this case, operation resumes within a few cycles after coming out of the STOP state.

hand, if the X bit is set, then processing will continue with the instruction immediately following the STOP instruction. Thus, operation will pick up after the STOP instruction in either case, with or without executing the XIRQ interrupt service routine first.

A.12   INSTRUCTION SET

Since we have discussed the 68HC11 register structure and instruction set in detail in Chap. 2, the purpose of this section is to list each instruction, to indicate the number of bytes and cycles which apply to each valid addressing mode for the instruction, and to show the effect of the instruction upon the bits of the condition code register. Since the instructions deal with the CPU registers, these are shown (again) in Fig. A.38. The instructions are then listed in Fig. A.39.

Examples of the instruction syntax required by a specific assembler are presented in Fig. A.40. Note in particular the syntax for the BSET, BCLR, BRSET, and BRCLR instructions. The operand field for the BRCLR actually includes three operands:

1. The address of the byte to be tested (SPSR), which is identified using direct addressing, and which becomes 29 in the assembled instruction.
2. The byte identifying which bits to test (10000000B), which actually represents immediate data, and which becomes 80 in the assembled instruction.
3. The branch address (WAIT), which employs relative addressing, and which becomes FC in the assembled instruction.

It is only the first of these operands for which optional addressing modes are available. They are identified in Fig. A.39 on the line for the BRCLR instruction by the entry under direct addressing and by the two entries under indexed addressing (using either X or Y as the index register).

A.13   CLOCK

The *clock* in the 68HC11 controls the timing of all operations going on within the chip. This includes not only the clocking of the CPU as it fetches and executes instructions, but also the clocking of other facilities on the chip which run independently of the CPU, such as the programmable timer.

The 68HC11 can operate with either an internal or an external clock source. Using the internal oscillator and an external crystal of 8.0 MHz as shown in Fig. A.41, it will operate at an internal frequency of 2.0 MHz. Being a CMOS chip with *static* registers (i.e., registers which do not have to have their contents refreshed periodically), the 68HC11 can operate at any frequency

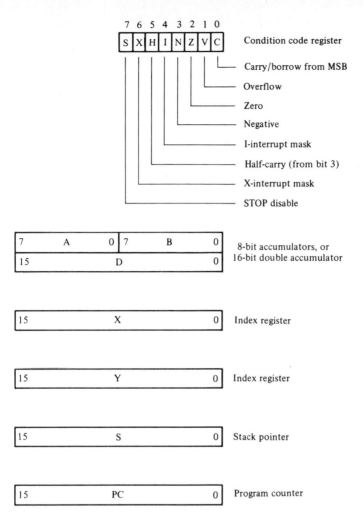

FIGURE A.38
Motorola 68HC11 register structure.

slower than this, all the way down to dc. That is, the clock can actually be stopped and then started again later without losing register contents. As mentioned earlier, the chip includes a clock monitor which can be used to detect if the clock ever drops below 10 kHz. If this capability is enabled, then a clock failure causes the chip to be reset.

With operations of the 68HC11 being clocked at the internal 2.0-MHz clock rate, all operations take integral multiples of the 0.50 μs internal cycle time. For example, the shortest instructions take two internal cycles, or 1.00 μs, to execute.

Bytes/cycles

Operation	Mnemonic	Description	Inherent	Immediate	Direct	Extended	Indexed X	Indexed Y	S	X	H	I	N	Z	V	C
Move, bits	BSET	M ← M OR mm			3/6		3/7	4/8	—	—	—	—	↔	↔	0	—
	BCLR	M ← M AND (m̄m̄)			3/6		3/7	4/8	—	—	—	—	↔	↔	0	—
Move, bytes	LDAA (LDAB)	A ← M		2/2	2/3	3/4	2/4	3/5	—	—	—	—	↔	↔	0	—
	STAA (STAB)	M ← A			2/3	3/4	2/4	3/5	—	—	—	—	↔	↔	0	—
	TAB	B ← A	1/2						—	—	—	—	↔	↔	0	—
	TBA	A ← B	1/2						—	—	—	—	↔	↔	0	—
	TAP	CC ← A	1/2						↔	→	↔	↔	↔	↔	↔	↔
	TPA	A ← CC	1/2						—	—	—	—	—	—	—	—
	CLRA (CLRB)	A ← 0	1/2						—	—	—	—	0	1	0	0
	CLR	M ← 0				3/6	2/6	3/7	—	—	—	—	0	1	0	0
	PSHA (PSHB)	Stack ← A	1/3						—	—	—	—	—	—	—	—
	PULA (PULB)	A ← Stack	1/4						—	—	—	—	—	—	—	—
Move, words	LDD	A ← M, B ← M+1		3/3	2/4	3/5	2/5	3/6	—	—	—	—	↔	↔	0	—
	STD	M ← A, M+1 ← B			2/4	3/5	2/5	3/6	—	—	—	—	↔	↔	0	—
	XGDX	D ↔ X	1/3						—	—	—	—	—	—	—	—
	XGDY	D ↔ Y	2/4						—	—	—	—	—	—	—	—
	LDX	X ← M:M+1		3/3	2/4	3/5	2/5	3/6	—	—	—	—	↔	↔	0	—
	STX	M:M+1 ← X			2/4	3/5	2/5	3/6	—	—	—	—	↔	↔	0	—
	LDY	Y ← M:M+1		4/4	3/5	4/6	3/6	3/6	—	—	—	—	↔	↔	0	—
	STY	M:M+1 ← Y			3/5	4/6	3/6	3/6	—	—	—	—	↔	↔	0	—
	LDS	S ← M:M+1		3/3	2/4	3/5	2/5	3/6	—	—	—	—	↔	↔	0	—
	STS	M:M+1 ← S			2/4	3/5	2/5	3/6	—	—	—	—	↔	↔	0	—
	PSHX	Stack ← X	1/4						—	—	—	—	—	—	—	—
	PULX	X ← Stack	1/5						—	—	—	—	—	—	—	—
	PSHY	Stack ← Y	2/5						—	—	—	—	—	—	—	—
	PULY	Y ← Stack	2/6						—	—	—	—	—	—	—	—
	TSX	X ← S+1	1/3						—	—	—	—	—	—	—	—
	TXS	S ← X−1	1/3						—	—	—	—	—	—	—	—
	TSY	Y ← S+1	2/4						—	—	—	—	—	—	—	—
	TYS	S ← Y−1	2/4						—	—	—	—	—	—	—	—
Increment	INCA (INCB)	A ← A+1	1/2						—	—	—	—	↔	↔	↔	—
	INC	M ← M+1				3/6	2/6	3/7	—	—	—	—	↔	↔	↔	—
	INX	X ← X+1	1/3						—	—	—	—	—	↔	—	—
	INY	Y ← Y+1	2/4						—	—	—	—	—	↔	—	—
	INS	S ← S+1	1/3						—	—	—	—	—	—	—	—
Decrement	DECA (DECB)	A ← A−1	1/2						—	—	—	—	↔	↔	↔	—
	DEC	M ← M−1				3/6	2/6	3/7	—	—	—	—	↔	↔	↔	—
	DEX	X ← X−1	1/3						—	—	—	—	—	↔	—	—
	DEY	Y ← Y−1	2/4						—	—	—	—	—	↔	—	—
	DES	S ← S−1	1/3						—	—	—	—	—	—	—	—

Condition Codes: S, X, H, I, N, Z, V, C. Indexed columns: X, Y.

Description	Mnemonic	Operation	INH	IMM	DIR	EXT	IND,X	IND,Y	S	X	H	I	N	Z	V	C
Complement	COMA (COMB)	A ← FFH-A	1/2						—	—	—	—	↕	↕	0	1
	COM	M ← FFH-M				3/6	2/6	3/7	—	—	—	—	↕	↕	0	1
Change CC bits	SEC	C ← 1	1/2						—	—	—	—	—	—	—	1
	CLC	C ← 0	1/2						—	—	—	—	—	—	—	0
	SEV	V ← 1	1/2						—	—	—	—	—	—	1	—
	CLV	V ← 0	1/2						—	—	—	—	—	—	0	—
	SEI	I ← 1 (disable interrupts)	1/2						—	—	—	1	—	—	—	—
	CLI	I ← 0 (enable interrupts)	1/2						—	—	—	0	—	—	—	—
Add, 8-bit	ADDA (ADDB)	A ← A+M		2/2	2/3	3/4	2/4	3/5	—	—	↕	—	↕	↕	↕	↕
	ABA	A ← A+B	1/2						—	—	↕	—	↕	↕	↕	↕
Add with carry, 8-bit	ADCA (ADCB)	A ← A+M+C		2/2	2/3	3/4	2/4	3/5	—	—	↕	—	↕	↕	↕	↕
Decimal adjust A	DAA	Adjust binary sum in A to BCD	1/2						—	—	—	—	↕	↕	↕	↕
Add, 16-bit	ADDD	D ← D+(M:M+1)		3/4	2/5	3/6	2/6	3/7	—	—	—	—	↕	↕	↕	↕
	ABX	X ← X+00:B	1/3						—	—	—	—	—	—	—	—
	ABY	Y ← Y+00:B	2/4						—	—	—	—	—	—	—	—
Subtract, 8-bit	SUBA (SUBB)	A ← A-M		2/2	2/3	3/4	2/4	3/5	—	—	—	—	↕	↕	↕	↕
	SBA	A ← A-B	1/2						—	—	—	—	↕	↕	↕	↕
	SBCA (SBCB)	A ← A-M-C		2/2	2/3	3/4	2/4	3/5	—	—	—	—	↕	↕	↕	↕
Negate	NEGA (NEGB)	A ← 0-A	1/2						—	—	—	—	↕	↕	↕	↕
	NEG	M ← 0-M				3/6	2/6	3/7	—	—	—	—	↕	↕	↕	↕
Subtract, 16-bit	SUBD	D ← D-(M:M+1)		3/4	2/5	3/6	2/6	3/7	—	—	—	—	↕	↕	↕	↕
Multiply, 8×8	MUL	D ← A*B	1/10						—	—	—	—	—	—	—	↕
Integer divide, 16/16	IDIV	X ← D/X; D ← r	1/41						—	—	—	—	—	↕	0	↕
Fractional divide, 16/16	FDIV	X ← D/X; D ← r	1/41						—	—	—	—	—	↕	↕	↕
AND	ANDA (ANDB)	A ← A AND M		2/2	2/3	3/4	2/4	3/5	—	—	—	—	↕	↕	0	—
OR (inclusive)	ORAA (ORAB)	A ← A OR M		2/2	2/3	3/4	2/4	3/5	—	—	—	—	↕	↕	0	—
Exclusive OR	EORA (EORB)	A ← A EOR A		2/2	2/3	3/4	2/4	3/5	—	—	—	—	↕	↕	0	—
Compare, 8-bit	CMPA (CMPB)	A-M		2/2	2/3	3/4	2/4	3/5	—	—	—	—	↕	↕	↕	↕
	CBA	A-B	1/2						—	—	—	—	↕	↕	↕	↕
	TSTA (TSTB)	A-0	1/2						—	—	—	—	↕	↕	0	0
	TST	M-0				3/6	2/6	3/7	—	—	—	—	↕	↕	0	0
	BITA (BITB)	A AND M		2/2	2/3	3/4	2/4	3/5	—	—	—	—	↕	↕	0	—
Compare, 16-bit	CPD	D-(M:M+1)		4/5	3/6	4/7	3/7		—	—	—	—	↕	↕	↕	↕
	CPX	X-(M:M+1)		3/4	2/5	3/6	2/6	3/7	—	—	—	—	↕	↕	↕	↕
	CPY	Y-(M:M+1)		4/5	3/6	4/7	3/7		—	—	—	—	↕	↕	↕	↕

FIGURE A.39

Motorola 68HC11 instruction set.

Bytes/cycles

Operation	Mnemonic	Description	Inherent	Immediate	Direct	Extended	Indexed X	Indexed Y	S	X	H	I	N	Z	V	C
Rotate	ROLA (ROLB) / ROL	C, A, B, or M (rotate diagram)	1/2			3/6	2/6	3/7	—	—	—	—	↔	↔	↔	↔
	RORA (RORB) / ROR	C, A, B, or M (rotate diagram)	1/2			3/6	2/6	3/7	—	—	—	—	↔	↔	↔	↔
Arithmetic shift	ASLA (ASLB) / ASL	C ← A, B, or M ← 0	1/2			3/6	2/6	3/7	—	—	—	—	↔	↔	↔	↔
	ASLD	C ← D ← 0	1/3						—	—	—	—	↔	↔	↔	↔
	ASRA (ASRB) / ASR	A, B, or M → C	1/2			3/6	2/6	3/7	—	—	—	—	↔	↔	↔	↔
Logic shift	LSLA (LSLB) / LSL	C ← A, B, or M ← 0	1/2			3/6	2/6	3/7	—	—	—	—	↔	↔	↔	↔
	LSLD	C ← D ← 0	1/3						—	—	—	—	↔	↔	↔	↔
	LSRA (LSRB) / LSR	0 → A, B, or M → C	1/2			3/6	2/6	3/7	—	—	—	—	0	↔	↔	↔
	LSRD	0 → D → C	1/3						—	—	—	—	0	↔	↔	↔

Operation	Mnemonic	Description	Relative	Immediate	Direct	Extended	Indexed X	Indexed Y	S	X	H	I	N	Z	V	C
Jump unconditionally	JMP	$PC \leftarrow M{:}M+1$				3/3	2/3	3/4	—	—	—	—	—	—	—	—
Branch unconditionally	BRA	$PC \leftarrow PC+M$	2/3						—	—	—	—	—	—	—	—
Branch never	BRN	$PC \leftarrow PC+2$	2/3						—	—	—	—	—	—	—	—

Branch conditionally on bit testing within a byte:

Operation	Mnemonic	Description	Relative	Immediate	Direct	Extended	Indexed X	Indexed Y	S	X	H	I	N	Z	V	C
if bits are set	BRSET	Branch if $(\overline{M})$ AND mm = 00H			4/6		4/7	5/8	—	—	—	—	—	—	—	—
if bits are clear	BRCLR	Branch if M AND mm = 00H			4/6		4/7	5/8	—	—	—	—	—	—	—	—

Branch conditionally on flag testing:

Condition	Mnemonic	Bytes/Cycles								
if C = 1	BCS	2/3	—	—	—	—	—	—	—	—
if C = 0	BCC	2/3	—	—	—	—	—	—	—	—
if Z = 1	BEQ	2/3	—	—	—	—	—	—	—	—
if Z = 0	BNE	2/3	—	—	—	—	—	—	—	—
if N = 1	BMI	2/3	—	—	—	—	—	—	—	—
if N = 0	BPL	2/3	—	—	—	—	—	—	—	—
if V = 1	BVS	2/3	—	—	—	—	—	—	—	—
if V = 0	BVC	2/3	—	—	—	—	—	—	—	—

Branch following a *signed* number comparison:

Condition	Mnemonic	Bytes/Cycles								
if <	BLT	2/3	—	—	—	—	—	—	—	—
if <=	BLE	2/3	—	—	—	—	—	—	—	—
if =	BEQ	2/3	—	—	—	—	—	—	—	—
if >=	BGE	2/3	—	—	—	—	—	—	—	—
if >	BGT	2/3	—	—	—	—	—	—	—	—
if <>	BNE	2/3	—	—	—	—	—	—	—	—

Branch following an *unsigned* number comparison:

Condition	Mnemonic	Bytes/Cycles								
if <	BLO	2/3	—	—	—	—	—	—	—	—
if <=	BLS	2/3	—	—	—	—	—	—	—	—
if =	BEQ	2/3	—	—	—	—	—	—	—	—
if >=	BHS	2/3	—	—	—	—	—	—	—	—
if >	BHI	2/3	—	—	—	—	—	—	—	—
if <>	BNE	2/3	—	—	—	—	—	—	—	—

Description	Mnemonic	Operation	Bytes/Cycles									
Jump to subroutine	JSR	Stack ← PC; PC ← M:M+1	2/5 3/6 2/6 3/7	—	—	—	—	—	—	—	—
Branch to subroutine	BSR	Stack ← PC; PC ← PC+M	2/6	—	—	—	—	—	—	—	—
Subroutine return	RTS	PC ← Stack	1/5	—	—	—	—	—	—	—	—
Software interrupt	SWI	Stack ← PC,Y,X,A,B,CC; 1 ← 1; PC ← M:M+1	1/14	—	—	—	1	—	—	—	—
Wait for interrupt	WAI	Stack ← PC,Y,X,A,B,CC; Halt	2/12	—	—	—	↔	↔	↔	↔	↔
Interrupt return	RTI	CC,B,A,X,Y,PC ← Stack	1/12	↔	↔	↔	→	↔	↔	↔	↔
Stop internal clock	STOP		1/2	—	—	—	—	—	—	—	—
No operation	NOP	PC ← PC+1	1/2	—	—	—	—	—	—	—	—

FIGURE A.39 (continued)

SOURCE FILE NAME: TEST.ASM

```
                    NOLIST                              ;Turn off listing
                    INCLUDE 6811REGS.HDR                ;File to define 68HC11 register names
                    LIST                                ;Turn listing back on again
0025        SMAL    EQU     25H
4455        BIG     EQU     4455H
E000                ORG     0E000H                      ;Beginning of ROM locations
E000 14030C         BSET    PORTC,00001100B             ;Set bits 2 and 3 of port C using direct addressing
E003 1D2501         BCLR    SMAL,X,00000001B            ;Indexed addressing, to clear bit 0 of the location
E006                                                    ;whose address is formed by temporarily adding 25H
E006                                                    ;to the content of X
E006 8625           LDAA    #SMAL                       ;Immediate addressing
E008 9625           LDAA    SMAL                        ;Direct addressing
E00A B64455         LDAA    BIG                         ;Extended addressing
E00D A625           LDAA    SMAL,X                      ;Indexed addressing relative to X
E00F 18A625         LDAA    SMAL,Y                      ;Indexed addressing relative to Y
E012 7C003B         INC     PPROG                       ;Extended addressing (INC does not use direct addressing)
E015 D31E           ADDD    TOC5                        ;Direct addressing
E017 3A             ABX                                 ;Add B to X, treating B as an unsigned number
E018 132980FC WAIT: BRCLR   SPSR,10000000B,WAIT         ;Hold up at this instruction as long as bit 7
E01C                                                    ;of SPSR is clear.  SPSR uses direct addressing.
E01C                                                    ;WAIT uses relative addressing.
E01C 1E0040F8       BRSET   0,X,01000000B,WAIT          ;Use indexed addressing to test bit 6 of the location
E020                                                    ;whose address is formed by temporarily adding 0 to the
E020                                                    ;content of X.  If bit 6 is set, then branch to WAIT.
E020 CE0005   AGAIN: LDX    #5                          ;Immediate addressing
E023 09             DEX                                 ;Inherent addressing
E024 26FD           BNE     AGAIN                       ;Relative addressing
0000                END
```

***** NO ERRORS DETECTED *****

FIGURE A.40

Example of instruction syntax for Avocet Systems 68HC11 assembler.

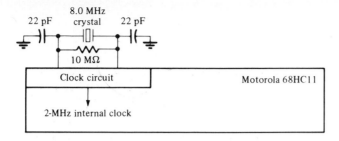

FIGURE A.41
Motorola 68HC11 clock circuitry.

A.14   PART NUMBERS, PACKAGING, AND PINOUT ALTERNATIVES

The 68HC11 which has been described here has the following two complete part numbers:

MC68HC11A8FN   52-lead quad pack
MC68HC11A8P   48-pin DIP pack

Motorola offers a part in which the ROM contains the BUFFALO monitor program described in Chap. 8:

MC68HC11A8FN1   52-lead quad pack
MC68HC11A8P1   48-pin DIP pack

Some other part numbers follow, leaving off the FN and P suffixes:

MC68HC11A0   ROM-less, EEPROM-less part
MC68HC11A1   ROM-less part
MC68HC811A2   2k EEPROM
MC68HC811A8   8.5k EEPROM
MC68HC11D4   4k ROM, 256-byte EEPROM, 192-byte RAM, no A/D
MC68HC811D4   4.3k EEPROM, 192-byte RAM, no A/D
MC68HC11E8   4 input capture, 512-byte RAM
MC68HC711E8   8k EPROM

The F version of the part will be packaged in a 68-pin quad pack, have 52 I/O lines, 1k RAM, 4-MHz *internal* clocking, a nonmultiplexed expansion bus, and on-chip generation of chip selects for external chips. The quad pack is also called a plastic-leaded chip carrier (PLCC).

The pin assignments for the 48-pin DIP pack and the 52-lead quad pack are listed in Fig. A.42. The orientation of the pins on each package is shown in Fig. A.43.

Pin name	52-lead PLCC	48-pin DIP	Single-chip function	Expanded-mode function
PA0/IC3	34	8	Port A input or Three programmable timer inputs	
PA1/IC2	33	7	Port A input	
PA2/IC1	32	6	Port A input	
PA3/OC5	31	5	Port A output or Five programmable timer outputs	
PA4/OC4	30	4	Port A output	
PA5/OC3	29	3	Port A output	
PA6/OC2	28	2	Port A output	
PA7/OC1	27	1	Port A I/O	
PB0	42	16	Output Port B	Upper address lines
PB1	41	15		
PB2	40	14		
PB3	39	13		
PB4	38	12		
PB5	37	11		
PB6	36	10		
PB7	35	9		
STRB / $R/\overline{W}$	6	28	Port B output strobe/ Port C output handshake	$R/\overline{W}$ (read/write)
STRA / AS	4	26	Port C input strobe or handshake	AS (address strobe)
PC0	9	31	I/O port C	Multiplexed lower address lines and data bus lines
PC1	10	32		
PC2	11	33		
PC3	12	34		
PC4	13	35		
PC5	14	36		
PC6	15	37		
PC7	16	38		

Signal			
PD0/RXD	20	42	I/O port D or SCI receive input
PD1/TXD	21	43	SCI transmit output
PD2/MISO	22	44	SPI master in, slave out
PD3/MOSI	23	45	SPI master out, slave in
PD4/SCK	24	46	SPI clock
PD5/SS	25	47	SPI slave select input
PE0/AN0	43	17	Input port E or A/D converter inputs
PE1/AN1	45	18	
PE2/AN2	47	19	
PE3/AN3	49	20	
PE4/AN4	44	—	
PE5/AN5	46	—	
PE6/AN6	48	—	
PE7/AN7	50	—	
VRH	52	22	High reference voltage input for A/D converter
VRL	51	21	Low reference voltage input for A/D converter
$\overline{\text{IRQ}}$	19	41	Interrupt input
$\overline{\text{XIRQ}}$	18	40	Nonmaskable interrupt input
$\overline{\text{RESET}}$	17	39	Reset chip
MODB / VKAM	2	24	Tie to +5 V to select single-chip mode — Tie to +5 V for expanded mode
MODA / LIR	3	25	Tie to ground to select single-chip mode — Pull up to +5 V for expanded mode; identifies first cycle of instruction
XTAL	8	30	8-MHz crystal connection
EXTAL	7	29	8-MHz crystal connection
E	5	27	2-MHz clock out — E (read, write timing signal)
VDD	26	48	+5-V power
VSS	1	23	Ground

FIGURE A.42
Motorola 68HC11 pin assignments.

537

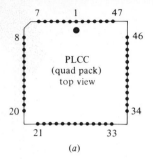

(a)

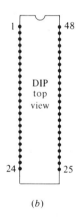

(b)

FIGURE A.43

Motorola 68HC11 pin orientations, viewed looking down on component side of PC board. (a) Plastic-leaded chip carrier (PLCC) package—also called a quad pack; (b) dual-in-line (DIP) package.

APPENDIX

B

INTEL 8096

B.1 OVERVIEW

When we use the term "Intel 8096" in this appendix, we are describing any one of Intel's second-generation 8096-family parts which carry a part number ending with BH. At the end of this appendix we list the variations on part number, designating parts with ROM, without ROM, with EPROM, with or without an A/D converter, and packaged in various ways. When compared with the first-generation parts, the BH parts add several performance refinements. They also include an EPROM version of the chip.

We first consider alternative ways to configure the 8096 (e.g., single-chip mode versus expanded mode). Next, we look at each of the on-chip resources, what options are available when using each one, how to initialize the resource, and how to interact with it during normal operation. We detail the instruction set, listing the internal clock cycles and bytes taken by each instruction for each addressing mode. Finally we list the pinouts associated with each available package and list the device options and part numbers.

B.2 CONFIGURING THE 8096

The 8096 can be operated in either the single-chip mode, or else two of its ports can be redefined to bring out the internal address bus and data bus. For the single-chip mode, the internal ROM or EPROM must necessarily be accessed. This choice is made by tying the $\overline{EA}$ (external access) pin high, as shown in Fig. B.1a. When $\overline{EA}$ is tied high, the internal ROM or EPROM is accessed during

539

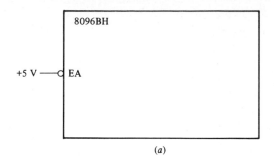

(a)

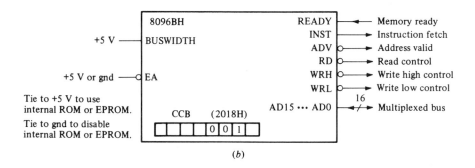

(b)

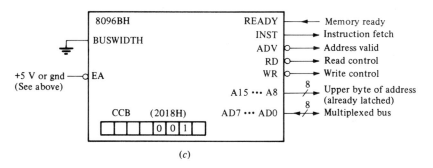

(c)

FIGURE B.1
Modes of Intel 8096 operation. (a) Single-chip mode; (b) expanded mode with 16-bit multiplexed bus; (c) expanded mode with 8-bit multiplexed bus.

instruction and data fetches from addresses 2080 to 3FFFH and for the interrupt vectors (located at addresses 2000 to 2011H). When operating in an expanded mode (Fig. B.1b and c) the internal ROM or EPROM can still be used by tying $\overline{EA}$ high. Alternatively (and necessarily, for a ROMless part), accesses to addresses 2000 to 2011H and 2080 to 3FFFH can be made to access off-chip memory by tying the $\overline{EA}$ pin low. With the $\overline{EA}$ pin tied low, nothing further needs to be done in order to access any address designated for external memory (i.e., most other addresses, as we will see in Fig. B.6). However, if the $\overline{EA}$

pin is tied high, then we have the option of using the internal ROM or EPROM together with external memory and devices, using one of the bus structures shown in Fig. B.1*b* and *c*.

One of the options made available by the BH series over the original 8096 family is the option to deal with either a 16-bit external data bus or else an 8-bit external data bus. The latter option permits expanding the 8096 with a single bytewide static RAM chip or with a single (bytewide) EPROM chip for program memory. The latter is particularly convenient for users who can either put their application program into a single EPROM, or who do not have the EPROM programming capability to separate their object code into even addresses and odd addresses, as required for the two (bytewide) EPROMs used with a 16-bit data bus. This choice of bus width is made in two places. When the 8096 comes out of reset, it reads the content of address 2018H of our user ROM or EPROM. This is called the chip configuration byte (CCB) and is defined in Fig. B.2. The

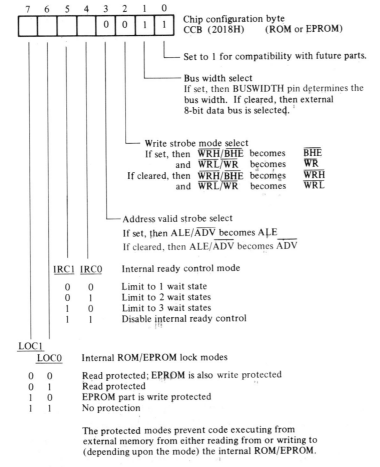

FIGURE B.2
The chip configuration byte (CCB).

8096 stores this byte in a chip configuration register (which is unaccessable by our software). Bit 1 works together with the external BUSWIDTH pin to determine the data bus width (when the $\overline{EA}$ pin is tied low). While the BUSWIDTH pin is shown in Fig. B.1 tied either high or low, it can actually be changed during each bus cycle of normal operation. For example, if it is tied to the A15 address line, then accesses to external addresses 8000 to FFFFH would use a 16-bit data bus while accesses to external addresses below this would use an 8-bit data bus. In either case, the full 16-bit address bus is brought out. However, when only an 8-bit data bus is brought out, the lines which bring out the upper half of the address bus do not have to be multiplexed. In this case, the designers of the chip have saved users the need for an external latch for the upper half of the address bus by latching the address internally.

Let us look at two other features which were introduced with the BH second-generation parts. The original 8096 parts gave the user of the expanded chip an ALE (address latch enable) output. This was used to latch the address. The BH parts permit this option but also offer the more versatile option illustrated in Figs. B.3 and B.4. The new option is selected with a 0 in bit 3 of CCB. As shown in Figs. B.3 and B.4, the $\overline{ADV}$ line remains high during any machine cycles which are not accessing external memory, but goes low during external accesses. Because of this, $\overline{ADV}$ can be used to simplify the decoding to enable external devices. In addition, during an external access, $\overline{ADV}$ drops low at precisely the correct time to latch the multiplexed address. Consequently, it can serve double duty, both helping with decoding and also latching the multiplexed address.

Another feature of the original 8096 parts operating in the expanded mode

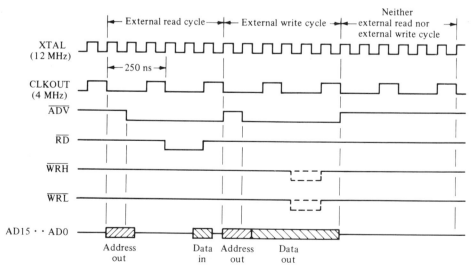

FIGURE B.3
Timing for expanded mode with 16-bit multiplexed bus.

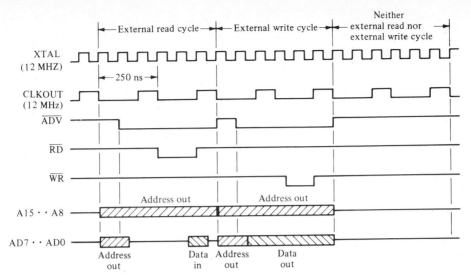

FIGURE B.4
Timing for expanded mode with 8-bit multiplexed bus.

was the need to decode a $\overline{\text{BHE}}$ (bus high enable) signal. This was used during writes to a byte at an odd address so that the lower byte on the 16-bit data bus could be left unchanged. Users of the original 8096 parts had to gate $\overline{\text{BHE}}$ together with a $\overline{\text{WR}}$ (write) signal to generate *two* write signals, one for chips connected to the upper half of the data bus and one for chips connected to the lower half of the data bus. The BH parts permit this option (for compatibility with the older parts) but include the fully decoded write signals (the $\overline{\text{WRH}}$ and $\overline{\text{WRL}}$ signals shown in Fig. B.3) as an alternative option. This latter option is selected with a 0 in bit 2 of CCB. As shown in Fig. B.3, these lines are normally high, going low only when an external write takes place to the upper byte, the lower byte, or both.

The expanded modes shown in Fig. B.1 include an INST output. This is a signal which takes on meaning when the $\overline{\text{RD}}$ (read) line is active, signaling that a read from an external device is taking place. *If* the read is an instruction fetch, then INST will be high. Otherwise it will be low during the read cycle. Users of logic analyzers and designers of 8096 emulators can use this signal to help sort out the activity on the external bus.

The READY control line shown in Fig. B.1 permits the 8096 to run at full speed for its internal accesses and yet to slow down somewhat for some or all of its external accesses. It is used in conjunction with bits 4 and 5 of CCB to introduce extra 250-ns* wait states into external read and write cycles. The options are spelled out by Figs. B.2 and B.5. If the READY line is tied high, as

---

* Assuming a 12-MHz crystal.

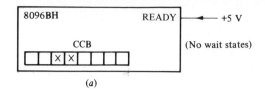

(a)

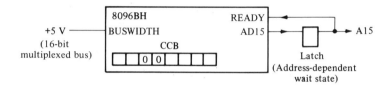

(b)

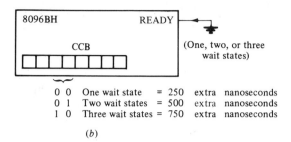

(c)

FIGURE B.5
Alternative uses of READY input. (a) Avoiding wait states entirely, for use with fast external parts; (b) use with slower external parts; (c) external accesses to addresses below 8000H get an extra wait state, whereas accesses above 8000H get no extra wait states.

shown in Fig. B.5a, then the CCB bits do not matter and no extra wait states are introduced into external reads or writes. On the other hand, if as shown in Fig. B.5b, the READY line is tied low (signifying that external devices are not ready), then this READY signal can be overridden by the CCB bits. Thus a 00 in bits 4 and 5 of CCB will now limit the delay to a single wait state. This is an excellent simplification to the scheme required in the original 8096 parts, where a one-shot was typically used to drive the READY line to achieve the same purpose.

As shown in Fig. B.5c, the READY line can be changed dynamically from cycle to cycle. If it is tied to the upper address line, then we can position external devices which can run at full speed in the 8000 to FFFF address range and slower external devices needing an extra wait state (or more) at lower addresses.

The final option introduced by CCB is the availability of a lock mode, selected by the coding of bits 6 and 7, as shown in Fig. B.2. A user can design a proprietary chip to carry out a sophisticated job and then build a business out of selling this solution to a general problem. Whether the software is in on-chip EPROM or on-chip ROM, it can be protected from copying by other users who manage to get the chip to execute code from external memory and then have that external program dump the internal memory to the serial port (for example). In a read-protected mode, only code executing from internal memory can read from memory addresses between 2020 to 3FFFH. In a write-protected mode, no code can write to memory addresses between 2000 and 3FFFH.

One problem arises with a memory-protection scheme such as this. If we purchase ROM-protected parts from Intel, then before we use them, we would like to test them. We can drive the $\overline{EA}$ line low and use our own program to test all the resources on the chip. However, this does not let us test the ROM contents. Intel supports the verification of ROM by including a 16-byte security key, located at addresses 2020 to 202FH. Before protected memory can be read, the chip must read external memory locations 4020 to 402FH and compare the contents with the internal security key. Accesses to protected memory will only be allowed if a match is found for all 16 bytes.

The memory space allocation for the 8096 is shown in Fig. B.6. This gives an overview of where resources are located in the memory space. The first 26 addresses form the register file, used to set up and access almost all of the on-chip resources. The allocation for this area is shown in Fig. B.7. The rest of page 0 is dedicated to internal RAM, for a total of 230 bytes of RAM. While the ROM or EPROM extends from 2000 to 3FFFH, Intel reserves addresses

2012 to 2017H

```
REGISTERS           EQU   0000H    ;Registers extend up to 0019H
RAM                 EQU   001AH    ;RAM extends up to 00FFH
EXTERNAL1           EQU   0100H    ;External memory space (up to 1FFDH)
PORT3               EQU   1FFEH
PORT4               EQU   1FFFH
ROM                 EQU   2000H    ;ROM, or EPROM, extends up to 3FFFH
INTERRUPT_VECTORS   EQU   2000H    ;Vectors extend up to 2011H
RESERVED            EQU   2012H    ;ROM up to 207FH is reserved for factory
                                   ; test code (except for CCB)
CCB                 EQU   2018H    ;Chip configuration byte
START               EQU   2080H    ;User program or data up to 3FFFH
EXTERNAL2           EQU   4000H    ;External memory space (up to FFFFH)
```

FIGURE B.6
Intel 8096 memory space allocation.

```
;*******************************************************************************
;           SYMBOLIC NAMES FOR THE I/O REGISTERS OF THE 8096
;*******************************************************************************
;
R0              EQU    00H:WORD   ; R     Zero register (reads as 0000H)
AD_COMMAND      EQU    02H:BYTE   ;   W   A/D command register
AD_RESULT_LO    EQU    02H:BYTE   ; R     A/D result, lo byte (byte read only)
AD_RESULT_HI    EQU    03H:BYTE   ; R     A/D result, hi byte (byte read only)
HSI_MODE        EQU    03H:BYTE   ;   W   HSI mode register
HSO_TIME        EQU    04H:WORD   ;   W   HSO time hi/lo (word write only)
HSI_TIME        EQU    04H:WORD   ; R     HSI time hi/lo (word read only)
HSO_COMMAND     EQU    06H:BYTE   ;   W   HSO command register
HSI_STATUS      EQU    06H:BYTE   ; R     HSI status register
SBUF            EQU    07H:BYTE   ; R/W   Receive buffer (read)
;                                         Transmit buffer (write)
INT_MASK        EQU    08H:BYTE   ; R/W   Interrupt mask register
INT_PENDING     EQU    09H:BYTE   ; R/W   Interrupt pending register
WATCHDOG        EQU    0AH:BYTE   ;   W   Watchdog timer register
TIMER1          EQU    0AH:WORD   ; R     Timer 1 hi/lo (word read only)
TIMER2          EQU    0CH:WORD   ; R     Timer 2 hi/lo (word read only)
BAUD_RATE       EQU    0EH:BYTE   ;   W   Baud rate control register
PORT0           EQU    0EH:BYTE   ; R     Port 0
PORT1           EQU    0FH:BYTE   ; R/W   Port 1
PORT2           EQU    10H:BYTE   ; R/W   Port 2
SP_CON          EQU    11H:BYTE   ;   W   Serial port control register
SP_STAT         EQU    11H:BYTE   ; R     Serial port status register
IOC0            EQU    15H:BYTE   ;   W   I/O control register 0
IOS0            EQU    15H:BYTE   ; R     I/O status register 0
IOC1            EQU    16H:BYTE   ;   W   I/O control register 1
IOS1            EQU    16H:BYTE   ; R     I/O status register 1
PWM_CONTROL     EQU    17H:BYTE   ;   W   Pulse width modulation control register
SP              EQU    18H:WORD   ; R/W   Stack pointer
```

FIGURE B.7
Intel 8096 allocation of page 0 addresses to registers.

and

$$2019 \text{ to } 207\text{FH}$$

for test purposes. Our interrupt vectors, CCB, and program code can fill out the remaining addresses.

While the 8096 includes instructions to read and write both bytewide and wordwide variables, some of the registers impose special requirements upon a user:

1. The following word-type (i.e., 16-bit) variables can only be read *as words*, and cannot be written to at all:

```
        TIMER1          TIMER2          HSI_TIME
```

2. The following word-type variable can only be written to, and then only with an instruction which deals with an entire word:

`HSO_TIME`

3. R0 really represents a hard-wired connection which always reads as 0000H. While it can be written to, it cannot be changed.

4. All of the other I/O registers can be accessed only as bytes. This even includes AD_RESULT_HI and AD_RESULT_LO. While it would seem that the 10-bit A/D converter output would be read as a word, such is not the case.

5. The register file addresses cannot be used as either the source address or the destination address for the multiply or divide instructions.

6. No register file address can be used as a base or index register for indexed or indirect addressing modes. The one exception to this rule arises with R0, which can be used as the base for long-indexed addressing. Since the word-type variable R0 is hard-wired to contain 0000H, this exception provides a way to access any location throughout the entire 65,536-byte memory space using its absolute address. For example,

```
LDB AL,1234H[R0]
```

will load AL with the content of address 1234H. In fact, most assemblers for the 8096 will convert

```
LDB AL,1234H
```

to the above form, making it appear that the 8096 includes an "extended" addressing mode when such is not actually the case.

B.3   GENERAL-PURPOSE I/O PORTS

While some of the ports on the 8096 serve double-duty, the alternative uses tend to be simple and easy to distinguish between. For example, Fig. B.8 shows

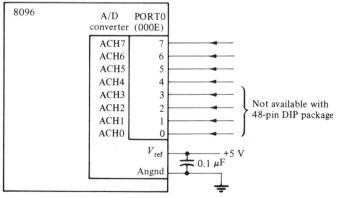

DC input current = 3 μA (max)

FIGURE B.8
Port 0 uses.

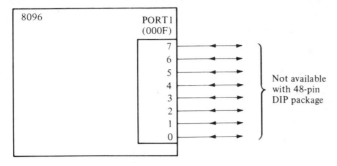

Output drive capability:
    0.8 mA at 0.45 V (max) when driving low.
    20 μA at 2.4 V (min) when driving high.

Input loading:
    100 μA (max) at 0.45 V when driving low.

Input/output selection:
    Write 1s to any bits which are to be used
    as inputs.
    (The weak internal pullup can then be
    overridden by the external device
    which drives the line.)

FIGURE B.9
Port 1 use as a quasi-bidirectional I/O port.

port 0, whose lines can serve as either general-purpose inputs or alternatively as inputs to the analog-to-digital converter facility.

Port 1 is a quasi-bidirectional I/O port, as shown in Fig. B.9. During writes to port 1, any lines which are being used as inputs must have a 1 written to them. The weak internal pullup is designed to be overridden by the external device which drives the line. While the output drive capability is suffcient to drive a 74LSxx input, a CMOS device driven by port 1 will require a pullup resistor* to +5 V in order to bring the output up well above the normal CMOS threshold voltage of 2.5 V. Port 1 inputs can be driven by either CMOS or TTL devices with no extra parts.

A problem can arise when using a bidirectional port which does not employ a separate data direction register to distinguish between inputs and outputs. Suppose, for example, that the lower 4 bits of port 1 are used as outputs while the upper 4 bits are used as inputs. Then the use of

```
ANDB        P1,#11111110B
```

would seem to be a proper way to clear bit 0 of port 1 while leaving the other bits unchanged. However, if bit 7 is being driven low by an external device when the ANDB instruction reads port 1, it will see the 0 on bit 7, it will AND it

---

* Perhaps 10 kΩ.

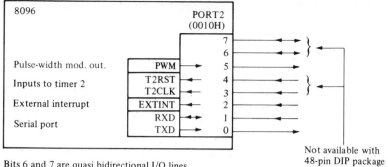

Bits 6 and 7 are quasi bidirectional I/O lines
with same behavior and drive characteristics as port 1 lines.

Drive capability of other outputs:

    2.0 mA at 0.45 V (max) when driving low.

    200 μA at 2.4 V (min) when driving high.

Input loading of port 2, bit 1 = 10 μA (max)

Input loading of port 2, bits 2, 3, 4 = 50 μA (max)

FIGURE B.10
Port 2 optional uses.

with a 1 (in the immediate operand), and finally it will write the resulting 0 back
to port 1. In this way, what had originally been set up as an input line can get a 0
written to it by the CPU. Thereafter, besides giving rise to contention on this bit
7 line between the 8096 and the external device, the input on bit 7 will always be
read as 0 (assuming that the port 1 output driver when driving low overrides the
external device's driver when it is driving high).

    The solution to this problem is to keep a copy of port 1 in RAM, perhaps
called P1_COPY. The following sequence of two instructions will then serve
the purpose intended above:

```
ANDB    P1_COPY,#11111110B
STB     P1_COPY,P1
```

    Port 2 includes four input lines, two output lines, and two quasi-bidirec-
tional I/O lines (which operate identically to those of port 1). One or more of six
of these lines can be dedicated to an alternative role, as shown in Fig. B.10. We
will discuss the selection of each of these alternatives when we deal with the
corresponding special-purpose facility on the chip.

    Ports 3 and 4 are shown in Fig. B.11. When used as ports, they have open-
drain outputs. By never writing anything but a 1 to a line, it can serve as an
input even as other lines serve as outputs. Each output line needs the addition
of a pullup resistor having a value of 15 kΩ or so. In the expanded mode, the
bus lines gain the ability to drive both high and low, forming the expansion bus
without the need for pullup resistors.

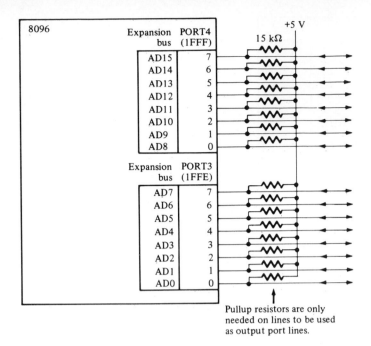

Pullup resistors are only
needed on lines to be used
as output port lines.

I/O port drive capability and loading:

Both ports have open-drain outputs when used as
bidirectional I/O parts.

Output drive: 0.8 mA at 0.45 V (max) when driving low.

Input loading: 10 μA (max)

Expansion bus drive capability and loading:

When used as an expansion bus, these lines
have strong internal pullup.

Output drive: 2.0 mA at 0.45 V (max) when driving low.
200 μA at 2.4 V (min) when driving high.

Input loading: 10 μA (max).

FIGURE B.11
Ports 3 and 4.

B.4   RESETS AND SELF-PROTECTION OPTIONS

Driving the reset pin low will reset the 8096 regardless of what it has been doing. The resulting reset state of the chip's resources is listed in Fig. B.12. Execution will begin at address 2080H.

The reset pin is a bidirectional line with a strong internal pullup. In addition to being driven by an external reset circuit, it can be driven by the internal watchdog timer circuit. Because of this, any external reset circuit must include

Register	Reset value
Port 1	11111111B
Port 2	110XXXX1B
Port 3	11111111B
Port 4	11111111B
PWM control	00H
Serial port (transmit)	Undefined
Serial port (receive)	Undefined
Baud rate register	Undefined
Serial port control	XXXX0XXXB
Serial port status	X00XXXXXB
A/D command	Undefined
A/D result	Undefined
Interrupt pending	Undefined
Interrupt mask	00000000B
Timer 1	0000H
Timer 2	0000H
Watchdog timer	0000H
HSI mode	11111111B
HSI status	00000000B
IOS0	00000000B
IOS1	00000000B
IOC0	X0X0X0X0B
IOC1	X0X0XXX1B
HSI FIFO	Empty
HSO CAM	Empty
HSO lines	000000B
PSW	0000H
Stack pointer	Undefined
Program counter	2080H

FIGURE B.12
Intel 8096 reset state.

an open-collector driver, as shown in Fig. B.13. The CMOS Schmitt trigger provides an extremely high input impedance, permitting a large RC time constant to be obtained with large R and reasonably sized C. It also provides some hysteresis so that noise on the power line while the RC circuit is charging will still cause only one exit from reset. Finally, it provides snap-action on its output, which is a desirable feature if other chips use the same reset signal. All chips will come out of the reset state at the same time. Furthermore, any chip needing a sharp rising edge on its reset input will get it.

This figure also shows all the miscellaneous lines to the 8096 which must be tied to either +5 V or to ground (except for those lines which are actively used, such as the READY or BUSWIDTH lines, already discussed).

While the 8096BH parts are not particularly low-powered microcontrollers,* they do offer a power-down option, shown being used in Fig. B.14. If V_{PD} is held above +4.5 V after V_{CC} drops to 0 V, then the data in the upper 16 bytes

* With a maximum power supply current of 200 mA, with all outputs disconnected.

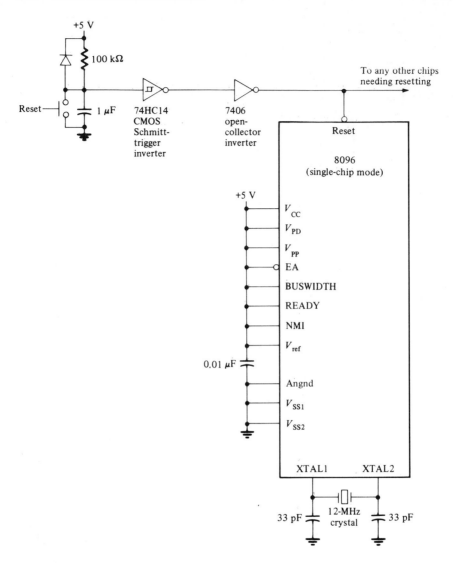

FIGURE B.13
Intel 8096 reset circuit and miscellaneous connections.

of RAM can be maintained with a current drain of less than 1 mA. The MAX690 battery switchover/reset generator chip supports the needed operation superbly. It provides a voltage threshold mechanism for bringing the chip out of reset at startup and for returning it to reset at power down. In addition, it can switch its V_{out} output from V_{CC} to V_{batt} automatically as V_{CC} drops below 4.65 V. With the 4.8-V NiCad rechargeable battery shown, the diode and resistor circuit is included so as to trickle-charge the battery while power is turned on.

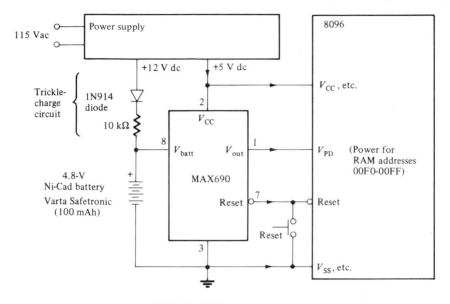

MAX690 CMOS battery switchover/reset generator

Maxim Integrated Products
8-pin DIP
When V_{CC} drops to 4.65 V,

- Reset output goes low
- V_{out} is connected to V_{batt} (with 0.1-V drop)

When V_{CC} rises above 4.65 V + hysteresis

- Reset output goes high
- V_{out} is connected to V_{CC} (with 0.2-V drop)

FIGURE B.14
Battery backup of power-down RAM.

Even if battery backup of 16 bytes of internal RAM is not worth the addition of a battery (and the attendant PC board space), the MAX690 is still an excellent reset circuit in that it is small (8-pin DIP), low in cost, provides an accurate 4.65-V threshold detector without a trimpot, and an active-low reset output with a weak pullup which can safely be overridden by a watchdog timer.

The 8096's watchdog timer is disabled at reset. It is not enabled until the following sequence is carried out:

```
LDB    WDT,#1EH
LDB    WDT,#0E1H
```

Thereafter, this same sequence must be executed no less often than every 16 ms to keep the watchdog timer from timing out and resetting the 8096.

If the 8096 ever goes awry and stops executing the intended code, presumably the watchdog timer will no longer see the above reset sequence.

Within 16 ms the 8096 will have its reset pin pulled low by the watchdog timer circuitry for 0.5 µs, resetting the chip and providing for the recovery back to normal operation.

B.5 INTERRUPTS

The 8096 supports the interrupt response to any of 22 different sources (up to 20 of which can all be enabled at the same time). Each interrupt source filters through a sequence of enabling conditions to determine whether it can actually interrupt CPU operation. Fundamental to interrupt operation is the program status word, or PSW, shown in Fig. B.15. The most-significant byte of PSW contains the CPU's flag bits. One of these, PSW's bit 9, is the global interrupt enable bit, I. This bit is cleared at reset time, disabling all interrupts initially. It is set under program control to enable any interrupts to the CPU. Thereafter it can be cleared by either of two instructions:

               DI   (disable interrupts)

or

               PUSHF   (push PSW onto stack and then clear PSW)

The I bit can be set by two other instructions:

               EI   (enable interrupts)

or

               POPF   (restore PSW from stack)

The latter instruction is often used at the end of an interrupt service routine to restore PSW to the same state which existed when the interrupt service routine was entered. This necessarily implies that POPF will restore the I bit to the 1 state in this case.

    The lower byte of PSW divides all interrupt sources into eight groups. Each group has its own interrupt mask bit in this lower byte. In fact, this byte can be separately read from and written to as the INT_MASK register, at

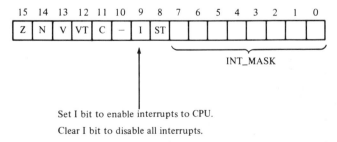

Set I bit to enable interrupts to CPU.

Clear I bit to disable all interrupts.

FIGURE B.15
Program status word, PSW.

address 0008H. As we shall see, this ability can be used at the beginning of an interrupt service routine, after a PUSHF instruction has cleared all of these bits, to reenable only the higher priority interrupt sources by setting only their bits in INT_MASK.

An overview of the interrupt sources is shown in Fig. B.16. As an example, consider "HSI data available." Figure B.16 states that the CPU will receive an interrupt when the high-speed input's FIFO is full, if three conditions are met:

1. Bit 7 of the IOC1 (I/O control register 1) is set.
2. Bit 2 of PSW (the program status word), which is also bit 2 of INT_MASK (address = 0008H) is set.
3. The I (interrupt enable) bit in the CPU is set.

When the FIFO-full interrupt occurs, two flags will be set:

1. Bit 2 of the INT_PENDING register. This is the flag which the CPU uses, together with INT_MASK and I, at the beginning of each instruction cycle to determine whether to vector off to an interrupt service routine.
2. Bit 6 of IOS1 (I/O status register 1). This flag exists so that *either* flag control *or* interrupt control can be used with the FIFO-full condition.

In contrast with the FIFO-full set of enabling conditions, some of the other interrupt sources achieve their local enabling by enabling the function itself. While this will become clearer as we examine each specific on-chip utility (e.g., the serial port), the transmit function of the serial port is an example. Setting bit 5 of IOC1 actually enables the transmit function itself, not just the interrupt which will occur when the transmission is complete.

Some interrupt sources require indirect testing to see if they have occurred. For example, when one or more of the interrupt sources in the software timers group causes an interrupt to occur, the interrupt service routine must poll *every one* of the sources which could have occurred. In the case of the four software timers, this is easy since each has its own flag bit in the IOS1 register. In addition, assuming that we sometimes want to take action when timer 2 is reset under interrupt control (e.g., setting up the next time when timer 2 is to be reset), we need to read TIMER2 itself and deduce whether it has indeed been reset. As another example, if we are using interrupt control to sample an analog signal with the A/D converter, then the above polling routine must read a timer and compare its time with the time which was used to initiate the A/D conversion (saved in a RAM variable). While this latter procedure seems awkward, it is actually easy to implement. The single instruction

```
CMP     TIMER1,SETTIME
```

will compare the actual timer 1 value with the RAM variable SETTIME and will leave the carry bit set (which, for the 8096, indicates that a borrow did *not*

Interrupt group	Interrupt source	Vector		Enabling in PSW register	INT_PENDING flag bit	Local enable bit	Local flag
		High byte	Low byte				
External interrupt	Port pin P2.2 (EXTINT)	200F	200E	I=1 PSW.7=1	INT_PENDING.7	IOC1.1=0	—
	Port pin P0.7 (ACH7)	200F	200E	I=1 PSW.7=1	INT_PENDING.7	IOC1.1=1	—
Serial port	Transmit ready	200D	200C	I=1 PSW.6=1	INT_PENDING.6	IOC1.5=1	SP_STAT.5
	Receive ready	200D	200C	I=1 PSW.6=1	INT_PENDING.6	SP_CON.3=1	SP_STAT.6
Software timers	Software timer 0	200B	200A	I=1 PSW.5=1	INT_PENDING.5	HSO_COMMAND.4=1	IOS1.0
	Software timer 1	200B	200A	I=1 PSW.5=1	INT_PENDING.5	HSO_COMMAND.4=1	IOS1.1
	Software timer 2	200B	200A	I=1 PSW.5=1	INT_PENDING.5	HSO_COMMAND.4=1	IOS1.2
	Software timer 3	200B	200A	I=1 PSW.5=1	INT_PENDING.5	HSO_COMMAND.4=1	IOS1.3
	Timer 2 reset (HSO-initiated)	200B	200A	I=1 PSW.5=1	INT_PENDING.5	HSO_COMMAND.4=1	—
	A/D conversion started (HSO-initiated)	200B	200A	I=1 PSW.5=1	INT_PENDING.5	HSO_COMMAND.4=1	—
High-speed input 0	HSI.0	2009	2008	I=1 PSW.4=1	INT_PENDING.4	—	HSI_STATUS.0

High-speed outputs	HSO.0	2007	2006	I=1	PSW.3=1	INT_PENDING.3	HSO_COMMAND.4=1	IOS0.0*
	HSO.1	2007	2006	I=1	PSW.3=1	INT_PENDING.3	HSO_COMMAND.4=1	IOS0.1*
	HSO.2	2007	2006	I=1	PSW.3=1	INT_PENDING.3	HSO_COMMAND.4=1	IOS0.2*
	HSO.3	2007	2006	I=1	PSW.3=1	INT_PENDING.3	HSO_COMMAND.4=1	IOS0.3*
	HSO.4	2007	2006	I=1	PSW.3=1	INT_PENDING.3	HSO_COMMAND.4=1	IOS0.4*
	HSO.5	2007	2006	I=1	PSW.3=1	INT_PENDING.3	HSO_COMMAND.4=1	IOS0.5*
HSI data available	HSI holding register full	2005	2004	I=1	PSW.2=1	INT_PENDING.2	IOC1.7=0	IOS1.7
	HSI FIFO full	2005	2004	I=1	PSW.2=1	INT_PENDING.2	IOC1.7=1	IOS1.6
A/D conv. complete	A/D conversion complete	2003	2002	I=1	PSW.1=1	INT_PENDING.1		AD_RESULT_LO.3 (=0 when idle)
Timer overflow	Timer 1 overflow	2001	2000	I=1	PSW.0=1	INT_PENDING.0	IOC1.2=1	IOS1.5
	Timer 2 overflow	2001	2000	I=1	PSW.0=1	INT_PENDING.0	IOC1.3=1	IOS1.4

* Read this bit (which is the present value of the HSO.*i* line) to determine if timed event has occurred.

FIGURE B.16
Intel 8096 interrupt sources, listed by group priority with highest priority listed at the top.

occur) if the A/D conversion has indeed been initiated and that new servicing is required.

A final example of the indirect interpretation of a flag arises during the polling which results after an interrupt in the high-speed output group occurs. The interrupt service routine may need to service one or more of the six high-speed output pins,

$$HSO.0 \quad . \quad . \quad . \quad HSO.5$$

For each one, it must keep track, in RAM, of what was supposed to happen and then read the output line to see if it did happen. We might maintain a variable called IOS0_DESIRED in which bits 0 through 5 are copies of what we have told the high-speed output to generate sometime in the future. Then the following instruction sequence uses the temporary variable, AL, in a test for each individual condition:

```
        LDB     AL,IOS0_DESIRED
        XORB    AL,IOS0         ;Matching bits produce 0
        JBS     AL,0,SKIP1      ;Skip if bit 0 is set
        BSR     SERVICE_HS00    ;Service HSO.0
SKIP1:  JBS     AL,1,SKIP2      ;Skip if bit 1 is set
        BSR     SERVICE_HS01    :Service HSO.1
SKIP2:  etc.
```

During the ongoing operation of the 8096, the CPU checks for interrupts after each instruction* is executed. If the I bit in the program status word (PSW) is set (enabling interrupts), then the CPU compares the bits of INT_MASK (the lower byte of the PSW) with the bits of INT_PENDING. If it finds one or more pairs of matching 1s, then an interrupt will occur. The CPU stacks the program counter and then loads the program counter with one of the eight vectors stored in addresses 2000 to 200FH. In picking which of the eight interrupt groups to service, it picks the highest enabled interrupting source, as listed in Fig. B.16. That is, it favors the interrupt associated with PSW.7 over that associated with PSW.6, etc. The CPU also clears the INT_PENDING bit corresponding to the selected interrupt service routine. Because of this, it is important to service *all* of the interrupt sources in that interrupt group before returning.

The INT_PENDING register (at address 0009H) offers some interesting opportunities. As discussed above, an interrupting source will automatically set the corresponding bit in INT_PENDING. That bit will be automatically cleared when the CPU services the interrupt. In addition, since the register can be both read from and written to, we can invoke any of the eight interrupt service routines at any time. For example, to invoke the A/D conversion complete interrupt service routine, we can execute the following:

```
        ORB     INT_PENDING,#00000010B
        ORB     INT_MASK,#00000010B
```

---

* Except EI, DI, PUSHF, POPF, and TRAP, each of which inhibits interrupts from being acknowledged until after the *next* instruction has been executed.

We can also eliminate a pending interrupt by clearing the associated bit of INT_PENDING. For example, within the high-speed output interrupt service routine, we need to poll each of six interrupt sources and possibly service each one, as discussed above. However, if the high-speed output circuitry generates a *new* output while we are in the interrupt service routine, then bit 3 of INT_PENDING will become set again. Depending upon where we are in the polling routine when this new interrupt condition occurs, we may service it now or else miss it and not handle it until after returning from the interrupt service routine and then responding to the new interrupt by reentering the interrupt service routine. Even if we have already handled the servicing of the new interrupt source, another interrupt from the high-speed output circuitry will be waiting for service when we execute the return-from-interrupt instruction. The subsequent polling routine may find that none of the six interrupt sources need servicing (because the new interrupting source has already been serviced).

As an alternative, *before* returning from the interrupt service routine, we might test bit 3 of INT_PENDING to see if it has been set again while its interrupt service routine has been executing. If the bit is set, then we can clear it with

```
ANDB INT_PENDING,#11110111B
```

and go through the entire polling routine again.

Note that we will not miss any interrupts by either scheme. However, the latter scheme does provide faster service for an interrupt source. In so doing, it decreases the worst-case latency for other lower-priority interrupt sources.

The designers of the 8096 have permitted us the flexibility of installing our own priority scheme in place of that of Fig. B.16. For example, suppose that we choose to give the A/D conversion complete interrupt the highest priority and the high-speed output interrupt the next highest priority. In this case, the A/D conversion complete interrupt service routine, ADCC_ISR, will take the following form:

```
ADCC_ISR:    PUSHF      ;Save the PSW, then clear PSW
             .          ;Service A/D converter (with I=0)
             .
             .
             POPF       ;Restore PSW
             RET        ;and return from interrupt
```

For the high-speed output interrupt service routine we have:

```
HSO-ISR:     PUSHF      ;Save the PSW, then clear PSW
             EI         ;Reenable interrupts after next instruction
             LDB INT_MASK,#00000010B ;Enable A/D converter ISR
             .          ;Service high-speed outputs
             .
             .
             POPF       ;Restore PSW
             RET        ;and return from interrupt
```

Each of the lower priority interrupt service routines reenables interrupts for these two sources as well as for any other sources to which we want to assign a higher priority. Within the lowest priority interrupt service routine, interrupts will be reenabled for all seven of the higher priority interrupt service routines.

This recasting of interrupt priorities depends upon the fact that when an interrupt service routine is entered, its first instruction will *always* be executed. When this is a PUSHF instruction, its execution clears PSW including the I bit, disabling further interrupts. Also, both PUSHF and the following EI instruction postpone any potential interrupt servicing until after the instruction which follows the EI instruction. This ensures that the instruction changing the INT_MASK register will be executed, changing priorities appropriately, and be followed by the reenabling of interrupts.

At the end of the interrupt service routine, the POPF instruction restores the original value of PSW, restoring the interrupt priority scheme to the same state it was in at the time this present interrupt occurred. Since interrupts are not acknowledged immediately following a POPF instruction, the RET instruction will be executed (clearing up the stack) before another interrupt is serviced.

The 8096 has a superbly fast approach to handling interrupts. A major feature stems from the availability of two- and three-operand instructions. They permit the interrupt service routines to take advantage of *context switching*. Upon entry, the only register which is automatically stacked is the program counter. The PSW is then stacked under program control, but many interrupt service routines need stack no other registers. RAM variables can be permanently dedicated to any counters and pointers which an interrupt service routine needs. The instructions of the interrupt service routine can use them *in place*, without moving them to CPU registers, which in fact do not exist. Consequently, the 8096 avoids a major source of overhead required by other microcontrollers when handling interrupts. The 8096 avoids the whole process of beginning an interrupt service routine by setting aside the contents of CPU registers and then loading these registers with the values needed by the interrupt service routine. It also avoids the need to restore the original contents to the CPU registers (other than the PSW) at the end of the interrupt service routine. In addition, the instructions support interrupt operations well, and they execute quickly. For example, if RX_POINTER is a RAM variable which points into a buffer used to store characters received from the serial port, then the single instruction

```
STB     SBUF,[RX_POINTER]+
```

reads SBUF (the receive data register of the serial port), stores this value in the RAM location pointed to by RX_POINTER, and then increments RX_POINTER, making it ready to handle the next interrupt. The entire operation takes only 2.0 µs, and no CPU accumulators or pointers are involved in carrying out the entire operation!

The 8096 also handles the *latency* of pending interrupts in an expeditious manner, where latency is defined to be the time from when an interrupt source requests service until it actually gets service. As we saw in Chap. 3, the best thing that can be done to reduce latency for lower-priority interrupting sources is to keep the interrupt service routines *short* for higher priority interrupt sources. The designers of the 8096 could hardly have done a better job in this respect.

To get a feel for the latency which can arise with the 8096, let us examine the worst-case latency of the highest priority interrupt source. For this examination, the times listed in Fig. B.17 are pertinent. Let us first consider the case in which the priority of interrupts is the same as that chosen by the Intel designers of the 8096. That is, we consider latency for an external interrupt. If *no* interrupt service routines are currently being executed (and if the I bit is set, enabling interrupts to the CPU), then all bits of the INT_MASK register are set, enabling all interrupt sources. The worst-case condition will occur if the CPU is executing an instruction which has a long execution time, such as the instruction to multiply two 16-bit words. As shown in Fig. B.17, the very longest instruction takes 10.75 μs. Even if we are concerned about worst-case latency at some specific point in our code and eliminate the very long instructions, we still have many instructions which may take as long as 4 μs to execute.

Since the 8096 requires 1 μs after an interrupting event before it can

---

Time from completion of an instruction until the 8096 is ready to execute the first instruction of an interrupt service routine.
5.25 μs

Time required between an interrupting event and the end of an instruction, in order for the 8096 to respond at the end of the instruction.
1.00 μs

Execution time of overhead instructions used in an interrupt service routine.

PUSHF	3.00 μs
EI	1.00 μs
LDB INT_MASK,#..	1.00 μs
POPF	2.25 μs
RET	3.00 μs

Execution time for the longest instructions.

NORML LONG,AL	10.75 μs (in worst-case)
MUL LONG,[POINTER]+	8.25 μs

Execution time for the longest instructions, if NORML, multiply, divide, and shift instructions are excluded.
4.00 μs

FIGURE B.17
Times which bear upon the servicing of 8096 interrupts (assuming a 12-MHz crystal and assuming operands are located within the 8096).

respond, the worst case arises if an external interrupt occurs 0.75 µs before the very longest instruction begins execution. After 11.50 µs have passed, the CPU will take the 5.25 µs needed to get to the interrupt service routine. The interrupt service routine begins with the PUSHF instruction discussed earlier, taking another 3.00 µs. Finally, we are ready to do the actual instructions needed to service the external interrupt. The worst-case latency is thus given by:

$$\text{Worst-case latency} = 0.75 + 10.75 + 5.25 + 3.00$$
$$= 19.75 \ \mu\text{s}$$

If we change the interrupt priority much from that provided by the Intel design-ers, a curious anomaly arises in the worst-case latency for the highest priority interrupt source. For the sake of this discussion, let us suppose that we want to reverse completely the inherent priorities of the eight interrupt groups, raising the timer overflow interrupts to highest priority and lowering the external inter-rupt to lowest priority. Now, the worst-case scenario would have a timer overflow interrupt occurring just 0.75 µs before the longest 8096 instruction began execution. Then sometime before the execution of this instruction has been completed,* the external interrupt occurs. An interrupt from each of the other six remaining interrupt groups also occurs shortly thereafter, so that in this worst-case scenario an interrupt from each of the eight interrupt groups becomes pending. Upon completion of the long instruction, the CPU will auto-matically vector to the external interrupt service routine (taking 5.25 µs to do this), where it will execute

```
PUSHF
EI
LDB INT_MASK,#01111111B
```

thus taking

$$5.25 + 3.00 + 1.00 + 1.00 = 10.25 \ \mu\text{s}$$

before interrupts are reenabled. At this point, the serial port interrupt source comes in, taking another 10.25 µs before reenabling interrupts. In this way, the worst-case latency for the timer overflow interrupt source which we have raised to highest priority becomes:

$$\text{Worst-case latency} = 0.75 + 10.75 + 7*(10.25) + (5.25 + 3.00)$$
$$= 91.50 \ \mu s \ !$$

This worst-case latency, which conceivably could happen if we completely reversed the priority of the interrupt sources, is extremely unlikely to occur in a real-life application. For one thing, we may not want to deviate very much from the original priority scheme. However, this example vividly illustrates one cost associated with priority rearrangements.

Before closing this section, there are two other interrupt sources to be

---

* Actually, more than 0.75 µs before the end of its execution.

considered. The 8096 includes a TRAP instruction which can serve as a software interrupt. It is unmaskable, so that when the CPU fetches this instruction, it will push the program counter onto the stack and vector to the interrupt service routine whose address is stored in memory at the TRAP instruction's vector location (2011,2010H). This instruction is used by the software for low-cost development boards to support single stepping and breakpoint implementation.

The 68-pin versions of the 8096-family of chips include an NMI (non-maskable interrupt) pin. A positive transition into this pin will clear the count in the watchdog timer, stack the program counter, and cause the CPU to fetch its next instruction from address 0000H in *external* memory. This feature is therefore of no use when the 8096 is used as a single-chip microcontroller. It has been included by Intel for use by their development systems, which reserve external memory addresses 0000 to 00FFH.

B.6 TIMERS AND HIGH-SPEED I/O

The 8096's timer 1, shown in Fig. B.18, is a 16-bit counter which is clocked every 2 μs (i.e., every eight internal clock cycles). It can be read from at any time but must never be written to. Furthermore, while 000BH contains the

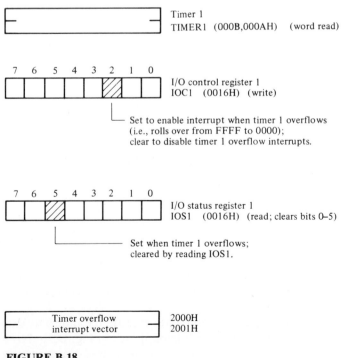

FIGURE B.18
Timer 1.

upper byte of timer 1 and 000AH contains its lower byte, the hardware accepts only reads of the entire 2-byte word, as for example

```
ADD HSO_TIME,TIMER1,#15
```

which reads TIMER1, adds 15 to this value, and stores the result in HSO_TIME.

Timer 1 is used in conjunction with the high-speed I/O system to be discussed shortly. Together, they make up the 8096's programmable timer capability, discussed in Chap. 3, which times input events and controls the timing for output events.

Timer 2, shown in Fig. B.19, is a 16-bit event counter. It must be clocked by a signal coming into the chip on either of two pins. Timer 2 is counted on both the rising edges *and* the falling edges of the input signal and the minimum time between edges is 2.0 μs. This corresponds to a square wave input having a maximum frequency of 250 kHz.

Either of two input pins can be used for the clock source for timer 2, as shown in Fig. B.19. If we want to keep track of the *time* when input transitions occur, then the high-speed input pin, HSI.1, becomes an appropriate choice. Otherwise T2CLK (i.e., bit 3 on port 2) can be selected, releasing HSI.1 for an alternative use. The choice between these two clock sources is made by setting or clearing bit 7 of the I/O control register 0 (IOC0), as shown in Fig. B.19.

Resetting is the only other control which can be exerted over timer 2. However, we have four choices for doing this:

1. Write a 1 to bit 1 of IOC0. This resets timer 2 (but does not hold it at reset).
2. Set up the high-speed output facility to reset timer 2. This permits timer 2 to be reset as a function of real time, as determined by timer 1. It also permits timer 2 to be reset when it (timer 2) has reached a certain count. In this way, a modulo-N counter of input events can be produced, perhaps to generate an interrupt and an output pulse each time that a rotating gear makes one complete revolution.
3. Receive a rising input edge on high-speed input pin HSI.0. This also permits us to determine the time when resetting took place, since, as we shall see shortly, the rising edge into HSI.0 can be used to capture the value of timer 1 at that instant.
4. Receive a rising input edge on T2RST. This choice might be used in place of HSI.0 if we do not actually need to know the time when resetting takes place. In this way, HSI.0 is released to carry out a timing function.

All of these options are controlled by what is written into I/O control register 0 (IOC0), and are spelled out by Fig. B.19.

The high-speed input (HSI) unit can be set up to capture the time at which input edges occur on any of the four HSI.*i* input lines shown in Fig. B.20. Note that two of the four lines are shared with two of the high-speed output unit

Timer 2
TIMER2   (000D, 000CH)   (word read)

Timer 2 is clocked by both rising and falling edges of selected input signal.

Minimum interval between input clock edges = 2.0 μs (with 12-MHz crystal)

Alternative clock sources:
— Port pin P2.3 (i.e., T2CLK; see Fig. B.10)
— High-speed input pin HSI.1

Alternative reset sources:
— Set bit 1 of IOC0 register
— Use high-speed output unit
— Rising edge on port pin P2.4 (i.e., T2RST; see Fig. B.10)
— Rising edge on high-speed input pin HSI.0

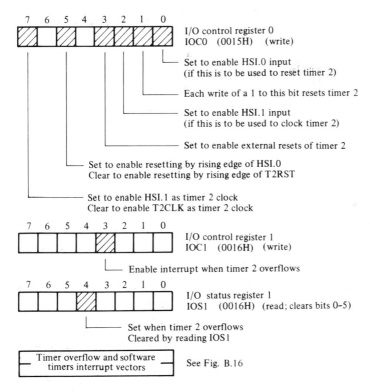

7 6 5 4 3 2 1 0

I/O control register 0
IOC0   (0015H)   (write)

Set to enable HSI.0 input
(if this is to be used to reset timer 2)

Each write of a 1 to this bit resets timer 2

Set to enable HSI.1 input
(if this is to be used to clock timer 2)

Set to enable external resets of timer 2

Set to enable resetting by rising edge of HSI.0
Clear to enable resetting by rising edge of T2RST

Set to enable HSI.1 as timer 2 clock
Clear to enable T2CLK as timer 2 clock

7 6 5 4 3 2 1 0

I/O control register 1
IOC1   (0016H)   (write)

Enable interrupt when timer 2 overflows

7 6 5 4 3 2 1 0

I/O status register 1
IOS1   (0016H)   (read; clears bits 0–5)

Set when timer 2 overflows
Cleared by reading IOS1

Timer overflow and software
timers interrupt vectors

See Fig. B.16

FIGURE B.19
Timer 2.

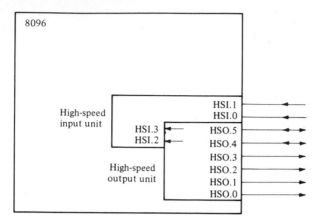

FIGURE B.20
High-speed I/O unit.

lines. As shown in Fig. B.21, input edges from each of the four lines are enabled into the HSI unit with four bits of the IOC0 register. Any of four possible input conditions can be sensed on each of these input lines:

> Every eighth rising edge
>
> Every rising edge
>
> Every falling edge
>
> Every rising and falling edge

The selected choice for each line is made by writing 2 bits to the HSI_MODE register.

Because each input to the HSI unit is likely to be used independently of the others, and because the IOC0 and HSI_MODE registers cannot be read, we generally need to keep a copy of each register in RAM and deal with them as illustrated by the following example:

```
ANDB    HSI_MODE_COPY,#11111100B   ;For HSI.0, sense
ORB     HSI_MODE_COPY,#00000010B   ; falling edges
STB     HSI_MODE_COPY,HSI_MODE     ;Update register
ORB     IOC0_COPY,#00000001B       ;Then enable
STB     IOC0_COPY,IOC0             ; HSI.0 input
```

When the selected input event occurs on any of the four lines, HSI.0, HSI.1, HSI.2, or HSI.3, the time is automatically read from timer 1 and put into a FIFO along with four identification bits. These four bits identify which one (or more) of the four inputs caused the input capture to occur. The FIFO has a width of $16 + 4 = 20$ bits for each entry and is seven entries deep. In addition, the oldest FIFO entry is moved out of the FIFO to the HSI holding register. Thus, the FIFO and HSI holding register combination can record up to *eight*

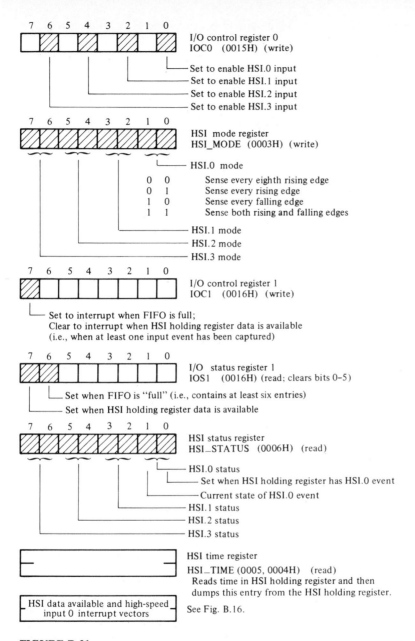

```
  7   6   5   4   3   2   1   0
┌───┬───┬───┬───┬───┬───┬───┬───┐
│   │▨▨│   │▨▨│   │▨▨│   │▨▨│   I/O control register 0
└───┴───┴───┴───┴───┴───┴───┴───┘   IOC0  (0015H)  (write)
```

Set to enable HSI.0 input
Set to enable HSI.1 input
Set to enable HSI.2 input
Set to enable HSI.3 input

```
  7   6   5   4   3   2   1   0
┌───┬───┬───┬───┬───┬───┬───┬───┐
│▨▨▨│▨▨▨│▨▨▨│▨▨▨│▨▨▨│▨▨▨│▨▨▨│▨▨▨│   HSI  mode register
└───┴───┴───┴───┴───┴───┴───┴───┘   HSI_MODE (0003H)  (write)
```

HSI.0 mode

0	0	Sense every eighth rising edge
0	1	Sense every rising edge
1	0	Sense every falling edge
1	1	Sense both rising and falling edges

HSI.1 mode
HSI.2 mode
HSI.3 mode

```
  7   6   5   4   3   2   1   0
┌───┬───┬───┬───┬───┬───┬───┬───┐
│▨▨│   │   │   │   │   │   │   │   I/O control register 1
└───┴───┴───┴───┴───┴───┴───┴───┘   IOC1  (0016H)  (write)
```

Set to interrupt when FIFO is full;
Clear to interrupt when HSI holding register data is available
(i.e., when at least one input event has been captured)

```
  7   6   5   4   3   2   1   0
┌───┬───┬───┬───┬───┬───┬───┬───┐
│▨▨│▨▨│   │   │   │   │   │   │   I/O  status register 1
└───┴───┴───┴───┴───┴───┴───┴───┘   IOS1   (0016H)  (read; clears bits 0-5)
```

Set when FIFO is "full" (i.e., contains at least six entries)
Set when HSI holding register data is available

```
  7   6   5   4   3   2   1   0
┌───┬───┬───┬───┬───┬───┬───┬───┐
│▨▨│▨▨│▨▨│▨▨│▨▨│▨▨│▨▨│▨▨│   HSI status register
└───┴───┴───┴───┴───┴───┴───┴───┘   HSI_STATUS (0006H)  (read)
```

HSI.0 status
Set when HSI holding register has HSI.0 event
Current state of HSI.0 event
HSI.1 status
HSI.2 status
HSI.3 status

```
┌─────────────────────────────┐
│─                           ─│   HSI time register
└─────────────────────────────┘   HSI_TIME (0005, 0004H)  (read)
```
Reads time in HSI holding register and then
dumps this entry from the HSI holding register.

```
┌─────────────────────────────┐
│HSI data available and high-speed│
│   input 0 interrupt vectors     │
└─────────────────────────────┘
```
See Fig. B.16.

FIGURE B.21
High-speed inputs.

input events before an overrun of the FIFO occurs. The HSI holding register can be read in two parts:

> HSI_STATUS, shown in Fig. B.21, interleaves the four identification bits from the HSI holding register with a reading of the present state of the actual HSI.*i* lines coming into the chip.
>
> HSI_TIME, also shown in Fig. B.21, has the 16-bit value which was put into the FIFO from timer 1 earlier.

Reading HSI_TIME is what triggers the HSI holding register to throw away its 20-bit value and replace it with the next oldest entry in the FIFO (if there is one). Reading HSI_TIME also clears bit 7 of IOS1, shown in Fig. B.21. This flag will remain cleared until another 20-bit entry ripples down to the HSI holding register. Consequently, bit 7 of IOS1 must always be tested first to see if there is an entry available in the HSI holding register. If there is, then it is read by first reading HSI_MODE followed by a read of HSI_TIME.

By putting a FIFO into their implementation of the input capture function of a programmable timer, the 8096 designers have dramatically increased the allowable latency permitted in the response of the CPU to interrupts resulting from input captures. If we set up to be interrupted when the HSI holding register contains an entry (by clearing bit 7 of IOC1), then we do not actually have to respond to that interrupt until the FIFO comes close to being overridden. Overriding the FIFO means that it and the HSI holding register are full at a time when the input capture circuitry is ready to put a new entry into the FIFO. If such an overrun occurs, the new entry will be thrown away.

Since there is no overrun flag for the FIFO to indicate that we have lost data, we should figure out how to detect this problem. In fact, if we are willing to think of the FIFO as holding a maximum of five entries, not seven, then we can use the FIFO-full flag (bit 6 of IOS1 in Fig. B.21) as an overrun error flag. This will work because the FIFO-full flag becomes set when the FIFO holds either six or seven entrys; that is, when the FIFO and HSI holding register together hold either seven or eight entries.

Reading IOS1 to test for FIFO data available and for an overrun calls for careful handling because the operation of reading actually clears bits 0 to 5 of the register. Furthermore, this register holds flags not only for the high-speed input unit but also for the high-speed output unit as well as overflow flags for each of the two timers. Again, the solution requires that a copy of IOS1 be maintained in RAM (and perhaps called IOS1_COPY). The *only* reading of IOS1 is done with the following operation:

```
ORB     IOS1_COPY,IOS1
```

This operation maintains any flags which have been set in the past, ready for subsequent testing. Now bit 7 can be tested with

```
JBS  IOS1_COPY,7,READ_FIFO
```

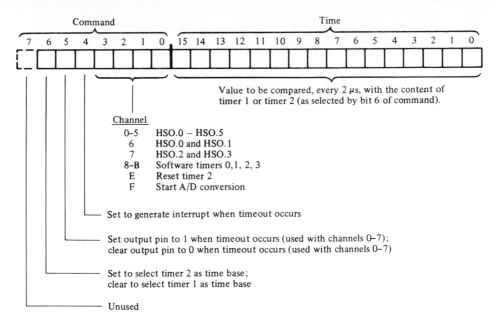

FIGURE B.22
Format of each CAM entry in the HSO unit.

which will jump to READ_FIFO if the HSI holding register holds an entry. If the jump is taken, then not only do we need to read HSI_STATUS and HSI_TIME to get the entry, but we also need to clear bit 7 of IOS1_COPY with

```
ANDB IOS1_COPY,#01111111B
```

and (perhaps) test bit 6 of IOS1_COPY for FIFO overrun.

The high-speed output (HSO) unit works with timer 1 to implement the output compare function described in Chap. 3. It works with timer 2 to implement the pulse accumulator function, discussed in Chap. 3 also. It works with the A/D converter to set up an arbitrary, and jitter-free,* sample rate. It does all of these functions, and more, without an overwhelming number of registers being required to support each function.

The heart of the high-speed output unit is a 23 × 8 CAM (content-addressable memory). Each of up to eight entries in the CAM is looked at once every 2 μs. Each entry takes the form shown in Fig. B.22. Sixteen bits of each entry hold a value which will be compared with the content of *either* timer 1 *or* timer 2 every 2 μs. Because timer 1 is clocked every 2 μs and timer 2 is not permitted to be clocked any faster than this, we are assured that a match will eventually

---

* In which the sampling interval is unvarying.

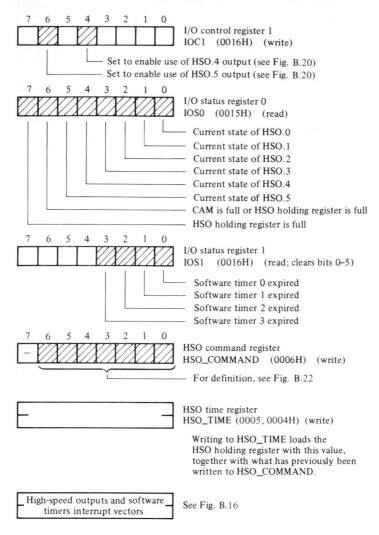

FIGURE B.23
High-speed outputs.

occur between each CAM entry and the selected timer.* When the match
occurs, the specified action will be invoked and then the entry is automatically
removed from the CAM.

The registers associated with the high-speed output unit are shown in Fig.
B.23. While the HSO.0, HSO.1, HSO.2 and HSO.3 pins on the chip are perma-

---

* An exception occurs if timer 2 is operated as a modulo-N counter, counting from 0 up to N and
then being reset back to 0. If the CAM entry requires timer 2 to hold a value above N, then the
compare will never occur. Not only is this entry in the CAM wasted, but it is never cleared out!

nently dedicated as timer outputs, the HSO.4 and HSO.5 pins need not be wasted if we do not need six timer outputs. As shown in Fig. B.20, either of these two lines can serve as timer inputs. However, if we *do* want to use HSO.4 and/or HSO.5, then we need to enable these outputs in the IOC1 register, as shown at the top of Fig. B.23.

Access is made to the CAM through the HSO holding register, shown in Fig. B.24. Before writing to the HSO holding register, we need to ensure that it is ready for an entry by checking bit 7 of IOS0, perhaps with

```
WAIT:     JBS  IOS0,7,WAIT
```

In contrast with most of the 8096's status registers, the IOS0 register does not include any bits which are automatically cleared when the register is read. Consequently, this test of a single bit in IOS0 does not have the side effect of destroying status information needed elsewhere. When the holding register is ready, the desired command is written to HSO_COMMAND. Then we write the desired time to HSO_TIME.

While the above scheme will load the CAM, often we need more than this. If we are loading the CAM with a time that is going to occur shortly, then we need to ensure that we get the entry not just in the HSO holding register but into the CAM itself (since the HSO holding register content cannot initiate a timed event). It is for this purpose that the extra flag, in bit 6 of IOS0, is included. When this flag is clear, we are assured that whatever we write to the

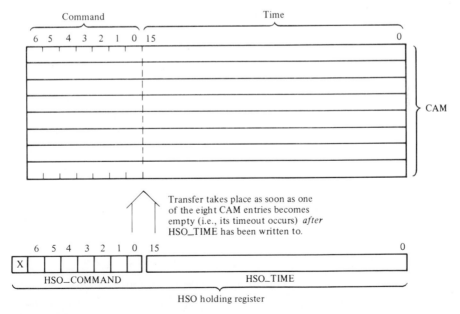

FIGURE B.24
Relationship between HSO_COMMAND, HSO_TIME, HSO holding register, and CAM.

HSO holding register will go immediately into the CAM. Actually, it is important to note that the earliest comparison we can obtain is *two counts* higher than whatever we read in timer 1 or timer 2. Thus, if we want to generate a 10-μs, positive-going pulse on HSO.3, we can use the following code:

```
            DI                                      ;Disable interrupts
WAIT3:      JBS     IOS0,6,WAIT3                    ;Wait for CAM
            LDB     HSO_COMMAND,#00100011           ;Set HSO.3 output
            ADD     HSO_TIME,TIMER1,#2              ; immediately
            LDB     HSO_COMMAND,#00000011           ;Clear HSO.3
            ADD     HSO_TIME,TIMER1,#(2+5-1)
            EI                                      ;Reenable interrupts
```

This code illustrates several important features. Interrupts are disabled because we cannot afford to find that bit 6 of the IOS0 register has been cleared by the timeout of a previously entered high-speed output command:

> Which generates an interrupt
>
> Which in turn loads another entry into the HSO holding register
>
> Which therefore fills the CAM again

The operand for the second ADD instruction is trying to make the second compare occur 10 μs, or 5 counts of timer 1, after the first compare (which was set up by the first ADD instruction). Since the first ADD instruction called for a compare to occur two counts after it read timer 1, we want to add 5 to this value. However, the intervening LDB and ADD instructions take a total of 2.50 μs to execute (as we see later in this appendix), so one count is subtracted from the operand. Because 2.50 μs is not evenly divisible by 2.00 (the time between increments of timer 1), the output pulsewidth will actually be either 10 or 12 μs, depending upon when the instructions are executed relative to the incrementing of timer 1.

If we care that the pulsewidth be exactly 10 μs, then we need only read timer 1 into a temporary variable (e.g., AX) and then form the operands for the two ADD instructions using this temporary variable. Our code would then look like:

```
            DI                                      ;Disable interrupts
WAIT3:      JBS     IOS0,6,WAIT3                    ;Wait for CAM
            LDB     HSO_COMMAND,#00100011  ;Set HSO.3 output
            LD      AX,TIMER1                       ;Read timer 1
            ADD     HSO_TIME,AX,#3
            LDB     HSO_COMMAND,#00000011  ;Clear HSO.3
            ADD     HSO_TIME,AX,#(3+5)
            EI                                      ;Reenable interrupts
```

One final point arises in connection with this example. When we disable interrupts because of the *critical region** introduced into our code, we increase

---

* In which we cannot afford to have the CPU field an interrupt.

the latency time for pending interrupts. This can be particularly troublesome if the CAM is full and if *none* of the comparisons in the CAM is scheduled to occur very soon.

As one example of the ongoing use of the high-speed output unit, consider that we want to augment the pulse-width-modulation output facility built into the 8096 with two more pulse-width-modulated outputs. Because channel 7 of the HSO command* operates on *both* HSO.2 and HSO.3, we do well to generate the pulse-width-modulated outputs on these two lines. Each time timer 1 reaches 0000H, we can set the outputs on these two lines to 1, and then use the timer 1 overflow interrupt to reload the CAM. Then the HSO.2 and HSO.3 outputs might be cleared when timer 1 matches RAM variables called PWM2 and PWM3. Each of these matches could also generate an interrupt, used to reload the CAM.

The power of the high-speed output's use of a CAM arises because we can obtain precise timing for so many different uses without having chip resources required for each one. On the other hand, no more than eight timer functions can be supported at any one time.

The software timers mentioned in Fig. B.22 are used to set a flag and (probably) generate an interrupt after a certain amount of time. For example, if we want to obtain real-time clock interrupts every 10 ms, then we might use software timer 0 to achieve this. The software timers' interrupt service routine (refer to Fig. B.16) would include the following lines:

```
            ORB   IOS1_COPY,IOS1        ;Capture 1s in IOS1
            JBC   IOS1_COPY,0,SKIP1     ;Skip if bit 0 not set
            SCALL RTI                   ;Service real-time interrupt
SKIP1:      JBC   IOS1_COPY,1,SKIP2     ;Check next bit
              .
              .
```

Then the RTI subroutine would take the following form:

```
RTI:        ANDB  IOS1_COPY,#11111110B  ;Clear copy of flag
            ADD   REAL_TIME,#5000       ;Update RAM variable
            DI                          ;Disable interrupts
RTI_1:      JBS   IOS0,7,RTI_1          ;Wait on HSO holding register
            LDB   HSO_COMMAND,#00011000B ;Software timer 0
            LDB   HSO_TIME,REAL_TIME
            EI                          ;Reenable interrupts
                                        ;Do jobs required every
              .                         ;10 ms
            RET                         ;Return from subroutine
```

This subroutine uses a 2-byte RAM variable called REAL_TIME to keep track of time, so as to generate a new interrupt every 10 ms.

As a final example, consider the use of the HSO unit, together with timer 2, to generate an interrupt once per revolution of a 120-tooth mechanical gear. Assume that a variable-reluctance pickup generates a wide pulse for each

---

* Refer to Fig. B.22.

tooth, that this pulse waveform is squared up with a Schmitt trigger, and that the resulting waveform drives the T2CLK clock input to timer 2 (refer to Fig. B.10). Since timer 2 will be clocked by *both* the rising and falling edges of the waveform, we want to make a modulo-240 counter, generating one interrupt per cycle of the counter. We will use software timer 1 for this purpose.

We need to initialize the registers associated with timer 2 appropriately (refer to Fig. B.19). Then the following test must be imbedded into the polling chain for the software timer interrupt service routine discussed above:

```
            ORB    IOS1_COPY,IOS1        ;Capture 1s in IOS1
            JBC    IOS1_COPY,0,ST_1      ;Software timer 0?
            SCALL  SOFTIME0              ;Service it
ST_1:       JBC    IOS1_COPY,1,ST_2      ;Software timer 1?
            SCALL  SOFTIME1              ;Service it
TO_2:       etc.

SOFTIME1:   ANDB   IOS1_COPY,#11111101B  ;Clear flag
            DI                           ;Disable interrupts
SOFT1:      JBS    IOS0,7,SOFT1          ;Wait for holding register
            LDB    HSO_COMMAND,#01011110B ;Reset timer 2
            LD     HSO_TIME,#240         ;Count 240 pulses
            EI                           ;Reenable interrupts
            RET                          ;Return from subroutine
```

Note that the SOFTIME1 subroutine sets up the HSO unit to watch timer 2. The instant that it reaches a count of 240, it will be reset to 0. Consequently, if we monitored timer 2, we would find it spending one "tooth-time" at each count, from 0 up to 239. It would spend no more than 2 μs at count 240, being prematurely reset back to 0 by the HSO unit, not by an edge on T2_CLK. That is, 240 input edges generate *exactly* one cycle of the counter. The interrupt service routine reinitializes the CAM with a new entry, to set up for the next time the gear has made a complete revolution.

B.7 PULSE-WIDTH-MODULATED OUTPUT

Bit 5 of port 2 can be a general purpose output, as shown back in Fig. B.10. Alternatively, it can generate a pulse-width-modulated output. The choice is made by the initialization of bit 0 in I/O control register 1, as shown in Fig. B.25. The pulse-width modulator includes an 8-bit PWM counter which is continuously clocked at the internal clock rate (4 MHz, given a 12-MHz crystal). This register is not accessible under program control. However, it is imbedded in circuitry which compares its output with the output of the pulse-width-modulation control register. The output line is set when the PWM counter equals 0. This same output line is cleared when the PWM counter equals the pulse-width-modulation control register. This simple action can be used to generate any of the waveforms shown in Fig. B.26, depending upon

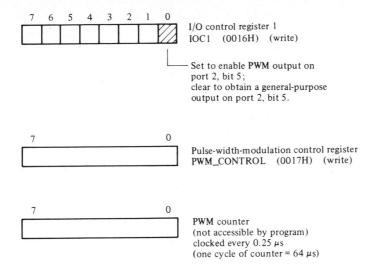

I/O control register 1
IOC1   (0016H)   (write)

Set to enable PWM output on
port 2, bit 5;
clear to obtain a general-purpose
output on port 2, bit 5.

Pulse-width-modulation control register
PWM_CONTROL   (0017H)   (write)

PWM counter
(not accessible by program)
clocked every 0.25 μs
(one cycle of counter = 64 μs)

Operation:
  Port 2, bit 5 is set when PWM counter = 00H.
  Port 2, bit 5 is cleared when PWM counter = content
    of PWM_CONTROL.
  If PWM_CONTROL = 00H, then output remains low.

FIGURE B.25
Pulse-width-modulated output.

what is written to PWM_CONTROL. The output is typically used to drive a CMOS buffer whose output drives all the way up to +5 V and down to 0 V. The output of the CMOS buffer, in turn, will have a dc component which is proportional to the content of PWM_CONTROL, in 256 steps ranging from 0 V up to 255/256 of +5 V.

B.8   ANALOG-TO-DIGITAL CONVERTER

The A/D converter on the second-generation 8096BH family of parts includes two major features not available on earlier parts.

  The input comes through a sample-and-hold circuit, permitting the collection of equally spaced samples, even when the input is changing during the conversion process.

  The conversion time has been shortened to 22 μs, permitting a maximum sampling rate of 45 kHz.

The connections between the A/D converter facility, the interrupt system, and the high-speed output unit permit the acquiring of sampled data inputs with great flexibility and rather low overhead.

PWM_CONTROL   Waveform

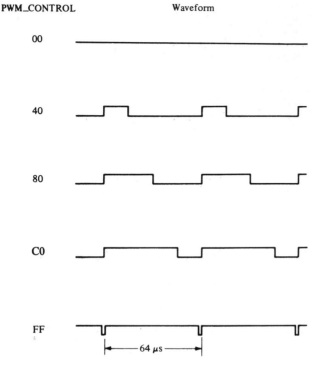

FIGURE B.26
PWM output waveforms.

The A/D converter shares pins with port 0, as shown in Fig. B.8. It also includes reference voltage inputs, V_{ref} and Angnd. Angnd should be connected to ground, while V_{ref} is typically tied to the V_{CC} supply. However, best results will be obtained if separate lines are run from the two power supply outputs (i.e., +5 V and 0 V) to these two connections. In this way, any voltage drop in the lines going to the V_{CC} and V_{SS} pins on the chip will not be seen by the A/D converter. The specification for the A/D converter *requires* that V_{ref} be within the range of 4.5 to 5.5 V. The analog inputs must remain within the range of Angnd and V_{ref}.

The conversion of an input, V_{in}, is determined from:

$$1023 \times \frac{V_{in} - \text{Angnd}}{V_{ref} - \text{Angnd}}$$

Thus, the input voltage is converted to a 10-bit binary number.

The registers associated with the A/D converter are shown in Fig. B.27. The A/D command register selects which of eight* input pins is to be sampled.

---

* Or four, for the 48-pin DIP part.

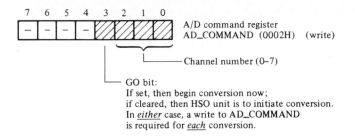

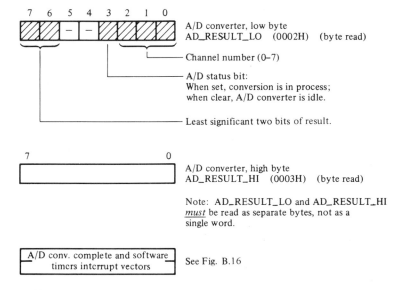

FIGURE B.27
A/D converter.

Each conversion requires a write to be made to this A/D command register. If the conversion is to begin immediately, then the GO bit must be set. Otherwise, with the GO bit written as a 0, the conversion will be triggered by a command, at a certain time, which has been loaded into the CAM of the high-speed output unit.

If we are using the A/D converter under interrupt control to obtain an ongoing sequence of samples from one of the inputs, then we have two choices for obtaining the interrupt. For relatively slow sampling, the A/D conversion complete interrupt can be used. It is only necessary that its worst-case latency be less than the sampling period. This is easy if the sampling period is something like 100 ms or 10 ms or even 1 ms.

For fast sampling, where the worst-case latency for the interrupt servicing is too long, the priority of the A/D converter can be raised. Even better, the

A/D conversion complete interrupt can be permanently disabled (by always keeping bit 1 of the INT_MASK register cleared), and the same high-speed output command which starts the conversion can be used to trigger an interrupt. If the conversion start time is stored in a 2-byte RAM variable, STARTTIME, then the interrupt will be fielded by the software timer interrupt service routine which can use the following instruction sequence to determine if the A/D converter is ready for servicing:

```
              ADD    STARTTIME,#11     ;Add conversion time
              CMP    TIMER1,STARTTIME  ;C=1 if done
              JNC    NOT_AD
              SCALL AD_SERVICE         ;Service the converter
NOT_AD:       SUB    STARTTIME,#11     ;Restore original value
```

For the fastest possible sampling, we must disable interrupts entirely (with a DI instruction) for the duration of the data gathering process. The high-speed output unit can provide equally spaced samples, and the testing of the A/D status bit (bit 3 of the AD_RESULT_LO register) used to read out the results. The A/D command register is double-buffered so that it can actually hold a second command while a first command is in progress. However, if the GO bit is set in the second command, then the first conversion will be aborted and the second conversion begun immediately. Consequently, for proper operation, the GO bit must be written as a 0 and the HSO unit used to trigger the conversion at a time when we know the conversion will be complete (e.g., more than 22 μs after the triggering of the last conversion). As an example, consider the SAMPLE subroutine below which has passed to it four parameters in RAM locations AH, AL, BX, and CX, where

AH contains the channel number (00H to 07H)

AL contains the number of samples to collect (0 to 255)

BX contains a (2-byte) pointer to the beginning of the buffer area where the samples are to be stored

CX contains the (2-byte) sampling period, expressed as a number of 2-μs ticks of timer 1 (11 or greater)

```
SAMPLE:      DI                                   ;Disable interrupts
SAMPLE1:     JBS   IOSO,6,SAMPLE1                 ;Wait until HSO is ready
             LD    DX,TIMER1                      ;Read timer into DX (RAM)
             ADD   DX,CX                          ;Set up first sample time
             LDB   AD_COMMAND,AH                  ;Arm converter
             LDB   HSO_COMMAND,#0FH               ;Prepare first conversion
             LD    HSO_TIME,DX
SAMPLE2:     ADD   DX,CX                          ;Set up next sample time
             LDB   AD_COMMAND,AH                  ;Arm converter
             LDB   HSO_COMMAND,#0FH               ;Prepare next conversion
             LD    HSO_TIME,DX
SAMPLE3:     JBC   AD_RESULT_LO,3,SAMPLE3         ;Conversion begun?
```

```
SAMPLE4:      JBS   AD_RESULT_LO,3,SAMPLE4  ;Conversion done?
              STB   AD_RESULT_LO,[BX]+       ;Store lower byte
              STB   AD_RESULT_HI,[BX]+       ;and then upper byte
              DJNZ  AL,SAMPLE2               ;Do again until done
              EI                            ;Reenable interrupts
              RET                           ; and return
```

Notice the opportunity made possible by the double buffering of the A/D command register. We can send the commands for the first *two* conversions before we have read out the result for the first one. In this way we will get down to SAMPLE4 and wait for the first output, knowing that when it is ready there will be no hurry to trigger another sample because one is already in the works. And another one will *always* be in the works while we wait at SAMPLE4.

B.9   SERIAL PORT

The serial port employs just two I/O pins on the 8096 chip. As shown in Fig. B.10, these are bits 0 and 1 of port 2, renamed TXD and RXD when the pins are dedicated to serial port use. Bit 0 of port 2 is always an output, whether used as TXD or an a general-purpose output pin (accessed by a write to PORT2). If the serial port is left unused, then bit 1 of port 2 is a general-purpose input pin (accessed by a read of PORT2). If the serial port is used as a UART, then bit 1 of port 2 serves as the RXD (receive data) input pin. On the other hand, if the serial port is used as an I/O serial port for I/O expansion, then bit 1 of port 2 serves as the data output pin when data is transmitted and serves as the data input pin when data is received.

Use as an I/O serial port permits the easy I/O expansion techniques discussed in Chaps. 6 and 7. An important characteristic of this implementation of the I/O serial port is that data is transmitted and received *least-significant bit first*. Thus, the least-significant bit of a data byte sent to an 8-bit shift register will control the last (most distant) bit of the shift register.

The registers associated with the serial port are shown in Fig. B.28. Bit 5 of I/O control register 1 (IOC1) must be set to assign bit 0 of port 2 to the TXD serial port function. The baud rate is selected by the writing of two consecutive *bytes* to a single address, 000EH, designated BAUD_RATE. As shown in Fig. B.29, the 16 bits are used in two parts:

The most-significant bit selects a clock source. While the 12-MHz crystal clock for the chip is the most likely clock source, the option exists to use T2CLK, one of the possible clock sources for timer 2.

The remaining bits, designated B, set up a divider for the selected clock source to obtain the desired baud rate. Even if a crystal clock frequency other than 12 MHz is used with the chip, this approach will still permit standard baud rates to be achieved.

```
   7  6  5  4  3  2  1  0
  ┌──┬──┬──┬──┬──┬──┬──┬──┐      I/O control register 1
  │  │  │▨▨│  │  │  │  │  │      IOC1   (0016H)  (write)
  └──┴──┴──┴──┴──┴──┴──┴──┘
           └── Set to enable TXD output
```

```
  ┌──────────────────────┐      Baud rate selection register
  │                      │      BAUD_RATE   (000EH)  (write)
  └──────────────────────┘
```

Write least-significant byte to BAUD_RATE;
then write most-significant byte to BAUD_RATE.

```
   7  6  5  4  3  2  1  0
  ┌──┬──┬──┬──┬──┬──┬──┬──┐      Serial port control register
  │ -│ -│ -│▨▨│▨▨│▨▨│▨▨│▨▨│      SP_CON   (0011H)  (write)
  └──┴──┴──┴──┴──┴──┴──┴──┘
```

Mode selection bits:

0	0	Mode 0: I/O serial port mode
0	1	Mode 1: normal UART mode (8 data bits)
1	0	Mode 2: } UART modes used for master-slave
1	1	Mode 3: } interconnections between several 8096s

PEN, parity enable bit (even parity)
If set, most-significant data bit position becomes
a parity bit;
if cleared, then parity is not employed.

REN, receive enable bit
Set to enable the receive function.

TB8, programs the ninth data bit (if not parity) during
transmission; cleared after transmission.

```
   7  6  5  4  3  2  1  0
  ┌──┬──┬──┬──┬──┬──┬──┬──┐      Serial port status register
  │▨▨│▨▨│▨▨│ -│ -│ -│ -│ -│      SP_STAT   (0011H)  (read)
  └──┴──┴──┴──┴──┴──┴──┴──┘
```

TI, transmit interrupt flag
Set when SBUF is available to transmit a byte;
cleared by reading SP_STAT.

RI, receive interrupt flag
Set when SBUF has received a byte;
cleared by reading SP_STAT.

RB8/RPE — ninth data bit received, if not parity;
— set on parity error, if parity function is enabled.

```
  ┌──────────────────────┐      Receive data register
  │                      │      SBUF   (0007H)  (read)
  └──────────────────────┘
```

```
  ┌──────────────────────┐      Transmit data register
  │                      │      SBUF   (0007H)  (write)
  └──────────────────────┘
```

```
  ┌──────────────────────┐      200CH
  │   Serial port        │      200DH
  │   interrupt vector   │
  └──────────────────────┘
```

FIGURE B.28
Serial port.

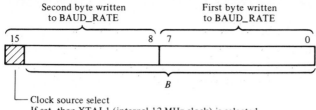

Clock source select
If set, then XTAL1 (internal 12-MHz clock) is selected.
If cleared, then T2CLK (external clock, 750 kHz maximum) selected.

Using XTAL1 and mode 0 (I/O serial port mode)

$$\text{Baud rate} = \frac{12\,\text{MHz}}{4 \times (B + 1)} \quad (\text{where } B \text{ must not be set to zero})$$

Example:
Maximum rate $(B = 1)$ gives $\frac{12}{4 \times (1 + 1)} = 1.5\,\text{MHz}$
Byte-transfer time = 8 bit times = 5.33 μs.
To set this up, write 01H, then 80H to BAUD_RATE.

Using XTAL1 and modes 1, 2, 3 (UART modes)

$$\text{Baud rate} = \frac{12\,\text{MHz}}{64 \times (B + 1)} \quad (\text{where } B \text{ must not be set to zero})$$

Examples:
For 9600 baud, write 13H, then 80H to BAUD_RATE
For 1200 baud, write 9BH, then 80H to BAUD_RATE
For 300 baud, write 70H, then 82H to BAUD_RATE

Using T2CLK and mode 0 (I/O serial port mode)

$$\text{Baud rate} = \frac{\text{T2CLK frequency}}{B} \quad (\text{where } B \text{ must not be set to zero})$$

Using T2CLK and modes 1, 2, 3 (UART modes)

$$\text{Baud rate} = \frac{\text{T2CLK frequency}}{16 \times B} \quad (\text{where } B \text{ must not be set to zero})$$

FIGURE B.29
Baud rate determination.

The actual baud rate for the port is related to the value of B as shown in Fig. B.29. For the I/O serial port, the maximum rate will probably be set up with

```
LDB   BAUD_RATE,#01H
LDB   BAUD_RATE,#80H
```

This maximum rate means that an 8-bit transfer will be completed in 5.33 μs. Under circumstances requiring that a slower rate be used (e.g., in the interface to an older, and therefore slow, CMOS chip like Motorola's MC14499 decoder/driver chip for four multiplexed seven-segment LED displays, discussed in Sec. 7.2), it needs to be pointed out that the pulse width of the clock is 333 ns, *regardless* of the baud rate. The slow chip may need to have clock pulses stretched, as was done in Fig. 7.6.

The serial port control register is used to select the serial port's mode of operation (e.g., the I/O serial port mode). Its REN bit (bit 3) is used to enable the receive function.

Data is transmitted by writing it to SBUF. When data has been received, it is read from SBUF. Although Intel uses the name SBUF for both of these functions, a read of SBUF (i.e., address 0007H) actually accesses a different register from that accessed by a write to SBUF.

Information for the use of the serial port as an I/O serial port is shown in Fig. B.30. Note that the clock output idles high and that data is valid, for both input and output, only during the rising edges of the clock.

We might choose to use the I/O serial port under flag control rather than interrupt control since a complete byte is transmitted in only 5.33 μs. To disable interrupts from the I/O serial port, we must clear bit 6 of the INT_MASK register, as discussed in conjunction with Fig. B.16. The transmit interrupt (TI) flag, bit 5 of SP_STAT, can still be used to indicate that the last byte written to SBUF has been automatically loaded into the output shift register and that now SBUF is ready for another byte.

To receive a byte of data via the I/O serial port, we need a way to trigger it to emit eight clock pulses and then collect the returned data into the receive data register (SBUF). As indicated in Fig. B.30, this can be done by clearing the REN bit, setting it again, and then waiting for the receive interrupt (RI) bit, bit 6 of SP_STAT, to become set. Alternatively, if we want to receive a sequence of bytes, we can set the REN bit just once and then test the RI bit not only to tell when each byte has been received into SBUF but also to trigger eight more clock pulses. This latter possibility raises a potential problem. By initiating the next transfer, this reading of the RI bit ignites a time bomb. At the end of 5.33 μs, the content of SBUF, which we presumably intend to read with the instruction which follows the successful testing of the RI bit, will be destroyed by the automatic loading of a new byte. If an interrupt is permitted to intervene, we will indeed lose the byte of data. The solution is to disable interrupts around this critical region with

```
          DI
WAIT:     JBC     SP_STAT,6,WAIT
          STB     SBUF,‹wherever›
          EI
```

Note that we do not have to disable interrupts during the *entire* transfer of N bytes, but only during the transfer of each byte. This will only add 7 μs, or so, to the latency of interrupt sources.

The use of the serial port as a UART for transmitting and receiving strings of ASCII characters is spelled out with Fig. B.31. This employs a 1-byte RAM

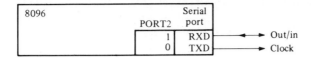

Data is transmitted least-significant bit (LSB) first, when
1. Bit 5 of IOC1 has previously been initialized to 1.
2. Mode 0, REN = 0, and baud rate have been set up.
3. A byte is written to SBUF.

The TI bit in SP_STAT will be set when SBUF is ready to send another byte.
Reading SP_STAT to test TI clears TI.

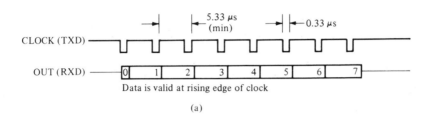

Data is received least-significant bit first, when
1. Mode 0 and baud rate have previously been set up.
2. If REN = 0 (REN = bit 3 of SP_CON), then set REN bit,
   If REN = 1 read SP_STAT. (This will also clear RI flag.)
   Either of these will initiate the output of eight clock pulses,
   and eight successive reads of IN (the RXD pin).

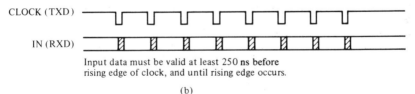

Input data must be valid at least 250 ns before
rising edge of clock, and until rising edge occurs.

(b)

FIGURE B.30
Serial port, mode 0 (I/O serial port mode). (a) Data transmission; (b) data reception.

variable called BUSY to maintain a 1-bit flag. This flag (bit 7) is used to let the serial port interrupt service routine inform the mainline program when the transmission of a string has been completed. In this way, the mainline program will not begin the transmission of a new string before the transmission of a previous string is completed. An alternative to this approach would be to pass any number of string pointers to the interrupt service routine through a queue, as discussed in Chap. 5.

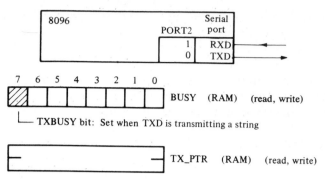

BUSY   (RAM)   (read, write)

└─ TXBUSY bit: Set when TXD is transmitting a string

TX_PTR   (RAM)   (read, write)

Initialization:

1. Set bit 5 of TOC1 to enable TXD.
2. Set up baud rate (see Fig. B.29).
3. Write 00001001B to SP_CON to enable RXD and mode 1.
4. Read SP_STAT to clear flags.
5. Clear the one-byte RAM variable, BUSY.
6. Initialize a "receive" queue.
7. Enable interrupts by setting bit 6 of INT_MASK
   and executing EI (enable interrupts) instruction.

Data format:

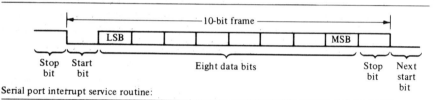

Serial port interrupt service routine:

1. Monitor SP_STAT by ORing it into a RAM variable, SP_STAT_COPY,
   since reading SP_STAT clears TI and RI flags.
2. Poll both of these flags; service if set; clear SP_STAT_COPY.
3. When a string-terminating character (e.g., 04H, the ASCII
   code for "End of transmission") is transmitted, clear TXBUSY.
4. Put each received character into the "receive" queue.

Initiating the transmission of a string:

1. Wait, if TXBUSY is set, until it becomes cleared by
   interrupt service routine.
2. Set TXBUSY.
3. Load TX_PTR to point to string to be transmitted.
4. Do a LDB SBUF, [TX_PTR]+ to initiate transmission.

FIGURE B.31
Serial port use as a UART for transmitting and receiving strings of ASCII characters.

A 2-byte RAM variable, TX_PTR, is used by the interrupt service routine. When the interrupt service routine reads the TI bit of SP_STAT and finds it set, it knows that SBUF is ready to transmit another character and that TX_PTR points to that character.

For this example, we simply have the receive part of the interrupt service routine put each received character into a queue, as discussed in Chap. 5. The mainline program can then pick up these characters at a more leisurely pace without missing any of them.

Figure B.31 shows the data format for each byte sent or received serially by the serial port operating in mode 1. Each character consists of a start bit, eight data bits, and a stop bit. If characters are not sent or received one right after another, then the line (TXD or RXD) idles high. In this way, when the next character does begin, the falling edge on the line as it begins the start bit will permit the receiver to resynchronize and read the bits of the character at the correct times.

For an application requiring the addition of a parity bit along with the data, we have two options. If the characters being sent and received are 7-bit ASCII characters, then mode 1 can be used as just described except that the serial port control register is preset to 00001101B, as shown in Fig. B.32a. This sets the parity enable bit, PEN. During a transmit operation, the most-significant bit of each byte written to SBUF will be replaced by an even-parity check bit. During a receive operation, if a parity error occurs, then bit 7 of the serial port status register will be set.

Using the serial port as a UART with parity and *eight* data bits, instead of seven, means doing exactly the same things as above except that the serial port is initialized into mode 3. This is illustrated in Fig. B.32b.

The final use of the serial port is to support communications between several 8096 chips organized into a master-slave configuration, as illustrated in Fig. B.33. In this figure, the AND gate serves as an active-low OR gate. Since inactive TXD outputs float high, the AND gate output will also be high when no slave is transmitting data back to the master. However, when the master tells one of the slaves to send it data, then that selected slave will drive its TXD line and the output of the AND gate will follow that signal.

The protocol used by the master to channel a multiple-byte message to a single slave is illustrated in Fig. B.34. The master operates its serial port in mode 3. In this mode a frame consists of a start bit, the eight data bits which come from what is written into SBUF, a ninth bit which comes from what is written into the TB8 bit (bit 4 of SP_CON), and a stop bit. The master sends multiple-byte messages consisting of an address frame followed by any number of data frames. The address frame has its ninth bit set whereas the data frames have their ninth bits cleared. The address frame also has its 8-bit data field set to the number of the slave to which the multiple-byte message is being sent. That is, a multiple-byte message being sent to slave 2 will have the number 100000010B in the nine-bit data field of the address frame.

(a) Data format: 1 start bit, 7 data bits, 1 parity bit, 1 stop bit

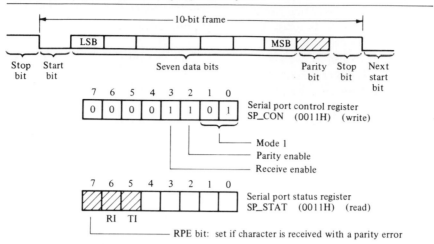

(b) Data format: 1 start bit, 8 data bits, 1 parity bit, 1 stop bit

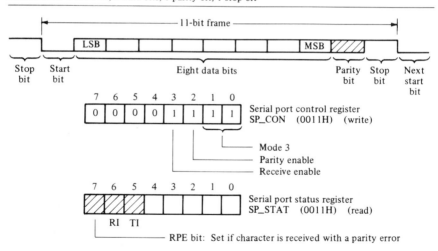

FIGURE B.32
UART use with parity. (*a*) Seven-data-bit frames; (*b*) eight-data-bit frames.

This protocol requires each slave to operate its serial port either in mode 2 or in mode 3 as it receives multiple-byte messages from the master. Both of these modes use a frame format consisting of a start bit, eight data bits, a ninth bit, and a stop bit. However, their response to the ninth bit differs. If the ninth bit is cleared, then an interrupt will occur *only* in a slave operating in mode 3. This is the mode which is used by an addressed slave to receive the data frames of a multiple-byte message. It gets an interrupt for each byte received. Unad-

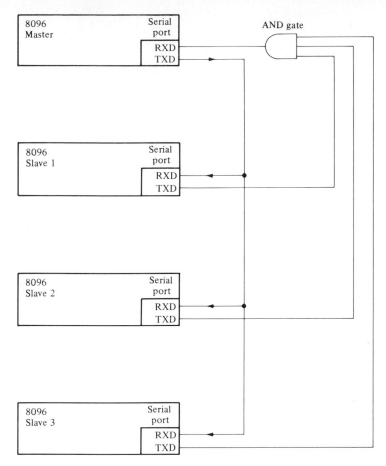

FIGURE B.33
Master-slave interconnection of several 8096 chips.

dressed slaves have set themselves up into Mode 2, specifically to avoid being interrupted by these data frames which are not addressed to them.

When the master has completed one multiple-byte message and is ready to begin another, it begins the new message with an address frame. *Every* slave, whether operating its serial port in mode 2 or 3, will be interrupted by this address frame. The slave to which the message is addressed switches to mode 3 to receive interrupts for the remaining bytes of the message. The other slaves switch to mode 2 so that they can go about their business without having to deal with the remaining bytes of a message which is not addressed to them.

B.10   EPROM PROGRAMMING

The 8096BH family of parts includes EPROM versions which can be programmed in any one of several ways. These parts all have part numbers of the

(a) Address-frame format:

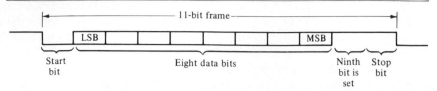

All slaves, whether in mode 2 or mode 3, will be interrupted.
The master sends, as data, the address (i.e., 1, 2, or 3)
of the slave to which the multiple-byte message is being sent.
The addressed slave sets its serial port for mode 3 operation.
The unaddressed slaves set their serial ports for mode 2 operation.

(b) Data-frame format:

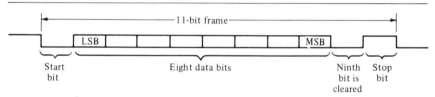

With the ninth bit cleared, this data-frame format will *not*
cause an interrupt in an unaddressed slave (operating in mode 2).
This data-frame format *will* cause an interrupt in the addressed
slave (operating in mode 3). It must check the RB8 bit (bit 7
of SP_STAT) with each reception so that it will know when a
new multiple-byte message is beginning.

FIGURE B.34
Master-to-slave frame formats. (*a*) First (address) frame of a multiple-byte message; (*b*) subsequent (data) frames of a multiple-byte message.

form 879XBH, where the "7" designates the EPROM capability and the "X" stands for a digit which specifies the package type (e.g., 68-pin ceramic pin-grid array) and whether the part includes an A/D converter which meets specifications.

In contrast with ROM versions of the part, which must be mask-programmed by Intel, the EPROM versions are user-programmed and can be erased by subjecting them to ultraviolet light. In this way, development can proceed with better and better versions of application code, until all desired features of the application code have been developed and until (hopefully) all bugs have been removed.

In Chap. 8 we saw the simplicity of circuitry needed if we first program a standard EPROM (e.g., an Intel 2764A, 8K × 8 EPROM). The circuit requires just one integrated circuit (in addition to the EPROM and the 879X part), a +12.75-V supply (in addition to the +5-V supply), and a few miscellaneous discrete parts. This uses an autoprogramming mode to program the 879X's EPROM.

In this section we look at two alternatives to the autoprogramming mode. The first alternative is used by a standard EPROM programmer which accepts a file from a computer and then uses that file to program the content of successive addresses in the EPROM. The second alternative permits "run-time" programming of the EPROM; that is, programming of the EPROM while it is running in the target application.

The 879X also supports a slave programming mode which permits up to 16 chips to be programmed in a "gang" programmer at the same time. Interested users will find that the 879X makes the programming of many chips at once not too different from the programming a single chip.

The circuit for an EPROM programmer is shown in Fig. B.35. After the

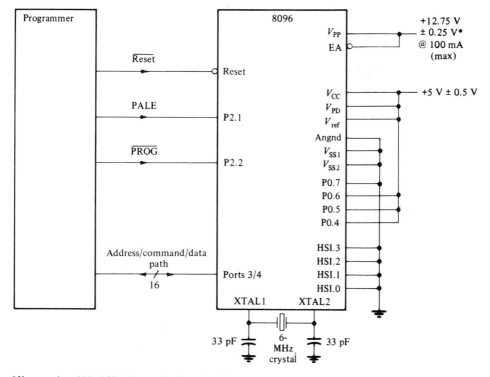

$*V_{PP}$ must be within 1 V of V_{CC} while $V_{CC} < 4.5$ V.

V_{PP} must be < 5 V until $V_{CC} > 4.5$ V.

V_{PP} must not have a low impedance path to ground while $V_{CC} > 4.5$ V.

FIGURE B.35
EPROM programming circuit.

programmer has put the 879X into the programming mode with the connections shown, together with a reset pulse, it controls the programming process with signals on the PALE (programming address latch enable) line and the $\overline{\text{PROG}}$ (programming pulse) line. It transfers commands, addresses, and data over the 16-bit address/command/data path shown in the figure.

Programming is carried out a word at a time using the algorithm of Fig. B.36. Each word is programmed with a succession of 25 separate programming cycles. Each of these cycles includes a 100-μs $\overline{\text{PROG}}$ pulse. The programmed word is then verified using the algorithm of Fig. B.37. The sequence of operations to program the entire EPROM and to verify the result takes about 30 s.

Command format:

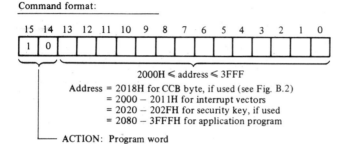

Timing diagram:

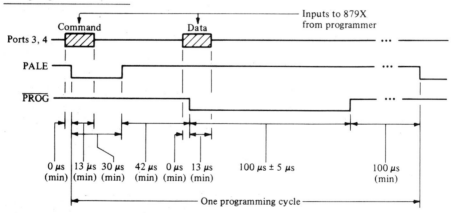

Modified Quick-Pulse programming algorithm

1. Program each two-byte EPROM location with 25 of the above programming cycles.
2. Verify that EPROM location is programmed correctly (see Fig. B.37).
3. If programming is correct, continue with next location; otherwise quit with failed part.
4. When all locations are programmed, verify each location again.
5. Entire 8K EPROM is programmed in 30 sec.

FIGURE B.36
EPROM programming algorithm.

Command format:

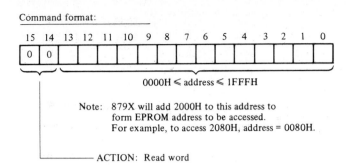

```
 15  14  13  12  11  10  9  8  7  6  5  4  3  2  1  0
┌───┬───┬──┬──┬──┬──┬──┬──┬──┬──┬──┬──┬──┬──┬──┬──┐
│ 0 │ 0 │  │  │  │  │  │  │  │  │  │  │  │  │  │  │
└───┴───┴──┴──┴──┴──┴──┴──┴──┴──┴──┴──┴──┴──┴──┴──┘
```

0000H ≤ address ≤ 1FFFH

Note: 879X will add 2000H to this address to
form EPROM address to be accessed.
For example, to access 2080H, address = 0080H.

ACTION: Read word

Timing diagram:

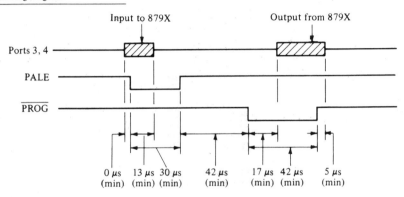

Input to 879X   Output from 879X

Ports 3, 4

PALE

PROG

0 μs   13 μs 30 μs   42 μs   17 μs 42 μs   5 μs
(min)   (min) (min)   (min)   (min) (min)   (min)

FIGURE B.37
EPROM verification.

The run-time programming mode permits an 879X part to be programmed in the target system. A voltage of +12.75 V must be applied to the V_{PP} pin. A circuit which permits the 879X part to control this voltage from an output port is shown in Fig. B.38. Since almost all 879X output lines are set to 1 at reset time (refer to Fig. B.12), this circuit drives V_{PP} to +5 V with OUT=1 at reset time. Then for run-time programming, OUT is lowered to 0 which causes V_{PP} to rise to +12.75 V. Also the $\overline{EA}$ pin must be at +5 V. Presumably it already is at +5 V since this is what it takes for the CPU in the chip to access its own EPROM. Since the erased EPROM contains all 1s, this run-time programming mode permits 1s to be changed to 0s, but not vice versa.

Once V_{PP} has been raised to +12.75 V, writes of a byte or a word to an EPROM address will continue to program that byte or word into that EPROM address until another read from, or write to, the EPROM occurs. Since the program in a single-chip microcontroller application resides in the EPROM, the trick is to avoid fetching any further instructions (from EPROM) for the duration of the programming pulse. A subroutine can be written which raises V_{PP} to +12.75 V, then executes the algorithm of Fig. B.36 in software to program just

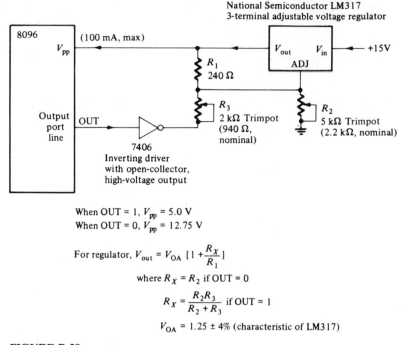

When OUT = 1, V_{pp} = 5.0 V
When OUT = 0, V_{pp} = 12.75 V

For regulator, $V_{out} = V_{OA} \left[1 + \dfrac{R_X}{R_1}\right]$

where $R_X = R_2$ if OUT = 0

$$R_X = \frac{R_2 R_3}{R_2 + R_3} \text{ if OUT = 1}$$

V_{OA} = 1.25 ± 4% (characteristic of LM317)

FIGURE B.38
Circuit for controlling V_{PP} from an output port.

1 byte, and then lowers V_{PP} back down to +5 V again. The subroutine can begin by copying the rest of its code into RAM and then executing a jump to this code. This will avoid accesses to EPROM except by the instruction which initiates the programming of a byte.

   Run-time programming adds an interesting capability to the 879X. However, it does so at the risk of destroying program code in the PROM if the CPU ever goes awry and just happens to execute the run-time programming subroutine with garbage parameters.

B.11   INSTRUCTION SET

The 8096 instruction set deals directly with operands in memory or in the registers, without requiring the intermediate loads to and stores from a CPU accumulator. In support of this, we have instructions which deal with up to three operands. These are listed in Fig. B.39 using generic names for the operands. The first letter of each operand stands for

Byte   *Word*   or   *Long word.*

The second letter stands for

Source   or   *Destination.*

Operation	MNE	Operand syntax	Description	Inherent	Immed.	Direct	Indirect — Normal	Indirect — Auto.	Indexed — Short	Indexed — Long	Z	N	C	V	VT	ST
Move, bytes	LDB	BD,BS*	BD ← BS		3/4	3/4	3/6	3/7	4/6	5/7	—	—	—	—	—	—
	STB	BS,BD*	BS → BD			3/4	3/7	3/8	4/7	5/8	—	—	—	—	—	—
	CLRB	BD	BD ← 0			2/4					1	0	—	0	—	—
Move, words	LD	WD,WS*	WD ← WS		4/5	3/4	3/6	3/7	4/6	5/7	—	—	—	—	—	—
	LDBSE	WD,BS*	WD ← BS, sign extend		3/4	3/4	3/6	3/7	4/6	5/7	—	0	—	0	—	—
	LDBZE	WD,BS*	WD ← BS, zero extend		3/4	3/4	3/6	3/7	4/6	5/7	—	0	—	0	—	—
	ST	WS,WD*	WS → WD			3/4	3/7	3/8	4/7	5/8	—	—	—	—	—	—
	CLR	WD	WD ← 0			2/4					1	0	—	0	—	—
	PUSH	WS*	WS → stack		3/12	2/12	2/15	2/16	3/15	4/16	—	—	—	—	—	—
	PUSHF	PSW	PSW → stack; 0 → PSW	1/12							0	0	0	0	0	0
	POP	WD*	WD ← stack			2/14	2/16	2/16	3/16	4/16	—	—	—	—	—	—
	POPF	PSW	PSW ← stack	1/9							↔	↔	↔	↔	↔	↔
Increment	INCB	BD	BD ← BD+1			2/4					↔	↔	↔	↔	←	—
	INC	WD	WD ← WD+1			2/4					↔	↔	↔	↔	←	—
Decrement	DECB	BD	BD ← BD−1			2/4					↔	↔	↔	↔	←	—
	DEC	WD	WD ← WD−1			2/4					↔	↔	↔	↔	←	—
Complement	NOTB	BD	BD ← FFH−BD			2/4					↔	↔	0	0	—	—
	NOT	WD	WD ← FFFFH−WD			2/4					↔	↔	0	0	—	—
Sign extend	EXTB	WD	WD ← extend low byte			2/4					↔	↔	0	0	—	—
	EXT	LD	LD ← extend low word			2/4					↔	↔	0	0	—	—
Change flag bits	SETC	C	C ← 1	1/4							—	—	1	—	—	—
	CLRC	C	C ← 0	1/4							—	—	0	—	—	—
	CLRVT	VT	VT ← 0	1/4							—	—	—	—	0	—
Enable interrupts	EI	I	I ← 1	1/4							—	—	—	—	—	—
Disable interrupts	DI	I	I ← 0	1/4							—	—	—	—	—	—
Add, 8-bit	ADDB	BD,BS*	BD ← BD+BS		3/4	3/4	3/6	3/7	4/6	5/7	↔	↔	↔	↔	←	↔
	ADDB	BD,BS1,BS2*	BD ← BS1+BS2		4/5	4/5	4/7	4/8	5/7	6/8	↔	↔	↔	↔	←	↔
Add with carry	ADDCB	BD,BS*	BD ← BD+BS+C		3/4	3/4	3/6	3/7	4/6	5/7	↔	↔	↔	↔	←	↔
Add, 16-bit	ADD	WD,WS*	WD ← WD+WS		4/5	3/4	3/6	3/7	4/6	5/7	↔	↔	↔	↔	←	↔
	ADD	WD,WS1,WS2*	WD ← WD+WS1+WS2		5/6	4/5	4/7	4/8	5/7	6/8	↔	↔	↔	↔	←	↔
Add with carry	ADDC	WD,WS*	WD ← WD+WS+C		4/5	3/4	3/6	3/7	4/6	5/7	↔	↔	↔	↔	←	—

FIGURE B.39

Intel 8096 instruction set.

				Bytes/cycles			Indirect		Indexed		Flags					
Operation	MNE	Operand syntax	Description	Inherent	Immed.	Direct	Normal	Auto.	Short	Long	Z	N	C	V	VT	ST
Subtract, 8-bit	SUBB	BD,BS*	BD ← BD−BS		3/4	3/4	3/6	3/7	4/6	5/7	↕	↕	↕	↕	←	—
	SUBB	BD,BS1,BS2*	BD ← BS1−BS2		4/5	4/5	4/7	4/8	5/7	6/8	↕	↕	↕	↕	←	—
Subtract with borrow	SUBCB	BD,BS*	BD ← BD−BS−(1−C)		3/4	3/4	3/6	3/7	4/6	5/7	↓	↕	↕	↕	←	—
Negate	NEGB	BD	BD ← 0−BD			2/4					↕	↕	↕	↕	←	—
Subtract, 16-bit	SUB	WD,WS*	WD ← WD−WS		4/5	3/4	3/6	3/7	4/6	5/7	↕	↕	↕	↕	←	—
	SUB	WD,WS1,WS2*	WD ← WS1−WS2		5/6	4/5	4/7	4/8	5/7	6/8	↕	↕	↕	↕	←	—
Subtract with borrow	SUBC	WD,WS*	WD ← WD−WS−(1−C)		4/5	3/4	3/6	3/7	4/6	5/7	↓	↕	↕	↕	←	—
Negate	NEG	WD	WD ← 0−WD			2/4					↕	↕	↕	↕	←	—
Compare	CMPB	BS1,BS2*	BS1−BS2		3/4	3/4	3/6	3/7	4/6	5/7	↕	↕	↕	↕	←	—
	CMP	WS1,WS2*	WS1−WS2		4/5	3/4	3/6	3/7	4/6	5/7	↕	↕	↕	↕	←	—
Multiply unsigned, 8×8	MULUB	WD,BS*	WD ← BD×BS		3/17	3/17	3/19	3/20	4/19	5/20	—	—	—	—	—	↕
	MULUB	WD,BS1,BS2*	WD ← BS1×BS2		4/18	4/18	4/20	4/21	5/20	6/21	—	—	—	—	—	↕
signed, 8×8	MULB	WD,BS*	WD ← BD×BS		4/21	4/21	4/23	4/24	5/23	6/24	—	—	—	—	—	↕
	MULB	WD,BS1,BS2*	WD ← BS1×BS2		5/22	5/22	5/24	5/25	6/24	7/25	—	—	—	—	—	↕
unsigned 16×16	MULU	LD,WS*	LD ← WD×WS		4/26	3/25	3/27	3/28	4/27	5/28	—	—	—	—	—	↕
	MULU	LD,WS1,WS2*	LD ← WS1×WS2		5/27	4/26	4/28	4/29	5/28	6/29	—	—	—	—	—	↕
signed, 16×16	MUL	LD,WS*	LD ← WD×WS		5/30	4/29	4/31	4/32	5/31	6/32	—	—	—	—	—	↕
	MUL	LD,WS1,WS2*	LD ← WS1×WS2		6/31	5/30	5/32	5/33	6/32	7/33	—	—	—	—	—	↕
Divide unsigned, 16/8	DIVUB	WD,BS*	WD(L) ← WD/BS; WD(H) ← WD mod BS		3/17	3/17	3/19	3/20	4/19	5/20	—	↕	—	↕	←	↕
signed, 16/8	DIVB	WD,BS*	WD(L) ← WD/BS; WD(H) ← WD mod BS		4/21	4/21	4/23	4/24	5/23	6/24	—	↕	—	↕	←	↕
unsigned 32/16	DIVU	LD,WS*	LD(L) ← LD/WS; LD(H) ← LD mod WS		4/26	3/25	3/27	3/28	4/27	5/28	—	↕	—	↕	←	↕
signed, 32/16	DIV	LD,WS*	LD(L) ← LD/WS; LD(H) ← LD mod WS		5/30	4/29	4/31	4/32	5/31	6/32	—	↕	—	↕	←	↕
AND	ANDB	BD,BS*	BD ← BD AND BS		3/4	3/4	3/6	3/7	4/6	5/7	↕	↕	0	0	—	—
	ANDB	BD,BS1,BS2*	BD ← BS1 AND BS2		4/5	4/5	4/7	4/8	5/7	6/8	↕	↕	0	0	—	—
	AND	WD,WS*	WD ← WD AND WS		4/5	3/4	3/6	3/7	4/6	5/7	↕	↕	0	0	—	—
	AND	WD,WS1,WS2*	WD ← WS1 AND WS2		5/6	4/5	4/7	4/8	5/7	6/8	↕	↕	0	0	—	—
OR (inclusive)	ORB	BD,BS*	BD ← BD OR BS		3/4	3/4	3/6	3/7	4/6	5/7	↕	↕	0	0	—	—
	OR	WD,WS*	WD ← WD OR WS		4/5	3/4	3/6	3/7	4/6	5/7	↕	↕	0	0	—	—
Exclusive OR	XORB	BD,BS*	BD ← BD XOR BS		3/4	3/4	3/6	3/7	4/6	5/7	↕	↕	0	0	—	—
	XOR	WD,WS*	WD ← WD XOR WS		4/5	3/4	3/6	3/7	4/6	5/7	↕	↕	0	0	—	—

Shift operations: Use an immediate operand to specify the number of shifts between 0 and 15. Alternatively, use the content of any page 0 address above 15 to specify this number of shifts.

Operation	Mnemonic	Operand	Effect	Bytes/cycles
Shift left	SHLB	BD,#shifts		3/7+shifts
	SHL	WD,#shifts	C ← [ ][ ]···[ ][ ] ← 0 (OPERAND)	3/7+shifts
	SHLL	LD,#shifts		3/7+shifts
	SHLB	BD,BS		3/7+shifts
	SHL	WD,BS		3/7+shifts
	SHLL	LD,BS		3/7+shifts
Logical shift right	SHRB	BD,#shifts		3/7+shifts
	SHR	WD,#shifts	0 → [ ][ ]···[ ][ ] → C (OPERAND)	3/7+shifts
	SHRL	LD,#shifts		3/7+shifts
	SHRB	BD,BS		3/7+shifts
	SHR	WD,BS		3/7+shifts
	SHRL	LD,BS		3/7+shifts
Arithmetic shift right	SHRAB	BD,#shifts		3/7+shifts
	SHRA	WD,#shifts	[ ]→[ ][ ]···[ ][ ] → C (OPERAND)	3/7+shifts
	SHRAL	LD,#shifts		3/7+shifts
	SHRAB	BD,BS		3/7+shifts
	SHRA	WD,BS		3/7+shifts
	SHRAL	LD,BS		3/7+shifts
Normalize long integer	NORML	LD,BD	Shift LD left until MSB=1; BD ← #shifts	3/11+shifts

				Bytes/cycles			
Operation	Mnemonic	Operand	Effect	Inherent	Relative	Direct	Direct, digit, relative
Branch unconditionally	SJMP	label	PC ← PC ± up to 1023		2/8		
	LJMP	label	PC ← PC ± up to 65535		3/8		
Jump, indirect	BR	[WS]	PC ← WS			2/8	
Decrement, branch if result <> 0	DJNZ	BD,label	BD ← BD−1; PC ← PC ± up to 127 if BD<>0			3/5,9	
Branch conditionally if selected bit number of BS is set (clear)							
Branch if bit set	JBS	BS,bit,label	PC ← PC ± up to 127 if test passes				3/5,8
Branch if bit clear	JBC	BS,bit,label					3/5,8
Branch conditionally on flag testing:							
if C=1	JC	label	PC ← PC ± up to 127 if test passes		2/4,8		
if C=0	JNC	label			2/4,8		
if Z=1	JE	label			2/4,8		
if Z=0	JNE	label			2/4,8		
if N=1	JLT	label			2/4,8		
if N=0	JGE	label			2/4,8		

FIGURE B.39 (*continued*)

595

Operation	MNE	Operand syntax	Description	Inherent	Relative	Direct	Direct, digit, relative	Z	N	C	V	VT	ST
if V=1	JV	label			2/4,8			—	—	—	—	—	—
if V=0	JNV	label			2/4,8			—	—	—	—	—	—
if VT=1	JVT	label			2/4,8			—	—	—	—	0	—
if VT=0	JNVT	label			2/4,8			—	—	—	—	0	—
if ST=1	JST	label			2/4,8			—	—	—	—	—	—
if ST=0	JNST	label			2/4,8			—	—	—	—	—	—
Branch following a *signed* number comparison:													
if <	JLT	label			2/4,8			—	—	—	—	—	—
if <=	JLE	label			2/4,8			—	—	—	—	—	—
if =	JE	label			2/4,8			—	—	—	—	—	—
if >=	JGE	label			2/4,8			—	—	—	—	—	—
if >	JGT	label			2/4,8			—	—	—	—	—	—
if <>	JNE	label			2/4,8			—	—	—	—	—	—
Branch following an *unsigned* number comparison:													
if <	JC	label			2/4,8			—	—	—	—	—	—
if <=	JNH	label			2/4,8			—	—	—	—	—	—
if =	JE	label			2/4,8			—	—	—	—	—	—
if >=	JNC	label			2/4,8			—	—	—	—	—	—
if >	JH	label			2/4,8			—	—	—	—	—	—
if <>	JNE	label			2/4,8			—	—	—	—	—	—
Branch to subroutine	SCALL	label	Stack ← PC; PC ← PC ± up to 1023		2/13			—	—	—	—	—	—
	LCALL	label	Stack ← PC; PC ← PC ± up to 65535			3/13							
Subroutine return and interrupt return	RET		PC ← Stack	1/12				—	—	—	—	—	—
Software interrupt	TRAP		Stack ← PC; PC ← [2011,2010H]	1/21				—	—	—	—	—	—
No operation	NOP		PC ← PC+1	1/4				—	—	—	—	—	—
	SKIP		PC ← PC+2	2/4				—	—	—	—	—	—
Reset system	RST	XXH	(see key)	1/16				0	0	0	0	0	0

KEY:

BS, BS1, BS2   Byte-type source operand having a page 0 address

WS, WS1, WS2	Word-type source operand having a page 0 even address
BD	Byte-type destination operand having a page 0 address
WD	Word-type destination operand having a page 0 even address
LD	Long-word-type destination operand having a page 0 address (divisible by 4)

BS*, WS*, BD*, WD* Operand which can employ any of the addressing modes listed

| 3/4 | 3-byte instruction which executes in 4 cycles (1.0 μs) |
| 2/4,8 | 2-byte branch instruction which executes in 4 cycles if the test fails, and 8 cycles if the test passes |

Flags: — No change
0 Flag cleared
1 Flag set
↕ Flag is set or cleared, as appropriate, for this instruction
↓ Flag may change from 0 to 1, if appropriate, but will not change from 1 to 0
↑ Flag may change from 1 to 0, if appropriate, but will not change from 0 to 1

Specific instructions:

LDBSE	Upper byte of destination word is loaded with FF or 00, depending upon sign bit of source byte
LDBZE	Upper byte of destination word is loaded with 00
EXTB WD	Upper byte of word is replaced by FF or 00, depending upon sign bit of lower byte
MULUB, MULB	BD is the lower byte of WD
MULU, MUL	WD is lower word of LD
DIVUB,DIVB	WD(L) represents lower byte of destination word
	WD(H) represents upper byte of destination word
	"mod" is the "modulo" operator which produces the remainder
JBS	Typical instruction syntax is JBS IOS0,6,NEXT
TRAP	Stacks the program counter and then loads it with the interrupt vector shown
SKIP XXH	This is a 2-byte instruction, for which the second byte can be anything and is ignored
RST	The PSW is cleared to 0; the PC is initialized to 2080H; the I/O registers are set to the same initial value which occurs at reset time; a pulse is put out on the reset pin.

Examples of addressing mode syntax:

LDB SBUF,#0FFH	Immediate addressing
LDB SBUF,TEMP	Direct addressing
LDB SBUF,[POINTER]	Normal indirect addressing
LDB SBUF,[POINTER]+	Indirect addressing with autoincrement
LDB SBUF,5[POINTER]	Short-indexed addressing
LDB SBUF,2000H[POINTER]	Long-indexed addressing
LDB SBUF,1234H[0]	Zero-register addressing (This is a special case of long-indexed addressing, used to access any location in the memory space. In this case, SBUF is loaded with the content of hex address 1234)

FIGURE B.39 (continued)

597

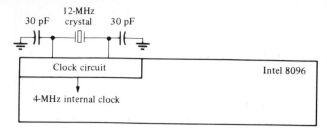

FIGURE B.40
Intel 8096 clock circuitry.

The asterisk indicates the only operand which can employ an addressing mode other than direct (i.e., page 0) addressing. Refer to the key presented at the end of the chart.

For instructions which deal with more than one operand, note the *sequence* of the operands in the assembly language representation. The destination operand is listed first for all instructions except the STB and ST instructions. If an instruction supports a variety of addressing modes, then it will be *only* the last operand which is supported in this way. The other operands must necessarily represent page 0 addresses (i.e., either the registers or the on-chip

		Without A/D	With A/D
ROMless	48-pin		8095BH
	68-pin	8096BH	8097BH
ROM	48-pin		8395BH
	68-pin	8396BH	8397BH
EPROM	48-pin		8795BH
	68-pin	8796BH	8797BH

P8X95BH = 48-pin plastic DIP
C8X95BH = 48-pin ceramic DIP
A8X96BH or A8X97BH = 68-pin ceramic PGA
N8X96BH or N8X97BH = 68-pin plastic PLCC
R8X96BH or R8X97BH = 68-pin ceramic LCC

FIGURE B.41
Intel 8096BH-family part numbers.

Pin name	68-lead PLCC	68-pin PGA/LCC	48-pin DIP	Single-chip function	Expanded-mode function
P0.0/ACH0	6	4	—	Port 0 or A/D converter input	
P0.1/ACH1	5	5	—		
P0.2/ACH2	7	3	—		
P0.3/ACH3	4	6	—		
P0.4/ACH4	11	67	43		
P0.5/ACH5	10	68	42		
P0.6/ACH6	8	2	40		
P0.7/ACH7	9	1	41		
VREF	13	65	45	Reference for A/D converter	
ANGND	12	66	44	Ground for A/D converter	
P1.0	19	59	—	Port 1	
P1.1	20	58	—		
P1.2	21	57	—		
P1.3	22	56	—		
P1.4	23	55	—		
P1.5	30	48	—		
P1.6	31	47	—		
P1.7	32	46	—		
P2.0/TXD	18	60	2	Port 2 or Serial port output	
P2.1/RXD	17	61	1	Serial port input	
P2.2/EXTINT	15	63	47	External interrupt input	
P2.3/T2CLK	44	34	—	Timer 2 clock input	
P2.4/T2RST	42	—	—	Timer 2 external reset input	
P2.5/PWM	39	39	13	Pulse-width modulator output	
P2.6	33	45	—		
P2.7	38	40	—		

FIGURE B.42
Intel 8096BH-family pin assignments.

Signal				Description
P3.0	60	18	32	
P3.1	59	19	31	
P3.2	58	20	30	
P3.3	57	21	29	
P3.4	56	22	28	
P3.5	55	23	27	
P3.6	54	24	26	
P3.7	53	25	25	Port 3 — Lower half of multiplexed bus
P4.0	52	26	24	
P4.1	51	27	23	
P4.2	50	28	22	
P4.3	49	29	21	
P4.4	48	30	20	
P4.5	47	31	19	
P4.6	46	32	18	
P4.7	45	33	17	Port 4 — Upper half of multiplexed bus
HSI.0	24	54	3	
HSI.1	25	53	4	
HSI.2/HSO.4	26	52	5	
HSI.3/HSO.5	27	51	6	
HSO.0	28	50	7	
HSO.1	29	49	8	
HSO.2	34	44	9	
HSO.3	35	43	10	High-speed inputs and outputs
RESET	16	62	48	Reset
XTAL1	67	11	36	12-MHz crystal connection
XTAL2	66	12	35	12-MHz crystal connection
CLKOUT	65	13	—	4-MHz clock out

Signal			Description
BUSWIDTH	64	14	—	Select 8- or 16-bit data bus
INST	63	15	—	External instruction fetch cycle
EA	2	8	39	Internal ROM enable/disable
ALE/ADV	62	16	34	External address valid
RD	61	17	33	External read
BHE/WRH	41	37	15	External write, upper byte
WR/WRL	40	38	14	External write, lower byte
READY	43	35	16	Input control for slow devices
NMI	3	7	—	Nonmaskable interrupt (development system support)
VCC	1	9	38	+5-V power
VPD	14	64	46	Standby power for 16 bytes of RAM
VSS1	68	10	11	Ground
VSS2	36	42	37	Ground
VPP	37	41	12	Programming power for EPROM parts

FIGURE B.42 (*continued*)

RAM). Many instructions permit a *source* operand to be accessed using these addressing modes. However, it helpful to notice that the only instructions which permit the destination operand to take on these addressing modes are

STB   and   ST   and   POP

The cycles given in the chart each take 0.25 μs when the 8096 is run with a 12-MHz crystal. The values given are correct for on-chip memory accesses. However, if either indirect or indexed addressing is used to access an off-chip address, then the correct times are five or six cycles longer than the values shown, depending upon the instruction. Also, if the stack is located in external memory then any instruction which involves stack use will require two to four extra cycles.

B.12   CLOCK

The 8096's clock controls the timing of all operations going on within the chip. An input clock frequency of between 6.0 and 12 MHz is required. This fre-

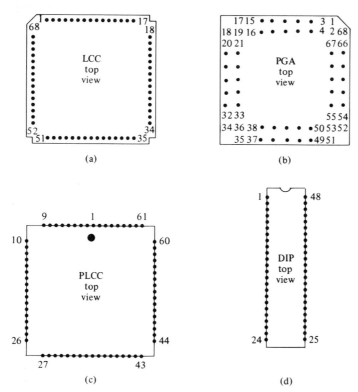

FIGURE B.43
Intel 8096BH-family pin orientations, viewed looking down on component side of PC board. (*a*) Leadless chip carrier (LCC); (*b*) pin grid array (PGA); (*c*) plastic-leaded chip carrier (PLCC); (*d*) dual-in-line (DIP) package.

quency is normally derived from an off-chip crystal working in conjunction with the on-chip oscillator, as shown in Fig. B.40. Alternatively, the clock input can be obtained from an off-chip oscillator. If this is done, it is important to meet pulse-width, voltage-level, and rise-and fall-time requirements for this input waveform.

Internally, the Intel 8096 creates a three-phase clock by counting down this clock input. With a 12-MHz crystal, internal operations are sequenced by three 4-MHz *internal* clock waveforms. This gives rise to an internal cycle time of 0.25 μs. This is the basic unit of time whereby events are carried out inside the Intel 8096.

B.13 PART NUMBERS, PACKAGING, AND PINOUT ALTERNATIVES

The part numbers for the 8096BH family of parts are shown in Fig. B.41. Corresponding to each package, we have the pinout shown in Fig. B.42. The top view for each chip is shown in Fig. B.43 to orient the pin numbers with the chip itself.

INDEX

Note: A number of items having their own entries are cross-indexed under Intel 8096 and Motorola 68HC11. The main entries for these devices are set off in **boldface** type.